Mobile Computing

ABOUT THE AUTHORS

ASOKE K. TALUKDER has been with the IT industry for about 25 years. He has held senior positions in different technology companies in India and abroad. In his last industry association, he was the founder-CTO of Cellnext, the pioneering wireless technology company in India offering technology and solutions in the domain of GSM, GPRS, SMS, MMS, Intelligent Networks, CDMA, and 3G. Since January 2003, he has been with the Indian Institute of Information Technology, Bangalore and is currently the DaimlerChrysler Chair and Associate Professor at IIITB. Asoke has worked in the USA, UK, Singapore, and India for companies like Microsoft, Oracle, Informix, Sequoia, NEC, Fujitsu-ICIM, Digital, iGate, Blue Star Infotech, ICL to name a few. A postgraduate in Physics from the University of Calcutta (1976), he set up the first X.25 network in India for the Department of Telecommunications way back in 1986. Later, he set up the first Java Centre in India in 1998. He was a key engineer for Oracle Parallel Server for Hewlett-Packard HP-FX fault tolerant computers as well as for the 64 bit Informix for DEC Alpha.

He is a recipient of many international awards for innovations and professional excellence including the ICIM Professional Excellence Award, ICL Services Trophy, ICL Chief Executive Excellence award, Atlas Club Excellence Award, etc. One of his ubiquitous middleware products was awarded the IBM Solutions Excellence Award in 2001; one other product on Java Card security was recipient of an award at the GSM World Congress in 2003.

ROOPA R. YAVAGAL completed her B.E. Computer Science from Basaveshwar Engineering College, Bagalkot, Karnataka in 1998. After her graduation, she joined the technical team at Cellnext and developed applications for cellular companies in India in the technology domain of GSM, GPRS, and CDMA. She did her M Tech from the Indian Institute of Information Technology (IIITB) in 2001. Her areas of interests include wireless application and service creation in wireless domain. She is currently working with Symphony Services, Bangalore.

Mobile Computing

Technology, Applications, and Service Creation

Asoke K. Talukder

DaimlerChrysler Chair

IIITB

International Institute of Information Technology

Bangalore, India

Roopa R. Yavagal

Symphony Services

Bangalore, India

New York   Chicago   San Francisco   Lisbon   London   Madrid
Mexico City   Milan   New Delhi   San Juan   Seoul
Singapore   Sydney   Toronto

The McGraw·Hill Companies

Cataloging-in-Publication Data is on file with the Library of Congress.

1 2 3 4 5 6 7 8 9 0   DOC/DOC   0 1 3 2 1 0 9 8 7 6

ISBN-13: 978-0-07-147733-8
ISBN-10: 0-07-147733-0

This book was first published in India in 2005 by Tata McGraw-Hill.

The sponsoring editor for this book was Stephen S. Chapman and the production supervisor was Richard C. Ruzycka. The art director for the cover was Brian Boucher.

Printed and bound by RR Donnelley.

This book was printed on acid-free paper.

McGraw-Hill books are available at special quantity discounts to use as premiums and sales pro-motions, or for use in corporate training programs. For more information, please write to the Director of Special Sales, Professional Publishing, McGraw-Hill, Two Penn Plaza, New York, NY 10121-2298. Or contact your local bookstore.

To my Parents—who taught me how to step into this world;

To my Gurus—who taught me how to face this world;

To my Wife and Daughter—who taught me how to love this world;

To all my Friends—who taught me how to live in this world.

—Asoke K Talukder

Preface

It has been known for centuries that knowledge is power. Pundits knew how to transform data into information and then into knowledge. They used data, information, and knowledge in a different context. What was not known for centuries is how to store data, information and knowledge in such a way that it could be made available to everybody! The convergence of information and communication technology (ICT—Information and Communication Technology) has created avenues to address all these challenges. With the help of ICT, it is possible to render information and knowledge to anybody, anywhere, anytime.

The last decade of the 20th century witnessed a lot of activities in ICT. The GSM was launched; www (World Wide Web) became popular; telecom industry saw lot of promises. Convergence of communication technology and information technology became evident. Investors started putting their money on ICT; the dot com boom happened. As an ICT person, I could not shy away from all this excitement. In 2000, we started a WAP (Wireless Application Protocol) portal company named Cellnext in the ICT space. Within a few months we realized that technology is an enabler and does not sell on its own. Within Cellnext, we quickly moved from WAP to GSM (Global System for Mobile Communications) and SMS (Short Message Service). Later we embraced IVR (Interactive Voice Response), GPRS (General Packet Radio Service), 3G, CDMA (Code Division Multiple Access) and MMS (Multimedia Mesaging Service). To put all these things together, we had to do a lot of research. Also, there was interest from academics to offer courses on some of these emerging technologies. I met Professor Dinesha and Professor Sadagopan, captains from IIIT-B (International Institute of Information Technology Bangalore). In 2001, I offered a one semester elective subject to second year M.Tech students of IIITB on WAP and SMS. Like technology, the course content changed over a period of time. In 2003, the elective WAP course evolved into a core course on Mobile Computing for the third semester M.Tech students. IIITB was one of the very few institutes offering Mobile Computing as a full semester course. This book is an outcome of lecture notes and topics of Mobile Computing course that I offered to the M. Tech students of IIIT-B the last few years.

Making information available from anywhere anytime is one set of challenges. Making information available all the time when the user is mobile and may be traveling on a train or a car on the freeway is another set of challenges! Mobile computing technology will address these challenges and enable the realization of global village with ubiquitous information, where people can seamlessly access any information from anywhere through any device while stationary or even at a state of mobility. There are a few books which cover wireless and mobile communications, but there are not many books that cover the service aspect of mobile applications. This book covers all the communication technologies starting from first generation to third generation cellular technologies, wired telecommunication technology, wireless LAN (WiFi), and wireless broadband (WiMax). It covers intelligent networks (IN) and emerging technologies like mobile IP, IPv6, and VoIP (Voice over IP).

The book is targeted to address a large cross-section of audience. This book can be used either as a textbook or a reference book. As this book gives a big picture of all the technologies from CTI (computer telephony interface) to 3G (third generation) including Bluetooth, IN, WiFi and WiMax, it can be used by non-technical people as a primer.

TO THE PROFESSIONAL

This book is designed to cover a broad range of topics suitable for an ICT professional. We believe that this book will be an excellent handbook for professionals who need to understand mobile computing. Each chapter covers one topics and they are more or less stand alone. Therefore, as a professional you can focus on any topic that interests you. If you need to understand some of wireless technologies to manage a project, this is the book for you. Even if you have not been in touch with some of the recent technologies, this book can help you to jumpstart. The book is suitable for application programmers, telecommunication professionals, or managers like you who want to know about technology and services in mobile computing.

TO THE TEACHER

The book is designed to provide a broad spectrum of subjects. Topics covered can be used for postgraduate studies in mobile computing. We have tried to keep a thread through the chapters; however, they are self-contained. For an undergraduate program, part of every chapter can be included. You may like to read through 'Organization of the Book' to be able to decide on the optimal mix of chapters in your course module.

TO THE STUDENT

We hope that this textbook will trigger your interest in the field of Mobile Computing and wireless services. We tried to organize this book as a textbook. However, we organized the chapters, topics, and the content in a fashion that it can also serve as a reference book or a handbook. This book is suitable for a computer science student who wants to develop a mobile application for computer savvy people, or a social science student who wants to design an application to eradicate the digital divide. The organization of the book has evolved from wire to wireless and telecom, Internet to convergence. Therefore, you may like to read the book in the sequence the chapters have been organized. However, we tried to keep the content of every chapter as independent as possible so that you can pick up a particular topic and understand the topic.

ORGANIZATION OF THE BOOK

The book is organized into 18 chapters:

Chapter 1 is the introduction to the book. This includes general introduction and builds the foundation for mobile computing. It covers various definitions and the significance of terms and technologies. As a part of the foundation, the chapter introduces various frameworks and explains how they fit into the complex jigsaw puzzle called mobile computing. How can mobile computing help businesses use information in an effective way! Also, it touches upon digital divide, services, and applications required for the masses at the "bottom of the pyramid". At the end, this chapter covers standard bodies and their roles. It also goes into detail of the scope of different standard bodies and their relevance in the context of mobile computing.

Chapter 2 captures the general architecture of the mobile computing and the first step towards making information ubiquitous through public network (Internet). This chapter deals with multi-tier architecture of application development and its significance. It discusses various types of middleware, their functions, roles, and how can they be used to implement mobile services. This chapter describes the philosophy behind context in a mobile application. It also goes into the detail of how to determine context and develop a context-aware system.

Chapter 3 introduces the concepts behind the telephony system. It deals with how to access information using a telephone as a client device. As telephones are available throughout the world, voice based applications can be considered as a legacy technology for today's mobile ubiquitous computing. This chapter introduces the philosophy behind voice based application development through CTI (Computer Telephony Interface/Computer telephony Integration) using IVR (Interactive Voice Response). It also covers Voice XML and Voice browsers.

Chapter 4 discusses many technologies. All these technologies are related to mobile computing; however they are yet to become mainstream technologies. These technologies are: Bluetooth, Radio Frequency Identifiers or RFID, Wireless Broadband or WiMax, Mobile and Cellular IP, IPv6 (Internet Protocol version 6) or the next generation IP. These technologies are very important and have every possibility to become mainstream technologies of tomorrow. Therefore, we have introduced these technologies as related technologies, but have not covered them in detail in this book.

Chapter 5 describes GSM technologies. It starts with the basic concepts of cellular networks. It describes the GSM architecture and different elements within the GSM network. It covers the role of these elements and how they interoperate to ensure seamless routing, handover, mobility management, and roaming within GSM networks. The chapter finishes with various security algorithms and security infrastructure within GSM.

Chapter 6 deals with SMS. It describes the SMS architecture in detail. It describes the SMS data technology over the air and over SMS gateways. It describes how SMS can be used as a data bearer to develop value added services and applications for the masses through SMS. It finishes with the SMPP (Short Message Peer-to-Peer) protocol and the GNU open source Kannel SMS gateway.

Chapter 7 moves up in the hierarchy from Second Generation (2G) to Two point Five Generation (2.5G), GSM to GPRS. This chapter starts with the GPRS architecture and various elements of the GPRS network. It describes the differences between GSM and GPRS. It describes in detail how GPRS deals with mobility issues with respect to data. It describes some mobile applications that are more suited for higher bandwidth GPRS networks.

Chapter 8 discusses the WAP and MMS technology. It starts with the description of WAP stack and WAP Application Environment. It covers WML (Wireless Markup Language) and how to develop applications using WML. It then describes WMLScript and Wireless Telephony Application Interface (WTAI). It then moves to Multimedia Messaging

Service. It introduces SMIL (Synchronized Multimedia Integration Language) and describes how to develop MMS applications using SMIL. The chapter finishes with DRM (Digital Writes Management) as defined by WAP forum.

Chapter 9 moves one step up further to discuss CDMA and 3G technology. This chapter begins with the concept of Spread Spectrum Technology and Code Division Multiple Access. IS-95 was the first cellular technology to use CDMA multiplexing technique. Though IS-95 is a 2nd generation network, it is described in this chapter. It then describes security, handoff, and roaming the IS-95 networks. It then goes into the 3G networks. It describes IMT-2000, CDMA-2000, UMTS, and WCDMA networks. The chapter concludes with description of 3G application framework and the concept of VHE (Virtual Home Network). Some of the new concepts of ubiquity, convergence, context awareness etc are describes as a part of 3G framework.

Chapter 10 discusses Wireless Local Area Network (WLAN) or WiFi technology. It describes where, why, how, and when to use Wireless LAN. It describes the scope of various standards within 802.11 family. It describes ad-hoc and infrastructure based wireless networks. It describes how CDMA and CSMA-CA (Carrier Sense Multiple Access—Collision Avoidance) is used in WLAN. The chapter covers aspects of mobility and roaming in WLAN. It then describes the deployment of WLAN. It very briefly touches upon sensor networks and how can WLAN be used for sensor networks. It then discusses WLAN security, what is available today, its vulnerabilities, and evolution of security standards to ensure higher data security over WLAN. The chapter finishes with a comparison between 3G and WiFi.

Chapter 11 covers Intelligent Networks and Interworking. This chapter introduces the SS#7 signaling network and describes how to use this network to develop Intelligent Network (IN) applications. It starts with the fundamentals of call processing and routing. It then explains when and why we call a network intelligent. It then goes into detail of the SS#7 network, different elements within a SS#7 network. It touches upon different application parts at the application layer of the SS#7 stack. This chapter finishes with some example applications in IN.

Chapter 12 introduces the philosophy and technique to develop technology neutral applications. In Chapters 3 thorough Chapter 10, we discussed various techniques for mobile applications. These techniques are specific to the specific technologies. This chapter is the introduction to Chapter 13 through Chapter 15 where we cover various platforms, tools and techniques to develop mobile applications.

Chapter 13 describes programming the PalmOS for the PDAs (Personal Digital Assistance) and wireless services. In this chapter we cover the tools and techniques to develop mobile applications for PalmOS based PDAs. It starts with PalmOS architecture and then moves forward on application development on PalmOS. We start with application development environment in PalmOS. It describes some of the aspects one needs to bear in mind while developing PalmOS applications. These include graphical user interface, forms, networking and communications, databases, and security. It then describes how to program PalmOS for multimedia and telephony interfaces.

Chapter 14 deals with what goes in a Symbian OS, which is becoming quite popular in communicators or high end mobile phones. This chapter starts with Symbian architecture and explains the application development environment in Symbian. It then describes how to develop applications in Symbian environment. It finishes the chapter with security considerations in Symbian.

Chapter 15 is all about Java for the wireless and mobile devices. It starts with Java 2 Micro Edition (J2ME). It describes the architecture of J2ME and goes in detail of how to develop applications using J2ME. This includes all aspects of J2ME starting from simple applications to multimedia, tickers, database, networking, communication using sockets. The chapter finishes with security considerations for J2ME.

Chapter 16 is about WindowsCE and using it for application development for small devices starting from mobile phones to PDAs. It describes the architecture of WindowsCE. It then describes the various development platforms available for application development in WindowsCE.

Chapter 17 talks about the Voice over Internet Protocol (VoIP). It deals with different protocol and technology for convergence of Internet with telecommunication network. It starts with H.323 and describes different elements and components within H.323. It then describes SIP (Session Initiation Protocol). It then describes all the technologies and protocols required to make the convergence of telecommunication and IP. It then describes application develop techniques and the environment within SIP.

Chapter 18 deals with the security issues in mobile computing. The majority of the information security challenges in desktop computing and mobile computing are common to both. Therefore, this chapter starts with vulnerability, exploits, and attacks. It then describes symmetric and public key encryption techniques. It then describes security protocols with specific to SSL (Secured Socket layer) and TLS (Transport Layer Security) and WTLS (Wireless TLS). It then covers the security framework for mobile environment to both and 3GPP (3rd Generation Partnership Program). The chapter ends with virus and worms in the wireless environment.

Though we have tried our best, it is possible that there are errors that have escaped our attention in the book. We would like to hear from you—your suggestions, feedback, criticism, or any other comments that will enable us improve the next edition of *Mobile Computing*. Please send your comments and suggestions to: asoke.talukder@iiitb.ac.in.

ASOKE K. TALUKDER

Acknowledgments

This book would not have been possible without the guidance and technical and personal support of a number of people. We would like to thank all those individuals, researchers, professionals, and technicians who are working in the domain of ICT. All of them have contributed to this book directly or indirectly. We went through many problems and challenges during my tenure with the industry. Many of the challenges that I faced have helped me to learn new things. Some of my former colleagues who have directly contributed to my work and whose names I would like to mention are: Nikhil Nanda, Rajan Swaroop, Nityananda, Atish Dasgupta, Rishi Pal, Manish Chaitanya, Rajkumar, Gururaj, Debmalya, Ashish Tara, Kurian John, Saugat Maitra, Ayush Sharma and Srihari P Mule.

I have read several books, and many articles on the web, and in various journals and magazines. They all have contributed to this work. I have acknowledged most of them as references and further readings. The omissions, if any, are inadvertent and not deliberate.

All trademarks and registered trademarks used in the book are the properties of their respective owners/companies.

I would like to thank Prof S. Sadagopan, Prof Prabhu, Prof Dinesha, Prof Debabrata Das of IIIT-B for their encouragement and support. I would like to acknowledge the contribution of Ms Roopa Yavagal, my co-author, who has written Chapters 12 to 15 of this book. I would like to thank Tata McGraw-Hill for publishing this book. Finally, I would like to thank my wife Kalyani and daughter Debi, without whose support this book would not have been complete. They not only inspired me to undertake the work of writing this book, but have also encouraged me to complete it.

ASOKE K. TALUKDER

Contents

CHAPTER 3 MOBILE COMPUTING THROUGH TELEPHONY   73

CHAPTER 4 EMERGING TECHNOLOGIES   99

Chapter 5 Global System for Mobile Communications (GSM) 137

Chapter 6 Short Message Service (SMS) 167

Chapter 7 General Packet Radio Service (GPRS) 203

CHAPTER 8   WIRELESS APPLICATION PROTOCOL (WAP)   225

CHAPTER 9   CDMA AND 3G   255

CHAPTER 10 WIRELESS LAN   297

CHAPTER 11 INTELLIGENT NETWORKS AND INTERWORKING   337

CHAPTER 12 CLIENT PROGRAMMING   365

CHAPTER 13 PROGRAMMING FOR THE PALM OS   379

CHAPTER 14 WIRELESS DEVICES WITH SYMBIAN OS   419

Chapter 18 Security Issues in Mobile Computing   591

Mobile Computing

CHAPTER 1

Introduction

1.1 MOBILITY OF BITS AND BYTES

It has been known for centuries that information is power. What was not known for centuries was how to store information and knowledge in such a way that it can be accessed by everybody from anywhere, anytime. Convergence of information technology and communication technology has created ways to address these challenges. Today we can have information access from anywhere all the time even at a state of mobility.

In the last two centuries, mobility has been redefined. Both physical and virtual objects are now mobile. Mobility of physical objects relate to movement of matters, whereas movements of virtual objects relate to movements of bits and bytes.

The foundation of mobility of information was laid by Joseph Henry, (1797–1878), who invented the electric motor and techniques for distant communication. In 1831, Henry demonstrated the potential of using electromagnetic phenomenon of electricity for long distance communication. He sent electric current over one mile of wire to activate an electromagnet, which caused a bell to ring. Later, Samuel F. B. Morse used this property of electricity to invent the telegraph. Morse transmitted his famous message "What hath God wrought?" from Washington to Baltimore over 40 miles in 1844. Then on March 10, 1876, in Boston, Massachusetts, Alexander Graham Bell laid the foundation of telephone by making the first voice call over wire–"Mr. Watson, come here, I want to see you".

On October 4, 1957 the USSR (Union of Soviet Socialist Republic now mainly Russia) launched the Sputnik. It was the first artificial earth satellite launched from Baikonur cosmodrome in Kazakhstan. This demonstrated the technological superiority of USSR. In response to this, the US formed the Advanced Research Projects Agency (ARPA) within the Department of Defense (DoD). The mandate for ARPA was to establish US lead in science and technology. ARPA funded different research projects to help research in computer networks. This laid the foundation of packet switched data networks. There were multiple flavors of packet switched networks in USA and in Europe. The important ones are TCP/IP and X.25. TCP/IP was driven by education and defense in the USA whereas X.25 was driven by European telecommunication industry and Governments. With the evolution of computers and the packet switched networks, movement of bits and bytes moved to a new state of maturity. Over last 175 years virtual reality evolved from ringing a bell to a mobile phone through mobile computing.

1.1.1 The convergence leading to ICT

The first step towards the convergence between telecommunication and IT happened in 1965 when AT&T used computers to do the switching in Electronic Switching System (ESS). On the other hand packet switch network was bringing communication closer to computers. The World Wide Web (WWW), which was started by Tim Berners-Lee in 1989 as a text processing software, brought these two faculties of technology together and established Internet as a powerful media. The Internet meets four primary needs of the society: communication, knowledge sharing, commerce, and entertainment. This convergence is called Information and Communications Technologies (ICT). Through ICT we are now moving towards an information-based society. ICT will address the need to access data, information, and knowledge from anywhere, anytime.

1.2 WIRELESS—THE BEGINNING

In 1947 researchers in AT&T Bell Labs conceived the idea of cellular phones. They realized that by using small service areas or cells they can reuse the frequency. This in turn can enhance the traffic capacity of mobile phones. AT&T requested the Federal Communication Commission (FCC) to allocate a large number of radio-spectrum frequencies so that widespread mobile telephone service would become feasible. FCC is a government agency in United States who regulates the usage and licensing of frequency bands. Every country has its regulatory agencies like FCC. In India the regulatory authority is TRAI (Telecom Regulatory Authority of India). FCC in USA is charged with regulating interstate and international communications by radio, television, wire, satellite and cable. Initially, FCC agreed to license a very small band to AT&T. This small frequency range made only 23 simultaneous phone conversations possible in one service area. With 23 channels there was no market incentive for either research or commercial deployment for AT&T. Though the idea of cellular was very much there in late forties, it did not take off.

1.2.1 Evolution of Wireless Networks

The first wireless network was commissioned in Germany in 1958. It was called A-Netz and used analog technology at 160 MHz. Only outgoing calls were possible in this network. That is to say that connection set-up was possible from the mobile station only. This system evolved into B-Netz operating at the same 160 MHz. In this new system, it was possible to receive an incoming call from a fixed telephone network, provided that location of the mobile station was known. This system was also available in Austria, the Netherlands, and Luxemburg. A-Netz was wireless but not a cellular network. Therefore, these systems (A-Netz and B-Netz) did not have any function, which permitted handover or change of base station. The B-Netz had 13,000 customers in West Germany and needed a big transmitter set, typically installable in cars.

In 1968, in USA, the FCC reconsidered its position on Cellular network concept. FCC agreed to allocate a larger frequency band for more number of mobile phones provided the technology to build a better mobile service be demonstrated. AT&T and Bell Labs proposed a cellular system to the FCC with many small, low-powered, broadcast towers, each covering a hexagonal 'cell' of a few kilometers in radius. Collectively these cells could cover a very large area. Each tower would use only a few of the total frequencies allocated to the system. As the phones traveled across the area, calls would be passed from tower to tower.

Besides AT&T and Bell Labs, other enterprises were also engaged in research in the wireless domain. In April 1973, Martin Cooper of Motorola invented the first mobile phone handset and made the first call from a portable phone to Joel Engel, his rival in AT&T Bell Labs. By 1977, AT&T and Bell Labs constructed a prototype of a public cellular network. In 1978, public trials of the cellular telephony system started in Chicago with over 2000 trial customers. In 1982, FCC finally authorized commercial cellular service for the USA. A year later in 1983, the first American commercial analog cellular service AMPS (Advanced Mobile Phone Service) was made commercially available in Chicago. This was the first cellular mobile network in the world.

While USA was experiencing the popularity of cellular phones, Japan and Europe were not lagging behind. In 1979, the first commercial cellular telephone system began operations in Tokyo. During the early 1980s, cellular phone experienced a very rapid growth in Europe, particularly in Scandinavia and the United Kingdom. There was decent growth of cellular phone in France and Germany as well. The message was quite clear by then that mobile technology was here to stay.

To take advantage of this growing market, each country in Europe developed its own analog mobile system and joined the bandwagon. These cellular systems developed by each country in Europe were mutually incompatible. These incompatibilities made the operation of the mobile equipment limited to national boundaries. Also, a mobile subscriber of one network cannot use the same device in another network in another country. Though the market was growing, these incompatible systems made the market very limited for equipment manufacturers. This became an increasingly unacceptable situation in a unified Europe.

To cope with these problems Europeans decided to evolve a standard for mobile phone technology. In 1982, the Conference of European Posts and Telegraphs (CEPT) formed a study group called the Groupe Spécial Mobile (GSM) to develop a standard for pan-European mobile system. In 1989, GSM responsibility was transferred to the European Telecommunication Standards Institute (ETSI), and GSM became a technical committee within ETSI. In 1990, phase I of the GSM specifications were published. Commercial service of GSM started in mid 1991. Although standardized in Europe, GSM became popular outside Europe as well. Therefore, to give a global flavor, GSM was renamed as 'Global System for Mobile Communications'. In the beginning of 1994, there were 1.3 million subscribers worldwide. This has grown to more than 1 billion by the end of February 2004 in over 200 countries. In October 2004, number of mobile subscribers in India crossed the number of fixed phones.

If we look at the critical success factor of GSM, we find quite a few technical and non-technical reasons for its tremendous success. These are:

- The developers of GSM sat together to arrive at a standard before they built the system. The advantage of standards is that they provide enough standardization to guarantee proper interoperability between different components of the system. GSM standards also facilitate the interworking between different vendors. Over 8000 pages of GSM recommendations ensure competitive innovation among suppliers.

- International roaming between networks. A subscriber from one network can seamlessly roam in another network and avail of services without any break.

- Emergence of SMS has spawned several applications within the GSM framework.

- The developers of GSM took considerable technological risk by choosing an unproven (in 80's) digital system. They had the confidence that advancements in compression algorithms and digital signal processing would allow the continual improvement of the system in terms of quality and cost.

1.2.2 Evolution of Wireless Data

Like the computers, the evolution of wireless technology has also been defined in generations. The first generation (1G) of wireless technology uses the analog technology. It uses FDMA (Frequency Division Multiple Access) technology for modulation; for example, AMPS (Advanced Mobile Phone Service) in US. The second generation or 2G technology uses digitized technology. It uses a combination of TDMA (Time Division Multiple Access) and FDMA technologies. An example is GSM. In 2G technology, voice is digitized over a circuit. In 1G and 2G networks, data is transacted over circuits. This technology is called Circuit Switched Data or CSD in short. Using modems, a data connection is established between the device and the network. This is similar to what happens in a dial-up network over analog telephones at home. The next phase in the evolution is 2.5G. In 2.5G technology, voice is digitized over a circuit. However, data in 2.5G is packetized. 2.5G uses the same encoding techniques as 2G does. GPRS networks is an example of 2.5G. The Third Generation or 3G wireless technology makes a quantum leap from a technology point of view. 3G uses Spread Spectrum techniques for media access and encoding. In 3G networks, both data and voice use packets. UMTS and CDMA2000 are examples of 3G networks.

While 1G, 2G, or 3G were making their marks in the metropolitan area wireless networks (MAN), wireless technology has been getting popular in local area networks (LAN) and personal area networks (PAN) as well. Wireless offers convenience and flexibility. With the success of wireless telephony and messaging services like paging, wireless communication is beginning to be applied to the realm of personal and business computing in the domain of local area networks. Wireless LANs are being deployed in homes, campuses, and commercial establishments. Wireless LANs are also being deployed in trains and commercial vehicles. The domain of wireless data networks today comprises of Wireless PAN (Bluetooth, Infrared), Wireless LAN (IEEE 802.11 family) and Wireless WAN (Wide Area Networks) (GSM, GPRS, 3G).

1.2.3 Evolution of Wireless LAN

In late 1980s, vendors started offering wireless products, which were to substitute the traditional wired LAN (Local Area Network) products. The idea was to use a wireless local area network to avoid the cost of installing LAN cabling and ease the task of relocation or otherwise modifying the network's structure. When the Wireless LAN (WLAN) was first introduced in the market, the cost per node was quite high and higher than the cost compared to its counterpart in the wired domain. However, as time progressed, the cost per node started dropping making wireless LAN quite attractive. Slowly WLAN started becoming popular and many companies started offering products. The question of interoperability between different wireless LAN products became critical. IEEE standard committee took the responsibility to form the standard for WLAN. As a result IEEE 802.11 series of standards emerged.

WLAN uses the unlicensed Industrial, Scientific, and Medical (ISM) band that different products can use as long as they comply with certain regulatory rules. These rules cover characteristics such as radiated power and the manner in which modulation occurs. WLAN is also known as Wireless Fidelity or WiFi in short. There are many products which use these unlicensed bands along with WLAN; examples could be cordless telephone, microwave oven etc. There are 3 bands within the ISM bands. These are 900-MHz ISM band, which ranges from 902 to 928 MHz; 2.4-GHz ISM band, which ranges from 2.4 to 2.4853 GHz; and the 5.4 GHz band, which range from 5.275 to 5.85 GHz. WLAN uses 2.4 GHz and 5.4 GHz bands. WLAN works both in infrastructure mode and ad hoc mode.

1.2.4 Evolution of Wireless PAN

Wireless technology offers convenience and flexibility. Some people will call this freedom from being entangled with the wire. The success of wireless technology in Cellular telephones or Wireless MAN (Metropolitan Area Network) made people look at using the technique in Wireless LAN and Wireless Personal Area Network (WPAN). Techniques for WPANs are infrared and radio waves. Most of the Laptop computers support communication through infrared, for which standards have been formulated by IrDA (Infrared Data Association–www.irda.org). Through WPAN, a PC can communicate with another IrDA device like another PC or a Personal Digital Assistant (PDA) or a Cellular phone.

The other best known PAN technology standard is Bluetooth. Bluetooth uses radio instead of infrared. It offers a peak over the air speed of about 1 Mbps over a short range of about 10 meters. The advantage of radio wave is that unlike infrared it does not need a line of sight. WPAN works in ad hoc mode only.

1.3 Mobile Computing

Mobile computing can be defined as a computing environment over physical mobility. The user of a mobile computing environment will be able to access data, information or other logical objects from any device in any network while on the move. Mobile computing system allows a user to perform a task from anywhere using a computing device in the public (the Web), corporate (business information) and personal information spaces (medical record, address book). While on the move, the preferred device will be a mobile device, while back at home or in the office the device could be a desktop computer. To make the mobile computing environment ubiquitous, it is necessary that the communication bearer is spread over both wired and wireless media. Be it for the mobile workforce, holidaymakers, enterprises, or rural population, the access to information and virtual objects through mobile computing are absolutely necessary for optimal use of resource and increased productivity.

Mobile computing is used in different contexts with different names. The most common names are:

- **Mobile Computing:** The computing environment is mobile and moves along with the user. This is similar to the telephone number of a GSM (Global System for Mobile communication) phone, which moves with the phone. The offline (local) and realtime (remote) computing environment will move with the user. In realtime mode user will be able to use all his remote data and services online.

- **Anywhere, Anytime Information:** This is the generic definition of ubiquity, where the information is available anywhere, all the time.

- **Virtual Home Environment:** Virtual Home Environment (VHE) is defined as an environment in a foreign network such that the mobile users can experience the same computing experience as they have in their home or corporate computing environment. For example, one would like to put ones room heater on when one is about 15 minutes away from home.

- **Nomadic Computing:** The computing environment is nomadic and moves along with the mobile user. This is true for both local and remote services.

- **Pervasive Computing:** A computing environment, which is pervasive in nature and can be made available in any environment.

- **Ubiquitous Computing:** A disappearing (nobody will notice its presence) every-place computing environment. User will be able to use both local and remote services.

- **Global Service Portability:** Making a service portable and available in every environment. Any service of any environment will be available globally.

- **Wearable Computers:** Wearable computers are those computers that may be adorned by humans like a hat, shoe or clothes (these are wearable accessories). Wearable computers need to have some additional attributes compared to standard mobile devices. Wearable computers are always on; operational while on move; hands free, context aware (with different types of sensors). Wearable computers need to be equipped with proactive attention and notifications. The ultimate wearable computers will have sensors implanted within the body and supposedly integrate with the human nervous system. These are part of new discipline of research categorized by "Cyborg" (Cyber Organism).

1.3.1 Mobile Computing Functions

We can define a computing environment as mobile if it supports one or more of the following characteristics:

- **User Mobility:** User should be able to move from one physical location to another location and use the same service. The service could be in the home network or a remote network. Example could be a user moves from London to New York and uses Internet to access the corporate application the same way the user uses in the home office.

- **Network Mobility:** User should be able to move from one network to another network and use the same service. Example could be a user moves from Hong Kong to New Delhi and uses the same GSM phone to access the corporate application through WAP (Wireless Application Protocol). In home network he uses this service over GPRS (General Packet Radio Service) whereas in Delhi he accesses it over the GSM network.

- **Bearer Mobility:** User should be able to move from one bearer to another and use the same service. Example could be a user was using a service through WAP bearer in his home network in Bangalore. He moves to Coimbatore, where WAP is not supported, he switch over to voice or SMS (Short Message Service) bearer to access the same application.

- **Device Mobility:** User should be able to move from one device to another and use the same service. Example could be sales representatives using their desktop computer in home office. During the day while they are on the street they would like to use their Palmtop to access the application.

- **Session Mobility:** A user session should be able to move from one user-agent environment to another. Example could be a user was using his service through a CDMA (Code Division Multiple Access) 1X network. The user entered into the basement to park the car and got disconnected from his CDMA network. User goes to home office and starts using the desktop. The unfinished session in the CDMA device moves from the mobile device to the desktop computer.

- **Service Mobility:** User should be able to move from one service to another. Example could be a user is writing a mail. To complete the mail user needs to refer to some other information. In a desktop PC, user simply opens another service (browser) and moves between them using the task bar. User should be able to switch amongst services in small footprint wireless devices like in the desktop.

- **Host Mobility:** The user device can be either a client or server. When it is a server or host, some of the complexities change. In case of host mobility the mobility of IP needs to be taken care of.

The mobile computing functions can be logically divided into following major segments (Figure 1.1):

1. **User with device:** The user device, this could be a fixed device like desktop computer in office or a portable device like mobile phone. Example: laptop computers, desktop computers, fixed telephone, mobile phones, digital TV with set-top box, palmtop computers, pocket PCs, two way pagers, handheld terminals, etc.

2. **Network:** Whenever a user is mobile, he will be using different networks at different places at different time. Example: GSM, CDMA, iMode, Ethernet, Wireless LAN, Bluetooth etc.

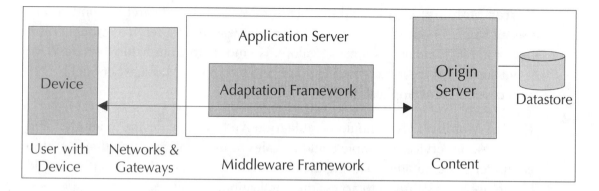

Figure 1.1 Mobile computing functions

3. **Gateway:** This is required to interface different transport bearers. These gateways convert one specific transport bearer to another transport bearer. Example: From a fixed phone (with voice interface) we access a service by pressing different keys on the telephone. These keys generate DTMF (Dual Tone Multi Frequency) signals. These analog signals are converted into digital data by the IVR (Interactive Voice Response) gateway to interface with a computer application. Other examples will be WAP gateway, SMS gateway etc.

4. **Middleware:** This is more of a function rather than a separate visible node. In the present context middleware handles the presentation and rendering of the content on a particular device. It will also handle the security and personalization for different users.

5. **Content:** This is the domain where the origin server and content is. This could be an application, system, or even an aggregation of systems. The content can be mass market, personal or corporate content. Origin server will have some means to accessing the database and the storage devices.

1.3.2 Mobile Computing Devices

The device for mobile computing can be either a computing device or a communication device. In computing device category it can be a desktop computer, laptop computer, or a palmtop computer. On the communication device side it can be a fixed line telephone, a mobile telephone or a digital TV. Usage of these devices are becoming more and more integrated into a task flow where fixed and mobile, computing and communication devices are used together. When computing technology is embedded into equipments, Human–Computer Interaction (HCI) plays a critical role in effectiveness, efficiency and user experience. This is particularly true as mobile information and communication devices are becoming smaller and more restricted with respect to information presentation, data entry and dialogue control. The human computer interface challenges are:

1. Interaction must be consistent from one device to another.

2. Interaction has to be appropriate for the particular device and environment in which the system is being used.

Note that the requirement does not call for identical metaphors and methods. The desktop computer allows for different interaction techniques than a palmtop computer or a digital TV. Using the keyboard and a mouse may be the obvious for the desktop computer. Using the pen may be appropriate for the palmtop or Tablet PC. Microphone and speaker may be appropriate for a fixed or mobile phone. A remote control on the other hand will be more desired for a digital TV.

1.4 DIALOGUE CONTROL

In any communication there are two types of user dialogues. These are long session-oriented transactions and short transaction. Going through a monolithic document page by page can be considered as a session-oriented transaction. Going to a particular page directly through an index can be considered as a short transaction. Selection of the transaction mode will depend on the type of device we use. A session may be helpful in case of services offered through computers with large screens and mouse. For devices with limited input/output like SMS for instance, short transactions may be desired.

Let us consider an example of bank balance enquiry over the Internet. In case of Internet banking through desktop computer, the user has to go through the following minimum dialogues:

1. Enter the URL of the bank site.

2. Enter the account number/password and Login into the application.

3. Select the balance enquiry dialogue and see the balance.

4. Logout from the internet banking.

The dialogue above is an example of session-oriented transaction. Using short transaction, the same objective can be met through one single dialogue. In short transaction, user sends a SMS message, say 'mybal', to the system and receives the information on balance. The application services all the 5 dialogue steps as one dialogue. In this case many steps like authentication, selection of transactions need to be performed in smarter ways. For example, the user authentication will be done through the mobile number of the user. It can be assumed that mobile devices are personal, therefore, authenticating the mobile phone implies authenticating the user account.

1.5 NETWORKS

Mobile computing will use different types of networks. These can be fixed telephone network, GSM, GPRS, ATM (Asynchronous Transfer Mode), Frame Relay, ISDN (Integrated Service Digital Network), CDMA, CDPD (Cellular Digital Packet data), DSL (Digital Subscriber Loop), Dial-up, WiFi (Wireless Fidelity), 802.11, Bluetooth, Ethernet, Broadband, etc.

1.5.1 Wireline Networks

This is a network, which is designed over wire or tangible conductors. This network is called fixed or wireline network. Fixed telephone networks over copper and fiber-optic will be part of this network family. Broadband networks over DSL or cable will also be part of wireline networks. Wireline network are generally public networks and cover wide areas. Though microwave or satellite networks do not use wire, when a telephone network uses microwave or satellite as a part of its infrastructure, it is considered part of wireline networks. When we connect to ISPs it is generally a wireline network. The Internet backbone is a wireline network as well.

1.5.2 Wireless Networks

Mobile networks are generally termed as wireless network. This includes wireless networks used by radio taxis, one way and two way pager, cellular phones. Example will be PCS (Personal Cellular System), AMPS (Advanced Mobile Phone System), GSM, CDMA, DoCoMo, GPRS etc. WiLL (Wireless in Local Loop) networks using different types of technologies are part of wireless networks as well. In a wireless network the last mile is wireless and works over radio interface. In a wireless network other than the radio interface rest of the network is wireline, this is generally called the PLMN (Public Land Mobile Network).

1.5.3 Ad-hoc Networks

In Latin, *ad hoc* literally means 'for this purpose only'. An ad-hoc (or spontaneous) network is a small area network, especially one with wireless or temporary plug-in connections. In these networks some of the devices are part of the network only for the duration of a communication session. An ad-hoc network is also formed when mobile, or portable devices, operate in close proximity of each other or with the rest of the network. When we beam a business card from our PDA (Personal Digital Assistant) to another, or use an IrDA (Infrared Data Association) port to print document from our laptop, we have formed an ad hoc network. The term 'ad hoc' has been applied to networks in which new devices can be quickly added using, for example, Bluetooth or wireless LAN (802.11x). In these networks devices communicate with the computer and other devices using wireless transmission. Typically based on short-range wireless technology, these networks don't require subscription services or carrier networks.

1.5.4 Bearers

For different type of networks, there are different types of transport bearers. These can be TCP/IP, http, protocols for or dialup connection. For GSM it could be SMS, USSD (Unstructured Supplementary Service Data) or WAP. For mobile or fixed phone, it will be Voice.

1.6 MIDDLEWARE AND GATEWAYS

Any software layered between a user application and operating system can be termed as middleware. Middleware examples are communication middleware, object oriented middleware, message oriented middleware, transaction processing middleware, database middleware, behavior management middleware, RPC middleware etc. There are some middleware components like behavior management middleware, which can be a layer between the client device and the application. In mobile computing context we need different types of middleware components and gateways at different layers of the architecture (Figure 1.2). These are:

1. Communication middleware

2. Transaction processing middleware

3. Behavior management middleware

4. Communication gateways.

1.6.1 Communication Middleware

The application will communicate with different nodes and services through different communication middleware. Different **connectors** for different services will fall in this category. Examples could be TN3270 for IBM mainframe services, or Javamail connector for IMAP or POP3 services

1.6.2 Transaction Processing Middleware

In many cases a service will offer session oriented dialogue (SoD). For a session we need to maintain a state over the stateless Internet. This is done through an application server. The user may be using a device, which demands a short transaction whereas the service at the backend offers a SoD. In such cases a separate middleware component will be required to convert a SoD to a short transaction. Management of the Web components will be handled by this middleware as well.

1.6.3 Behavior Management Middleware

For different devices we need different types of rendering. We can have applications, which are developed specially for different types of rendering. For example, we can have one application for Web, another for WAP, and a different one for SMS. On the contrary, we may choose to have a middleware, which will manage entire device specific rendering at the run time. This middleware will identify the device properly and handle all the behavior related stuff independent of the application. The system may be required to have some context awareness. All these will be handled by behavior management middleware.

1.6.4 Communication Gateways

Between the device and the middleware there will be network of networks. Gateways are deployed when there are different transport bearers or networks with dissimilar protocols. For example, we need an IVR gateway to interface voice with a computer, or an WAP gateway to access internet over a mobile phone.

The following diagram (Figure 1.2) depicts a schematic diagram of services in a mobile computing environment where services from enterprise to a vending machine can be used from different devices.

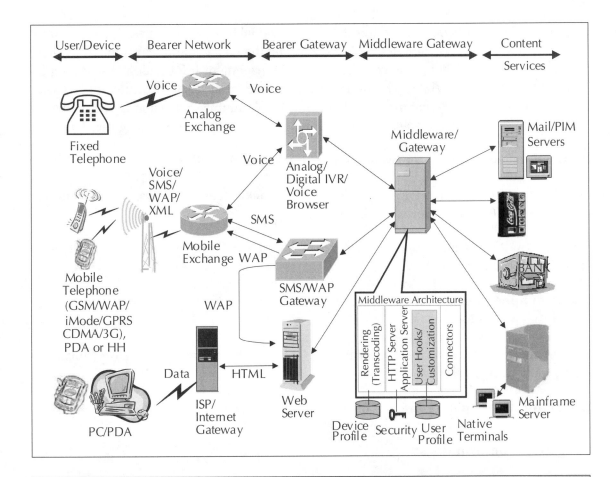

Figure 1.2 A Schematic Representation of a Mobile Computing Environment

1.7 APPLICATION AND SERVICES (CONTENTS)

Data and information, through mobile computing services are required by all people regardless of the fact that they are mobile are not. Mobile users will include people like mobile executives, sales people, service engineers, road warriors, farmers in the field, milkman, newspaper boy, courier or pizza delivery boy. Logically everybody is a mobile user in some respect in some part of the lifestyle. For people, who are stationary, mobile computing is necessary in the off office hours. For example we may need to do a bank transaction from home at night or respond to an urgent mail while at home.

There can be many applications and services for the mobile computing space. These application or services run on origin server. These are also known as content servers.

Contents will primarily be lifestyle specific. An individual has different lifestyles in different social environment. Also, lifestyles do change during the course of the day. One individual can be an executive needing the corporate MIS (Management Information System) application during the day. While at home the same individual at leisure can use applications for youth lifestyle or entertainment. The list of possible mobile applications can never be complete. From lifestyle perspective they can be grouped into different categories like:

Personal–belongs to the user (wallet, life-tool, medical records, diary).

Perishable–time sensitive and relevance passes quickly (general news, breaking news, weather, sports, business news, stock quotes).

Transaction Oriented–transactions need to be closed (bank transactions, utility bill payment, mobile shopping).

Location Specific–information related to current geographical location (street direction map, restaurant guide).

Corporate–corporate business information (mail, ERP, inventory, directory, business alerts, reminders).

Entertainment–applications for fun, entertainment.

Here are some examples:

News: This is a very big basket of applications having different types of news. News could be political, current affair, breaking news, business news, sports news, community news etc. While people are in move, they can always be connected to their culture, community through news using mobile computing.

Youth: This is a very high growth market with different applications to suit the style and lifestyle of the youth. These are primarily messaging based application like person-to-person messaging, chat, forums, dating etc.

Weather: There are different types of applications and services where mobile computing can make a difference. If we look at very closely when a person is on a vacation, and driving from one city to another, access to weather information can sometime save lives. In this case location aware weather information is desirable. Notification services on weather is a very sought after application. GPS-based system to locate a person (with GPS receiver) can sometime save lives in case of natural calamity.

Corporate application: Standard corporate information is one of the most desirable information set for mobile workers. This will include corporate mail, address book, appointments, MIS applications, corporate Intranet, corporate ERP etc.

Sales Force Automation: This group will offer many applications. This will cater the large population of road worriers or sales personnel. Applications will include sales order booking, inventory enquiry, shipment tracking, logistics related applications etc. These applications will be very effective over wireless devices.

m-broker: Getting correct and timely information related to different stocks are very important. Also, online trading of stocks while on move is quite critical for certain lifestyle. Stock ticker, stock alerts, stock quote, and stock trading can be made ubiquitous so that users can check their portfolio and play an active role in this marketplace.

Telebanking: We need to access our banks for different transactions. Earlier people used to go to the bank, but things are changing where banks are coming to customers through telebanking. If telebanking can be made ubiquitous it helps everybody, both the customer and the bank. Many banks in India are today offering banking over Internet (web), voice and mobile phones through SMS.

m-shopping: Application to do different types of shopping using mobile devices like Palmtop, PocketPC, mobile phone etc. Buying of a soft drink/soda from a vending machine in an airport or a movie theatre using a mobile phone may be very handy especially when we do not have change.

Micropayment-based application: Micropayments involve transactions where the amount of money involved is not very high, may be maximum 1000 rupees ($ 25) or so. Micropayment using mobile phones can help rural people to do business in much effective way.

Interactive games: Many mobile network operators have started offering different types of contest and interactive games to be played using mobile phone. The applications could be similar to quiz, housie etc.

Interactive TV shows: Many TV companies around the world use email, SMS and voice as a bearer for interactive TV. In these shows viewers are encouraged to participate by asking questions, share opinions or even answer to different quizzes.

Digital/Interactive TV: These are interactive TV programs through digital TV using set-top boxes and Internet. Video on demand, community programs, health care, and shopping applications are quite popular for this media.

Experts on call: Application system for experts. Experts use these services to schedule their time and business with clients; clients use this to schedule business with the expert. A typical example could be to fix up an appointment with the tax consultant.

GPS-based systems: Applications related to location tracking come under this category. This could be a simple service like tracking a vehicle. Another example could be tracking an individual who got stuck due to bad weather while on a tracking trip. Fleet management companies and context aware systems need these.

Remote monitoring: This is important for children at home where parents monitor where their children are or what are they doing. Also, monitoring and controlling of home appliances will be part of this application.

Entertainment: This contains a very large basket of applications starting from horoscope to jokes. Many people in part of Asia decide their day based on the planetary positions and horoscope information.

Directory services: This includes information related movie theatre, public telephone, restaurant guide, public information system and Yellow pages.

Sports: This service offers online sports update. In India live cricket score is the most popular mobile computing application. Live cricket score is available in India with many service providers through Web, Voice, SMS and WAP.

Maps/navigation guide: This is an application, which has lot of demand for traveling individuals. These services need to be location aware to guide the user to use the most optimum path to reach a destination. The directions given by these applications also take the traffic congestion, one way, etc. into consideration.

Virtual office: There are many people who are self-employed and do not have a physical office. Mobile office and virtual office is very useful for them. They can check their mails, schedule appointments etc. while they are mobile. Insurance agents and many other professions need these types of services.

m-exchange for industries: Exchange for manufacturing industry accessible from mobile device can be a very cost effective solution for small/cottage industries. It may not be possible for a cottage industry to invest on a computer. However, accessing an exchange for a manufacturing company through a SMS may be affordable.

m-exchange for agricultural produce: Exchange for farmers on different type of agricultural products can be very useful for countries like India. If farmers can get information about where to get good price for their product, it helps both farmers and consumers.There is a system www.echoupal.com to do exactly this. Think of this available over mobile phone.

Applications for speech/hearing challenged people: Telecommunication always meant communicating using voice. There are people who cannot speak or hear. These include people with disability and senior citizens who lost speech due to old age or following a stroke. Text-based communication can help rehabilitate some of these disabled individuals.

Agricultural information: Think about a case where a farmer receives an alert in his local language through his mobile phone and immediately knows that the moisture content in air is 74%. He can then decide how much to water his harvest. This can save his money, can save the harvest (excess water is sometime harmful), and save the scarce water resource. Simputer, a PDA like portable device, with its voice interface can change the economics of rural India with this kind of applications.

Corporate knowledge-based applications: Many corporations today have knowledgebase. Making this ubiquitous can reduce cost and increase productivity.

Community knowledge-based applications: Knowledge is power. Like in a corporation, knowledge is equally important for a community. Making knowledge ubiquitous always help the society at a large.

Distance learning: Applications related to distance learning may be very desirable for countries with digital divide. For virtual schools in Asia, or Africa, it is possible to have very good faculty from outside these continents.

Digital library: Internet converted the world into a global village and global library. Accessing this library from anywhere anytime will only help reducing the digital divide. These libraries will have support of local language and search. Digital library is easy and cheaper to commission.

Telemedicine & healthcare: Making telemedicine and healthcare application ubiquitous may save many lives. For example, someone is travelling and having a chest pain and needs immediate attention. The patient has been taken to a doctor in a remote town. Access to patient record in home location can help in faster diagnosis. Reminder service for medicine or checkups can be very useful service. In rural India, virtual clinics can help many people who otherwise do not have access to medical care.

Micro-credit schemes: Micro-credit has a distinct role to play in the micro-economy of a country. Grameen Bank with all its applications in Bangladesh is the best example of micro-credit.

Environmental protection and management: Ubiquity is a must for applications on environmental protection and management. Applications related to industrial hygiene will be part of this family.

e-governance: These applications are very important to bridge the digital divide. The Bhoomi project of Karnataka government has computerized twenty million land records of 0.67 million farmers living in 30,000 villages in the state. Many such projects in the Government can be made electronic, resulting into better and faster access to information managed by the government.

Virtual laboratories: There are many laboratories and knowledge repositories around the world. These types of applications make the facility of these laboratories available across the boundary of culture and countries.

Community forums: There are different social and community meetings. In the case of India, Panchayats can be made electronic. These may help increase the involvement of more people to participate in community developments.

Law enforcements: Most of the time law enforcement staffs are on the streets. They need access to different types of services through wireless means. These may be access to criminal records, information related to vehicles, or even a picture of the accident site taken through a MMS phone can help the insurance companies to resolve the claim faster.

Job facilitator: These could be either proactive alerts or information related to jobs and employment opportunities.

Telemetric applications: Almost every industry and sphere of life has the need for telemetric applications. Examples could be monitoring and control in manufacturing industry; vehicle tracking; meter reading; health care and emergency services; vending machine monitoring; research (telemetric orthodontic); control and service request for different emergency services for utilities like power plants etc.

Downloads: Different types of downloads starting from ringing tone to pictures are part of this family. In many countries this type of applications are very popular. It is estimated that the market for ringing tone is more than 1 billion dollars.

Alerts & Notifications: This can be either business or personal alerts. Simple examples could be breaking news alerts from a newspaper. Complex examples of alert could be for a doctor when the patient is in critical condition. In India many mobile operators are offering cricket alerts. In this service, subscribers receive score information every 15 minutes; about every wicket fall!

1.8 DEVELOPING MOBILE COMPUTING APPLICATIONS

Any portal system today supports user mobility. If I have an Internet mail account like hotmail or yahoo, I can access my mail from anywhere. I need a desktop or laptop computer to access my mailbox. I may not be able to access the same mail through some other device like a fixed phone. There are a number of factors that make mobile computing different from desktop computing. As a result of mobility the attributes associated with devices, network, and users are constantly changing. These changes imply that context and behavior of applications needs to be adapted to suit the current environment. The context and behavior adaptation is required to provide a service that is tailored to the user's present situation. There are several ways in which context and behavior can be adapted. One way is to build applications without any context or behavior awareness. Context and behavior adaptation will be handled by a behavior management middleware at the runtime. Other option is to build different applications specific to different context and behavior patterns. There could be systems in the organization, which was originally developed 15 years ago for some direct connected terminals like VT52. Due to change in the market expectation these systems need to be made mobile. Complexities involved in making an existing application mobile versus developing a new mobile system will be different. For a new application it is possible to embed the behavior within the application. However, for a long life system or a legacy application the content behavior adaptation will need to be done externally.

1.8.1 New Mobile Applications

Let us assume that in a bank, some new applications need to be built for e-Commerce. The bank wants to offer banking through voice (telephone) and Web (Internet). Assuming that the bank already has a computerized system in place, the bank will develop two new applications. One will handle the telephone interface through Interactive Voice Response (IVR) and the other through Web. At a later point in time, if the bank decides to offer SMS and WAP, they will develop two new applications to support SMS and WAP interfaces respectively. To protect the investment and quick adaptation, the bank may decide to use transaction processing middleware and RPC middleware. All these are possible only if it is a fresh applications development.

1.8.2 Making Legacy Application Mobile

How do we make a long life existing legacy application mobile? We define an application as legacy if it has one or more of the following characteristics:

1. The application has moved into the sustenance phase in the software development lifecycle.

2. An application, which cannot be modified. This could be due to unavailability of the original development platforms, unavailability of original source code or unavailability of expertise to make necessary changes.

3. Products and packaged software where enterprise does not have any control. This could be due to high cost of ownership for new upgrade or the vendor does not have any plan to support the new requirement.

Let us assume that an enterprise has licensed an ERP system from an external vendor. The enterprise wants to offer a notification of yesterday's sales figures to some select executives at 9:30 AM everyday morning through SMS. The ERP vendor plans to offer similar function in their next release six months down the line. The license fee for the next upgrade will be very expensive. Another example is that a wireless network operator wants to offer enterprise mails through its network. In all such cases the adaptation will be done without changing the base product. This requires a framework that attempts to perform most of the adaptation dynamically. Content and behavior management will be managed real-time through a behavior management middleware.

1.9 SECURITY IN MOBILE COMPUTING

The security issues in mobile computing environment pose a special challenge. This is because we have to offer services over the air using networks over which we do not have any control. All the infrastructure and technology designed by GSM and other forums are primarily to increase the revenue of the network operators. This makes the technology complex and very much dependent on the network operator. For example, the SMS technology is operator centric; WAP requires WAP gateway. These gateways are installed in the operator's network and managed by the operator. The security policy implemented by the network operator depends on operator's priority and revenue generation potential and not on the need of the content provider.

In a mobile computing environment user can move from one network to another, one device to another, one bearer to another. Therefore, theoretically the security implementations need to be device independent, network independent, bearer independent, so on and so forth. The requirement is to arrive at a security model, which can offer a homogenous end-to-end security.

1.10 STANDARDS—WHY IS IT NECESSARY?

Standards are documented agreements containing technical specifications or other precise criteria to be used consistently as rules, guidelines or definitions of characteristics. Standards ensure that materials, products, processes and services fit for their defined and agreed purpose. A standard begins as a technical contribution, which is supported by a number of interested parties to the extent that they indicate their willingness to participate in the standard's development. Standards are available for experts to challenge, examine and validate. No industry in today's world can truly claim to be completely independent of components, products, rules of application that have been developed in other sectors. Without standards, interoperability of goods and services will not be possible.

When the proposed standard or technical document is near completion, the formulating Engineering Committee circulates the draft of the document for a ballot. The purpose of this ballot is to identify any unresolved issues and to establish consensus within the formulating group. Every effort is made to address and resolve comments received.

The opposite of standard is proprietary. Proprietary systems for similar technologies are seen as technical barriers to trade and competition. Today's free-market economies increasingly encourage diverse sources of supply and provide opportunities for expanding markets. On the technology front, fair competition needs to be based on identifiable, clearly defined common references that are recognized from one country to the other, and from one region to the next. An industry-wide standard, internationally recognized, developed by consensus among trading partners, serves as the language of trade.

There are some fundamental differences between how USA and Europe adapt technology. In USA, market force and time to market drive the technology. Interoperability always has been the primary issue in Europe. Therefore, in Europe, standards drive adaptation of technology. This is one of the reasons why USA has more proprietary systems compared to Europe.

1.10.1 Who makes the standards

There are many institutes that generate and provide standards across the world. There are standard bodies at the regional or country level; also, there are bodies at the international level. Based on the area of operations, standard bodies are formed by the governments, professional institutes or industry consortiums. These standard bodies sometime also function as regulators. In India there is a standard body under the Government of India, which is called Bureau of Indian Standard or simply BIS (www.bis.org.in). A standards process include following steps:

1. Consensus on a proposed standard by a group or 'consensus body' that includes representatives from materially affected and interested parties.

2. Broad-based public review and comment on draft standards.

3. Consideration of and response to comments submitted by voting members of the relevant consensus body and by public review commenters.

4. Incorporation of approved changes into a draft standard.

5. Right to appeal by any participant that believes that due process principles were not sufficiently respected during the standards development in accordance with the ANSI-accredited procedures of the standards developer.

1.11 STANDARD BODIES

International Organization for Standardization

The International Organization for Standardization (**ISO**) (http://www.iso.ch) is a worldwide federation of national standards bodies from more than 140 countries, one from each country. ISO is a non-governmental organization established in 1947. The mission of ISO is to promote the development of standardization and related activities in the world with a view to facilitating the international exchange of goods and services, and to developing cooperation in the spheres of intellectual, scientific, technological and economic activity. Though ISO is commonly believed as the acronym for International Standard Organization, in fact the word 'ISO' is derived from the Greek isos, meaning 'equal'. Sometimes ISO makes its own standard, sometime it adapts standards from its member organizations and makes it an international standard. One of the most widely known standard from ISO is ISO9000. ISO9000 relates to Software quality. The famous 7-layer model for Open System Interconnection (OSI) is ISO standard (ISO7498). For Information security ISO has come up with the recommendation ISO17799.

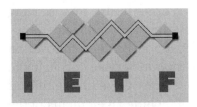

The Internet Engineering Task Force (**IETF**) (http://www.ietf.org) is the standard-making body for Internet and related technologies. IETF is an open international community of network designers, operators, vendors and researchers concerned with the evolution of the Internet architecture and the smooth operation of the Internet. It is open to any individual. The actual technical work of the IETF is done in its working groups. Working groups are organized into several areas by topic (e.g., routing, transport, security, etc.). The Internet Assigned Numbers Authority (IANA) is the central coordinator for the assignment of unique IP address. The IANA is chartered by the Internet Society (ISOC) to act as the clearinghouse to assign and coordinate the use of numerous Internet protocol parameters. Standards defined by IETF are called Request For Comment or RFC. The standard for email is defined in RFC821 (Simple Mail Transfer Protocol or SMTP); RFC2616 describes the version 1.1 of Hypertext Transfer Protocol (HTTP/1.1).

ETSI (the European Telecommunications Standards Institute) (http://www.etsi.org) is an organization whose mission is to produce the telecommunications standards that will be used for decades to come throughout Europe and possibly beyond. ETSI unites members from countries inside and outside of Europe, and represents regulators, network operators, manufacturers, service providers, research bodies and users. ETSI plays a major role in developing a wide range of standards and other technical documentation as Europe's contribution to world-wide standardization in telecommunications, broadcasting and information technology. ETSI's prime objective is to support global harmonization by providing a forum in which all the key players can contribute actively. ETSI is officially recognized by the European Commission. GSM Standard is created, maintained and managed by a committee within ETSI. GSM standards document GSM 01.04 (ETR 350): 'Digital cellular telecommunications system (Phase 2+); Abbreviations and acronyms'. GSM 12.13 standard defines the interface Digital cellular telecommunications system (Phase 2+); Man-Machine Interface (MMI) of the Mobile Station (MS) (GSM 02.30 version 7.1.0 Release 1998).

OMA and **WAP** Forum (http://www.wapforum.org) (http://www.openmobilealliance.org). The Open Mobile Alliance (OMA) has been established by the consolidation of the WAP Forum and the Open Mobile

Architecture initiative. It intends to grow the market for the entire industry by removing barriers to interoperability and supporting a seamless and easy-to-use mobile experience for end users. The Open Mobile Alliance encourages competition through innovation and differentiation, while ensuring the interoperability of mobile service through the entire value chain. The supporters of the Open Mobile Alliance recognize the significant industry benefits of creating a standards organization that will include all elements of the wireless value chain, and contribute to timely and efficient introduction of services and applications to the market. WAP and MMS standards are created, maintained and managed by OMA.

ITU (International Telecommunication Union) (www.itu.int) is an organization within the United Nations System. It was founded on the principle of cooperation between governments and the private sector. With a membership encompassing telecommunication policy-makers and regulators, network operators, equipment manufacturers, hardware and software developers, regional standards-making organizations and financing institutions. ITU's activities, policies and strategic direction are determined and shaped by the industry it serves. ITU has three Sectors of the Union; they are Radio communication (ITU-R), Telecommunication Standardization (ITU-T), and Telecommunication Development (ITU-D). Their activities cover all aspects of telecommunication, from setting standards that facilitate seamless interworking of equipment and systems to adopting operational procedures for the wireless services and designing programmes to improve telecommunication infrastructure. **ITU Telecommunication Standardization Sector (ITU-T)'s** mission is to ensure an efficient and on-time production of high quality standards (Recommendations) covering all fields of telecommunications. **ITU-T** was founded in 1993, replacing the former International Telegraph and Telephone Consultative Committee (CCITT) whose origins go back to 1865. Any telephone in this world has a unique number (technically known as Global Title). These numbering schemes are defined through the ITU-T standards E.164.

The IEEE Standards Association (**IEEE-SA**) (http://standards.ieee.org) is an organization that produces standards, which are developed and used internationally. While the IEEE-SA focuses considerable resources on the long-respected full consensus standards process carried out by the standards committees and IEEE Societies, the IEEE-SA pioneers new and innovative programs to increase the value of IEEE standards to members, industry, and the global society. IEEE-SA members continue to set the pace for the development of standards products, technical reports and documentation that ensure sound engineering practices worldwide. IEEE-SA demonstrates strong support of an industry-led consensus process for the development of standards and operating procedures and guidelines. Standards for Wireless LAN are created, maintained and managed by IEEE. These are defined through different 802.11 standards.

Electronic Industries Alliance

The Electronic Industries Alliance (**EIA**) (http://www.eia.org) is a national trade organization within USA that includes the full spectrum of U.S. electronics industry. The Alliance is a partnership of electronic and high-tech associations and companies whose mission is promoting the market development and competitiveness of the U.S. high-tech industry through domestic and international policy efforts. EIA comprises companies whose products and services range from the smallest electronic components to the most complex systems used by defense, space and industry, including the full range of consumer electronic products. The progressive structure of the Alliance enables each sector association to preserve unique autonomy while uniting in common cause under EIA. One of the most commonly used EIA standard is EIA RS-232. This is a standard for 25-pin connector between a computer and a modem.

The World Wide Web Consortium (**W3C**) (http://www.w3.org) develops interoperable technologies (specifications, guidelines, software, and tools) to lead the Web to its full potential. W3C is a forum for information, commerce, communication and collective understanding. By promoting interoperability and encouraging an open forum for discussion, W3C is commited to leading the technical evolution of the Web. To meet the growing expectations of users and the increasing power of machines, W3C is already laying the foundations for the next generation of the Web. W3C's technologies will help make the Web a robust, scalable and adaptive infrastructure for a world of information. W3C contributes to efforts to standardize Web technologies by producing specifications (called 'Recommendations') that describe the building blocks of the Web. W3C recommendations include HTML, XML, CSS (Cascading Style Sheet), Web Services, DOM (Document Object Model), MathML (Maths Markup Language), PNG (Portable Network Graphics), SGV (Scalable Vector Graphics), RDF (Resource Description Framework), P3P (Platform for Privacy Preferences) etc.

3GPP (http://www.3gpp.org) is to produce globally applicable technical specifications and technical reports for 3rd Generation Mobile System based on evolved GSM core networks and the radio access technologies that they support—i.e., Universal Terrestrial Radio Access (UTRA) both

Frequency Division Duplex (FDD) and Time Division Duplex (TDD) modes. The scope was subsequently amended to include the maintenance and development of the Global System for Mobile communication (GSM) technical specifications and technical reports including evolved radio access technologies (e.g. General Packet Radio Service (GPRS) and Enhanced Data rates for GSM Evolution (EDGE)).

ANSI (www.ansi.org) The American National Standards Institute (ANSI) is the national standard organization in the United States. In many instances, U.S. standards are taken forward to ISO and IEC (International Electrotechnical Commission), where they are adopted in whole or in part as international standards. For this reason, ANSI plays an important part in creating international standards that support the worldwide sale of products, which prevent regions from using local standards to favor local industries. ANSI Standard X3.4-1968 defines 'American National Standard Code for Information Interchange (ASCII)' character set. ASCII character set is used in almost every modern computer today. The same standard has also been adapted as ISO 8859-1 standard.

UMTS (www.umts-forum.org) Universal Mobile Telecommunications System (UMTS) represents an evolution in terms of services and data speeds from today's 'second generation' mobile networks like GSM. As a key member of the 'global family' of third generation (3G) mobile technologies identified by the ITU, UMTS is the natural evolutionary choice for operators of GSM networks. Using fresh radio spectrum to support increased numbers of customers in line with industry forecasts of demand for data services over the next decade and beyond, UMTS is synonymous with a choice of WCDMA radio access technology that has already been selected by many licensees worldwide. UMTS-Forum is the standards-making body for WCDMA (Wideband Code Division Multiple Access) and UMTS technology.

Bluetooth (http://www.bluetooth.com) Bluetooth wireless technology is a worldwide specification for a small-form factor, low-cost radio solution that provides links between mobile computers, mobile phones, other portable handheld devices, and connectivity to the Internet. The standards and specification for Bluetooth are developed, published and promoted by the Bluetooth Special Interest Group.

CDMA Development Group (http://www.cdg.org). The CDMA Development Group (CDG) is an international consortium of companies who have joined together to lead the adoption and evolution of CDMA wireless systems around the world. The CDG is comprised of the world's leading CDMA service providers and manufacturers. By working together, the members will help ensure interoperability among systems, while expediting the availability of CDMA technology to consumers.

PKCS (http://www.rsasecurity.com/rsalabs/pkcs/). The Public-Key Cryptography Standards are specifications produced by RSA Laboratories in cooperation with secure systems developers worldwide for the purpose of accelerating the deployment of public-key cryptography. First published in 1991 as a result of meetings with a small group of early adopters of public-key technology, the PKCS documents have become widely referenced and implemented. Contributions from the PKCS series have become part of many formal and de facto standards, including ANSI X9 documents, PKIX, SET, S/MIME and SSL (Secure Sockets Layer).

PAM Forum (http://www.pamforum.org). In the world of ubiquitous computing, knowing the position and context of a device is very important. The Presence and Availability Management (PAM) Forum is an independent consortium with a goal to accelerate the commercial deployment of targeted presence and availability applications and services that respect users' preferences, permissions and privacy. Working in partnership with industry participants, the PAM Forum will define a framework for the various standards and specifications needed for context/location aware applications.

Parlay Group (http://www.parlay.org). The Parlay Group is a multi-vendor consortium formed to develop open, technology-independent application programming interfaces (APIs). Parlay integrates intelligent network (IN) services with IT applications via a secure, measured, and billable interface. By releasing developers from underlying code, networks and environments, Parlay APIs allow for innovation within the enterprise. These new, portable, network-independent applications are connecting the IT and telecom worlds, generating new revenue streams for network operators, application service providers (ASPs), and independent software vendors (ISVs). Using Parlay APIs, one will be able to develop applications, which rely on network-related data like service provisioning. Parlay will also help develop location/context aware applications and services.

1.12 PLAYERS IN THE WIRELESS SPACE

In a wireless network there are many stakeholders. These are:

1. Regulatory authorities

2. The operator or the service provider

3. The user or the subscriber

4. Equipment vendors (network equipment and user device)

5. Research organizations.

In most part of the world, the radio spectrums are regulated. Generally, a license is required to use a part of this spectrum. There are certain bands like ISM (Industry, Scientific and Medical) used by cordless telephones or microwave ovens or 802.11 which are unregulated. This means that if one develops equipment which needs to use these bands, then one need not go to the government and ask for a permission to use them. However, for the regulated bands, one has to get a license before it can be used. GSM, CDMA etc. use frequency bands, which are regulated. Therefore, a network company offering these services needs to get a clearance from the government. Governments generally auction these spectrums to different network operators. The spectrums for 3G networks were auctioned in Europe for about 100 billion Dollars. In India, the whole country was divided into metros and circles. The average license fee for GSM for these circles was in the tune of hundreds of crores of rupees.

Once the license is obtained, the network operator needs to conduct a detailed survey of the region with a plan for the cell sites. Cell site survey is very important and critical for any wireless network. Cell site survey is logically the design of the architecture of the network. During this phase the network operator determines the location of the base station and the positioning of the cell. The location of a wireless base station tower will be determined by many factors; examples could be subscriber density, hills and other obstacles. Similarly for wireless LAN or WiFi, the location of AP (Access Points) will be determined by the layout of the building floor, concrete, glass walls etc. A site survey is necessary for a wireless LAN as well before the APs are installed.

Cellular network operators need to create the infrastructure. There are a few equipment manufacturers who supply the hardware to the network operators. These hardware will be MSC (Mobile Switching Centre), BSC (Base Station Controller), BTS (Base Transceiver Station), and the Cells. Cells and BTSs are spread across the region; however, MSCs and BSCs are generally installed under one roof commonly known as switching room. Some of the leading manufacturers of these hardware are Ericsson and Nokia in Europe; Motorola and Lucent in the USA; Samsung in Asia.

To use a cellular network, we need a handset or device. There are different types of handsets available in the market today. All these devices offer voice and SMS as minimum, and range up to fancy handsets which offer WAP, MMS, J2ME or even digital cameras. Some of the leading suppliers of these handsets are Nokia, Sony Ericsson, Motorola, Samsung, LG etc. An Indian company is planning to manufacture mobile phones and enter this market.

In GSM world, all these handsets contain a small piece of card known as SIM (Subscriber Identity Module). These are technically Smart Cards. These are processor cards with a small memory and an independent processor. The size of the memory ranges from 8K bytes to 64K bytes. They contain some secured data installed by the network operator related to the subscriber and the network. Some of the leading suppliers of SIM card are Gemplus, Schlumberger, Orga etc.

When a person wants to subscribe to a cellular phone, he contacts a cellular operator. The subscriber is then registered with the network as a prepaid or postpaid subscriber. In a GSM network, the operator issues a SIM card to the subscriber with all the relevant information. The subscriber buys a handset and installs the SIM card inside the handset. A provisioning and activation needs to be done within the network for this new subscriber. During the provisioning, some of the databases within the operator will be updated. Once the databases for authentication and billing are completed, the subscriber is activated. Following the activation, the subscriber can use the network for making or receiving calls. In the case of a WiLL/CDMA network like Reliance Infocomm, there is no SIM card within the handset. However, all the provisioning information are embedded within the CDMA handset.

REFERENCES/FURTHER READING

1. Asoke K Talukder, Mobile Computing–Impact in Our Life, Harnessing and Managing Knowledge, edited by C. R. Chakravarthy, L. M. Patnaik, T. Sabapathy, M. L. Ravi, Tata McGraw-Hill, p 13, 2002.

2. David Lewin, Susan Sweet, The economic benefits of mobile services in India, A case study for the GSM Association, CLM28, Version 1, OVUM, January 2005.

3. Africa: The Impact of Mobile Phones, Moving the debate forward, The Vodafone Policy Paper Series, Number 2, March 2005.

4. Ashok Jhunjhunwala, Bhaskar Ramamurthi, and Timothy A. Gonsalves, The Role of Technology in Telecom Expansion in India, IEEE Communications Magazine, November 1998.

5. Ken Banks and Richard Burge, Mobile Phones: An Appropriate Tool For Conservation And Development. Fauna & Flora International, Cambridge, UK, 2004.

6. Dean Eyers, Telecom Markets and the Recession: An Imperfect Storm, Gartner Report, AV-14-9944, 27 November 2001.

7. John B. Horrigan, Senior Researcher, Lee Rainie, Director, Getting Serious Online, Pew Internet & American Life Project Report, March 2002.

8. The Path towards UMTS–Technologies for the Information Society, UMTS Forum Report # 2, 1998, http://www.umts-forum.org.

9. The Future Mobile Market Global trends and developments with a focus on Western Europe UMTS Forum Report # 8, March 1999, http://www.umts-forum.org.

10. Enabling UMTS Third Generation Services and Applications, UMTS Forum Report # 11, October 2000, http://www.umts-forum.org.

11. UMTS Report, An Investment Perspective, Durlacher Reasearch www.durlacher.com.

12. Alistair McGrath, Christophe de Hauwer, Tim Willey, and Adam Mantzos, Industry Analysis Wireless Data The World in Your Hand, Arthur Andersen Technology, Media and Communications, October, 2000.

13. Milojicic D., Douglis F., Wheeler R. (Ed), Mobility Processes, Computers, and Agents, Addison-Wesley, 1999.

REVIEW QUESTIONS

Q1: What are the essential functional differences between 1st Generation, 2nd Generation, and 3rd Generation of networks?

Q2: Describe what do you understand by Wireless PAN, Wireless LAN, Wireless MAN?

Q3: What is an ISM band? Why is it called free band?

Q4: What are the characteristic of a mobile computing environment?

Q5: Give an example of five mobile computing applications?

Q6: What are the advantages and disadvantages of standards? Name the standard committees responsible for 3G?

CHAPTER 2

Mobile Computing Architecture

2.1 HISTORY OF COMPUTERS

Nothing has changed the world around us the way digital technology and computers have. Computers have entered every aspect of our life and the environment around us. The origin of computers can be traced back to thousands of years. Though different forms of computers were in existence for centuries, the real transformation happened with electronic or digital computers. The development of electronic computer started during the Second World War. In 1941 German engineer Konrad Zuse had developed a computer called Z3 to design airplanes and missiles. In 1943, the British developed a computer called Colossus for cryptanalysis to decode encrypted messages transacted by Germans. With a team of engineers in 1944, Howard H. Aiken developed the Harvard–IBM Automatic Sequence Controlled Calculator Mark I, or Mark I for short. This is considered as the early general-purpose computer. In 1945 John von Neumann introduced the concept of stored program. Another general-purpose computer development spurred by the War was the Electronic Numerical Integrator and Computer, better known as ENIAC, developed by John Presper Eckert and John W. Mauchly in 1946. In 1947, the invention of the transistor by John Bardeen, Walter H. Brattain, and William Shockley at Bell Labs changed the development scenario of digital computers. The transistor replaced the large, energy-hungry vacuum tube in first generation computers. Jack Kilby, an engineer with Texas Instruments, developed the integrated circuit (IC) in 1958. IC combined all the essential electronic components (inductor, resistor, capacitor etc.) onto a small silicon disc, which was made from quartz. By the 1980s, very large scale integration (VLSI) squeezed hundreds of thousands of components onto a chip. VLSI led the development of third generation computers. All these early computers contained all the components we find today in any modern-day computers like printers, persistent storage, memory, operating systems and stored programs. However, one aspect of modern-day computers was missing in these machines–that was the networking aspect of today's computers.

2.2 History of Internet

Following the successful launch of Sputnik in 1957 by the Russians, USA felt the need of research in certain focused areas. Therefore, Advance Research Project Agency (ARPA) was formed to fund Science and Technology projects and position USA as a leader in technology. Internet represents one of the best examples of the benefits of sustained investment on research and development through ARPA. Beginning with the early research in packet switching, the government, industry and academia have been partners in evolving and deploying the exciting Internet technology. People in almost all parts of life starting from education, IT, telecommunications, business, and society at large have felt the influence of this pervasive information infrastructure. Today, almost anybody and everybody on the street uses terms like 'faculty@iiitb.ac.in' or 'http://www.isoc.org'.

In the early sixties, Leonard Kleinrock developed the basic principles of packet switching at MIT. During the same period Paul Baran in a series of RAND Corporation reports recommended several ways to accomplish packet switch network as well. In 1965 Lawrence G. Roberts in association with Thomas Merrill, connected the TX-2 computer in Massachusetts to the Q-32 in California with a low speed dial-up telephone line creating the first computer network. In 1971 Ray Tomlinson at BBN wrote the software to send and read simple electronic mail. In October 1972 demonstration of the ARPANET was done at the International Computer Communication Conference (ICCC). This was the first public demonstration of this new network technology to the public. It was also in 1972 that the initial 'hot' application, electronic mail, was introduced. In 1973 work began on the Transmission Control Protocol (TCP) at a Stanford University laboratory headed by Vincent Cerf.

In 1986, the U.S. NSF (National Science Foundation) initiated the development of the NSFNET which provides a major backbone communication service for the Internet. In Europe, major international backbones such as NORDUNET and others provide connectivity to a large number of networks. Internet slowly evolved as the universal network of networks, which connects almost every data networks of the world with a reach spread over the whole earth. It can be debated as to what the definition and scope of this global network is. On 24 October 1995, the Federal Networking Council (FNC) unanimously passed a resolution to officially define the term Internet. According to this resolution, the definition of Internet is 'Internet refers to the global information system that (i) is logically linked together by a globally unique address space based on the Internet Protocol (IP) or its subsequent extensions/follow-ons; (ii) is able to support communications using the Transmission Control Protocol/Internet Protocol (TCP/IP) suite or its subsequent extensions/follow-ons, and/or other IP-compatible protocols; and (iii) provides, uses or makes accessible, either publicly or privately, high level services layered on the communications and related infrastructure described herein.'

Vannevar Bush through his July 1945 essay 'As We May Think', described a theoretical machine he called a 'memex', which was to enhance human memory by allowing the user to store and retrieve documents linked by associations. This can be considered as the early hypertext. During 1960s, Doug Engelbart prototyped an 'oNLine System' (NLS) that does hypertext browsing, editing, etc. He invented the mouse for this purpose. In 1991 Tim Berners-Lee invented HTML (Hyper Text Markup Language) and HTTP (Hyper Text Transport Protocol). Tim wrote a client program and named it as 'WorldWideWeb', which finally became the 'www' (World Wide Web), almost synonymous with Internet. We would like to differentiate all these technologies by different names. We will use Web for the http, www technology, internet for the interworking with the network of networks, and Internet for the internet managed by IETF (Internet Engineering Task Force).

2.3 Internet—The Ubiquitous Network

For any content to be available anywhere, we need a ubiquitous network that will carry this content. As of today, there are two networks, which are ubiquitous. One is the telecommunication network and the other is the Internet network. Both these networks are in real terms network of networks. Different networks have been joined together using a common protocol (glue). In loose terms it can be stated that SS#7 is the glue for telecommunication network whereas TCP/IP is the glue for Internet. We need one of these networks to transport content from one place to another.

We have three types of basic contents: audio, video and text. Some of these contents can tolerate little delay in delivery whereas some cannot. Packet switched networks like Internet are better-suited for contents which can tolerate little delay. Telecommunication or circuit switch networks are better-suited for realtime contents that cannot tolerate delays. A ubiquitous application needs to use these networks for taking the content from one place to another. A network can be divided into three main segments viz., Core, Edge and Access.

Core: As the name signifies, core is the part of the network that is the backbone. This is the innermost part of the network. The primary functions for the core network is to deliver traffic efficiently at the least cost. Core looks at the traffic more from the bit stream point of view. Long-distance operators and backbone operators own core networks. This part of the network deals with transmission media and transfer points.

Edge: As the name suggests, this is at the edge of the network. These are generally managed and owned by ISPs (Internet Service Providers) or local switches and exchanges. Edge looks at the traffic more from the service point of view. It is also responsible for the distribution of the traffic.

Access: This part of the network services the end point or the device by which the service will be accessed. This deals with the last mile of transmission. This part is either through a wireline or the wireless. From the mobile computing point of view, this will be mostly through the wireless.

Internet is a network of networks and is available universally. In the last few years, the popularity of web-based applications has made more and more services available through the Internet. This had a snowball effect encouraging more networks and more contents to be added to the Web. Therefore, Internet is the preferred bearer network for audio, video or text contents that can tolerate delay. Internet supports many protocols. However, for ubiquitous access, web-based application is desirable. A web-based application in the Internet uses HTTP protocol and works like a request/response service. This is similar to conventional client/server application. The fundamental difference between a web application and a conventional client/server paradigm is that in the case of conventional client/server application, the user facing client interface contains part of the business logic. However, in the case of web applications, the client will be a thin client without any business logic. The thin client or the agent software in the client device will relate only to the rendering functions. Such user agents will be a web browser like Mozila, Internet Explorer or Netscape Navigator.

The types of client devices that can access the Internet are rapidly expanding. These client devices are networked either through the wireless or through a wireline. The server on the contrary, is likely to be connected to the access network through wired LAN. In addition to standard computers of different shapes and sizes, client devices can be Personal Digital Assistants (PDA) such as the PalmPilot, Sharp Zaurus, or iPaq; hand-held personal computers such as the EPOC, Symbian, Psion and numerous Windows-CE machines; mobile phones with GPRS/WAP and 3G capability such as Nokia, Sony Ericsson etc; Internet-capable phones such as the Smartphone (cellular) and Screenphone (wired); set-top boxes such as WebTV etc. Even good old voice-based telephone can be used as the client device. Voice-activated Internet browser will be very useful for visually challenged people. To fulfill the promise of universal access to the Internet, devices with very diverse capabilities need to be catered to. For the wireless, the devices range from the small footprint mobile phone to the large footprint laptop computers.

2.4 ARCHITECTURE FOR MOBILE COMPUTING

In mainframe computers many mission critical systems use Transaction Processing (TP) environment. At the core of a TP system, there is TP monitor software. In a TP system all the terminals (VDU–Visual Display Terminal, POS–Point of Sale Terminal, Printers etc.) are terminal resources (objects). There are different processing tasks, which process different transactions or messages; these are processing resources (objects). Finally there are database resources. A TP monitor manages terminal resources, database objects and coordinates with the user to pick up the right processing task to service business transactions. The TP monitor manages all these objects and connects them through policies and rules. A TP monitor also provides functions such as queuing, application execution, database staging, and journaling. When the world moved from large expensive centralized mainframes to economic distributed systems, technology moved towards two-tier conventional client/server architecture. With growth in cheaper computing power and penetration of Internet-based networked systems, technology is moving back to centralized server-based architecture. The TP monitor architecture is having a reincarnation in the form of three-tier software architecture.

In the early days of mainframes, the TP monitor and many other interfaces were proprietary. Even the networked interfaces to different terminals were vendor-specific and proprietary. The most successful early TP system was the reservation system for the American Airlines. This was over a Univac computer using U100 protocol. For IBM TP environment, which runs on OS/390 known as CICS (Customer Information Control System), the network interface was through SNA. In India when BSNL (earlier known as DoT–Department of Telecom) launched the 197 telephone directory enquiry system in 1986, it was on TPMS (Transaction Processing Management System) running on ICL mainframe running VME operating system. The network interface was over X.25 interface.

The network-centric mobile computing architecture uses a three-tier architecture Figure 2.1. In the three-tier architecture, the first layer is the User Interface or Presentation Tier. This layer deals with user facing device handling and rendering. This tier includes a user system interface where user services (such as session, text input, dialog and display management) reside. The second tier is the Process Management or Application Tier. This layer is for application programs or process management where business logic and rules are executed. This layer is capable of accommodating hundreds of users.

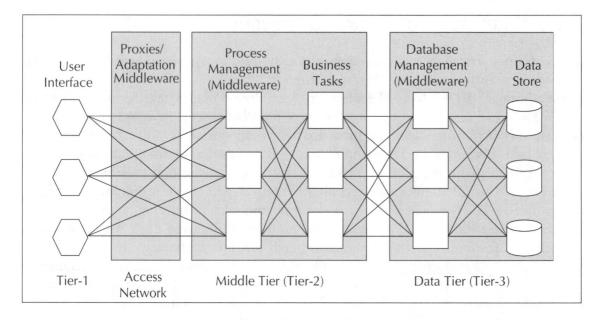

Figure 2.1 Three-tier Architecture for Mobile Computing

In addition, the middle process management tier controls transactions and asynchronous queuing to ensure reliable completion of transactions. The third and final tier is the Database Management or Data Tier. This layer is for database access and management. The three-tier architecture is better suited for an effective networked client/server design. It provides increased *performance*, *flexibility*, *maintainability*, *reusability*, and *scalability*, while hiding the complexity of distributed processing from the user. All these characteristics have made three-tier architectures a popular choice for Internet applications and net-centric information systems. Centralized process logic makes administration and change management easier by localizing changes in a central place and using it throughout the systems.

2.5 THREE-TIER ARCHITECTURE

To design a system for mobile computing, we need to keep in mind that the system will be used through any network, any bearer, any agent and any device. To have a universal access, it is desirable that the server is connected to a ubiquitous network like the Internet. To have an access from any device, a web browser is desirable. The reason is simple; web browsers are ubiquitous, we get a browser in any computer. The browser agent can be Internet Explorer or Netscape Navigator or Mozila or any other standard agent. Also, the system should preferably be context aware. We will discuss context awareness later.

We have introduced the concept of three-tier architecture. We have also discussed why it is necessary to go for Internet and three-tier architecture for mobile computing. The important question is what a mobile three-tier application actually should consist of. Figure 2.2 depicts a three-tier architecture for a mobile computing environment. These tiers are presentation tier, application tier and data tier. Depending upon the situation, these layers can be further sublayered.

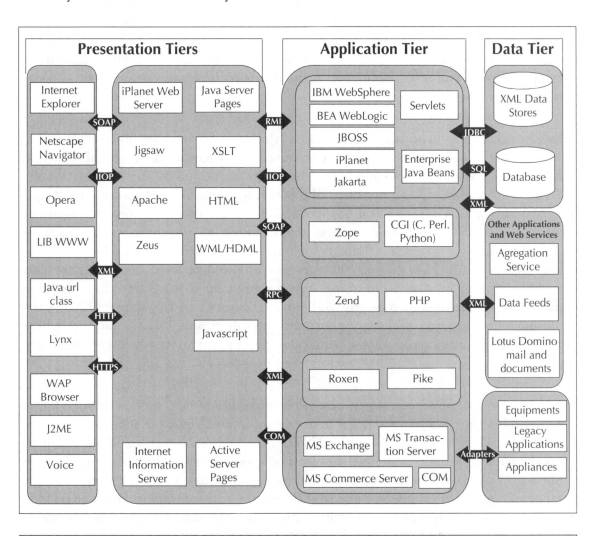

Figure 2.2 The mobile computing architecture

2.5.1 Presentation (Tier-1)

This is the user facing system in the first tier. This is the layer of agent applications and systems. These applications run on the client device and offer all the user interfaces. This tier is responsible for presenting the information to the end user. Humans generally use visual and audio means to receive information from machines (with some exceptions like vibrator in mobile phones). Humans also use keyboard (laptop computers, cell phones), pen (tablet PC, palmtops), touch screen (kiosks), or voice (telephone) to feed the data to the system. In the case of the visual, the presentation of information will be through a screen. Therefore, the visual presentation will relate to rendering on a screen. 'Presentation Tier' includes web browsers (like Mozila, lynx, Internet Explorer and Netscape Navigator), WAP browsers and customized client programs. A mobile computing agent needs to be context-aware and device-independent.

In general, the agent software in the client device is an Internet browser. In some cases, the agent software is an applet running on a browser or a virtual machine (Java Virtual Machine for example). The functions performed by these agent systems can range from relatively simple tasks like accessing some other application through http API, to sophisticated applications like realtime sales and inventory management across multiple vendors. Some of these agents work as a web scraper. In a web scraper, the agent embeds functionality of http browser and functions like an automated web browser. The scraper picks up part of the data from the web page and filters off the remaining data according to some predefined template. These applications can be in Business to Business (B2B) space, Business to Consumer (B2C) space or Business to Employee (B2E) space, or machine to machine (M2M) space. Applications can range from e-commerce, workflow, supply chain management to legacy applications.

There are agent software in the Internet that access the remote service through telnet interface. There are different flavors of telnet agents in use. These are standard telnet for Unix servers; TN3270 for IBM OS/390; TN5250 for IBM AS/400 or VT3K for HP3000. For some applications, we may need an agent with embedded telnet protocol. This will work like an automated telnet agent (virtual terminal) similar to a web scraper. These types of user agents or programs work as M2M interface or software robots. These kinds of agents are used quite frequently to make legacy applications mobile. Also, such systems are used in telecommunication world as mediation servers within the OSS (Operation and Support Subsystem).

2.5.2 Application Tier (Tier-2)

The application tier or middle tier is the 'engine' of a ubiquitous application. It performs the business logic of processing user input, obtaining data, and making decisions. In certain cases, this layer will do the transcoding of data for appropriate rendering in the Presentation Tier. The Application Tier may include technology like CGI's, Java, JSP, .NET services, PHP or ColdFusion, deployed in products like Apache, WebSphere, WebLogic, iPlanet, Pramati, JBOSS or ZEND. The application tier is presentation and database-independent.

In a mobile computing environment, in addition to the business logic there are quite a few additional management functions that need to be performed. These functions relate to decisions on rendering, network management, security, datastore access etc. Most of these functions are implemented using different middleware software. A middleware framework is defined as a layer of software, which sits in the middle between the operating system and the user facing software. Stimulated by the growth of network-based applications and systems, middleware technologies are gaining an increasing importance in the netcentric computing. In case of a netcentric architecture, a middleware framework sits between an agent and the business logic. Middleware covers a wide range of software systems, including distributed objects and components, message-oriented communication, database connectors, mobile application support, transaction drivers, etc. Middleware can also be considered as a software gateway connecting two independent open objects.

It is very difficult to define how many types of middleware are there. A very good description of middleware is available in Carnegie Mellon University Software Engineering Institute (http://www.sei.cmu.edu/str/descriptions/middleware.html), which readers can refer to.

We can group middleware into following major categories.

1. Message-Oriented Middleware

2. Transaction Processing Middleware

3. Database Middleware

4. Communication Middleware

5. Distributed Object and Components

6. Transcoding Middleware.

Message-oriented Middleware (MOM)

Message-oriented Middleware is a middleware framework that loosely connects different applications through asynchronous exchange of messages. A MOM works over a networked environment without having to know what platform or processor the other application is resident on. The message can contain formatted data, requests for action, or unsolicited response. The MOM system provides a message queue between any two interoperating applications. If the destination process is out of service or busy, the message is held in a temporary storage location until it can be processed. MOM is generally asynchronous, peer-to-peer, and works in publish/subscribe fashion. In publish/subscriber model one or many objects subscribe to an event. As the event occurs, it will be published by the asynchronous loosely coupled object. The MOM will notify the subscribers about this event. However, most implementations of MOM support synchronous (request/response) message passing as well. MOM is most appropriate for event-driven applications. When an event occurs, the publisher application hands off to the messaging middleware application the responsibility of notifying subscribers that the event has happened. In a netcentric environment, MOM can work as the integration platform for different applications. Example of MOM are Message Queue from IBM known as MQ Series. The equivalent from Java is JMS (Java Message Service).

Transaction Processing (TP) Middleware

Transaction Processing Middleware provides tools and an environment for developing transaction-based distributed applications. An ideal TP system will be able to input data into the system at the point of information source and the output of the system is delivered at the point of information sink. In an ideal TP system, the device for input and output can potentially be different (Figure 2.3). Also, the output can be an unsolicited message for a device. TP is used in data management, network access, security systems, delivery order processing, airline reservations, customer service, etc. to name a few. TP systems are generally capable of providing services to thousands of clients in a distributed client/server environment. CICS (Customer Information Control System) is one of the early. TP application systems on IBM mainframe computers.

TP middleware maps numerous client requests through application-service routines to different application tasks. In addition to these processing tasks, TP middleware includes numerous management features, such as restarting failed processes, dynamic load balancing and enforcing consistency of distributed data. TP middleware is independent of the database architecture. TP middleware optimizes the use of resources by multiplexing many client

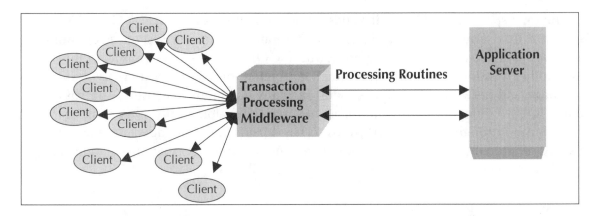

Figure 2.3 Transaction Processing Middleware

functions onto a much smaller set of application-service routines. This also helps in reducing the response time. TP middleware provides a highly active system that includes services for delivery-order processing, terminal and forms management, data management, network access, authorization, and security. In the Java world and net-centric systems, transaction processing is done through J2EE application server through entity beans and session beans.

Communication Middleware

Communication Middleware is used to connect one application to another application through some communication middleware, like connecting one application to another application through telnet. These types of middleware are quite useful in the telecommunication world. There are many elements in the core telecommunication network where the user interface is through telnet. A mediation server automates the telnet protocol to communicate to these nodes in the network. Another example could be to integrate legacy applications through proprietary communication protocols like TN5250 or TN3270.

Distributed Object and Components

An example of distributed objects and components is CORBA (Common Object Request Broker Architecture). CORBA is an open distributed object computing infrastructure being standardized by the Object Management Group (http://www.omg.org). CORBA simplifies many common network programming tasks used in a netcentric application environment. These are object registration, object location, and activation; request demultiplexing; framing and error-handling; parameter marshalling and demarshalling; and operation dispatching. CORBA is vendor-independent infrastructure. A CORBA-based program from any vendor on almost any computer, operating system, programming language and network, can interoperate with a CORBA-based program from the same or another vendor, on almost any other computer, operating system, programming language and network. CORBA is useful in many situations because of the easy way that CORBA integrates machines from so many vendors, with sizes ranging from mainframes through minis and desktops to hand-helds and embedded systems. One of its most important, as well as the most frequent, uses is in servers that must handle a large number of clients, at high hit rates, with high reliability.

Transcoding Middleware

Transcoding Middleware is used to transcode one format of data to another format to suit the need of the client. For example, if we want to access a web site through a mobile phone supporting WAP, we need to transcode the HTML page to WML page so that the mobile phone can access it. Another example could be accessing a map from a PDA. The same map, which can be shown in a computer, needs to be reduced in size to fit the PDA screen. Technically transcoding is used for content adaptation to fit the need of the device. Content adaptation is also required to meet the network bandwidth needs. For example, some frames in a video clip need to be dropped for a low bandwidth network. Content adaptation used to be done through proprietary protocols. To allow interoperability, IETF has accepted the Internet Content Adaptation Protocol (ICAP). ICAP is now standardized and described in RFC3507.

Internet Content Adaptation Protocol (ICAP)

Popular web servers are required to deliver content to millions of users connected at ever-increasing bandwidths. Progressively, contents is being accessed through different devices and agents. A majority of these services have been designed keeping the desktop user in mind. Some of them are also available for other types of protocols. For example, there are a few sites that offer contents in HTML and WML to service desktop and WAP phones. However, the model of centralized services that are responsible for all aspects of every client's request seems to be reaching the end of its useful life. ICAP, the Internet Content Adaptation Protocol, is a protocol aimed at providing simple object-based content vectoring for HTTP services. ICAP is a lightweight protocol to do transcoding on HTTP messages. This is similar to executing a 'remote procedure call' on a HTTP request. The protocol allows ICAP clients to pass HTTP messages to ICAP servers for some sort of transformation. The server executes its transformation service on messages and sends back responses to the client, usually with modified messages. The adapted messages may be either HTTP requests or HTTP responses. For example, before a document is displayed in the agent, it is checked for virus.

There are two major components in ICAP architecture:

1. What is the semantics for the transformation? How do I ask for content adaptation?
2. How is policy of the transformation managed? What kind of adaptation do I ask for and from where? How do I define and manage the adaptation?

ICAP works at the edge part of the network as depicted in Figure 2.4. It is difficult, if not impossible, to define the devices users may like to use to access content within Internet. Customized edge delivery of Internet content will help to improve user experience. When applications are delivered from an edge device, end users find the applications execute more quickly and are more reliable. Typical data flow in an ICAP environment is depicted in Figure 2.4 and is described below.

1. A user agent makes a request to an ICAP-capable surrogate (ICAP client) for an object on an origin server.
2. The surrogate sends the request to the ICAP server.
3. The ICAP server executes the ICAP resource's service on the request and sends the possibly modified request, or a response to the request back to the ICAP client.

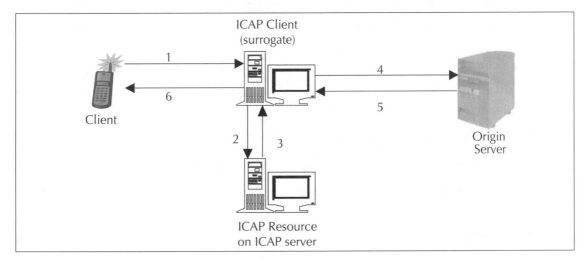

Figure 2.4 Typical data flow in an ICAP environment

4. The surrogate sends the request, possibly different from the original client request, to the origin server.

5. The origin server responds to the request.

6. The surrogate sends the reply (from either the ICAP server or the origin server) to the client.

It is envisioned that in future, ICAP servers may be available to provide some of the following services:

- Suite content delivery based on network bandwidth.
- Suite content delivery based on device characteristics.
- Language translation based on the user's preference.
- Virus checking for the requested content.
- Content filtering based on the sensor rating like PG (Parental Guidance), R (Restricted).
- Local realtime advertisement insertion like television
- Local realtime advertisement elimination for premium subscribers
- Wireless protocol translation
- Anonymous Web usage profiling for a dating service
- Transcoding or image enhancement
- Image magnification for the elderly
- Image size reduction based on device display characteristics
- Intelligent Video condensation by dropping frames
- Digest production/batch download of Web content
- Content filtering based on copyright or digital signature
- Peer-to-Peer compression and encryption of data

Web services

As the need for peer-to-peer, application-to-application communication and interoperability grows, the use of Web services on the Internet will also grow. Web Services provide a standard means of communication and information exchange among different software applications, running on a variety of platforms or frameworks. *Web service is a software system identified by a URI, whose public interfaces and bindings are defined using XML (eXtensible Markup Language).* Its definition can be discovered by other software systems connected to the network. Using XML based messages these systems may then interact with the Web service in a manner prescribed by its definition.

The basic architecture includes Web service technologies capable of:

- Exchanging messages.
- Describing Web services.
- Publishing and discovering Web service descriptions.

The Web services architecture defines the standards for exchange of messages between service requester and the service provider. Service providers are responsible for publishing a description of the services they provide. Requesters must be able to find and discover the descriptions of the services.

Software agents in the basic architecture can take on one or all of the following roles:

- Service requester–requests the execution of a Web service.
- Service provider–processes a Web service request.
- Discovery agency–agency through which a Web service description is published and made discoverable.

The interactions involve the publish, find and bind operations. A service is invoked after the description is found, since the service description is required to establish a binding.

2.5.3 Data Tier (Tier-3)

The Data Tier is used to store data needed by the application and acts as a repository for both temporary and permanent data. The data could be stored in any form of datastore or database. These can range from sophisticated relational database, legacy hierarchical database, to even simple text files. The data can also be stored in XML format for interoperability with other system and datasources. A legacy application can also be considered as a data source or a document through a communication middleware.

Database Middleware

We have discussed that for a mobile computing environment, the business logic should be independent of the device capability. Likewise, though not essential, it is advised that business logic should be independent of the database. Database independence helps the maintenance of the system better. Database middleware allows the business logic to be independent and transparent of the database technology and the database vendor. Database middleware runs between the application program and the database. These are sometimes called database connectors as well. Example of such middleware will be ODBC, JDBC, etc. Using these middleware, the application will be able to access data from any data source. Data sources can be text files, flat file, spreadsheets, or a network, relational, indexed, hierarchical, XML database, object database, etc. from vendors like Oracle, SQL, Sybase, etc.

SyncML

SyncML protocol is an emerging standard for synchronization of data access from different nodes. When we moved from conventional client/server model of computing to netcentric model of computing, we moved from distributed computing to centralized computing with networked access. The greatest benefit of this model is that resources are managed at a centralized level. All the popular mobile devices like handheld computers, mobile phones, pagers and laptops work in an occasionally connected computing mode and access these centralized resources from time to time. In an occasionally connected mode, some data are cached in the local device and accessed frequently. The ability to access and update information on the fly is key to the pervasive nature of mobile computing. Examples are emails and personal information like appointments, address book, calendar, diary, etc. Storing and accessing the phone numbers of people from the phone address book is more user-friendly compared to accessing the same from a server. However, managing appointments database is easier in a server, though cacheing the same on the mobile client is critical. Users will cache emails into the device for reference. We take notes or draft a mail in the mobile device. For workflow applications, data synchronization plays a significant role. The data in the mobile device and the server need to synchronize. Today vendors use proprietary technology for performing data synchronization. SyncML protocol is the emerging standard for synchronization of data across different nodes. SyncML is a new industry initiative to develop and promote a single, common data synchronization protocol that can be used industry-wide.

The ability to use applications and information on a mobile device, then to synchronize any updates with the applications and information back at the office or on the network, is key to the utility and popularity of mobile computing. The SyncML protocol supports naming and identification of records, common protocol commands to synchronize local and network data. It supports identification and resolution of synchronization conflicts. The protocol works over all networks used by mobile devices, both wireless and wireline. Since wireless networks employ different transport protocols and media, a SyncML will work smoothly and efficiently over:

- HTTP 1 (i.e. the Internet)

- WSP (the Wireless Session Protocol, part of the WAP protocol suite)

- OBEX (Object Exchange Protocol, i.e. Bluetooth, IrDA and other local connectivity)

- SMTP, POP3 and IMAP

- Pure TCP/IP networks

- Proprietary wireless communication protocols.

2.6 DESIGN CONSIDERATIONS FOR MOBILE COMPUTING

The mobile computing environment needs to be context-independent as well as context-sensitive. Context information is the information related to the surrounding environment of an actor in that environment. The term 'context' means, all the information that help determine the state of an object (or actor). This object can be a person, a device, a place, a physical or computational object, the surrounding environment or any other entity being tracked by the system. In a mobile computing environment, context data is captured so that decisions can be made about how to adapt content or behavior to suit this context. Mobility implies that attributes associated with devices and users will change constantly. These changes mean that content and behavior of applications should be adapted to suit the current situation. There are many ways in which content and behavior can be adapted. Following are some examples:

1. **Content with context awareness** Build each application with context awareness. There are different services for different client context (devices). For example a bank decides to offer mobile banking application through Internet, PDA and mobile phone using WAP. These services are different and are http://www.mybank.com/inet.html, http://www.mybank.com/palm.html and http://www.mybank.com/wap.wml, respectively. The service http://www.mybank.com/inet.html assumes that the user will use computers to access this

service. Therefore it is safe to offer big pages with text box, drop down menu. Also, it is fine to add a few animated pictures for the new product the bank is launching. We know that http://www.mybank.com/palm.html is a service for a PalmOS PDA. As the display size is small, we design the screen to be compact for the PDA and do not offer the same product animation. For the WAP service at http://www.mybank.com/wap.wml, we do a completely different user interface; we make all drop down options available through the option button in the mobile phone and remove all the graphics and animations.

2. **Content switch on context** Another way is to provide intelligence for the adaptation of content within the service. This adaptation happens transparent to the client. In this case the service is the same for Internet, PDA and WAP. All access the bank's service through http://www.mybank.com/. A intelligent piece of code identifies the agent to decide what type of device or context it is. This intelligent code does the adaptation at run time based upon the agent in hand. The simplest way to do this is to look at the User-Agent value at the HTTP header and decide whether to route the request to http://mybank.com/inet.html or http://www.mybank.com/palm.html or http://www.mybank.com/wap.wml.

3. **Content transcoding on context.** Another way is to provide an underlying middleware platform that performs the adaptation of the content based on the context and behavior of the device. This adaptation happens transparent to the client and the application. The middleware platform is intelligent enough to identify the context either from the http parameters or additional customized parameters. In this case the service may be in html or XML, the middleware platform transcode the code from html (or XML) to html, and wml on the fly. It can also do the trasncoding based on policy so that the html generated for a computer is different from a PDA.

Following sections describe different types of context that can enhance the usability, reliability and security of the service. Figure 2.5 depicts the old web and the web of the future for mobile computing.

2.6.1 Client Context Manager

When we humans interact with other persons, we always make use of the implicit situational information of the surrounding environment. We interpret the context of the current situation and react appropriately. For example, we can go close to a lion in a zoo, but definitely not in the wild. Or, a person discussing some confidential matter with another person observes the gestures and voice tone of the other person and reacts in an

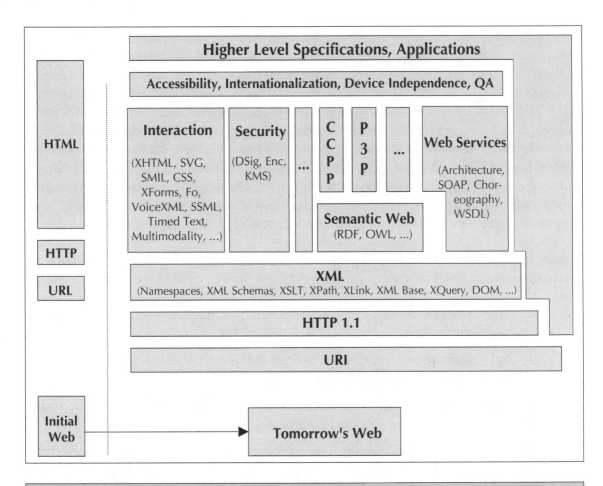

Higher Level Specifications, Applications

Accessibility, Internationalization, Device Independence, QA

HTML

Interaction

(XHTML, SVG, SMIL, CSS, XForms, Fo, VoiceXML, SSML, Timed Text, Multimodality, ...)

Security

(DSig, Enc, KMS)

...

C C P P

P 3 P

...

Semantic Web
(RDF, OWL, ...)

Web Services

(Architecture, SOAP, Choreography, WSDL)

HTTP

URL

XML
(Namespaces, XML Schemas, XSLT, XPath, XLink, XML Base, XQuery, DOM, ...)

HTTP 1.1

URI

Initial Web

Tomorrow's Web

Figure 2.5 The content architecture with respect to mobile computing

appropriate manner or changes the subject if someone shows up suddenly. When we use content through a PC within the four walls of the organization, we do not have any problem. A majority of the applications can safely assume that the context is the enterprise LAN. It can be assumed that the environment is secured; it can also be assumed that the user will be using the systems in a particular fashion using the browser standardized by the company. These applications are developed keeping the large screen (for mainly PC) and browsers in mind. A mobile computing application, on the other hand, needs to operate in dynamic conditions. This is due to various device characteristics and network conditions. This demands a reactive platform that can make decisions about how to respond to changes to device capability, user preferences, enterprise policy, network policy and many other environmental factors. Context can be used as the basis by which an adaptation manager or algorithm decides to modify content or application behavior. We therefore need a Client Context Manager to gather and maintain information pertaining to the client device, user, network and the environment surrounding each mobile device. All these information will be provided by a set of *Awareness Modules.* Awareness modules are sensors of various kinds. These sensors can be hardware sensors or software sensors or a combination of these. A hardware sensor can be used to identify the precise location of a user; whereas, a software sensor can be used to determine the type of the user agent. These awareness modules can be in the device, network, or even in the middleware. We use the term middleware in a very generic context. A middleware can be a functional module in the content server, a proxy or an independent system. For example, an awareness module in the device will provide information about its capabilities. Another example could be a location manager that tracks the location and orientation of the mobile device.

Almost any information available at the time of an interaction can be seen as context information. Some examples are:

1. **Identity:** The device will be in a position to communicate its identity without any ambiguity.

2. **Spatial information:** Information related to the surrounding space. This relates to location, orientation, speed, elevation and acceleration.

3. **Temporal information:** Information related to time. This will be time of the day, date, timezone and season of the year.

4. **Environmental information:** This is related to the environmental surroundings. This will include temperature, air quality, moisture, wind speed, natural light or noise level. This also includes information related to the network and network capabilities.

5. **Social situation:** Information related to the social environment. This will include who you are with, and people that are nearby; whether the user is in a meeting or in a party.

6. **Resources that are nearby:** This will relate to the other accessible resources in the nearby surroundings like accessible devices, hosts or other information sinks.

7. **Availability of resources:** This will relate to information about the device in use. This will include battery power, processing power, persistence store, display, capabilities related to I/O (Input/Output) and bandwidth.

8. **Physiological measurements:** This relates to the physiological state of the user. This includes information like blood pressure, heart rate, respiration rate, muscle activity and tone of voice.

9. **Activity:** This relates to the activity state of the user. This includes information like talking, reading, walking and running.

10. **Schedules and agendas:** This relates to the schedules and agendas of the user.

A system is context-aware if it can extract, interpret and use context-related information to adapt its functionality to the current context. The challenge for such systems lies in the complexity of capturing, representing, filtering and interpreting contextual data. To capture context information generally some sensors are required. This context information needs to be represented in a machine-understandable format, so that applications can use this information. In addition to being able to obtain the context-information, applications must include some 'intelligence' to process the information and deduce the meaning. These requirements lead us to three aspects of context management:

1. **Context sensing:** The way in which context data is obtained.

2. **Context representation:** The way in which context information is stored and transported.

3. **Context interpretation:** The way in which meaning is obtained from the context representation.

W3C has proposed a standard for context information. This standard is called Composite Capabilities/Preference Profiles (CC/PP), for describing device capabilities and user preferences. All these context information are collated and made available to the management components.

Composite Capabilities/Preference Profiles (CC/PP)

Composite Capabilities/Preference Profiles (CC/PP) is a proposed W3C standard for describing device capabilities and user preferences. Special attention has been paid to wireless devices such as mobile phones and PDAs. In practice, the CC/PP model is based on RDF (Resource Description Framework) and can be serialized using XML.

A CC/PP profile contains a number of attribute names and associated values that are used by an application to determine the appropriate form of a resource to deliver to a client. This is to help a client or proxy/middleware to describe their capabilities to an origin server or other sender of resource data. It is anticipated that different applications will use different vocabularies to specify application-specific properties within the scope of CC/PP. However, for different applications to interoperate, some common vocabulary is needed. The CC/PP standard defines all these. Following is an example of a device RDF in CC/PP terminology.

```xml
<?xml version="1.0"?>
<!-- Checked by SiRPAC 1.16, 18-Jan-2001 -->
<rdf:RDF xmlns:rdf="http://www.w3.org/1999/02/22-rdf-syntax-ns#"
         xmlns:ccpp="http://www.w3.org/2000/07/04-ccpp#">

  <rdf:Description rdf:about="MyProfile">
    <ccpp:component>
      <rdf:Description rdf:about="TerminalHardware">
        <rdf:type rdf:resource="HardwarePlatform" />
        <display>320x200</display>
      </rdf:Description>
    </ccpp:component>

    <ccpp:component>
      <rdf:Description rdf:about="TerminalSoftware">
        <rdf:type rdf:resource="SoftwarePlatform" />
        <name>EPOC</name>
        <version>2.0</version>
        <vendor>Symbian</vendor>
      </rdf:Description>
    </ccpp:component>

    <ccpp:component>
      <rdf:Description rdf:about="TerminalBrowser">
        <rdf:type rdf:resource="BrowserUA" />
        <name>Mozilla</name>
        <version>5.0</version>
        <vendor>Symbian</vendor>
        <htmlVersionsSupported>
          <rdf:Bag>
            <rdf:li>3.0</rdf:li>
            <rdf:li>4.0</rdf:li>
          </rdf:Bag>
        </htmlVersionsSupported>
```

```
    </rdf:Description>
   </ccpp:component>

  </rdf:Description>
 </rdf:RDF>
```

CC/PP is designed in such a way that an origin server or proxy can perform some sort of content to device matching. CC/PP is designed to suit an adaptation algorithm. The sequence of steps in the general case would look something like the following (Figure 2.6):

1. Device sends serialized profile model with request for content.

2. Origin server receives serialized RDF profile and converts it into an in-memory model.

3. The profile for the requested document is retrieved and an in-memory model is created.

4. The device profile model is matched against the document profile model.

5. A suitable representation of the document is chosen. At this stage the document to be returned can be chosen from a number of different versions of the same document (content switch on context) or it can be dynamically generated (content transcoding on context).

6. Document is returned to device and presented.

If a document or application is specific about how it should be displayed, or if there are several versions of the document or application for different devices, then the Adaptation Manager can ask the Client Context Manager for detailed context information. Client context manager will enquire with relevant Awareness Module and extract the necessary

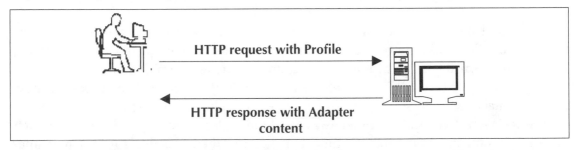

Figure 2.6 The simplest use of CC/PP

context information. This fine-grained approach allows a high level of adaptation to take place. In cases where the document does not provide profile information, or the profile is limited in description, the Adaptation Manager can obtain a general context class from the Context Manager and perform some limited adaptation. For example, some adaptation can still take place where the location of the user is important. The Policy Manager can specify some rules about how adaptation should take place when a user is at a certain location, regardless of the information provided in an application or document profile.

Policy Manager

The Policy Manager is responsible for controlling policies related to mobility. A policy is a set of rules; these rules need to be followed under different conditions. Introduction of mobility within an enterprise brings with it different types of challenges that are not normally seen in traditional computing environments. When we consider mobility, it is assumed that the data or information will be visible from outside the four walls of the enterprise. Organizations generally have policies regarding the disclosure of information. For example, documents from certain systems can be printed only on certain printers in the organization. Some hard copy documents may be viewed only at the office of the CEO. These kinds of policies must be transferable to a mobile computing environment. Mobile computing Policy Manager will be able to define policy for documents/services and assign roles to users. Each role will have permissions, prohibitions and obligations associated with it. Each policy will have access rights associated with respect to read, write, execute. A policy in combination with role and current context information will be able to determine what actions a user is allowed to perform, or what actions a user is obligated to perform.

Semantic Web

As mentioned earlier, policies are sets of rules. When we drive in the street we are expected to follow the right of way. In a party there are some etiquettes to be followed. We humans learn these rules, policies, laws, and etiquettes from documents or experienced people. This is to help us to behave correctly in the society. The question is how to make a machine understand policies and make them behave in the expected fashion? Data in the Web is generally hidden away in HTML files, how do we determine which content is useful in some contexts, but often not in others. Facilities to put machine-understandable data on the Web are becoming a necessity. The Semantic Web is targeted to address this need. The idea of having data on the web defined and linked in a way that it can be used by machines not just for display purposes, but for automation, security, filtering, integration and reuse of data across various applications.

Semantic Web technologies are still very much in their infancy. It is believed that a large number of Semantic Web applications can be used for a variety of different tasks, increasing the modularity of applications on the Web. The Semantic Web is generally built on syntaxes which use URIs to represent data, usually in tuple-based structures: i.e. many tuples of URI data that can be held in databases, or interchanged on the world Wide Web using a set of particular syntaxes developed especially for the task. These syntaxes are called 'Resource Description Framework (RDF)' syntaxes.

Security Manager

The Security Manager provides secure connection between the client device and the origin server. Depending on the security policies of an organization, if the security requirements are not met, some content may not be viewable. Security manager will ensure security with respect to:

- **Confidentiality**–the message being transacted needs to be confidential. Nobody will be able to see it.

- **Integrity**–the message being transacted needs to be tamper-resistant. Nobody will be able to change any part of the message.

- **Availability**–the system will be available. Nobody will be able to stop the service.

- **Non-repudiation**–the users of the system can be identified. Nobody after using the system can claim otherwise.

- **Trust**–there are complex issues of knowing what resources, services or agents to trust. The system will be trusted.

Confidentiality is managed by encryption. Using encryption techniques we change the message to some other message so that it cannot be understood. There are different types of encryption algorithms and standards. In a defined environment like enterprise LAN or a VPN (Virtual Private Network), we can standardize some encryption algorithm like 128 bits AES to be used. However, in a ubiquitous environment, the environment is unpredictable with ad hoc groups of devices. Also, the networks and their security level cannot be guaranteed all the time. Integrity can be managed using different hashing algorithms. Availability relates to peripheral security related to Web server, firewall etc. The non-repudiation can be managed with digital signature. For trust we may need to establish some sort of third-party recommendation system. Third-party rating system can also help establish trust. The security manager needs to manage all these aspects.

Platform for Privacy Preference Project (P3P)

The Platform for Privacy Preference Project (P3P) is an emerging standard defined by W3C. P3P enables web sites to express their privacy practices in a standardized format so that it can be retrieved and interpreted by user agents. With P3P, users need not read the privacy policies they visit; instead, key information about the content of the web site can be conveyed to the user. Any discrepancies between a site's practices and the user's preferences can be flagged as well. The goal of P3P is to increase user trust and confidence in the Web.

P3P provides a technical mechanism to inform users about privacy policies about the site. This will help users to decide whether to release personal information or not. However, P3P does not provide any mechanism for ensuring that sites act according to their policies. P3P is intended to be complementary to both legislative and self-regulatory programs that can help enforce web site policies.

Adaptability Manager

The Adaptability Manager is responsible for adapting content, behavior and other aspects according to context and policy. The Adaptability Manager may take any number of actions depending on the information passed to it by the Context Manager. This information may or may not be in the form of RDF. The most obvious action to perform is to transcode content so that it may be viewed on a particular device. Other actions might include appending location-specific information to documents.

Content Adaptation and Transcoding

In a ubiquitous situation, services are used from any device through any network. Therefore, the content should be able to adapt to these dynamic situations. The adaptation may be static or dynamic.

Content adaptation can be performed either at the content level at the server end or at the agent level in the client device. Content adaptation can be done at an intermediate level in a middleware framework as well. To do a good job of content adaptation, we need to go beyond the header. We need to consider the requirements of the entire Web page or relationships between its various components in different media. It also needs to look at adaptation within the scope of the same and a different modality. Modes can be audio, video, voice, image or text. We are differentiating between audio and voice by the characteristics that audio is a sound clip as an object like the audio part of a multimedia lecture, whereas voice is realtime and synthesized from some other form or representation. Content adaptation needs to consider the following attributes.

1. **Physical capabilities of the device,** viz., screen size i.e., width and height in pixels, color and bits/pixel.

2. **Logical capabilities of the device** for displaying video, image and playing audio.

3. Effective **Network bandwidth**.

4. **Payload** can be defined as the total amounts of bits that can be delivered to the agent for the static parts. For streaming media this will be the initial buffer space required before the media starts playing. For storage constrained devices, the payload will be defined as the storage space.

Transcoding can be classified as the following:

- **Spatial transcoding** is transcoding in space or dimension. In this transcoding technique a standard frame is downscaled and reduced. The frame is changed from one size to a different size to suit the target device.

- **Temporal transcoding** copes with a reduction of number of frames in the time scale. This technique downscales the number of transferred frames to suit the target device and the network bandwidth.

- **Color transcoding** is sometimes requested for monochrome clients. Using less bits for pixel can reduce bandwidth and sometime modify the perception of images.

- **Code transcoding** is used to change coding from one standard to another. One such example could be compression of the data or transcode a BMP file to WBMP for wireless device.

- **Object or semantic transcoding** comprises some different techniques based on computer vision techniques. The goal is to extract semantically valuable objects from the scene and transfer them with the lower amount of compression in order to maintain both details and speed.

Server side content adaptation can be achieved through the concept of InfoPyramid. InfoPyramid creates context-aware content through static transcoding. The transcoding is done off-line at the content creation time. InfoPyramid is used to store multiple resolutions and modalities of the transcoded content, along with any associated meta-data. For server side adaptation, each atomic item of the document is analysed to determine its resource requirements. The types of resources considered are those that may differentiate different client devices. The resource requirements is determined by the following attributes.

1. Static content size in bits.

2. Display size such as height, width and area.

3. Streaming bit-rate.

4. Color requirements.

5. Compression formats.

6. Hardware requirements, such as display for images, support for audio and video.

This is very useful for enterprises whose users are likely to use the service from different networks and devices. For example, a bank or a courier company which has its customer base across the world and is likely to use the service from any device from any network. When the Web server receives a user request, it determines the capabilities of the requesting client device. A customization module (context-sensitive content switch) dynamically selects the page from the InfoPyramids. The selection is based on the resolutions or modalities that best meet the client capabilities. This selected content is then rendered in a suitable delivery format for delivery to the client. This type of transcoding is most suitable for enterprises where the content type is known.

In case of client-side adaptation, the adaptation is done by the agent application. The agent application does the adaptation based on its capabilities. For example, let us assume that the client device does not support color: therefore, a color image received by the agent will be displayed as black and white image. Client-side adaptation can be quite effective for static images. However, it may not be very effective for streaming payload delivery.

The other technique of transcoding is through a middleware. One big benefit of the middleware approach is that it is totally transparent to the device and the content. Content providers do not have to change the way they author or serve content. However, there are a number of drawbacks to this approach:

1. Content providers have no control over how their content will appear to different clients.

2. There may be legal issues arising from copyright that may preclude or severely limit the transcoding by proxies.

3. HTML tags mainly provide formatting information rather than semantic information.

4. Transcoding sometimes-could be difficult to apply to many media types such as video and audio.

5. Developing a general purpose transcoding engine is very difficult if not impossible.

Transcoding through middleware is transparent to both device and content. Therefore, this transcoding technique has to be very robust and universal. That is why this transcoding technique is the most difficult to engineer. It is most desirable for content aggregators and value added service providers.

Content Rating and Filtering

Any city in the world has regions well marked like business district, residential area, shopping complex, so on and so forth. In Bangalore, for example, Commercial Street, Koramangala, Shivaji Market signify commercial/shopping area, residential area and market place respectively. Looking at the name of a web site or the document header, can we make some judgment about the content? This is necessary for content filtering and personalization. If we want to make sure that children at home are not accessing some restricted material, how do we do this? In a bookstore, adult magazines are displayed on the topmost shelf so that children cannot reach them. Children below 18 are not allowed to buy cigarettes or alcohol from a shop. In Internet, everything is freely accessible. How do we enforce such social discipline in the electronic world?

W3C has proposed a standard called PICS (Platform for Internet Content Selection) for rating of web content. Filtering of the content can take place depending on this rating. PICS specification is a set of technical specifications for labels (meta-data) that help software and rating services to work together. Rating and labeling services choose their own criteria for proper identification and filtering of the content. Since rating will always involve some amount of subjective judgment, it is left to the service provider to define the ratings. Rating can be through self-labeling or third party labeling of content. In a third party labeling some independent rating agency can be used. The rating of Internet sites was originally designed to help parents and teachers control what children access on the Internet, but it also facilitates other uses for labels, including code signing and privacy.

The RSACI (Recreational Software Advisory Council–Internet) has a PICS-compliant rating system called Resaca. Web pages that have been rated with the Resaca system contain labels recognized by many popular browsers like Netscape and Internet Explorer. Resaca uses four categories–violence, nudity, sex, and language–and a number for each category indicating the degree or level of potentially offensive content. Each number can range from 0, meaning the page contains no potentially offensive content, to 4, meaning the page contains the highest levels of potentially offensive content. For example, a page with a Resaca language level of 0 contains no offensive language or slangs. A page with a language level of 4 contains crude, vulgar language or extreme hate speech. When an end-user asks to see a particular URL, the software filter fetches the document but also makes an inquiry to the label bureau to ask for labels that describe that URL. Depending on what the labels say, the filter may block access to that URL. PICS labels can describe anything that can be named with a URL. That includes FTP and Gopher. E-mail messages do not normally have URLs, but messages from discussion lists that are archived on the Web do have URLs and can thus be labeled. A label can include a cryptographic signature. This mechanism lets the user check that the label was authorized by the service provider.

While the motivation for PICS was concern over children accessing inappropriate materials, it is a general 'meta-data' system, meaning that labels can provide any kind of descriptive information about Internet materials. For example, a labeling vocabulary could indicate the literary quality of an item rather than its appropriateness for children. Most immediately, PICS labels could help in finding particularly desirable materials, and this is the main motivation for the ongoing work on a next generation label format that can include arbitrary text strings. More generally, the W3C is working to extend Web meta-data capabilities generally and is applying them specifically in the following areas:

1. Digital Signature–coupling the ability to make assertions with a cryptographic signature block that ensures integrity and authenticity.

2. Intellectual Property Rights Management–using a meta-data system to label Web resources with respect to their authors, owners and rights management information.

3. Privacy (P3)–using a meta-data system to allow sites to make assertions about their privacy practices and for users to express their preferences for the type of interaction they want to have with those sites.

4. Personalization–based on some policy, the content can be personalized to suit the need of the user and the service.

Regardless of content control, meta-data systems such as PICS are going to be an important part of the Web, because they enable more sophisticated commerce (build and manage trust relationships), communication, indexing, and searching services. Content filtering can take place either at the client end or at the middleware proxy end.

Content Aggregation

Over a period, the dynamics associated with the content has changed considerably. Earlier there was a requester requesting for content and a responder responding to the content requested. The game was simple with only two players, the requester and the responder. These contents were corporate content or content for the mass (primarily web sites). There was no concept of charging for the content. Today there is a concept of OEM (Original Equipment Manufacturer) in content. There are some organizations who creates contents like an OEM. There are other ASPs (Application Service Providers), MVNOs (Mobile Virtual Network Operators), and content aggregators who source content from these OEMs and provide the content as a value added service to different individuals, content providers, and network operators.

In the current scenario, there are primarily four parties involved; they are end user (EU), the content provider (CP), the content aggregator (CA), and the ISP (Internet Service Provider) or the wireless or wireline network operator (NO). The network operator will have routers, cache, gateways and other nodes to offer the service. In this scheme anybody can become a requester or a responder. There could be different parameters, which will determine the content. These parameters are of two types static and dynamic. The static adaptation parameters are those which can be received before the service begins. The content is adapted, based on this parameter. The dynamic adaptation parameters are those which are required with every request. For example, a user may initiate a request for a MPEG stream. The NO will transcode the stream to suit the bandwidth of the end user and delivers the same to the user. However, through a dynamic parameter, the user can specify a different parameter for transcoding.

From the content aggregator's perspective we may classify the service into two categories:

1. Single service request–this works at user level and works for only one user. For example, a user may request the proxy server at the NO to translate the page into Hindi and then deliver the same to the user. In this case, the end user buys the content and the translation service.

2. Group service request–this works for a group of users. This type of request is initiated either at the CA level or the NO level. For example, the content aggregator has some arrangement for advertisement. The content aggregator examines all the HTML pages and inserts an advertisement at an appropriate place.

Seamless Communication

The basic premise of a ubiquitous system is that the system will be available and accessible from anywhere, anytime and through any network or device. A user will be able to access the system after moving from one place to another place (foreign place). The user will also be able to access the system while on the move (traveling mode). Mobile healthcare professionals for example, may need to seamlessly switch between different modes of communication when they move from house to outdoors etc. A corporate user requires a similar kind of facility as well. Also, what is necessary is, during the movement, the session needs to continue. If we take the example of healthcare scenario, some data and information are exchanged between the patient and the hospital. While the patient is moved from home, to ambulance, to a helicopter, to the hospital, the information exchange has to continue without any interruption.

Seamless communication will combine seamless handoffs and seamless roaming. Handoff is the process by which the connection to the network (point of attachment) is moved from one base station (access point) to another base station within the same network. Whereas, roaming will involve the point of attachment moving from one base station of one network to another base station of another network. The basic challenge in handoff is that it has to work while a session is in progress. Cellular technology with respect to voice has reached a level of maturity where a seamless voice communication is possible through handoff and roaming. The data technology is yet to mature to provide a similar level of service. In some parts of the world, handoff is termed as handover.

The seamless communication offers users to roam across different wireless networks. Roaming generally works within homogeneous networks, like GSM to GSM or CDMA2000 to CDMA2000. True seamless roaming will include handoff and roaming in a heterogeneous hybrid network. User will move from a WiFi to 3G to wired LAN to GSM while the session is in progress. Users will be able to communicate using whatever wireless device is currently at hand. Thus, GPRS-enabled cell phones, PDAs and laptops will be able to roam and communicate freely and access the Internet across both WLANs and WWANs.

In a seamless roaming, the following aspects need to be maintained and managed in a seamless fashion without any disruption of service:

1. Authentication across network boundaries.
2. Authorization across network boundaries.
3. Billing and charging data collection.
4. End-to-end data security across roaming.
5. Handoff between wireless access points.
6. Roaming between networks.
7. Session migration.
8. IP mobility.

The task of managing authentication between client devices and networks, often involving multiple login names and passwords, will become automatic and invisible to the user, as will the configuration of various settings and preferences that accumulate with client devices.

Autonomous Computing

The world is heading for a software complexity crisis. Software systems are becoming bigger and more complex. Systems and applications range millions of lines of code and require skilled IP professionals to install, configure, tune and maintain. New approaches are needed to provide flexible and adaptable software and hardware both for mobile devices and the intelligent environment. Ease of use will have some effect on acceptance of a ubiquitous system. The scale of these ubiquitous systems necessitates 'autonomic' systems. The purpose of autonomous system is to free users and system administrators from the details of system operation and maintenance complexity. Also, the system will run 24/7. The essence of autonomous system is self-management, which is a combination of the following functions:

1. **Self-configurable:** An autonomous system will configure itself automatically in accordance with high-level policies. This will suit the functional requirement of the user.
2. **Self-optimizing:** An autonomous system will continuously look for ways to improve its operation with respect to resource, cost and performance. This will

mean that an autonomous system will keep on tuning hundreds of tunable parameters to suit the user and the environment.

3. **Self-healing:** An autonomous system will heal detect, diagnose and repair localized problems resulting from bugs or failures. These failures could be the result of either software or hardware failure.

4. **Self-protecting:** An autonomous system will be self-protecting. This will be from two aspects. It will defend itself from external attacks; also, it will not propagate or cascade failure to other parts of the system.

5. **Self-upgradeable:** An autonomous system will be able to grow and upgrade itself within the control of the above properties.

Design tools and theories may be needed to support large-scale autonomic computing for small devices.

2.6.2 Context Aware Systems

The role of a Context Manager is to maintain information pertaining to location, mobile devices, network, users, the environment around each mobile device and any other context information deemed relevant. Following is a description of these information and relevance in the mobile computing environment.

- **Location Information:** This feature helps us to identify the location of the user/device. This can be achieved in either of the two ways. One is through the device and the other is through the network. From the device, the best way to find the location is through GPS (Global Positioning Systems). GPS-based systems can offer location information to a precision of 10 feet radius. Also, the location of the base station with which the device is associated can help us to get the location information. In certain networks, GSM for example, the base station location can be obtained from the device through the CID (Cell ID) value. From the network side the location of the device can be determined through timing advance technology. However, this information relates to a point when a successful call was made. Base-station-based location information is likely to be correct to the precision of 100 feet radius.

- **Device Information:** This feature helps us to know the characteristics of the device. This is required to determine the resource capability and the user interface capability. In a mobile computing environment the user will move from device to device. Therefore, it is essential to know the device context. The device information can be obtained from the device and from the network. Through the User-Agent parameter of http protocol we can get some information about the device. As this information is provided by the browser in the device, the information is very generic. This does not give the device properties like color, pixel capability, display size etc. From the network side, the information about the device can be obtained from the EIR (Equipment Identity Register) database of the network. In all the wireless networks (GSM, GPRS, UMTS, 3G) we have the EIR. However, we do not have any concept of EIR in wireless LAN or WiFi.

- **Network information:** In a mobile computing environment, the user moves from network to network. Sometime they are even heterogeneous in nature. Network information is required to identify the capability of the network. Capability information will include security infrastructure, services offered by the networks etc. For example, while roaming a user moves from a GPRS network to a GSM network. Therefore, the rendering may need an adaptation from WAP to SMS. In the future, some of these will be done through programmable networks.

- **User information:** This information is required to identify the user correctly. From the security point of view, the system needs to ensure that the user is a genuine user who he claims to be. We need to ensure that nobody else is impersonating. This information can be validated through authentication independent of device or network. However, user preferences information need to be obtained from the network. For charging the user properly we need to refer to some subscriber information available in the network.

- **Environment information:** This includes ambient surrounding awareness. We need to know the temperature, elevation, moisture, and other ambient-related information. These information are necessary for sensor-based networks.

For general mobile-computing environment we need location information, network information, user information, and the device information. We also notice that for a majority of the parameters we need to access the information available in different databases within the network. These information are being available through different network interfaces of intelligent networks. These interfaces are Softswitch (http://www.softswitch.org), JAIN (Java API for IN http://java.sun.com/products/jain), Parlay (http://www.parlay.org), and TINA (www.tinac.com). These are explained in Chapter 11

GPS

Global Positioning System (GPS) is a system that gives us the exact position on the Earth. GPS is funded by and controlled by the US Department of Defense. There are GPS satellites orbiting the Earth, which transmit signals that can be detected by anyone with a GPS receiver. Using the receiver, we can determine the location of the receiver. GPS has 3 parts: the space segment, the user segment, and the control segment.

The space segment consists of 24 satellites, each in its own orbit 11,000 nautical miles above the Earth. The GPS satellites each take 12 hours to orbit the Earth. Each satellite is equipped with an accurate clock to let it broadcast signals coupled with a precise time message.

The user segment consists of receivers, which can be in the users' hand, embedded in a mobile device or mounted in a vehicle. The user segment receives the satellite signal which travels at the speed of light. Even at this speed, the signal takes a measurable amount of time to reach the receiver. The difference between the time the signal is sent and the time it is received, multiplied by the speed of light, enables the receiver to calculate the distance to the satellite. To measure precise latitude, longitude and altitude, the receiver measures the time it took for the signals from four separate satellites to get to the receiver. If we know our exact distance from a satellite in space, we know we are somewhere on the surface of an imaginary sphere with radius equal to the distance to the satellite radius. If we know our exact distance from four satellites, we know precisely where we are on the surface of the each.

2.7 MOBILE COMPUTING THROUGH INTERNET

We discussed that a network can be divided into three major functional areas, namely, core, edge and the access. Likewise, we can divide a ubiquitous network into three functional areas. Out of the three, the core and the edge are likely to be Internet and internet. By internet we define a network which is a combination of various networks and interworks with one another, whereas Internet with the uppercase I is the Internet as we know. For mobile and ubiquitous computing, the access network will be both wireless and wired networks. In the case of wireless access network, it could range from infrared, Bluetooth, WiFi, GSM, GPRS, IS-95, CDMA etc. For wired, it is expected to be some kind of LAN. In the case of wired network the bandwidth is higher, stable and the device is likely to be a workstation with a large memory and display. Also, such devices are not constrained by the limited battery power.

When the user-facing device is a wired device, the complexity and challenges are far less. However, some of the constraints for wireless can still apply in the case of wired devices and networks. Therefore, from the mobile computing client point of view, consideration for wired device will be the same as a wireless client.

2.8 Making Existing Applications Mobile-Enabled

There are many applications that are now being used within the intranet or the corporate network, that need to be made ubiquitous. These are different productivity tools like e-mail or messaging applications, workflow systems etc. Information systems for partners and vendors and employees like sales force automation etc. will also fall within this category. These applications need to be made ubiquitous and mobile-computing capable. There are many ways by which this can be achieved.

1. **Enhance existing application** take the current application. Enhance the application to support mobile computing.

2. **Rent an application from an ASP** there are many organizations who develop ubiquitous application and rent the same at a fee.

3. **Write a new application** develop a new application to meet the new business requirement of the mobile computing.

4. **Buy a packaged solution** there are many companies who are offering packaged solutions for various business areas starting from manufacturing to sales and marketing. Buy and install one of these which will also address the mobile computing needs of the enterprise.

5. **Bridge the gap through middleware** use different middleware techniques to face-lift and mobile-computing-enable the existing application.

One of these techniques, or any combinations can be used to make an application ubiquitous. If the enterprise has a source code for the application, enhancement of the existing application may be a choice. Writing a new application by taking care of all the aspects described above may also be a possibility. Buying a package or renting a solution from an ASP can also be a preferred path for some business situations.

Many of these applications might have been developed inhouse, but may not be in a position to be enhanced. Some might have been purchased as products. A product developed by outside agency cannot be enhanced or changed as desired. In many of such situations, mobile computing enabling can be done through middleware. The combination of communication middleware and application middleware can be used to make an application mobile. Let us assume that the enterprise has its sales and distribution application running in SAP in IBM AS/400 system. The enterprise wants this system to be wireless-enabled for its mobile sales force. Using TN5250 communication middleware, the application can be abstracted as an object. Through a transaction processing middleware and APIs, the SAP application can be used as a document. By using a transcoding middleware, the application can be wireless-enabled and used through WAP, J2ME or even SMS (Short Message Service). Through middleware, some additional security features can be added.

REFERENCES/FURTHER READING

1. History of Internet: http://www.isoc.org/internet/history/.

2. Brief History of Internet: http://www.isoc.org/internet/history/brief.shtml.

3. Internet Timeline: http://www.zakon.org/robert/internet/timeline/.

4. Milojicic D., Douglis F., Wheeler R. (Ed), Mobility Processes, Computers, and Agents, Addison-Wesley, 1999.

5. Mari Korkea-aho, Context-Aware Applications Survey, http://www.hut.fi/~mkorkeaa/doc/context-aware.html.

6. VB45: Vannevar Bush: As we may think, Atlantic Monthly: http://www.theatlantic.com-/unbound/flashbks/computer/bushf.htm.

7. Jeffry O. Kephart, David M. Chess, The Vision of Automatic Computing; IEEE Computer Magazine, P41-50, January 2003.

8. Mosaic: http://archive.ncsa.uiuc.edu/SDG/Software/Mosaic/NCSAMosaicHome.html.

9. Rakesh Mohan, John R. Smith, and Chung-Sheng Li, Adapting Multimedia Internet Content for Universal Access; IEEE Transactionson Multimedia, Vol. 1, No. 1, March 1999, PP. 104–114.

10. Javed I. Khan and Yihua He, Ubiquitous Internet Application Services on Sharable Infrastructure:Technical Report 2002-03-02, Internetworking and Media Communications Research Laboratories, Deptt. of Computer Science, Kent State University; http://medianet.kent.edu/technicalreports.html.

11. SyncML: http://www.openmobilealliance.org/syncml/.

12. Internet Content Rating Association: http://www.icra.org.

13. Platform for Internet Content Selection: http://www.w3.org/2000/03/PICS-FAQ/.

REVIEW QUESTIONS

Q1: Describe the significance of core, edge, and access network. What are their functions?

Q2: What are the different tiers in three-tier architecture? Describe the functions of these tiers?

Q3: Explain how can an ISP implement a system using ICAP where some web sites are inaccessible during certain time of the day?

Q4: What do you understand by context? Why is context important? To develop a navigational system for a car, what types of context information will be necessary?

Q5: You have been asked to develop a location aware restaurant guide system for the Restaurant Foundation of India. Describe 4 main functions of this system. Describe how will you implement these 4 functions?

Q6: What is seamless communication? How can seamless communication help in an emergency service rescue operation?

CHAPTER 3

Mobile Computing through Telephony

3.1 EVOLUTION OF TELEPHONY

The first telephone system developed by Alexandra Graham Bell allowed a two-way voice communication between two individuals in two locations on either side of a wire. We (known as **calling** or **A** party, the person who makes the call) speak into one unit of the phone at one end of the wire and someone else hears our voice at another location (known as **called** or **B** party, the person who responds to the call) at the other end of the wire, instantly in real-time. During the long era of analog telephony, the purpose of interconnecting two subscribers was to establish a physical connection between their respective telephone devices. This is achieved by establishing a physical circuit between two parties (A party and B party). In early days, each telephone was connected to a central place (the exchange) and from this exchange the operator would manually connect the call to another subscriber. Whenever a subscriber turned the crank of the telephone, a ringing signal sounded at the operator's switchboard. Upon answering the signal, the operator was asked to connect the call to the other subscriber, which the operator did manually. The operator was required to make a note of who placed the call, whom the call was for, and when it started and ended. This information made it possible to charge the caller for the call, the classic billing and charging information. If we wanted to make a call to someone outside our own local exchange, say to the neighboring exchange, an operator at our exchange would call an operator at the adjacent exchange and then ask the other operator to connect through to the desired subscriber. If we wanted to call someone much further away we had to book a trunk call. In the case of a trunk call, the call would have to be set up with a whole chain of operators, each one calling the next and so on.

We can say that the market forces of the early 1890s prompted the development of the first automatic telephone exchange. It was called the 'Strowger switch', after its originator

Almon B. Strowger. Strowger did not invent the idea of automatic switching; it was first invented in 1879 by Connolly and McTigthe. Strowger was the first person to put it to commercial use. Almon B. Strowger was an undertaker in Kansas City, USA. The story goes that there was another local competing undertaker whose wife was a telephone operator at the local (manual) telephone exchange. Whenever any caller used to request to be connected to Strowger, calls were deliberately put through to his competitor, who in fact was the husband of this operator. This made Strowger devise a system to eliminate the human factor of the whole equation! Strowger developed a system of automatic switching using an electromechanical switch based around electromagnets and pawls. The first version of automatic exchange was installed in 1892 al La Porte, Indiana, USA.

In 1912, the Swedish engineer Gotthief Betulander patented an automatic switching system based on a grid. This type of exchange was also electromechanical and was called crossbar exchange. In 1960 the first Electronic Switching System (ESS) was developed by AT&T and commissioned for testing. Finally, on 30 May 1965, the first commercial electric central office was put into operation at Succasunna, New Jersey. The ESS required a staggering four thousand man-years of work at Bell Labs. In 1976, Bell Labs developed the 4ESS toll switch for the long-distance voice network. This was the first digital circuit switch. The idea behind a digital switch was that the analog voice is digitized before it is given to a switch for switching. The 1960s and 1970s saw the advent of telephone exchanges that were controlled by processors and software (digital computers). These were called stored program control exchanges. The primary objective of a sophisticated telephone exchange is still the same as that of the manual exchange a century ago. These are to detect the A-subscriber's (calling party) call attempt, connect him to the correct B-subscriber (called party), and to save data about the call for the purpose of billing.

The digital revolution in telephony started with the introduction of electronic switches. The next major milestone was achieved in 1962 when the carrier system was made digital. Work on digital transmission began back in the 1920s, when the Bell System researcher Harry Nyquist determined that it was possible to encode an analog signal in digital form if the analog signal was sampled at twice its frequency. Each sample could then be encoded and transmitted. There would be enough information in the encoded signal for the original voice signal to be reconstructed into an understandable analog signal at the receiving end. Assuming that the audio voice band is 0 to 4000 Hzs, we start with a 4 KHz analog voice channel. Then we take a snapshot of the voice signal's amplitude at 1/8000th of a second (every at double the frequency of 4 KHz). Then we convert the measured amplitude to a number (the **quantization** process) that is represented by 8 bits. This type of digitization is called Pulse Code Modulation (PCM). Thus, PCM requires 64 Kb/s of digital bandwidth (8 KHz * 8 bits). Alex H. Reeves, working for Western Electric Company in Paris, first conceived PCM in 1937. Bell Laboratory scientists first introduced digital transmission using PCM in 1962. The Bell Laboratory system was named T1 with a transmission rate of 1.544 Mb/s carrying 24 channels of 64 Kb/s each. In Europe, a similar system was called E1, where it had a bandwidth of 2 Mb/s and carried 32 channels of 64 Kb/s. The developments in transmission techniques have been advancing largely to reduce network costs. We have witnessed an evolution from systems employing open-wire lines to multiplexed, analog systems using coaxial or radio links, on to digital fiber-optic systems with a capacity of tens of Gbit/s per fiber pair. The first commercial optical systems came on the scene in 1980.

In a manual switching system, an operator would be able to inform the caller of the current status of the call. In manual exchanges, the operator's intelligence was a control system separate from the switching mechanism. An operator, alerted to an incoming call:

- Listens to and remembers the desired number.
- Finds the right way to connect the caller's line to the line being called.
- Checks if the desired line is free.
- Makes the connection.
- Notes down the call details: time of call, duration of call, calling number and called number.

Having removed the need for an operator in the automated exchange, a system was necessary to indicate the progress of the call to the caller. A series of distinct tones were generated by a machine called **Ring Generator**. The tones produced were as follows:

- **Dial Tone (DT).** This is a signal applied to the line after the calling party (**A** party) has lifted his handset and the switching equipment has allocated him an available outlet (a circuit) for this call to proceed.
- **Busy Tone (BT).** Busy tone indicated either that the called subscriber (**B** party) is already off-hook (busy) or that the route to the called subscriber is congested.
- **Ring Tone (RT).** When a circuit between **A** party and the **B** party is established, the telephone rings at **B** party's end and a ring tome is generated for the **A** party.

A normal telephone system is called Public Switched Telephone Network (PSTN). PSTN nodes can be subdivided into three main categories: local exchanges (also known as End Office), transit exchanges (also known as Local Access Tandem) and international exchanges (also known as Interexchange Carrier). Local exchanges are used for the connection of subscribers. Transit exchanges switch traffic within and between different geographical areas. International exchanges, and other gateway-type exchanges switch traffic to telecommunication networks in foreign countries and other networks. A physical wire (also known as local loop) is laid from the local exchange to the telephone device at each subscriber's place. This is traditionally also known as the **last mile**. In case of a wireless network like GSM or WiLL (Wireless in Local Loop), there is no wire from the local exchange to the telephone. The communication between the local exchange and the telephone device is managed over the wireless radio interface. In India, there are network operators who are offering basic or fixed telephone, WiLL and GSM.

3.2 MULTIPLE ACCESS PROCEDURES

In a PSTN network, a separate physical wire is used to connect the subscriber's telephone with the switch. Therefore, multiple users can have speech communication at the same time without causing any interference to each other. The scene is different in the case of wireless communication. Radio channel, used in a wireless network, is shared by multiple subscribers. Unless we control simultaneous access of the radio channel (by multiple users), collisions can occur. In a connection-oriented communication, collision is undesirable. Therefore, every mobile subscriber must be assigned a dedicated communication channel on demand. This is achieved by using different multiplexing techniques (Figure 3.1).

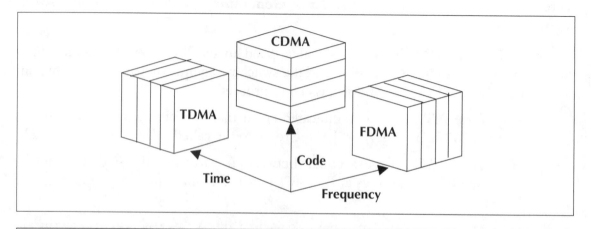

Figure 3.1 Multiple access procedures

3.2.1 Frequency Division Multiple Access

Frequency Division Multiple Access (FDMA) is one of the most common multiplexing procedures. The available frequency band is divided into channels of equal bandwidth so that each communication is carried on a different frequency. This multiplexing technique is used in all the first generation analog mobile networks like Advanced Mobile Phone System (AMPS) in the USA and Total Access Communication System (TACS) in the UK.

3.2.2 Time Division Multiple Access

Time Division Multiple Access (TDMA) is a more expensive technique compared to FDMA as it needs precise synchronization between the transmitter and the receiver. The TDMA technique is used in digital mobile communication. In a TDMA system, the whole frequency bandwidth is subdivided into sub-bands using FDMA techniques. TDMA technique is then used in each of these sub-bands to offer multiple access. GSM uses such a combination of FDMA and TDMA. A frequency range of 25 MHz holds 124 single channel of 200 KHz each. Each of these frequency channels contains 8 TDMA conversation channel.

3.2.3 Code Division Multiple Access

Code Division Multiple Access (CDMA) is a broadband system. CDMA uses spread spectrum technique where each subscriber uses the whole system bandwidth. Unlike the FDMA or TDMA where a frequency or time slot is assigned exclusively to a subscriber, in CDMA all subscribers in a cell use the same frequency band simultaneously. To separate the signals, each subscriber is assigned an orthogonal code called 'chip'.

3.2.4 Space Division Multiple Access

Along with TDMA, FDMA, and CDMA, we need to make use of the space effectively. Space division multiple access (SDMA) is a technique where we use different part of the space for multiplexing. SDMA is used in radio transmission and is more useful in satellite communications to optimize the use of radio spectrum by using directional properties of antennas. In SDMA, antennas are highly directional, allowing duplicate frequencies to be used at the same time for multiple surface zones on earth. SDMA requires careful choice of zones for each transmitter, and also requires precise antenna alignment.

3.3 MOBILE COMPUTING THROUGH TELEPHONE

One of the early examples of mobile computing was accessing applications and services through voice interface. This technology was generally referred to as Computer Telephony Interface (CTI). Different banks around the world were offering telephone banking for quite sometime using this technology. In a telephone banking application, the user calls a number and then does his banking transaction through a fixed telephone. In this application the telephone does many functions of a bank teller. Input to this system is a telephone keyboard and output is a synthesized voice. These applications can be used from anywhere in the world. The only issue in this case is the cost of a call. Let us take the example of a bank, which has branches only in Bangalore (like some co-operative banks in India). Let us assume that this bank offers telephone banking facility only in Bangalore city in India. The service number for the bank is +91(80)2692265 (+91 80 2MYBANK). Assuming I am in Bangalore, it costs me a local call to check the balance in my account. When I am traveling and want to check my account detail from elsewhere, say Delhi, I make a call to the Bangalore number +91802692265 from Delhi and pay a long distance charge. Let us now assume that the bank has gone for a VPN (Virtual Private Network) between Delhi and Bangalore. The bank now offers telephone banking services in Delhi through a service number +91 (11) 26813241. This will enable me to use the same service in Delhi at the cost of a local call. The only challenge is that the number in Delhi is different from that of Bangalore. In such cases the bank customers are required to remember multiple service numbers for mobile computing.

The telephone companies soon came up with a brilliant idea to solve this problem of multiple numbers by offering 800 services using Intelligent Networks (IN) technology. This is also commonly known as **Toll Free** numbers. In this technology only one number like 1-800-2MYBANK is published. This number is not attached to any specific exchange or any specific city. When a subscriber calls this number an optimal routing is done and the call is connected to the nearest service center. The advantage is that users remember only one number. They can call the same number from anywhere. They also need not worry about the distance of the call as these numbers are generally toll free. Toll free means that the call is charged to the B party instead of the A party. In India this service is available as 1-600 service. For example, to shop through TV we dial 1-600-117247. No matter where we are making the call in India from, we will be connected to

'Shop 24 Seven' at Mumbai. If we dial 1-600-111100 from anywhere in India we will be connected to Microsoft office in Delhi.

To make this type of mobile computing work through voice interfaces, we use Interactive Voice Response (IVR). In USA and Japan IVRs are commonly known as Voice Response Unit (VRU). The technical name for this technology is CT (Computer Telephony) or CTI (Computer Telephony Interface or Computer Telephony Integration). IVR software can be hosted on a Windows-NT, Linux, or other computers with the voice cards. There are many companies who manufacture voice cards; however, one of the most popular card vendors is from Intel/Dialogic. IVR works as the gateway between a voice-based telephone system and a computer system. Multiple telephone lines are connected to the voice card through appropriate telecom interfaces (E1 or an analog telephone extension). When a caller dials the IVR number, a ring tone is received by the voice card within the IVR. The voice card answers the call and establishes a connection between the caller and the IVR application. The caller uses the telephone keyboard to input data. Figure 3.2 depicts an IVR infrastructure. The switch can be either a PSTN exchange or a local PBX in the office. For PSTN switch, the voice card will have E1 interface whereas for a PBX, the voice card will have analog interface. The IVR will have all the gateway-related functions. The server will host the business application.

A telephone keyboard has 12 keys (viz., 1, 2, 3, 4, 5, 6, 7, 8, 9, 0, *, and #). The English alphabetic characters are also mapped on these 12 keys. They are mapped as follows.

1. Alphabet A, B, C on key 2
2. Alphabet D, E, F on key 3

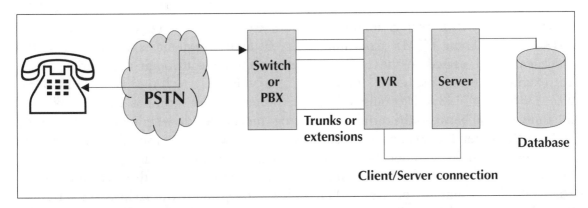

Figure 3.2 The IVR architecture

3. Alphabet G, H, I on key 4

4. Alphabet J, K, L on key 5

5. Alphabet M, N, O on key 6

6. Alphabet P, Q, R, S on key 7

7. Alphabet T, U, V on key 8

8. Alphabet W, X, Y, Z on key 9

It is possible to enter alphabetic data through the telephone keyboard by pressing a key in multiple successions. For example DELHI will be entered as 3-3 (D), 3-3-3 (E), 5-5-5-5 (L), 4-4-4 (H), 4-4-4-4 (I). These key inputs are received by the voice card as DTMF (Dual Tone Multi Frequency) inputs generated through combination of frequencies. Following is the table (Table 3.1) of these frequencies:

Table 3.1 DTMF frequencies			
	1209 Hz	1336 Hz	1477 Hz
697 Hz	1	2/ABC	3/DEF
770 Hz	4/GHI	5/JKL	6/MNO
852 Hz	7/PQRS	8/TUV	9/WXYZ
941 Hz	*	0	#

If we press key 1, it will generate a frequency 697 + 1209 Hz. Likewise 0 will be 941+1336 Hz. These DTMF signals are different audio frequencies interpreted by the voice card and passed to the IVR program as numbers through appropriate APIs. For example the user presses '2' three times. The voice card will receive 697+1336 Hz-697+1336 Hz-697+1336 Hz. This will be interpreted by a program as 2-2-2. Looking at the time interval between the numbers, the program can decide whether the user entered '222' or 'B'. When the application needs to send an output to the user, the standard data is converted into voice either through synthesizing voice files or through TTS (Text To Speech). In a cheque-printing software we print the amount in both words and figure. For example an amount of 'Rs. 320,145.00' will be printed on a cheque as 'Rupees three lacs twenty thousand one hundred forty-five only'. Within the cheque-printing application, one function converted the numeric number 320145 into text. Likewise in the case of IVR application, we assemble a series of prerecorded voice prompts to generate the equivalent sound response. In this case we assemble voice data 'three' 'lacs' 'twenty' 'thousand' 'one' 'hundred' 'forty' 'five' 'only' and then give the voice card to play. We can generate the same voice response by giving the number 320,145 to the TTS interface to convert the text into speech and play through the IVR. TTS is a interface software which takes text and numbers as input and generates equivalent sounds at the run time. There are different TTS available for different languages. In India there are companies who have Hindi TTS software. Hindi TTS software takes Devanagari text stream as input and generates the voice as if someone is reading the same text.

3.3.1 Overview of the Voice Software

Voice technology encompasses the processing and manipulation of an audio signal in a Computer Telephony (CT) system. It supports filtering, analyzing, recording, digitizing, compressing, storing, expanding and replaying of audio voice. A CT system also includes the ability to receive, recognize and generate specific telephone and network tones. This fundamental technology is at the core of most IVR systems. Voice products also offer Digital Signal Processing (DSP) technology and signal processing algorithms, for building the core of any converged communications system. Most of the voice cards come with industry-standard Peripheral Component Interface (PCI) bus expansion boards. The PCI interface makes it possible to integrate these voice products into Windows or Linux systems quite easily (Figure 3.3).

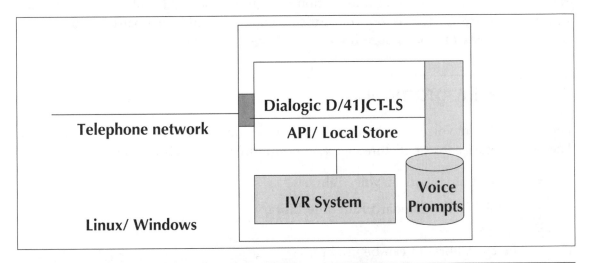

Figure 3.3 Inside an IVR

One of the most popular voice cards used for small office interface is D/41JCT-LS from Dialogic. Dialogic (part of Intel) products are the de-facto industry standard. The D/41JCT-LS board is a four-port analog converged communications voice, fax, and software-based speech recognition board. This board is ideal for building enterprise unified messaging and interactive voice response (IVR) applications. The D/41JCT-LS provides four telephone line interface circuits for direct connection to analog loop start lines through RJ11 (the standard telephone jack used in homes) interface. D/41JCT-LS possesses dual-processor architecture, comprising a digital signal processor (DSP) and a general-purpose microprocessor, which handles all telephony signaling and performs DTMF (touchtone) and audio/voice signal processing tasks. A voice card also has some on-board memory and with voice store-and-forward feature.

3.3.2 Voice Driver and API

In this section we describe Dialogic Voice Driver APIs. Dialogic is now part of Intel and one of the leading vendors on voice-based hardware. Many IVR vendors around the world use Dialogic cards from Intel in their IVR systems. Voice driver in an IVR system is used to communicate and control the voice hardware on the IVR system. This section describes Dialogic APIs. Voice card from some other vendor will have similar type of APIs. A voice driver can make calls, answer calls, identify caller id, play and record sound from the phone line, detect DTMF signals (touch-tones) dialed by the caller. It can tear down a call, detect when the caller has hung up. It also offers APIs to record the transaction details. Transaction information is required for audit trail and for charging. Voice boards are treated as **board devices**, channels within a board are treated as **channel devices or board sub-devices** by the voice driver.

3.3.3 IVR Programming

There are different voice libraries provided by Dialogic to interface with the voice driver. The voice libraries for single-threaded and multi-threaded applications include:

- libdxxmt.lib–the main Voice Library
- libsrlmt.lib–the Standard Run-time Library

These C function libraries can be used to:

- Utilize all the voice board features of call management

- Write applications using a **Single-threaded Asynchronous** or **Multi-threaded** paradigm
- Configure devices
- Handle events that occur on the devices
- Return device information
- Gather call transaction details.

The Standard Run-time Library provides a set of common system functions that are device independent and are applicable to all Dialogic devices.

3.3.4 Single-threaded Asynchronous Programming Model

Single-threaded asynchronous programming enables a single program to control multiple voice channels within a single thread. This allows the development of complex applications where multiple tasks must be coordinated simultaneously. The asynchronous programming model supports both polled and callback event management.

3.3.5 Multi-threaded Synchronous Programming Model

The multi-threaded synchronous programming model uses functions that block application execution until the function completes. This model requires that the application control each channel from a separate thread or process. The operating system can put individual device threads to sleep while allowing threads that control other Dialogic devices to continue their actions unabated. When a Dialogic function is completed, the operating system wakes up the function's thread so that processing continues. This model enables the IVR system to assign distinct applications to different channels dynamically in real time.

Voice APIs

To use the voice board, Dialogic provides different APIs. All Dialogic APIs are prefixed with dx_; this helps to identify them easily. APIs are available for device management, configuration function, input output functions, play and record functions, tone detection functions, tone generation functions, call control functions etc. Following are some of the important functions in Dialogic voice card, which are used quite often to develop a mobile computing application.

dx_open( )	– open a voice channel
dx_close( )	– close a voice channel
dx_wtcallid( )	– waits for rings and reports Caller ID
dx_getdig( )	– get digits from channel digit buffer (for reading the key input)
dx_play( )	– plays recorded voice data
dx_playvox( )	– play a single vox file
dx_playwav( )	– play a single wave file
dx_rec( )	– record voice data
dx_recvox( )	– records voice data to a single vox file
dx_recwav( )	– records voice data to a single wave file
dx_dial( )	– dial an ASCII string of digits.

3.4 DEVELOPING AN IVR APPLICATION

Like any other application development, computer telephony/IVR application development also requires definition of the user interface. The user interface in IVR application is called the **Call Flow**. In a call flow we define how the call will be managed. Also, we note down the precise prompts that are played as an output. As described earlier, these prompts are generally pre-recorded by people with professional voice. Let us take a simple example of ticket booking in a theatre. In this application, the user dials a service number and enters a phone number. The operator calls the user back and accepts the booking request. The extra step of call back is done for security reason. Figure 3.4 depicts the call flow for this application.

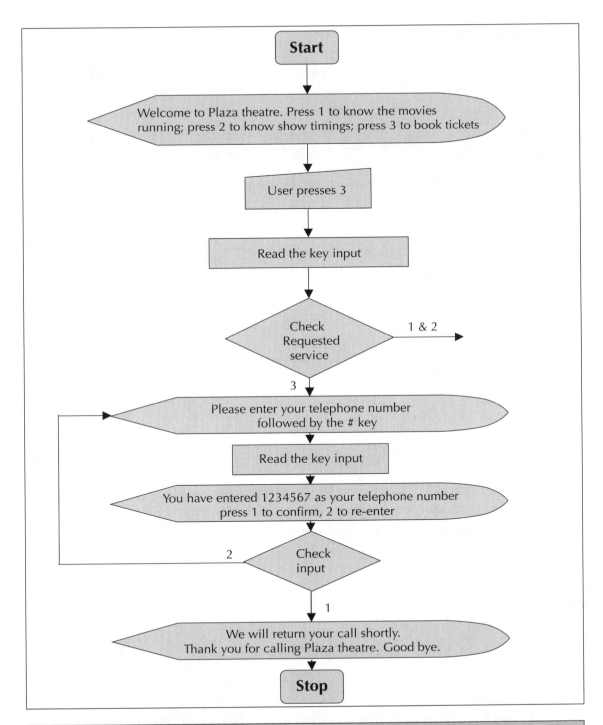

Figure 3.4 Call flow for a theatre ticket booking

Example IVR application

Like any other program, in an IVR program we need to open the user interface device (voice board in the case of IVR) and the data store. Table 3.2 lists an example for one such IVR program.

Table 3.2 An IVR example

```
1    /* Play a voice file. Terminate on receiving 4 digits or at end of file*/
2    #include <fcntl.h>
3    #include <srllib.h>
4    #include <dxxxlib.h>
5    #include <windows.h>
6    main()
7    {
8        int chdev;
9        DX_IOTT iott;
10       DV_TPT tpt;
11       DV_DIGIT dig;
12       .
13       .
14       /* Open the device using dx_open( ).
15       Get channel device descriptor in * chdev. */
16       if ((chdev = dx_open("dxxxB1C1",NULL)) == -1)
17           {
18           /* process error */
19           }
20       /* set up DX_IOTT */
21       iott.io_type = IO_DEV|IO_EOT;
22       iott.io_bufp = 0;
23       iott.io_offset = 0;
24       iott.io_length = -1; /* play till end of file */
25       if ((iott.io_fhandle =
```

```
26    dx_fileopen("prompt.vox", O_RDONLY|O_BINARY)) == -1)
27    {
28    /* process error */
29    }
30    /* set up DV_TPT */
31    dx_clrtpt(tpt,3);
32    tpt[0].tp_type = IO_CONT;
33    tpt[0].tp_termno = DX_MAXDTMF; /* Maximum number of digits */
34    tpt[0].tp_length = 4; /* terminate on 4 digits */
35    tpt[0].tp_flags = TF_MAXDTMF; /* terminate if already in buf. */
36    tpt[1].tp_type = IO_CONT;
37    tpt[1].tp_termno = DX_LCOFF; /* LC off termination */
38    tpt[1].tp_length = 3; /* Use 30 ms (10 ms resolution * timer) */
39    tpt[1].tp_flags = TF_LCOFF|TF_10MS; /* level triggered, clear
40        history, * 10 ms resolution */
41    tpt[2].tp_type = IO_EOT;
42    tpt[2].tp_termno = DX_MAXTIME; /* Function Time */
43    tpt[2].tp_length = 100; /* 10 seconds (100 ms resolution * timer) */
44    /* clear previously entered digits */
45    if (dx_clrdigbuf(chdev) == -1)
46        {
47        /* process error */
48        }
49      /* Now play the file */
50    if (dx_play(chdev,&iott,&tpt,EV_SYNC) == -1)
51        {
52        /* process error */
53        }
54    /* get digit using dx_getdig( ) and continue processing. */
55    /* Set up the DV_TPT and get the digits */
56    if ((numdigs = dx_getdig(chdev,tpt, &digp, EV_SYNC))== -1)
```

```
57              {
58              /* process error */
59              }
60          for (cnt=0; cnt < numdigs; cnt++)
61          {
62              printf("\nDigit received = %c, digit type = %d",
63              digp.dg_value[cnt], digp.dg_type[cnt]);
64          }
65          /* go to next state */
66          .
67          .
68          .
69      }
```

Line 16 is to open a channel for use in a Dialogic card. It is necessary to open the channel before any type of access of the same.

Line 50 is to play the voice file, which was prerecorded with voice 'Hello World'. Pre-recorded voice files are recordings of normal voice and stored in digitized form. This can be done using normal telephone speaker and Dialogic card. However, in a majority of cases, this is done in a professional studio using professional people with a good voice.

In line 56 we read the digits entered through the telephone keypad.

3.5 VOICE XML

In mobile computing through telephone, the IVR is connected to the server through client/server architecture. It is also possible to host the IVR and the application on the same system. In the last few years, mobile computing through voice has come a long way. Today Internet (http) is used in addition to client/server interface between the IVR and the server. This increases the flexibility in the whole mobile-computing architecture. Http is used for voice portals as well. In the case of a voice portal, a user uses an Internet site through voice interface. For all these advanced features, VoiceXML has been introduced. Recent IVRs are equipped with DSP (Digital Signal Processing) and are capable of recognizing voice. The output is synthesized voice through TTS (Text to Speech).

The Voice eXtensible Markup Language (VoiceXML) is an XML-based markup language for creating distributed voice applications. VoiceXML is designed for creating audio dialogs that feature synthesized speech, digitized audio, recognition of spoken voice and DTMF key input. Using VoiceXML, we can create Web-based voice applications that users can access through telephone.

VoiceXML supports dialogs that feature :

- Spoken input
- DTMF (telephone key) input
- Recording of spoken input
- Synthesized speech output ('text-to-speech')
- Recorded audio output
- Dialog flow control
- Scoping of input

Architectural Model

The architectural model for VoiceXML is depicted in Figure 3.5. It has the following components:

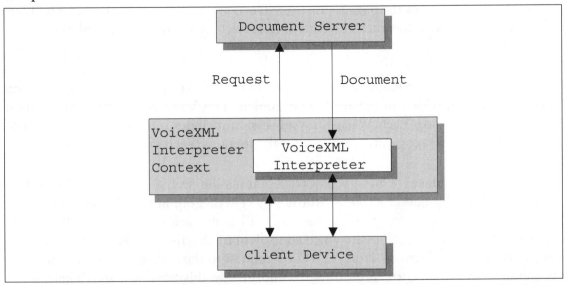

Figure 3.5 Architectural Model

A **Document Server** (e.g. a web server) services requests from a client application. The client side of the application runs on a **VoiceXML Interpreter**, and is accessed through the **VoiceXML interpreter context**. The server delivers VoiceXML documents, which are processed by the VoiceXML Interpreter. The VoiceXML Interpreter Context is responsible for special actions on voice escape phrases.

For instance, in an interactive voice response application, the VoiceXML interpreter context may be responsible for detecting an incoming call, acquiring the initial VoiceXML document, and answering the call, while the VoiceXML interpreter manages the dialog after answer. The implementation platform generates events in response to user actions (e.g. spoken or character input received, disconnect) and system events (e.g. timer expiration).

3.5.1 How Voice XML Fits into Web Environment?

All of us are familiar with the web as it works today. We use a visual GUI web browser (such as Netscape Communicator or Internet Explorer), which renders and interprets http requests to present information to the user (text, graphics, audio, multimedia, etc.). When the user makes a selection (for example, a click on a hyperlink), the web browser sends an HTTP request to the web server. The web server responds by locating the new page and returns the page to the user. The content server may also have to interact with a back-end infrastructure (database, servlets, etc.) to obtain and return the requested information.

The Voice Browser extends this paradigm. In Figure 3.6 a telephone and a **Voice Server** have been added to the web environment. The Voice Server manages several Voice Browser sessions. Each Voice Browser session includes one instance of the Voice Browser, the speech recognition engine, and the text-to-speech engine.

VoiceXML introduces a new way of presenting the web information. Instead of presenting the information visually (through HTML, graphics and text), the Voice Browser presents the information to the caller in audio using VoiceXML. When the caller says something (which is the voice equivalent of clicking on something to make a selection), the Voice Browser sends an HTTP request to the web server, which accesses the same back-end infrastructure, to return information this time in audio. This type of portal is known as voice portal. Voice portal is very useful in a hands free situation while driving.

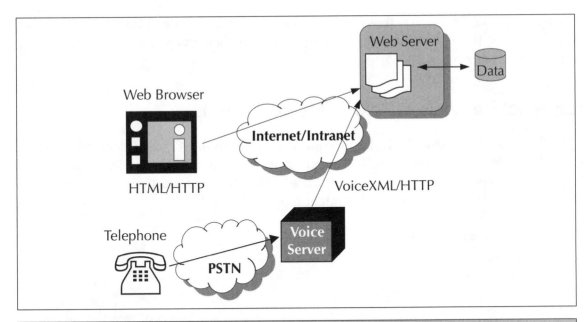

Figure 3.6 Voice browser and voice portal over VoiceXML Architecture

The Voice Browser

An audio Voice Browser is similar to a visual web browser like Netscape Communicator or Microsoft Internet Explorer. Through voice browser, we interact with a web server using our voice and a telephone. Instead of rendering and interpreting a HTML document (like a GUI browser), the Voice Browser renders and interprets VoiceXML documents. Instead of clicking a mouse and using keyboard, we use our voice and a telephone (and even the phone keypad) to access web information and services.

Dialogs

A VoiceXML application defines a series of dialogs between a user and a computer. Each VoiceXML document forms a conversational finite state machine. The user is always in one conversational state, or dialog, at a time. Each dialog determines the next dialog to which to transition. Transitions are specified using URIs, which define the next document and dialog to use.

There are two types of dialogs that can be implemented in VoiceXML:

- forms
- menus.

Forms define an interaction that collects values for a set of fields. **Menus**, on the other hand, present the user with choices or options and then transition to another dialog based on the choice.

Essential Elements of Voice XML Documents

The first line of any VoiceXML application must contain the <?xml> element. The second line must contain the <vxml> element. And each VoiceXML <tag>, must have an associated </tag>. The very last line of VoiceXML document must be the </vxml> tag. So, at a minimum, a VoiceXML document looks like this:

```
<?xml version="1.0"?>
<vxml version="1.0">
.


.


.


Interesting stuff goes here
.


.


.

</vxml>
```

In between the <vxml> and the </vxml> tags we put all the really interesting stuff, the VoiceXML code that defines the dialog with the user.

Prompts

In a VoiceXML application, we present information to the user through audio prompts. These prompts can either be prerecorded audio, or they can be synthesized speech (TTS). We use the <prompt> element in VoiceXML to generate TTS. Any text within the body of a <prompt> element is spoken. In the following example, the text-to-speech software will read 'Would you like coffee, tea, milk or nothing?' to the caller:

```
<prompt>
    Would you like coffee, tea, milk or nothing
</prompt>
```

Grammars

Each dialog has one or more speech and/or DTMF grammars associated with it.

In VoiceXML, we use the **\<grammar\>** element to define what the caller can say to the application at any given time. There are three different types of grammars supported in VoiceXML:

- Inline
- External
- Built-in

Inline grammars are those that are defined right in the VoiceXML code. For example

```
<grammar>
     credit card    |    credit    |    tuition    |    tuition bill
</grammar>
```

In an inline grammar, the words and phrases that a caller is allowed to say are defined within the body of the \<grammar\> element. Each word or phrase is separated by a vertical bar ("|") symbol. This symbol essentially means "or." So, in the previous example, the caller can say either "credit card" or "credit" or "tuition" or "tuition bill".

External grammars are those that are specified outside of the VoiceXML document in another file and are referenced from within the VoiceXML code. We use the \<grammar\> element to specify an external grammar, too. For example:

```
<grammar>src="names.gram" type="application/x-jsgf"</grammar>
```

In this example, the grammar is defined in 'names.gram' file.

Form

Form is one of the ways of developing a dialog with the caller in VoiceXML. Forms are central to VoiceXML. A VoiceXML form is a process to present information and gather input from the caller. A form is, basically, a collection of one or more fields that the caller fills in by saying something. A VoiceXML form is a very similar concept to a paper or online form, except that in the case of VoiceXML, we cannot see the field and instead of typing or writing in a field, we say something to fill it in.

In VoiceXML, we define a form using the **<form>** element and fields within the form using the **<field>** element. Here is a simple Voice Form.

```
<?xml version="1.0"?>
<vxml version="1.0">
<form id="add_funds">
        <field name="amount" type="currency">
            <prompt>How much?</prompt>
        </field>
        <field>
            <prompt>Charge   to   credit   card   or   tuition
            bill?</prompt>
            <grammar> credit card | credit | tuition | tuition
            bill</grammar>
        </field>
</form>
</vxml>
```

This form has an **id** of 'add_funds', and it contains two **fields**. The first field asks the user how much money to add to the meal account ('How much') and is expecting the user to say an amount as currency (e.g., 'one thousand rupees'). The second field asks for the type of transaction ('charge to credit card or tuition bill') and is expecting the caller to say either 'credit card', 'credit', 'tuition', or 'tuition bill'.

As we can see, **fields** define the information the application needs from the caller. Fields tell the caller what to say, and they also define the words and phrases that the caller can say (or the keys that can be pressed). Based on the caller's input–in other words, what the caller says or which keys were pressed–the application takes an appropriate action. When the user provides a valid response, the field is considered FILLED and the application can then do something with this information.

Events

VoiceXML provides a form-filling mechanism for handling 'normal' user input. In addition, VoiceXML defines a mechanism for handling events not covered by the form mechanism. Events are thrown by the platform under a variety of circumstances, such as when the user does not respond, doesn't respond intelligibly, requests help, etc.

Links

A *link* supports mixed initiative. It specifies a grammar that is active whenever the user is in the scope of the link. If user input matches the link's grammar, control transfers to the link's destination URI. A <link> can be used to throw an event to go to a destination URI.

VoiceXML Elements

Element	Purpose Page
<assign>	Assign a variable a value.
<audio>	Play an audio clip within a prompt.
<block>	A container of (non-interactive) executable code.
<break>	JSML element to insert a pause in output.
<catch>	Catch an event.
<choice>	Define a menu item.
<clear>	Clear one or more form item variables.
<disconnect>	Disconnect a session.
<div>	JSML element to classify a region of text as a particular type.
<dtmf>	Specify a touch-tone key grammar.
<else>	Used in < if > elements.
<elseif>	Used in < if > elements.
<emp>	JSML element to change the emphasis of speech output.
<enumerate>	Shorthand for enumerating the choices in a menu.
<error>	Catch an error event.
<exit>	Exit a session.
<field>	Declares an input field in a form.
<filled>	An action executed when fields are filled.
<form>	A dialog for presenting information and collecting data.

`<goto>`	Go to another dialog in the same or different document.
`<grammar>`	Specify a speech recognition grammar.
`<help>`	Catch a help event.
`<if>`	Simple conditional logic.
`<initial>`	Declares initial logic upon entry into a (mixed-initiative) form.
`<link>`	Specify a transition common to all dialogs in the link's scope.
`<menu>`	A dialog for choosing amongst alternative destinations.
`<meta>`	Define a meta data item as a name/value pair.
`<noinput>`	Catch a noinput event.
`<nomatch>`	Catch a nomatch event.
`<object>`	Interact with a custom extension.
`<option>`	Specify an option in a <field>.
`<param>`	Parameter in <object> or <subdialog>.
`<prompt>`	Queue TTS and audio output to the user.
`<property>`	Control implementation platform settings.
`<pros>`	JSML element to change the prosody of speech output.
`<record>`	Record an audio sample.
`<reprompt>`	Play a field prompt when a field is re-visited after an event.
`<return>`	Return from a subdialog.
`<sayas>`	JSML element to modify how a word or phrase is spoken.
`<script>`	Specify a block of ECMAScript client-side scripting logic.
`<subdialog>`	Invoke another dialog as a subdialog of the current.
`<submit>`	Submit values to a document server.
`<throw>`	Throw an event.
`<transfer>`	Transfer the caller to another destination.
`<value>`	Insert the value of a expression in a prompt.
`<var>`	Declare a variable.
`<vxml>`	Top-level element in each VoiceXML document.

3.6 TELEPHONY APPLICATION PROGRAMMING INTERFACE (TAPI)

In the previous sections we have discussed how to program a Dialogic card and develop voice based applications and services. However, there are quite a few higher level frameworks available where a developer can develop voice based services without going too deep into it. TAPI (Telephony Application Programming Interface) is one such example. There is another related standard for speech called Speech Application Programming Interface (SAPI). Developed jointly by Intel and Microsoft, TAPI and SAPI are two standards that can be used when developing voice telephony applications. Using TAPI, programmers can take advantage of different telephone systems, including ordinary PSTN, ISDN, and PBX (Private Branch Exchange) without having to understand all their details. Use of these API will save the programmer the pain of trying to program hardware directly. Through TAPI and SAPI a program can "talk" over telephones or video phones to people or phone-connected resources. Through TAPI one will be able to:

- Simple user interfaces to setup calls. This can be calling someone by clicking on their picture or other images

- Use simple graphical interface to set up a conference call and then attend the call at the scheduled time

- See who you're talking to

- Attach voice greeting with an email. This will allow the receiver to listen to this greeting while opening the email

- Set groups and security measures such that a service can receive phone calls from certain numbers (but not from others)

- Send and receive faxes

- Same set of TAPI APIs are available in many smart phones. This facilitates accessing telephony interfaces from a mobile phone along with from a desktop computer

In addition to the interface for applications, TAPI includes an interface for convergence of both traditional PSTN telephony and IP telephony. IP telephony or VoIP (Voice over IP) is an emerging set of technologies that enables voice, data, and video collaboration over Internet protocol. VoIP is discussed in detail in chapter 17.

REFERENCES/FURTHER READING

1. Fundamentals of Telecommunications, The International Engineering Consortium, http://www.iec.org.

2. William C. Y. Lee, Mobile Cellular Telecommunications Analogue and Digital Systems, McGraw-Hill, 2000.

3. Marion Cole, Introduction to Telecommunications Voice, data, and the Internet, Pearson Education Asia, 2001.

4. Jochen Schiller, Mobile Communications, Pearson Education Asia, 2001.

5. Aronsson's Telecom History Timeline, http://www.aronsson.se/hist.html.

6. Voice XML Forum: http://www.voicexml.org.

7. IBM, WebSphere Voice Server for DirectTalk, User's Guide, 2001.

8. IBM WebSphere Voice Server, Software Developers Kit (SDK), Programmer's Guide, 2000.

9. Dialogic card references: http://www.intel.com.

10. Voice-enabled e-Business Unlocking e-Business Opportunities, Intel Corporation.

11. TAPI: http://www.microsoft.com.

12. Enterprise Computer Telephony Forum (ECTF): http://www.ectf.org.

REVIEW QUESTIONS

Q1: Describe what is multiple access? Why is multiple access important? Describe FDMA, TDMA, CDMA, and SDMA with the application areas and examples?

Q2: What are the steps you need to follow during the design of an application development using voice?

Q3: You have been asked by a bank to develop an account enquiry system over telephone. Design the architecture of such a system?

Q4: Draw a call flow diagram for authentication of a user using CTI?

Q5: Design a hands free voice based email client application for the sales persons of the company?

CHAPTER 4

Emerging Technologies

4.1 INTRODUCTION

The mainstream wireless technologies have been discussed in chapters 5 through 10. We will discuss wireless networks and application development using cellular networks like GSM, SMS, GPRS, WAP, CDMA, and 3G in chapter 5 through 9. We will also discuss wireless local area network (WLAN or WiFi) in chapter 10. In this chapter however, will we discuss some technologies which are not yet in the mainstream but are potential candidates for the same. These technologies are included here to make the mobile computing story complete. These include technologies like Bluetooth (802.15.1a), Radio frequency identifier (RFID), Wireless metropolitan area network or wireless broadband (WiMax-802.16), Mobile IP, IPv6, and Java Card. Bluetooth is a technology in the personal area network (PAN). RFID is emerging as a leading technology in the logistics, manufacturing, and retail industry. Wireless broadband is expected to be a mainstream technology very soon. Mobile IP allows data handoff over different sub-networks. IPv6 is the next generation internet protocol. Java Card technology is emerging as a forerunner in the security and personal identity domain. Therefore, we introduce all these technologies in this chapter.

4.2 BLUETOOTH

Bluetooth was the nickname of a Danish king Harald Blåtand, who unified Denmark and Norway in the 10th century. The concept behind Bluetooth wireless technology was unifying the telecom and computing industries. Bluetooth technology allows users to make ad hoc wireless connections between devices like mobile phones, desktop or notebook computers without any cable. Devices carrying Bluetooth-enabled chips can easily transfer data at a speed of about 720 Kbps within 50 meters (150 feet) of range or beyond through walls, clothing and even luggage bags.

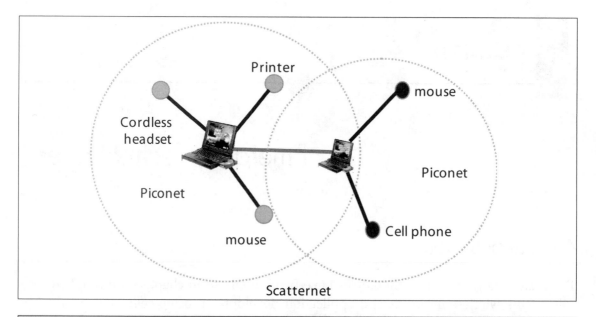

Figure 4.1 Blutooth scatternet as a combination of Piconets

4.2.1 Bluetooth Protocol

The Bluetooth radio is built into a small microchip and operates in a globally available frequency band ensuring interoperability worldwide. Bluetooth uses the unlicensed 2.4 GHz ISM (Industrial Scientific and Medical) frequency band. There are 79 available Bluetooth channels spaced 1 MHz apart from 2.402 GHz to 2.480 GHz. The Bluetooth standard is managed and maintained by Bluetooth Special Interest Group (www.bluetooth.com). IEEE has also adapted Bluetooth as the 802.15.1a standard. Bluetooth allows power levels starting from 1mW covering 10cm to 100mW covering upto 100 meters. These power levels are suitable for short device zone to personal area networks within a home. Bluetooth supports both unicast (point-to-point) and multicast (point-to-multi-point) connections. Bluetooth protocol uses the concept of master and slave. In a master-slave protocol a device cannot talk as and when they desire. They need to wait till the time the master allows them to talk. The master and slaves together form a piconet. Up to seven 'slave' devices can be set to communicate with a 'master'. Several of these **piconets** can be linked together to form a larger network in an ad hoc manner. The topology can be thought as a flexible, multiple piconet structure. This network of piconets is called **scatternet** (Figure 4.1). A scatternet is formed when a device from one piconet also acts as a member of another piconet. In this scheme, a device being master in one piconet can simultaneously be a slave in the other one.

Bluetooth protocol is a combination of different protocols. The Bluetooth Core protocols plus the Bluetooth radio protocols are required by most of Bluetooth devices, while the rest of the protocols are used by different applications as needed. At the physical layer Bluetooth uses spread spectrum technologies. It uses both direct sequence and frequency hopping spread spectrum technologies. Bluetooth uses connectionless (ACL–Asynchronous Connectionless Link) and connection-oriented (SCO–Synchronous Connection-oriented Link) links. Together, the Cable Replacement layer, the Telephony Control layer, and the Adopted protocol layer form application-oriented protocols enabling applications to run over the Bluetooth Core protocols.

4.2.2 Bluetooth Protocol Stack

Bluetooth protocol stack can be thought of combination of multiple application specific stacks as depicted in Figure 4.2. Different applications run over one or more vertical slices from this protocol stack. These are RFCOMM (Radio Frequency COMMunication), TCS Binary (Telephony Control Specification), and SDP (Service Discovery Protocol). Each application environments use a common data link and physical layer. RFCOMM and the TCS binary (Telephony Control Specification) protocol are based on the ETSI TS 07.10 and the ITU-T Recommendation Q.931 respectively. Some applications have some relationship with other protocols, e.g., L2CAP (Logical Link Control and Adaptation Protocol) or TCS may use LMP (Link Manager Protocol) to control the link manager.

Bluetooth protocol stack can be divided into four basic layers according to their functions. These are:

- **Bluetooth Core Protocols** this comprises of Baseband, Link Manager Protocol (LMP), Logical Link Control and Adaptation Protocol (L2CAP), and Service Discovery Protocol (SDP).

  o **Baseband** The Baseband and Link Control layer enables the physical RF link between Bluetooth units forming a piconet. This layer uses inquiry and paging procedures to synchronize the transmission with different Bluetooth devices. Using SCO and ACL link different packets can be multiplexed over the same RF link. ACL packets are used for data only, while the SCO packet can contain audio only or a combination of audio and data. All audio and data packets can be provided with different levels of CRC (Cyclic Redundancy Code) or FEC (Forward Error Correction) for error detection/correction.

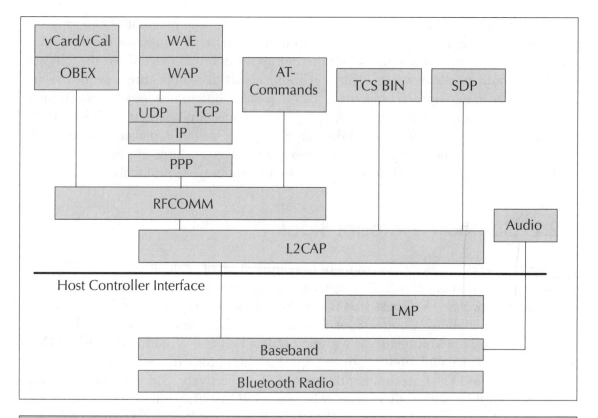

Figure 4.2 Blutooth Protocol stack

o **Link Manager Protocol (LMP)** When two Bluetooth devices come within each other's radio range, link managers of either device discover each other. LMP then engages itself in peer-to-peer message exchange. These messages perform various security functions starting from authentication to encryption. LMP layer performs generation and exchange of encryption keys as well. This layer performs the link setup and negotiation of baseband packet size. LMP also controls the power modes, connection state, and duty cycles of Bluetooth devices in a piconet.

o **Logical Link Control and Adaptation Protocol (L2CAP)** this layer is responsible for segmentation of large packets and the reassembly of fragmented packets. L2CAP is also responsible for multiplexing of Bluetooth packets from different applications.

o **Service Discovery Protocol (SDP)** The Service Discovery Protocol (SDP) enables a Bluetooth device to join a piconet. Using SDP a device inquires what services are available in a piconet and how to access them. SDP uses a client-server model where the server has a list of services defined through service records. One service record in a server describes the characteristics of one service. In a Bluetooth device there can be only one SDP server. If a device provides multiple services, one SDP server acts on behalf of all of them. Similarly multiple applications in a device may use a single SDP client to query servers for service records. A Bluetooth device in an inquiry mode broadcasts ID packets on 32 frequency channels of the Inquiry Hopping Sequence. It sends two ID packets every 625 μs and then listens for responses the following 625 μs. At this stage the unique identity of the devices called Bluetooth globalID is exchanged. A globalID indicates a device's profile along with capability functions. Upon matching of the device profile a connection is set up and devices exchange data. When a connection is set up, the paging device becomes the master and the paged device becomes the slave. A Bluetooth device may operate both as a server and as a client at the same time forming a scatternet. They can also switch from master to slave and vice versa. The master slave switch can take between 4:375 and 41:875 ms. In a piconet, a master device can be a laptop or PDA, while slaves devices could be printers, mouse, cellular phones etc.

- **Cable Replacement Protocol** this protocol stack has only one member viz., Radio Frequency Communication (RFCOMM).

  o RFCOMM this is a serial line communication protocol and is based on ETSI 07.10 specification. The "cable replacement" protocol emulates RS-232 control and data signals over Bluetooth baseband protocol.

- **Telephony Control Protocol** this comprises of two protocol stacks viz., Telephony Control Specification Binary (TCS BIN), and the AT-Commands.

  o **Telephony Control protocol Binary** TCS Binary or TCS BIN is a bit-oriented protocol. TCS BIN defines the call control signaling protocol for set up of speech and data calls between Bluetooth devices. It also defines mobility management procedures for handling groups of Bluetooth TCS devices. TCS Binary is based on the ITU-T Recommendation Q.931.

  o **AT-Commands** this protocol defines a set of AT-commands by which a mobile phone can be used and controlled as a modem for fax and data transfers. AT (short form of attention) commands are used from a computer or DTE (Data Terminal Equipment) to control a modem or DCE (Data Circuit terminating Equipment). AT-commands in Bluetooth are based on ITU-T Recommendation V.250 and GSM 07.07.

- **Adopted Protocols** This has many protocol stacks like Point-to-Point Protocol (PPP), TCP/IP Protocol, OBEX (Object Exchange Protocol), Wireless Application Protocol (WAP), vCard, vCalendar, Infrared Mobile Communication (IrMC), etc.

  o **PPP** Bluetooth offers PPP over RFCOMM to accomplish point-to-point connections. Point-to-Point Protocol is the means of taking IP packets to/from the PPP layer and placing them onto the LAN.

  o **TCP/IP** Protocol is used for communication across the Internet. TCP/IP stacks are used in numerous devices including printers, handheld computers, and mobile handsets. Access to these protocols is operating system independent, although traditionally realized using a socket programming interface model. TCP/IP/PPP is used for the all Internet Bridge usage scenarios. UDP/IP/PPP is also available as transport for WAP.

  o **OBEX Protocol** OBEX is a session protocol developed by the Infrared Data Association (IrDA) to exchange objects. OBEX, provides the functionality of HTTP in a much lighter fashion. The OBEX protocol defines a folder-listing object, which can be used to browse the contents of folders on remote devices.

  o **Content Formats** vCard and vCalendar specifications define the format of an electronic business card and personal calendar entries developed by the Versit consortium, These are now maintained by the Internet Mail Consortium. Other content formats, supported by OBEX, are vMessage and vNote. These content formats are used to exchange messages and notes. They are defined in the IrMC (IrDA Mobile Communication) specification. IrMC also defines a format for synchronization of data between devices.

4.2.3 Bluetooth Security

In a wireless environment where every bit is on the air, security concerns are high. Bluetooth offers security infrastructure starting from authentication, key exchange, to encryption. In addition to encryption, a frequency-hopping scheme with 1600 hops/sec is employed. All of this make the system difficult to eavesdrop. At the lowest levels of the protocol stack, Bluetooth uses the publicly available cipher algorithm known as SAFER+ to authenticate a device's identity. In addition to these basic security functions, different application verticals use their own security infrastructure at the application layer.

4.2.4 Bluetooth Application Models

Each application model in Bluetooth is realized through a Profile. Profiles define the protocols and protocol features supporting a particular usage model.

- **File Transfer** The file transfer usage model offers the ability to transfer data objects from one device (e.g., PC, smart-phone, or PDA) to another. Object types include .xls, .ppt, .wav, .jpg,.doc files, folders or directories or streaming media formats. Also, this model offers a possibility to browse the contents of the folders on a remote device.

- **Internet Bridge** In this usage model, mobile phone or cordless modem acts as modem to the PC, providing dial-up networking and fax capabilities without need for physical connection to the PC.

- **LAN Access** In this usage model multiple data terminals use a LAN access point (LAP) as a wireless connection to an Ethernet LAN. Once connected, the terminals operate as if they were connected directly to the LAN.

- **Synchronization** The synchronization usage model provides a device-to-device (phone, PDA, computer, etc.) synchronization of data. Examples could be PIM (personal information management) information, typically phonebook, calendar, message, and note information.

- **Headset** The headset can be wirelessly connected for the purpose of acting as a remote device's audio input and output interface. This is very convenient for hands free cellular phone usage in automobiles.

4.3 RADIO FREQUENCY IDENTIFICATION (RFID)

RFID is a radio transponder carrying an ID (Identification) that can be read through radio frequency (RF) interfaces. These transponders are commonly known as RFID tags or simply tags. To assign an identity to an object, a tag is attached to the object. Data within the tag provides identification for the object. The object could be an entity in a manufacturing shop, goods in transit, item in a retail store, a vehicle in a parking lot, a pet, or a book in a library. Biologists had been using RFID for sometime to track animals for the purpose of studying animal behavior and conservation. The earliest use of RFID was for tracking farm animals. A RFID system comprises of different functional areas like:

1. Means of reading or interrogating the data in the tag

2. Mechanism to filter some of the data

3. Means to communicate the data in the tag with a host computer

4. Means for updating or entering customized data into the tag.

RFID tags are categorized on three basic criteria. These are based on frequency, application area and the power level.

- **On Frequency:** There are 6 basic frequencies on which RFID operates. These are 132.4 KHz, 13.56 MHz, 433 MHz, 918 MHz, 2.4 GHz and 5.8 GHz. Low-frequency (30 KHz to 500 KHz) systems have short reading ranges and lower system costs. Tags in this frequency range are slow in data transfer and suitable for slow-moving objects. They are most commonly used in security access, asset tracking and animal identification applications. High-frequency (850 MHz to 950 MHz and 2.4 GHz to 2.5 GHz) systems offer long read ranges and high data transfer speeds. High reading speed is required for fast moving objects like railway wagon tracking and identification of vehicles on freeways for automated toll collection. The higher the frequency the higher the data transfer rates.

- **On Application:** RFIDs are also grouped according to application and usage. Speed of the object and distance to read determines the type of tag to be used. RFID used for livestock will be different from the tag used in railroad. The significant advantage of all types of RFID systems is the contactless, non-line-of-sight nature of the technology. Tags can be read through a variety of substances such as snow, fog, paint, plastic, wall, container and other challenging conditions, where barcodes or other optical means of reading are not effective. RFID tags can also be read at high speeds. In these cases RFIDs can respond within 100 milliseconds. A RFID tag contains two segments of memory. One segment is a factory-set and used to uniquely identify a tag. The other segment is usable by the application. Application specific data can be written or stored in this portion of the tag. The read/write capability of a RFID system is an advantage in interactive applications such as work-in-process or maintenance tracking. Compared to the barcode, RFID is a costlier technology. However, RFID has become indispensable for a wide range of automated data collection and identification applications that would not be possible otherwise.

- **Power-based grouping:** RFIDs can be grouped into two types based on power requirements. These are active and passive tags. Passive tags are generally in low frequency range, whereas tags at higher frequency range can be either active or passive.

- **Active RFID tags:** Active tags are powered by an internal battery and are typically read/write. The life of an active tag is limited by the life of the battery. The data within an active tag can be rewritten or modified. An active tag's memory can vary from a few bytes to 1MB. The battery-supplied power of an active tag generally gives it a longer read range. The trade off is, greater the size the greater cost, and a limited operational life. Depending upon the battery type and temperatures, the life of such tags could be 10 years. Some active tags can also be smart and do not send their information all the time. In a typical read/write RFID system, a tag might give a machine a set of instructions, and the machine would then report its performance to the tag. This encoded data would then become part of the tagged part's history. This data can be details about the port of transit with dates.

- **Passive RFID tags:** Passive tags operate without a power source of its own. A passive tag obtains operating power from the reader's antenna. The data within a passive tag is read only and generally cannot be changed during operation. Passive tags are lighter, less expensive and offer a virtually unlimited operational lifetime. The trade off is that they have shorter read ranges than active tags and require a higher-powered reader. Passive tags contain data usually 32 to 128 bits long.

RFID tags are of different shapes and sizes. Animal tracking tags are inserted beneath the skin and are as small as a pencil lead. Tags can be screw-shaped to identify trees or wooden logs. In stores, plastic tags are attached to merchandise and used as anti-theft device. Heavy-duty large tags are used to track containers or heavy machinery. The reader emits radio waves in any range from one centimeter to 25 meters or more. When an RFID tag passes through the electromagnetic zone of the reader, it detects the reader's activation signal. The reader decodes the data encoded in the tag's integrated circuit and the data is passed to the host computer for processing. A basic RFID system consist of three components:

- A transponder programmed with unique information (RFID tag)

- A transceiver with decoder (a reader)

- An antenna or coil.

The antenna emits radio signals to read data from or write data into the tag. Antennas control data acquisition and communication. An antenna is fitted with the transceiver to become a reader. Close proximity passive tags rely on electromagnetic or inductive coupling techniques (Figure 4.3a). Whereas, active tags are based upon propagating electromagnetic waves techniques (Figure 4.3b). Coupling is via 'antenna' structures forming

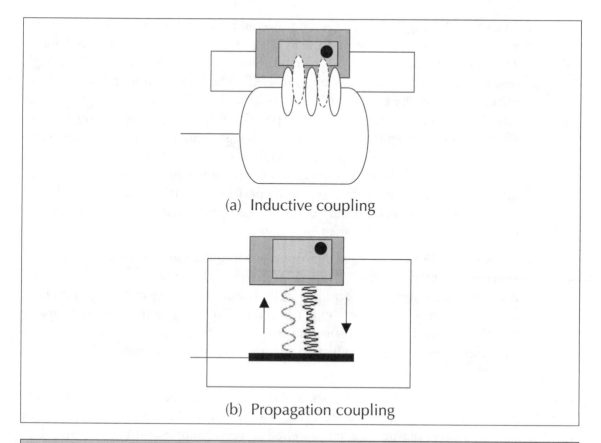

(a) Inductive coupling

(b) Propagation coupling

Figure 4.3 Passive and active RFID

an integral feature in both tags and readers. While the term antenna is generally considered more appropriate for propagating systems, it is also loosely applied to inductive systems.

4.3.1 Areas of Application for RFID

Potential applications for RFID may be identified in virtually every sector of industry, commerce and services where data is to be collected. The attributes of RFID are complementary to other data-capture technologies and therefore able to satisfy particular application requirements that cannot be adequately accommodated by alternative technologies. Principal areas of application for RFID that can be currently identified include:

- Transportation and Logistics
- Manufacturing and Processing

- Security
- Animal tagging
- Store in an enterprise
- Retail store
- Community library
- Time and attendance
- Postal tracking
- Airline baggage reconciliation
- Road toll management.

The lack of standards has been a deterrent for the growth of the RFID industry. Standards are essential for interoperability and growth of a technology. Standardization helped GSM and barcode technology to be widely accepted. Therefore, a number of organizations in US and Europe are working to address this issue. ANSI's X3T6 group is currently developing a draft document-based systems' operation at a carrier frequency of 2.45 GHz. ISO has already adopted international RFID standards for animal tracking, ISO 11784 and 11785.

4.4 Wireless Broadband (WiMax)

Wireless technologies are proliferating in a major way into the first-mile (as computer people call it) or last-mile (as communication people call it) subscriber access, as opposed to twisted-pair local loop. These technologies are generally referred to as (WLL–wireless local loop) or WiLL (wireless in local loop). Wireless local loop is also known as fixed-wireless system. The world is moving towards a convergence of voice, data and video. This convergence will demand interoperability and high data rate. Keeping this in mind, the IEEE 802 committee set up the 802.16 working group in 1999 to develop wireless broadband or WirelessMAN (wireless metropolitan area network) standards. Wireless MAN offers an alternative to high bandwidth wireline access networks like fiber optic, cable modems and DSL (Digital Subscriber Line). Figure 4.4 depicts a WirelessMAN architecture.

The release of WirelessMAN (IEEE 802.16) standards in April 2002 has paved the way for the entry of broadband wireless access as a new bearer to link homes and businesses with core telecommunications networks. WirelessMAN provides network access to buildings through exterior antennas communicating with radio base stations. The technology is

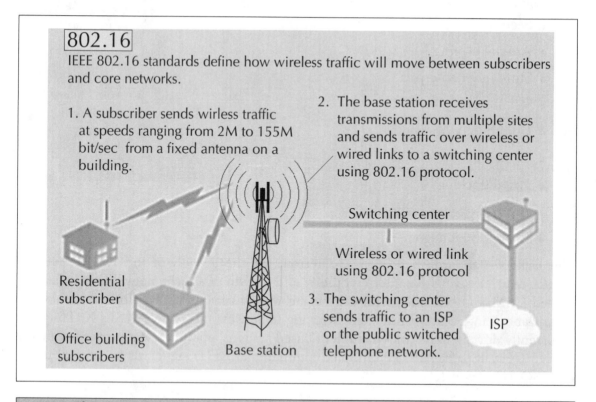

Figure 4.4 The WirelessMAN (wireless metropolitan area network)

expected to provide less expensive access with more ubiquitous broadband access with integrated data, voice and video services. One of the most attractive aspects of wireless broadband technology is that networks can be created in just weeks by deploying a small number of base stations on buildings or poles to create high-capacity wireless access systems. In a wired set up, one physical wire will connect the device with the network. Also, we need to keep many wires reserved for future growth. Therefore, the initial investment in wired infrastructure is very high. Wireless network can grow as the demand increases. At any point in time the number of active users are always a fraction of the number of subscribers. In a wireless environment the number of channels is always low compared to the number of subscribers. This makes wireless technologies very attractive to the service providers.

IEEE 802.16 standardizes the air interface and related functions associated with WLL. Three working groups have been chartered to produce following standards:

- IEEE 802.16.1–Air interface for 10 to 66 GHz.

- IEEE 802.16.2–Coexistence of broadband wireless access systems.

- IEEE 802.16.3–Air interface for licensed frequencies, 2 to 11 GHz.

- Extensive radio spectrum is available in frequency bands from 10 to 66 GHz worldwide. In a business scenario, 802.16 can serve as a backbone for 802.11 networks. Other possibilities are using 802.16 within the enterprise along with 802.11a, b or g.

IEEE 802.16 standards are concerned with the air interface between a subscriber's transceiver station and a base transceiver station. The 802.16 standards are organized into a three-layer architecture.

- The physical layer: This layer specifies the frequency band, the modulation scheme, error-correction techniques, synchronization between transmitter and receiver, data rate and the multiplexing structure.

- The MAC (Media Access Control) layer: This layer is responsible for transmitting data in frames and controlling access to the shared wireless medium through media access control (MAC) layer. The MAC protocol defines how and when a base station or subscriber station may initiate transmission on the channel.

- Above the MAC layer is a convergence layer that provides functions specific to the service being provided. For IEEE 802.16.1, bearer services include digital audio/video multicast, digital telephony, ATM, Internet access, wireless trunks in telephone networks and frame relay.

4.4.1 Physical Layer

To support duplexing, 802.16 adapted a burst design that allows both time-division duplexing (TDD) and frequency-division duplexing (FDD). In TDD the uplink and downlink share a channel but do not transmit simultaneously. In the case of FDD the uplink and downlink operate on separate channels and sometimes simultaneously. Support for half-duplex FDD subscriber stations is also supported in 802.16. Both TDD and FDD alternatives support adaptive burst profiles in which modulation and coding options may be dynamically assigned on a burst-by-burst basis.

The 2–11 GHz bands, both licensed and unlicensed, are used in 802.16. Design of the 2–11 GHz physical layer is driven by the need for non-line-of-sight operation. The draft currently specifies that compliant systems implement one of three air interface specifications, each of which provides for interoperability. The 802.16 standard specifies three physical layers for services:

- WirelessMAN-SC2: This uses a single-carrier modulation format. This is to support existing networks and protocols.

- WirelessMAN-OFDM: This uses orthogonal frequency-division multiplexing with a 256-point transform. Access is by TDMA. This air interface is mandatory for license-exempt bands.

- WirelessMAN-OFDMA: This uses orthogonal frequency-division multiple access with a 2048-point transform. In this system, multiple access is provided by addressing a sub-set of the multiple carriers to individual receivers.

4.4.2 802.16 Medium Access Control

The IEEE 802.16 MAC protocol was designed for point-to-multipoint broadband wireless access. It addresses the need for very high bit rates, both uplink (to the base station) and downlink (from the base station). To support, a variety of services like multimedia and voice, the 802.16 MAC is equipped to accommodate both continuous and bursty traffic. To facilitate the more demanding physical environment and different service requirements of the frequencies between 2 and 11 GHz, the 802.16 project is upgrading the MAC to provide automatic repeat request (ARQ) and support for mesh, rather than only point-to-multipoint, network architectures.

4.4.3 Broadband Applications

Wireless broadband allows higher data rates in homes and offices. Therefore all the user applications in home and offices are potential candidates for wireless broadband. These include standards Ethernet LAN or WiFi. However, along with the existing applications a new brand of applications are also being thought about. One such system is mobile cellular system.

4.4.4 Broadband Mobile Cellular System

During different discussions on systems and architecture of mobile computing, we talked about mobility with the network being static. In mobile cellular system the cellular network itself will be mobile. A cellular system like 3G can provide high data rate. WirelessMAN is also geared up to support high data rate. However, these high data rates

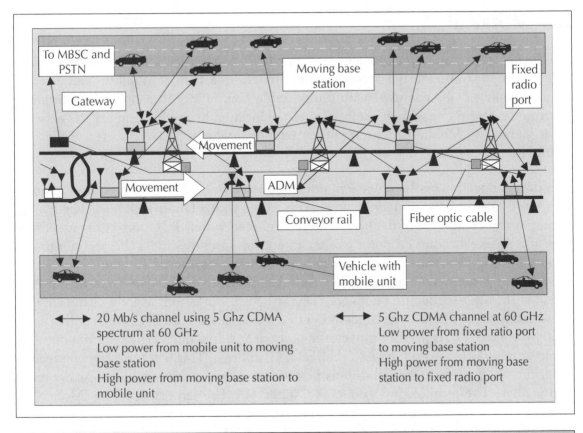

Figure 4.5 Mobile broadband communication system with moving BTS

are possible with low speed mobility. Scientists are now thinking in terms of high-speed mobility specially designed for high-speed telematics application.

Figure 4.5 depicts one such mobile communication system to support high-speed mobility. This is achieved by installing moving base stations and fixed radio ports uniformly distributed along the median of the roadway. The moving base stations allow communication links to be established between the mobile units traveling on the roadway and a fixed communication network through the fixed radio ports. The small-cell (picocell) architecture of the proposed system enables the use of extremely lightweight low-power mobile units that can be used almost anywhere. In this architecture the picocell will move in the direction of the moving vehicle so that the relative speed between them are low. This proposed infrastructure is suitable for high-speed multilane highways in cities. The proposed system will be able to communicate to devices traveling at speeds up to and in excess of 150 KMph.

4.5 MOBILE IP

Mobile computing in a true sense will be able to provide an environment where a user will be able to continuously access data and services in a state of mobility. Mobile computing should not be confused with the portable computing. In a portable computing environment we move with the device from one location to another and use the network while stationary. Mobile computing offers seamless computing facility even if the user changes the network.

A data connection between two end-points through TCP/IP network requires a source IP address, source TCP port and a target IP address with a target TCP port. The combination of one IP address of the host system combined with a TCP port as the identification of a service becomes a point of attachment for an end-point. TCP port number is application-specific and remains constant. IP address, on the other hand, is network-specific and varies from network to network. IP addresses are assigned to a host from a set, of addresses assigned to a network. This structure works well as long as the client is static and is using a desktop computer. Let us assume that the user is mobile and is using a laptop with WiFi. As the user moves, the point of attachment will change from one subnet to another subnet resulting in a change of IP address. This will force the connection to terminate. Therefore, the question is how do we allow mobility while a data connection is alive. The technology to do so is 'Mobile IP'. The term 'mobile' in 'Mobile IP' signifies that, while a user is connected to applications across the Internet and the user's point of attachment changes dynamically, all connections are maintained despite the change in underlying network properties. This is similar to the handoff/roaming situation in cellular network. In a cellular network, when a user is mobile, the point of attachment (base station) changes. However, in spite of such changes the user is able to continue the conversation.

4.5.1 How does Mobile IP work?

Internet Protocal routes packets from a source endpoint to a destination endpoint through various routers. An IP address of a host can be considered to be a combination of network address (most significant 24 bits) and the node address (least significant 8 bits). Let us assume a 'C' class IP address 203.197.175.123 to be of the mail server of iiitb (email.iiitb.ac.in). We can assume that the first 24 bits 203.197.175 is the address of the network and the last 8 bits containing 123 is the address of the host. The **network portion** of an IP address is used by routers to deliver the packet to the last router in the chain to which the target computer is attached. This last router then uses the **host**

portion (123 in this example) of the IP address to deliver the IP packet to the destination computer. In addition to the IP addresses of the hosts, for a meaningful communication we need the TCP or UDP (User Datagram Protocal) port of the applications. The port number is used by the host to deliver the packet to the appropriate application.

A TCP connection is identified by a quadruplet that contains the IP address and port number of the sender endpoint along with the IP address and port number of the receiving endpoint. To ensure that an active TCP connection is not terminated while the user is mobile, it is essential that all of these four identities remain constant. The TCP ports are application specific and generally constant. However, the IP address changes from subnet to subnet. Therefore, to fix this problem mobile IP allows the mobile node to use two IP addresses. These IP addresses are called **home address** and **care-of address**. The home address is static and known to everybody as the identity of the host. The care-of address changes at each new point of attachment and can be thought of as the mobile node's location specific address. This is similar to the concept of HLR (Home Location Register) and VLR (Visitor Location Register) in cellular networks. When the mobile node is roaming and is attached to a foreign network, the home agent receives all the packets for the mobile node and arranges to forward them to the mobile node's current point of attachment. The network node that is responsible for forwarding and managing this transparency is known as the **home agent**.

Whenever the mobile node moves, it registers its new care-of address with its home agent. The home agent forwards the packet to the foreign network using the care-of address. The delivery requires that the packet header is modified so that the care-of address becomes the destination IP address. This new header (Figure 4.8) encapsulates the original packet, causing the mobile node's home address to have no impact on the encapsulated packet's routing. This phenomenon is called **tunneling**. Figure 4.6 shows in general terms how Mobile IP deals with the problem of dynamic IP addresses.

Let us take an example of IP datagrams being exchanged over a TCP connection between the mobile node (A) and another host (server X in Figure 4.6), the following steps occur:

- Server X wants to transmit an IP datagram to node A. The home address of A is advertised and known to X. X does not know whether A is in the home network or somewhere else. Therefore, X sends the packet to A with A's home address as the destination IP address in the IP header. The IP datagram is routed to A's home network.

- At the A's home network, the incoming IP datagram is intercepted by the home agent. The home agent discovers that A is in a foreign network.

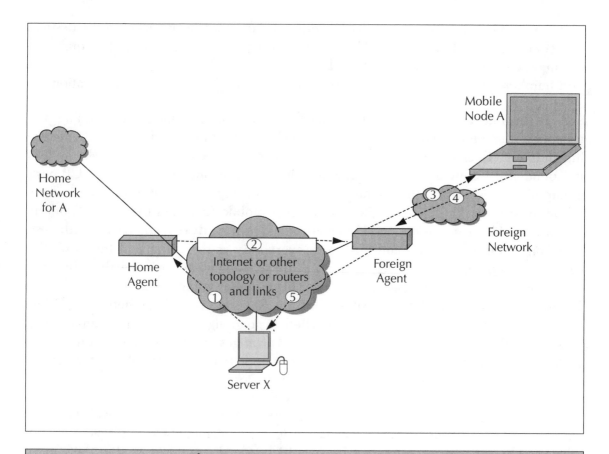

Figure 4.6 Mobile IP architecture

A care-of-address has been allocated to A by this foreign network and available with the home agent. The home agent encapsulates the entire datagram inside a new IP datagram, with A's care-of address in the IP header. This new datagram with the care-of-address as the destination address is retransmitted by the home agent.

- At the foreign network, the incoming IP datagram is intercepted by the **foreign agent**. The foreign agent is the counterpart of the home agent in the foreign network. The foreign agent strips off the outer IP header, and delivers the original datagram to A.

- A intends to respond to this message and sends traffic to X. In this example, X is not mobile; therefore X has a fixed IP address. For routing A's IP datagram to X, each datagram is sent to some router in the foreign network. Typically, this router is the foreign agent. A uses X's IP static address as the destination address in the IP header.

- The IP datagram from A to X travels directly across the network, using X's IP address as the destination address.

To support the operations illustrated in the example above, mobile IP needs to support three basic capabilities:

- **Discovery:** A mobile node uses a discovery procedure to identify prospective home agents and foreign agents.

- **Registration:** A mobile node uses a registration procedure to inform its home agent of its care-of address.

- **Tunneling:** Tunneling procedure is used to forward IP datagrams from a home address to a care-of address.

4.5.2 Discovery

The Mobile IP discovery procedure has been built on top of an existing ICMP router discovery and advertisement procedure as specified in RFC 1256. Using these procedures a router can detect whether a new mobile node has entered into its network. Also, using this procedure the mobile node determines whether it is in a foreign network. For the purpose of discovery, a router or an agent periodically issues a router advertisement ICMP message. The mobile node on receiving this advertisement packet compares the network portion of the router IP address with the network portion of its own IP address allocated by the home network. If these network portions do not match, then the mobile node knows that it is in a foreign network. A router advertisement can carry information about default routers and information about one or more care-of addresses. If a mobile node needs a care-of address without waiting for the agent advertisement, the mobile node can broadcast a solicitation that will be answered by any foreign agent.

4.5.3 Registration

Once a mobile node obtained a care-of-address from the foreign network, the same needs to be registered with the home agent. The mobile node sends a registration request to the home agent with the care-of address information. When the home agent receives this request, it updates its routing table and sends a registration reply back to the mobile node.

Authentication: As a part of registration, the mobile host needs to be authenticated. Using 128-bit secret key and the MD5 hashing algorithm, a digital signature is generated. Each mobile node and home agent shares a common secret. This secret makes the digital signature unique and allows the agent to authenticate the mobile node. At the end of the registration a triplet containing the home address, care-of address and registration lifetime is maintained in the home agent. This is called a **binding** for the mobile node. The home agent maintains this association until the registration lifetime expires. The registration process involves the following four steps:

- The mobile node requests for forwarding service from the foreign network by sending a registration request to the foreign agent.

- The foreign agent relays this registration request to the home agent of that mobile node.

- The home agent either accepts or rejects the request and sends a registration reply to the foreign agent.

- The foreign agent relays this reply to the mobile node.

We have assumed that the foreign agent will allocate the care-of address. However, it is possible that a mobile node move to a network that has no foreign agents or on which all foreign agents are busy. As an alternative therefore, the mobile node may act as its own foreign agent by using a **colocated** care-of address. A colocated care-of address is an IP address obtained by the mobile node that is associated with the foreign network. If the mobile node is using a colocated care-of address, then the registers happens directly with its home agent.

4.5.4 Tunneling

Figure 4.7 shows the tunneling operations in Mobile IP. In the mobile IP, **IP-within-IP** encapsulation mechanism is used. Using IP-within-IP, the home agent, adds a new IP header called **tunnel header**. The new tunnel header uses the mobile node's care-of address as the tunnel destination IP address. The tunnel source IP address is the home agent's IP address. The tunnel header uses 4 as the protocol number (Figure 4.8), indicating that the next protocol header is again an IP header. In IP-within-IP, the entire original IP header is preserved as the first part of the payload of the tunnel header. The foreign agent after receiving the packet, drops the tunnel header and delivers the rest to the mobile node.

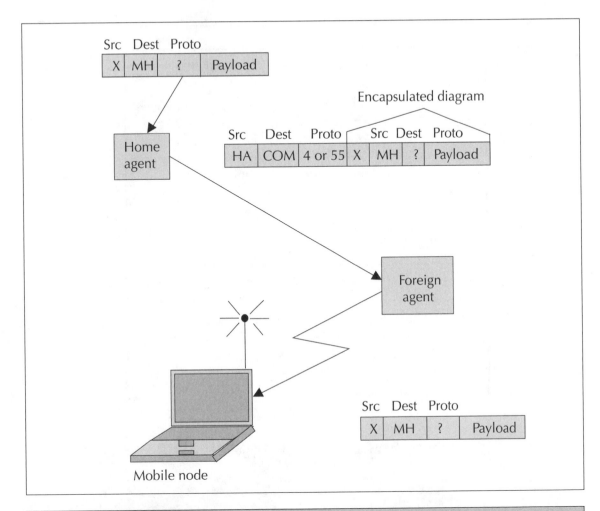

Figure 4.7 Tunneling operations in Mobile IP

When a mobile node is roaming in a foreign network, the home agent must be able to intercept all IP datagram packets sent to the mobile node so that these datagrams can be forwarded via tunneling. The home agent, therefore, needs to inform other nodes in the home network that all IP datagrams with the destination address of the mobile node should be delivered to the home agent. In essence, the home agent steals the identity of the mobile node in order to capture packets destined for that node that are transmitted across the home network. For this purpose ARP (Address Resolution Protocol) is used to notify all nodes in the home network.

Let us take the example of Figure 4.6. The original IP datagram from X to A has a source address as IP address of X and a destination address as the home IP address of A.

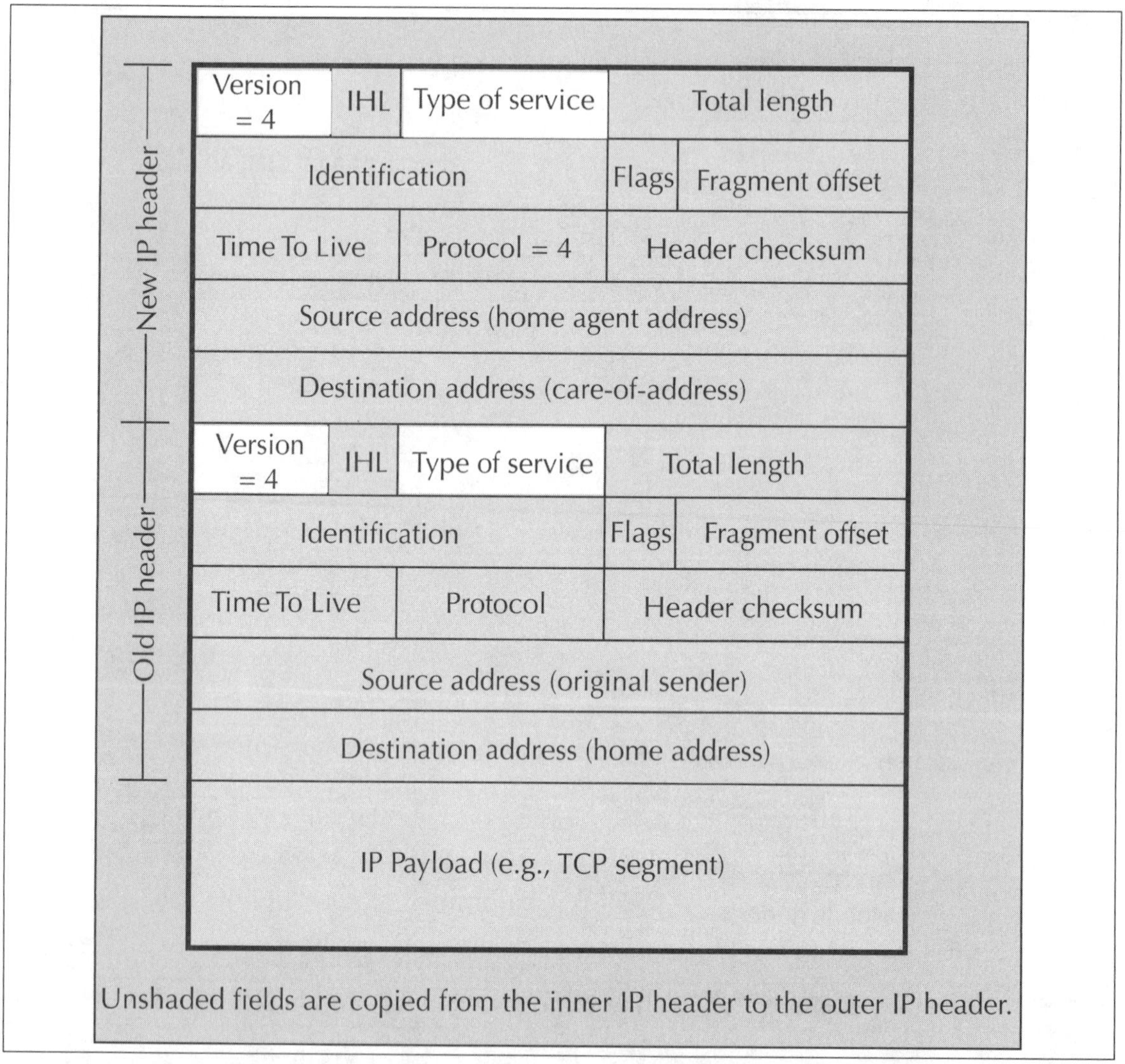

Version = 4	IHL	Type of service	Total length	
Identification			Flags	Fragment offset
Time To Live		Protocol = 4	Header checksum	
Source address (home agent address)				
Destination address (care-of-address)				

New IP header

Version = 4	IHL	Type of service	Total length	
Identification			Flags	Fragment offset
Time To Live		Protocol	Header checksum	
Source address (original sender)				
Destination address (home address)				
IP Payload (e.g., TCP segment)				

Old IP header

Unshaded fields are copied from the inner IP header to the outer IP header.

Figure 4.8 The IP headers in mobile IP (IP encapsulation)

The datagram is routed through the Internet to A's home network, where it is intercepted by the home agent. The home agent encapsulates the incoming datagram with an outer IP header. This outer header includes a source address same as the IP address of the home agent and a destination address equal to the care-of-address. As the care-of-address has the network portion of the foreign network, the packet will find its way directly to the mobile host. When this new datagram reaches the host in the foreign network, it strips off the outer IP header to extract the original datagram. From this stripped off packet it also finds out the original sender. This is necessary for the host to know who has sent the packet so that the response reaches the right destination.

4.5.5 Cellular IP

The primary design goal for mobile IP protocols is to allow a host to change its point of access during data transfer without being disconnected or needing to be reconfigured. An important design goal for mobile host protocols is to support handoffs without significant disturbance to ongoing data transmission. A change of access point while connectivity is maintained is called a handoff.

To manage mobility, generally a "two tier addressing" scheme is used. One address is for a fixed location which is known to all; other one is for a dynamic location which changes as the user moves. In case of GSM this is done through Home Location Register and Visitor Location Register. Same is true in Mobile IP, where a mobile host is associated with two IP addresses: a fixed home address that serves as the host-identifier; and a care-of-address that reflects its current point of attachment. The mobile IP architecture comprises three functions:

1. A database that contains the most up-to-date mapping between the two address spaces (home address to care-of-address)
2. The translation of the host identifier to the actual destination address
3. Agents ensuring that the source and destination packets for arriving and outgoing packets are updated properly so that routing of packets are proper

Whenever the mobile host moves to a new subnet managed by a different foreign agent, the dynamic care-of-address will change. This changed care-of-address needs to be communicated to the home agent. This process works for slowly moving hosts. For a high speed mobile host, the rate of update of the addresses needs to match the rate of change of addresses. Otherwise, packets will be forwarded to the wrong (old) address. Mobile IP fails to update the addressed properly for high speed mobility. **Cellular IP** (Figure 4.9), a new host mobility protocol has been designed to address this issue.

In a Cellular IP, none of the nodes know the exact location of a mobile host. Packets addressed to a mobile host are routed to its current base station on a hop-by-hop basis where each node only needs to know on which of its outgoing ports to forward packets. This limited routing information (referred as mapping) is local to the node and does not assume that nodes have any knowledge of the wireless network topology. Mappings are created and updated based on the packets transmitted by mobile hosts.

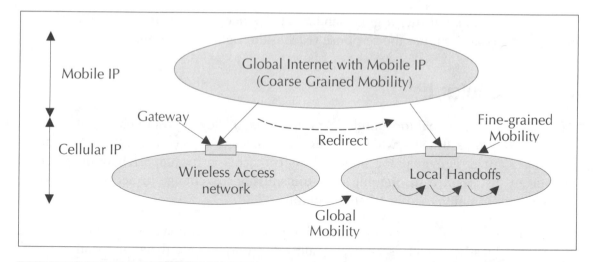

Figure 4.9 Relationships between Mobile IP and Cellular IP

Cellular IP uses two parallel structures of mappings through **Paging Caches** (PC) and **Routing Caches** (RC). PCs maintain mappings for stationary and idle (not in data communication state) hosts; whereas, RC maintains mappings for mobile hosts. Mapping entries in PC have a large timeout interval, in the order of seconds or minutes. RCs maintain mappings for mobile hosts currently receiving data or expecting to receive data. For RC mappings, the timeout are in the packet time scale. Figure 4.10 illustrates the relationship between PCs and RCs. While idle at location 1, the mobile host X keeps PCs up-to-date by transmitting dummy packets at a low frequency (step 1 in Figure 4.10). Let us assume that the host is mobile and moved to location 2 without transacting any data. The PC mapping for X now points to location 2. While at location 2, there are data packets to be routed to the mobile host X, the PC mappings are used to find the host (step 2). As there is data transmission, the mapping database to be used will be the RC. As long as data packets keep arriving, the host maintains RC mappings, either by its outgoing data packets or through the transmission of dummy packets (step 3).

Idle mobile hosts periodically generate short control packets, called **paging-update packets**. These are sent to the nearest available base station. The paging-update packets travel in the access network from the base station toward the gateway router, on a hop-by-hop basis. Handoff in Cellular IP is always initiated by the mobile host. As the host approaches a new base station, it redirects its data packets from the old to the new base station. First few redirected packets will automatically configure a new path of RC mappings for the host to the new base station. For a time equal to the timeout of

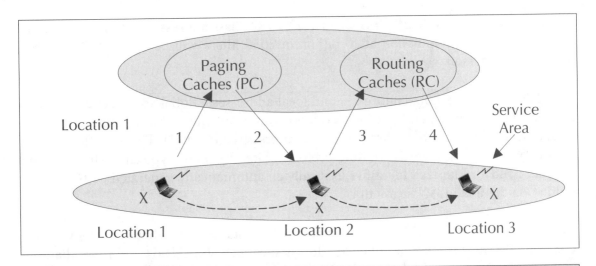

Figure 4.10 Cellular IP Paging and Routing

RC mappings, packets addressed to the mobile host will be delivered at both old and new base stations.

4.6 INTERNET PROTOCOL VERSION 6 (IPv6)

Internet offers access to information sources worldwide. We access Internet through increasing variety of wireless devices offering IP connectivity, such as PDAs, palmtops, handhelds, laptops, and digital cellular phones. The explosion in the number of devices connected to the Internet, combined with projections for the future, made scientists think seriously whether the 32-bit address space of TCP/IP is sufficient. IP version 6 (IPv6), the successor to today's IP version 4 protocol (IPv4), dramatically expands the available address space. Internet Engineering Task Force (IETF) has produced a comprehensive set of specifications (RFC 1287, 1752, 1886, 1971, 1993, 2292, 2373, 2460, 2473 etc.) that define the next-generation IP protocol originally known as 'IPNg,' now renamed as 'IPv6'. IPv6 addresses both a short-term and long-term concern for network owners, service providers and users.

4.6.1 Address Space

IPv6 uses 128 bit addresses for each packet, creating a virtually infinite number of IP addresses (approx. 3.4*10**38 IP addresses), as opposed to 3758096384 IPv4 addresses

(2**31 A Class address + 2**30 B Class + 2**29 C Class address). This also means that if we set the world population at 10 billion in 2050, there will be 3.4*10**27 addresses available per person.

In IPv6, there are global addresses and local addresses. Global addresses are used for routing of global Internet. Link local addresses are available within a subnet. IPv6 uses hierarchical addressing with three-level of addresses (Figure 4.11). This includes a Public Topology (the 48 bit external routing prefix), a Site Topology (typically a 16 bit subnet number), and an Interface Identifier (typically an automatically generated 64 bit number unique on the local LAN segment).

End-user-sites get their address prefix from an ISP that provides them the IPv6 service. General IPv6 host is given a linklocal address such as fe80::EUI-64 and more than one global address such as global-prefix::EUI-64. It has 64bit length and made by IEEE EUI-64 format. Interface ID is used to specific Interface in the same link. Interface ID

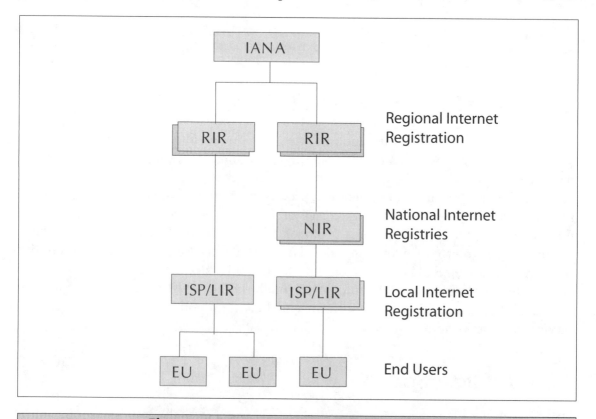

Figure 4.11 Hierarchical addressing of IPv6

is generated to use Interface's link layer address. An Ethernet MAC address for a device is 48 bits long, Interface ID is created by adding 2 octet "0xfffe" in it's center. Like 02:60:8c:de:7:79 becomes 260:8cff:fede:779.

4.6.2 IPv6 Security

One of the biggest differences between IPv6 and IPv4 is that all IPv6 nodes are expected to implement strong authentication and encryption features to improve Internet security. IPv6 comes native with a security protocol called IP Security (IPSec). Many vendors adapted IPSec as a part of IPv4. IPSec protocol is a standards-based method of providing privacy, integrity and authenticity to information transferred across IP networks.

IPSec combines several different security technologies into a complete system to provide confidentiality, integrity and authenticity. In particular, IPSec uses:

- Diffie-Hellman key exchange mechanism for deriving key between peers on a public network.

- Public key cryptography to guarantee the identity of the two parties and avoid man-in-the-middle attacks.

- Bulk encryption algorithms, such as 3DES, for encrypting the data.

- Keyed hash algorithms, such as HMAC, combined with traditional hash algorithms such as MD5 or SHA for providing packet authentication.

- Digital certificates signed by a certificate authority to act as digital ID cards.

- IPSec provides IP network-layer encryption.

4.6.3 Packet payload

Each IPv6 packet payload is attached a tag which can be customized to enable a better quality in the packet flow, or by a price of other class, such as non-real time quality of service or 'real-time' service. This feature does not exist natively in IPv4, although a part of payload could be used for the same, reducing unique information amount carried by the packet.

Information is packetized into IPv6 packets, with the corresponding levels of control. A neighbour discovery feature (care-of-address, and stateless Prefix or Stateful DHCPv6) will in principle allow the device carrying these packets to configure itself for a consistent dialogue with other devices or software interfaces. The same can be done with IPv4 packets, but with the intervention of humans or specific tools and services and only for selected information and software architectures.

4.6.4 Migrating from IPv4 to IPv6

The Migration from IPv4 to IPv6 is quite an involved task. This includes the following:

1. Migration of the network components to be able to support IPv6 packets. As there is no change at the physical layer between IPv4 and IPv6, network components like hub or switch need not change. As there is a change in the packet header the routers need to be upgraded. However, using IP tunneling IPv6 packets can propagate over an IPv4 envelope. Existing routers can support IP tunneling.

2. Migration of the computing nodes in the network: this will need the operating system upgrades so that they support IPv6 along with IPv4. Upgraded systems will have both IPv4 and IPv6 stacks. Therefore, both the IPv4 and IPv6 applications can run without any difficulty.

3. Migration of networking applications in both client and server systems: this requires porting of the applications from IPv4 to IPv6 environment.

Migration of Windows System

The Microsoft Windows 9x families do not support IPv6. Windows XP and Windows Server 2003 support IPv6 natively. Windows 2000 Professional can be upgraded to support IPv6. IPv6 in Windows support different tools and dlls. These are:

wship6.dll: The Winsock helper dynamically linked library for the INET6 address family

wininet.dll: Winsock INET6 libraries

ftp.exe: This is the IPv6 ftp client and server application

telnet.exe: This is the IPv6 telnet client application

tlntsvr.exe: Telnet server

ipv6.exe: This tool retrieves and displays configuration information about the IPv6 protocol. This tool is used to view the state of interfaces, the neighbor caches, the binding cache, the destination cache, and the route table. This utility can also be used to manually configure interfaces, addresses, and route table entries.

ping6.exe: This tool is equivalent to the current IPv4 ping.exe tool. It sends ICMPv6 Echo Request messages, waiting for the corresponding ICMPv6 Echo Reply messages and then displaying information on round trip times.

tracert6.exe: This tool is equivalent to the current IPv4 tracert.exe tool. It sends ICMPv6 Echo Request messages with monotonically increasing values of the Hop Limit field to discover the path traveled by IPv6 packets between a source and destination.

ttcp.exe: This tool is used to send TCP segment data or UDP messages between two nodes. ttcp.exe supports both IPv4 and IPv6.

6to4cfg.exe: This tool is used to configure IPv6 connectivity over an IPv4 network.

ipsec6.exe: This tool is used to configure policies and security associations for IPv6 IPSec traffic.

checkv4.exe: This tool is used to scan source code files to identify code that needs to be changed to support IPv6. This is similar to the *lint* command in Unix.

Migration of Linux System

Linux kernel 2.4.x either supports IPv6 directly or can be upgraded to support IPv6. All versions after Red Hat Linux 7.1 support IPv6 directly. The kernel needs to be built and properly configured for IPv6. Different tools are available in Linux as a part of v6 installation. These are:

#ping6: equivalent of ping in IPv4

#traceroute6: equivalent of tracerout in IPv4

#tracepath6: equivalent of tracepath in IPv4

#tcpdump: equivalent of tcpdump in IPv4

#proto: displays only tunneled IPv6-in-IPv4 traffic

#ip6: displays all native IPv6 traffic including ICMPv6

#icmp6: displays only native ICMPv6 traffic.

4.6.5 Migration of Applications

There are many networking software and systems around the world that use TCP/IP socket. One of the changes we need to do in all these applications is to allow the larger address space for the destination endpoint. This is similar to the classic Y2K problem of last century, where the space provided for a date field had to be enlarged. Moreover, in the case of IPv6, the header has been changed according to the practical need of the present-day applications as well as to facilitate high speed routing. The difference is that Y2K was a time-bound project and had to be finished before 1 January 2000. However, in case of IPv6, there is no such compulsion other than all applications in 3G network will have to be IPv6 as 3G will not support IPv4.

Figures 12a and 12b are the snapshots of a telnet application after migration to IPv6 environment.

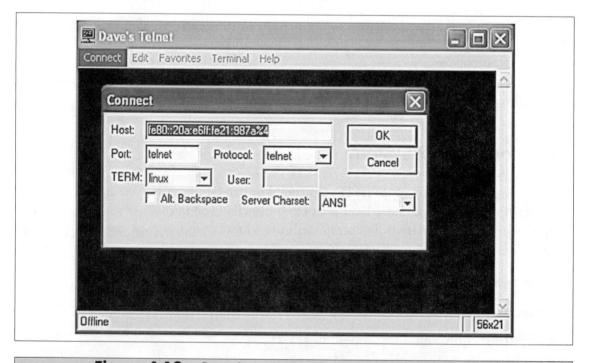

Figure 4.12a Snapshot of telnet connection request in IPv6
(Windows XP through dtelnet)

Figure 4.12b Snapshot of telnet client in IPv6

4.6.6 Interconnecting IPv6 Networks

Till all the routers/system become IPv6-compatible, the interconnection between IPv6 networks can be accomplished by tunneling. Tunneling is one of the key deployment strategies for both service providers as well as enterprises during the period of IPv4 and IPv6 coexistence. Tunneling service providers can offer an end-to-end IPv6 service without major upgrades to the infrastructure and without impacting current IPv4 services.

A variety of tunnel mechanisms are available. These mechanisms include:

1. Manually created tunnels such as IPv6 manually configured tunnels (RFC 2893).

2. IPv6 over IPv4 tunnels.

3. Semiautomatic tunnel mechanisms such as that employed by tunnel broker services.

4. Fully automatic tunnel mechanisms such as IPv4-compatible and 6 to 4.

The dual-stack routers run both IPv4 and IPv6 protocols simultaneously and thus can interoperate directly with both IPv4 and IPv6 end systems and routers.

4.6.7 Mobile IP with IPv6

IPv6 includes many features for streamlining mobility support that are missing in IP version 4, including Stateless Address Autoconfiguration and Neighbor Discovery. IPv6 with hierarchical addressing scheme will be able to manage IP mobility much efficiently. IPv6 also attempts to simplify the process of renumbering, which could be critical to the future routability of the Internet traffic. Mobility Support in IPv6, as proposed by the Mobile IP working group, follows the design for Mobile IPv4. It retains the ideas of a home network, home agent and the use of encapsulation to deliver packets from the home network to the mobile node's current point of attachment. While discovery of a care-of-address is still required, a mobile node can configure its a care-of address by using Stateless Address Autoconfiguration and Neighbor Discovery. Thus, foreign agents are not required to support mobility in IPv6.

4.7 JAVA CARD

Java Card is a smart card with Java framework. Smart card was developed in 1974, by Roland Moreno. Smart card is a plastic card with intelligence and memory. Smart cards are becoming popular as identity module and wireless security devices. In many countries driving licenses are being issued on smart cards. The SIM card on a GSM mobile phone is a smart card as well. The importance of smart card made ISO to standardize all its interfaces. These are done through ISO 7816 standards. These ISO standards define the physical characteristic of the card (ISO 7816-1: Physical Characteristics), locations and dimensions of the contacts (7816-2:: Dimensions and Locations of the Contacts), signals and transmission interfaces (7816-3:: Electronic Signals and Transmission Protocols), and command interfaces (7816-4:: Interindustry Commands for Interchange). A smart card is embedded with either (i) a microprocessor and a memory chip or (ii) only a memory chip with non-programmable logic. A microprocessor card can have an intelligent program resident within the card which can add, delete, and otherwise manipulate information on the card. A memory card on contrast, can store some information for some pre-defined operation. Smart cards are capable of carrying data, functions, and information on the card. Therefore, unlike memory strip cards, they do not require access to remote databases at the time of the transaction.

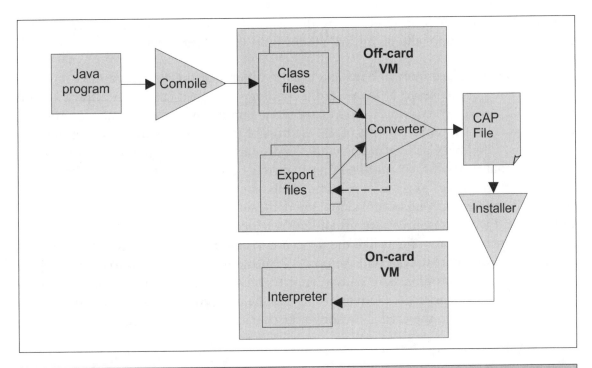

Figure 4.13 Architecture of Java card applications development process

Microprocessor based smart cards which used to be used for some specific application areas are becoming quite common. Smart cards have now emerged as multi function cards. To allow interoperability, Java was chosen as the vehicle for interoperability. All the microprocessor based smart cards now offer Java API framework on the smart card. This is why smart cards with Java framework are also called Java Cards. 3GPP has decided to use Java Card as the standard for USIM and ICC (Integrated Circuit cards). Java Card technology preserves many of the benefits of the Java programming languages such as: productivity, security, robustness, tools, and portability. For Java card, the Java Virtual Machine (JVM), the language definition, and the core packages have been made more compact to bring Java technology to the resource constrained smart cards.

A smart card of a GSM SIM card supporting Java Card functionalities may typically have 8 or 16 bit microprocessor running at speed between 5 MHz to 40 MHz with 32K to 128K bytes of EEPROM (Electronically Erasable Programmable Read Only Memory). Though Java card works in a master/slave mode, using the proactive SIM technology of GSM Phase 2+, it is possible for the application on the SIM card to get activated in an automated fashion. Also, Java card technology supports OTA (Over The Air) downloads. In OTA download, a Java applet (through SMS) can be downloaded by the network operator proactively or by the user interactively over the wireless media. Applications written for the Java Card platform are referred to as applets.

The development framework in Java card is different from that on a desktop computer. The major challenge of Java Card technology on smart card is to fit Java system software in a resource constraint smart card while conserving enough space for applications. Java Card supports a subset of the features of Java language available on desktop computers. The Java Card virtual machine on a smart card is split into two parts (Figure 4.13): one that runs off-card and the other that runs on-card. Many processing tasks that are not constrained to execute at runtime, such as class loading, bytecode verification, resolution and linking, and optimization, are dedicated to the virtual machine that is running off-card where resources are usually not a concern. Thle e on-card components of Java Card include components like the Java card virtual machine (JCVM), the Java card runtime environment (JCRE), and the Java API. Task of the compiler is to convert a Java source into Java class files. The converter will convert class files into a format downloadable into the smart card. Converter ensures the byte code validity before the application is installed into the card. The converter checks the classes off-card for,

- Well formedness

- Java Card subset violations

- Static variable initialization

- Reference resolution

- Byte code optimization

- Storage Allocation.

- The java card interpreter

- Executes the applets

- Controls run-time resources

- Enforces runtime security

Following conversion by the off-card VM into CAP (Converted APlet) format, the applet is transferred into the card using the installer. The applet is selected for execution by the JCRE. JCRE is made up of the on-card virtual machine and the Java Card API classes. JCRE performs additional runtime security checks through applet firewall. Applet firewall partitions the objects stored into separate protected object spaces, called contexts. Applet firewall controls the access to shareable interfaces of these objects. The JCVM is a scaled down version of standard JVM (Java Virtual Machine). Elements of standard Java not supported in JCVM are,

- Security manager
- Dynamic class loading
- Bytecode verifier
- Threads
- Garbage collection
- Multi dimensional arrays
- Char and strings
- Floating point operation
- Object serialization
- Object cloning

As mentioned above Java applications for a Java Cards are called Applets. Java Card applets should not be confused with Java applets on the Internet. A Java Card applet is not intended to run within an Internet browser environment. The reason for choosing the name applet is that Java Card applets can be loaded into the Java Card runtime environment after the card has been manufactured. That is, unlike applications in many embedded systems, Java Card applets do not need to be burned into the ROM during manufacture.

REFERENCES/FURTHER READING

1. Perkins, C., "Mobile IP," IEEE Communications Magazine, May 1997.

2. Perkins, C., "Mobile Networking through Mobile IP," IEEE Internet Computing, January-February 1998, p58.

3. Andras G. Valko, Cellular IP: A New Approach to Internet Host Mobility, ACM SIGCOMM Computer Communication Review, pp50–65.

4. Perkins, C., Mobile IP: Design Principles and Practices, ISBN 0-201-63469-4, Prentice Hall PTR, 1998.

5. Georgios Karagiannis, Mobile IP, Ericsson Open Report # 3/0362-FCP NB 102 88 Uen, 13 July, 1999.

6. AndrOs G. Valko, Cellular IP: A New Approach to Internet Host Mobility, ACM SIGCOMM Computer Communication Review.

7. RFC 2005, Applicability Statement for IP Mobility Support.

8. D. Cong & M. Hamlen, C. Perkins, The Definitions of Managed Objects for IP Mobility Support, RFC 2006.

9. Charles D. Gavrilovich, Jr., Gray Cary Ware & Freidenrich LLP, Broadband Communication on the Highways of Tomorrow, IEEE Communications Magazine, April 2001.

10. Carl Eklund, Roger B. Marks, Kenneth L. Stanwood and Stanley Wang, IEEE Standard 802.16:

11. A Technical Overview of the WirelessMAN Air Interface for Broadband Wireless Access, IEEE Communications Magazine, June 2002.

12. William Stallings, IEEE 802.16 for broadband wireless, Network World, 09/03/01, RFC1825: Security.

13. Nathan J. Muller, Bluetooth Demystified, Tata McGraw-Hill 2001.

14. C. S. R. Prabhu, A. Prathap Reddi, Bluetooth Technology and Its Applications with Java and J2ME, Prentice-Hall of India, 2004.

15. Georgios Karagiannis, Mobile IP, Ericsson State of the Art Report # 3/0362-FCP NB 102 88 Uen, 1999.

16. Gil Held, Data Over Wireless Networks Bluetooth, WAP, & Wireless LANs, McGraw-Hill, 2001.

17. RFC 2002: IP Mobility Support–http://www.faqs.org/rfcs/rfc2002.html.

18. Guidelines For 64-Bit Global Identifier (EUI-64) Registration Authority: http://standards.ieee.org/regauth/oui/tutorials/EUI64.html.

19. IEEE 802.16 for Broadband: http://www.nwfusion.com/news/tech/2001/0903-tech.html.

20. Official Bluetooth site: http://www.bluetooth.com.

21. Association for Automatic Identification and Mobility: http://www.aimglobal.org.

22. Zhiqun Chen, Technology for Smart Cards: Architecture and Programmer's Guide, Sun, Addison-Wisley, 2000.

23. Java card Forum: http://www.javacardforum.org/.

REVIEW QUESTIONS

Q1: Describe the protocol stack of Bluetooth?

Q2: How does a new Bluetooth device discover a Bluetooth network? For inter-operability, the system needs to be open. Describe the security principles in Bluetooth?

Q3: What is active RFID? Describe two applications of active RFID. How is active RFID different from passive RFID? Describe two applications of passive RFID?

Q4: What is WiMax (Wireless broadband)? How is it different from WiFi (Wireless LAN)?

Q5: Explain how does Mobile IP work? What are the challenges with mobile IP with respect to high speed mobility? How does Cellular IP solve some of these challenges?

Q6: Explain three limitations of IPv4 that are overcome by IPv6. You have a communication application that uses sockets in IPv4, what are the steps you need to follow to port this application from IPv4 to IPv6?

Q7: You need to develop a secured healthcare application. What information will you keep in the Java card and what will be in the backend server? How will you secure these information on the Java Card?

CHAPTER 5

Global System for Mobile Communications (GSM)

5.1 GLOBAL SYSTEM FOR MOBILE COMMUNICATIONS

GSM is much more than just the acronym for Global System for Mobile Communication. It signifies an extremely successful technology and bearer for mobile communication system. GSM today covers 71% of all the digital wireless market. The mobile telephone has graduated from being a status symbol to being a useful appliance. People use it not only in business but also in everyday personal life. Its principal use is for wireless telephony, and messaging through SMS. It also supports facsimile and data communication.

GSM is based on a set of standards, formulated in the early 1980s (see Table 5.1 for the GSM timeline). In 1982, the Conference of European Posts and Telegraphs (CEPT) formed a study group called the Groupe Spécial Mobile (GSM) to study and develop a pan-European mobile system, which was later rechristened as Global System for Mobile Communication. See Chapter 1 for cellular network evolution and standards. The proposed GSM system had to meet certain business objectives. These are:

- Support for international roaming
- Good speech quality
- Ability to support handheld terminals
- Low terminal and service cost
- Spectral efficiency
- Support for a range of new services and facilities
- ISDN compatibility.

Due to its innovative technologies and strengths, GSM rapidly became truly global. Many of the new standardization initiatives came from outside Europe. Depending on locally available frequency bands, different air interfaces were defined. These are 900 MHz, 1800 MHz and 1900 MHz. However, architecture, protocols, signaling and roaming are identical in all networks independent of the operating frequency bands.

Table 5.1 GSM History timeline

Year	Event
1982	Groupe Spécial Mobile (GSM) established
1987	Essential elements of wireless transmission specified
1989	GSM become an ETSI technical committee
1990	Phase 1 GSM 900 specification (designed 1987 through 1990) frozen
1991	First GSM network launched
1993	First roaming agreement came into effect
1994	Data transmission capability launched
1995	Phase 2 launched. Fax and SMS roaming services offered
2002	SMS volume crosses 24 billions/year, 750 millions subscribers

GSM uses a combination of FDMA (Frequency Division Multiple Access) and TDMA (Time Division Multiple Access). See Section 3.2 for definition of these multiple access procedures. The GSM system has an allocation of 50 MHz (890–915 MHz and 935–960 MHz) bandwidth in the 900 MHz frequency band. Using FDMA, this band is divided into 124 (125 channels, 1 not used) channels each with a carrier bandwidth of 200 KHz. Using TDMA, each of these channels is then further divided into 8 time slots. Therefore, with the combination of FDMA and TDMA we can realize a maximum of 992 channels for transmit and receive. In order to be able to serve hundreds of thousands of users, the frequency must be reused. This is done through cells.

The frequency reuse concept led to the development of cellular technology as originally conceived by AT&T and Bell Labs way back in 1947. The essential characteristics of this reuse are as follows:

- The area to be covered is subdivided into radio zones or cells (Figure 5.1). Though in reality these cells could be of any shape, for convenient modeling purposes these are modeled as hexagons. Base stations are positioned at the center of these cells.

- Each cell i receives a subset of frequencies fb_i from the total set assigned to the respective mobile network. To avoid any type of co-channel interference, two neighboring cells never use the same frequencies.

- Only at a distance of D (known as frequency reuse distance), the same frequency from the set fb_i can be reused. Cells with distance D from cell i, can be assigned one or all the frequencies from the set fb_i belonging to cell i.

- When moving from one cell to another during an ongoing conversation, an automatic channel change occurs. This phenomenon is called handover. Handover maintains an active speech and data connection over cell boundaries.

The regular repetition of frequencies in cells result in a clustering of cells. The clusters generated in this way can consume the whole frequency band. The size of a cluster is defined by k, the number of cells in the cluster. This also defines the frequency reuse distance D. Figure 5.1 shows an example of cluster size of 4.

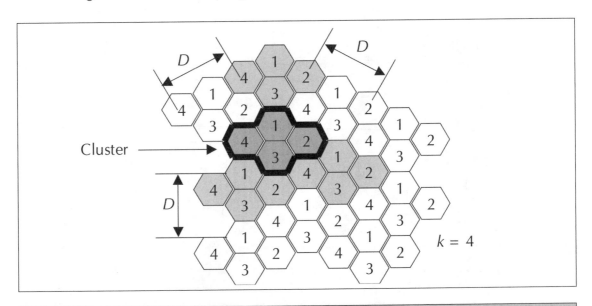

Figure 5.1 Cell clusters in GSM

5.2 GSM ARCHITECTURE

GSM networks are structured in hierarchic fashion (Figure 5.2). It consists at the minimum one administrative region assigned to one MSC (Mobile Switching Centre). The administrative region is commonly known as PLMN (Public Land Mobile Network). Each administrative region is subdivided into one or many Location Area (LA). One LA consists of many cell groups. Each cell group is assigned to one BSC (Base Station Controller). For each LA there will be at least one BSC. Cells in one BSC can belong to different LAs.

Cells are formed by the radio areas covered by a BTS (Base Transceiver Station) (Figure 5.3). Several BTSs are controlled by one BSC. Traffic from the MS (Mobile Station) is routed through MSC. Calls originating from or terminating in a fixed network or other mobile networks is handled by the GMSC (Gateway MSC). Figure 5.3 depicts the architecture of a GSM PLMN from technology point of view, whereas Figure 5.4 depicts the same architecture from operational point of view.

For all subscribers registered with a cellular network operator, permanent data such as the service profile is stored in the Home Location Register (HLR). The data relate to the

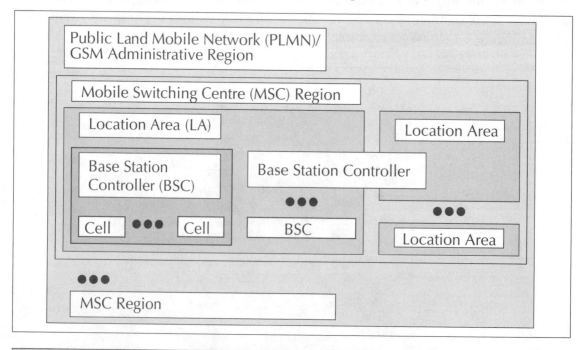

Figure 5.2 GSM System Hierarchy

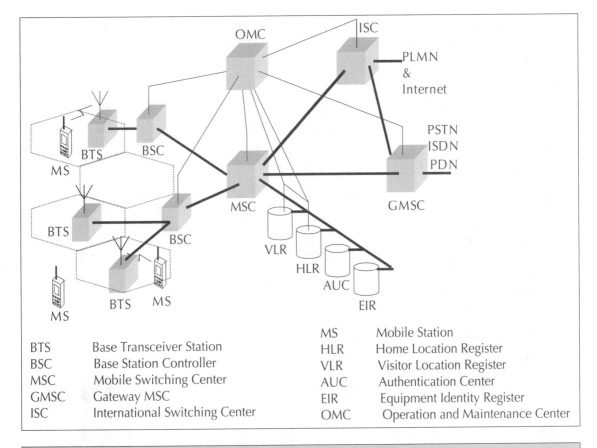

Figure 5.3 Architecture of GSM

following information:

- Authentication information like International Mobile Subscriber Identity (IMSI)

- Identification information like name, address, etc. of the subscriber

- Identification information like Mobile Subscriber ISDN (MSISDN) etc.

- Billing information like prepaid or postpaid

- Operator selected denial of service to a subscriber

- Handling of supplementary services like for CFU (Call Forwarding Unconditional), CFB (Call Forwarding Busy), CFNR (Call Forwarding Not Reachable) or CFNA (Call Forwarding Not Answered)

- Storage of SMS Service Center (SC) number in case the mobile is not connectable so that whenever the mobile is connectable, a paging signal is sent to the SC

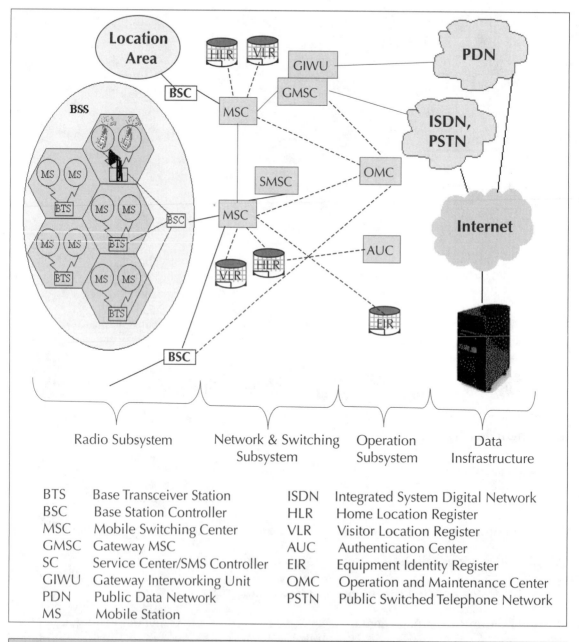

Figure 5.4 System Architecture of GSM

- Provisioning information like whether long distance and international calls allowed or not

- Provisioning information like whether roaming is enabled or not

- Information related to auxiliary services like Voice mail, data, fax services etc.

- Information related to auxiliary services like CLI (Caller Line Identification) etc.

- Information related to supplementary services for call routing. In GSM network one can customize the personal profile to the extent that while the subscriber is roaming in a foreign PLMN, incoming calls can be barred. Also, outgoing international calls can be barred etc.

There is some variable information, which could also be part of the HLR. This includes the pointer to the VLR, location area of the subscriber, Power OFF status of the handset etc.

5.3 GSM Entities

The GSM technical specifications define different entities that form the GSM network by defining their functions and interface requirements. The GSM network can be divided into four main groups (Figure 5.4):

- The Mobile Station (MS). This includes the Mobile Equipment (ME) and the Subscriber Identity Module (SIM).

- The Base Station Subsystem (BSS). This includes the Base Transceiver Station (BTS) and the Base Station Controller (BSC).

- The Network and Switching Subsystem (NSS). This includes Mobile Switching Center (MSC), Home Location Register (HLR), Visitor Location Register (VLR), Equipment Identity Register (EIR), and the Authentication Center (AUC).

- The Operation and Support Subsystem (OSS). This includes the Operation and Maintenance Center (OMC).

5.3.1 Mobile Station

Mobile Station is the technical name of the mobile or the cellular phone. In early days mobile phones were a little bulky and were sometimes installed in cars like other equipments. Even the handheld terminals were quite big. Though the phones have become smaller and lighter, they are still called Mobile Stations. MS consists of two main elements:

- The mobile equipment or the mobile device. In other words, this is the phone without the SIM card.
- The Subscriber Identity Module (SIM).

There are different types of terminals distinguished principally by their power and application. The handheld GSM terminals have experienced the highest evolution. The weight and volume of these terminals are continuously decreasing. The life of a battery between charging is also increasing. The evolution of technologies allowed decrease of this power to 0.8 W.

The SIM is installed in every GSM phone and identifies the terminal. Without the SIM card, the terminal is not operational. The SIM cards used in GSM phones are smart processor cards. These cards posses a processor and a small memory. By inserting the SIM card into the terminal, the user can have access to all the subscribed services. The SIM card contains the International Mobile Subscriber Identity (IMSI) used to identify the subscriber to the system, a secret key for authentication, and other security information. Another advantage of the SIM card is the mobility of the users. In fact, the only element that personalizes a terminal is the SIM card. Therefore, the user can have access to its subscribed services in any terminal using his or her SIM card. The SIM card may be protected against unauthorized use by a password or personal identity number. Typically SIM cards contain 32 K bytes of memory. Part of the memory in the SIM card is available to the user for storing address book and SMS messages. Applications are developed and stored in SIM cards using SAT (SIM Application Toolkit). SAT is something similar to Assembly languages of computers and is proprietary to the SIM vendor. Nowadays Java Smart cards are coming to the market. In Java Smart card, the applications are written in Java language and are portable across SIM cards from different vendors.

5.3.2 The Base Station Subsystem

The BSS (Base Station Subsystem) connects the Mobile Station and the NSS (Network and Switching Subsystem). It is in charge of the transmission and reception for the last mile. The BSS can be divided into two parts:

- The Base Transceiver Station (BTS) or Base Station in short.
- The Base Station Controller (BSC).

The Base Transceiver Station corresponds to the transceivers and antennas used in each cell of the network. In a large urban area, a large number of BTSs are potentially deployed. A BTS is usually placed in the center of a cell. Its transmitting power defines the size of a cell. The BTS houses the radio transmitter and the receivers that define a cell and handles the radio-link protocols with the Mobile Station. Each BTS has between one and sixteen transceivers depending on the density of users in the cell.

Base Station Controller is the connection between the BTS and the Mobile service Switching Center (MSC). The BSC manages the radio resources for one or more BTSs. It handles handovers, radio-channel setup, control of radio frequency power levels of the BTSs, exchange function, and the frequency hopping.

5.3.3 The Network and Switching Subsystem

The central component of the Network Subsystem is the Mobile Switching Center (MSC). It does multiple functions:

- It acts like a normal switching node for mobile subscribers of the same network (connection between mobile phone to mobile phone within the same network).

- It acts like a normal switching node for the PSTN fixed telephone (connection between mobile phone to fixed phone).

- It acts like a normal switching node for ISDN.

- It provides all the functionality needed to handle a mobile subscriber, such as registration, authentication, location updating, handovers and call routing.

- It includes databases needed in order to store information to manage the mobility of a roaming subscriber.

These different services are provided in conjunction with several functional entities, which together form the Network Subsystem. The signaling between functional entities in the Network Subsystem uses Signaling System Number 7 (SS7). SS7 is used for trunk signaling in ISDN and widely used in today's public networks. SS7 is also used for SMS, prepaid, roaming and other intelligent network functions.

The MSC together with Home Location Register (HLR) and Visitor Location Register (VLR) databases, provide the call-routing and roaming capabilities of GSM. The HLR is considered a very important database that stores information of subscribers belonging to the covering area of a MSC. Although a HLR may be implemented as a distributed database, there is logically only one HLR per GSM network. The HLR contains all the administrative information of each subscriber registered in the corresponding GSM network. This includes information like current location of the mobile, all the service provisioning information and authentication data. When a phone is powered off, this information is stored in the HLR. The location of the mobile is typically in the form of the signaling address of the VLR associated with the mobile station. HLR is always fixed and stored in the home network, whereas the VLR logically moves with the subscriber.

The VLR can be considered a temporary copy of some of the important information stored in the HLR. VLR is similar to a cache, whereas HLR is the persistent storage. The VLR contains selected administrative information borrowed from the HLR, necessary for call control and provisioning of the subscribed services. This is true for each mobile currently located in the geographical area controlled by a VLR. GSM standards define interfaces to HLR; however, there is no interface standard for VLR. Although each functional entity can be implemented as an independent unit, all manufacturers of switching equipment implement the VLR as an integral part of the MSC, so that the geographical area controlled by the MSC corresponds to that controlled by the VLR. Note that MSC contains no information about a particular mobile station–this information is stored in location registers. When a subscriber enters the covering area of a new MSC, the VLR associated with this MSC will request information about the new subscriber from its corresponding HLR in the home network. For example if a subscriber of a GSM network in Bangalore is roaming in Delhi, the HLR data of the subscriber will remain in Bangalore with the home network, however, the VLR data will be copied to the roaming network in Delhi. The VLR will then have enough information in order to assure the subscribed services without needing to refer to the HLR each time a communication is established. Though the visiting network in Delhi will provide the services, the billing for the services will be done by the home network in Bangalore.

Within the NSS there is a component called Gateway MSC (GMSC) that is associated with the MSC. A gateway is a node interconnecting two networks. The GMSC is the interface between the mobile cellular network and the PSTN. It is in charge of routing calls from the fixed network towards a GSM user and vice versa. The GMSC is often implemented in the same node as the MSC. Like the GMSC, there is another node called GIWU (GSM Interworking Unit). The GIWU corresponds to an interface to various networks for data communications. During these communications, the transmission of speech and data can be alternated.

5.3.4 The Operation and Support Subsystem (OSS)

As the name suggests, Operations and Support Subsystem (OSS) controls and monitors the GSM system. The OSS is connected to the different components of the NSS and to the BSC. It is also in charge of controlling the traffic load of the BSS. However, the increasing number of base stations, due to the development of cellular radio networks, has resulted in some of the maintenance tasks being transferred to the BTS. This transfer decreases considerably the costs of the maintenance of the system. Provisioning information for different services is managed in this layer.

Equipment Identity Register (EIR) is a database that contains a list of all valid mobile equipment within the network, where each mobile station is identified by its International Mobile Equipment Identity (IMEI). EIR contains a list of IMEIs of all valid terminals. An IMEI is marked as invalid if it has been reported stolen or is not type approved. The EIR allows the MSC to forbid calls from this stolen or unauthorized terminals.

Authentication Center (AUC) is responsible for the authentication of a subscriber. This is a protected database and stores a copy of the secret key stored in each subscriber's SIM card. These data help to verify the user's identity.

5.3.5 Message Centre

Short message or SMS is one of the most popular services within GSM. SMS is a data service and allows a user to enter text message up to 160 characters in length when 7-bit English characters are used. It is 140 octets when 8-bit characters (some European alphabets or binary data) are used, and 70 characters in length when non-Latin alphabets such as Arabic, Chinese or Hindi are used (70 characters of 16-bit Unicode). SMS is a proactive bearer and is an **always ON** network. Message center is also referred to as Service Centre (SC) or SMS Controller (SMSC). SMSC is a system within the core GSM network, which works as a the store and forward system for SMS messages. Refer to Figure 5.5 for SMS architecture.

There are two types of SMS, SMMT (Short Message Mobile Terminated Point-to-Point), and SMMO (Short Message Mobile Originated Point-to-Point). SMMT is an incoming short message from the network and is terminated in the MS (phone or Mobile Station). SMMO is an outgoing message, originated in the MS, and forwarded to the network for delivery. For an outgoing message, the SMS is sent from the phone to SC

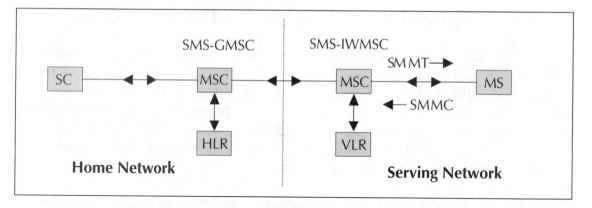

Figure 5.5 The network structure for the short message transfer

via the VLR and the Interworking MSC (IWMSC). For incoming SMS message the path is from SC to the MS via the HLR and the Gateway MSC (GMSC). Please see chapter 6 for SMS and related technologies.

5.4 Call Routing in GSM

Human interface is analog. However, the advancement in digital technology makes it very convenient to handle information in digital fashion. In GSM there are many complex technologies used between the human analog interface in the mobile and the digital network (Figure 5.6).

Digitizer and source coding: The user speech is digitized at 8 KHz sampling rate using Regular Pulse Excited–Linear Predictive Coder (RPE–LPC) with a Long Term Predictor loop. In this technique, information from previous samples is used to predict the current sample. Each sample is then represented in signed 13-bit linear PCM value. This digitized data is passed to the coder with frames of 160 samples. The encoder compresses these 160 samples into 260-bits GSM frames resulting in one second of speech compressed into 1625 bytes and achieving a rate of 13 Kbits/sec.

Channel coding: This step introduces redundancy information into the data for error detection and possible error correction. The gross bit rate after channel coding is 22.8 kbps (or 456 bits every 20 ms). These 456 bits are divided into eight 57-bit blocks, and the result is interleaved amongst eight successive time slot bursts for protection against burst transmission errors.

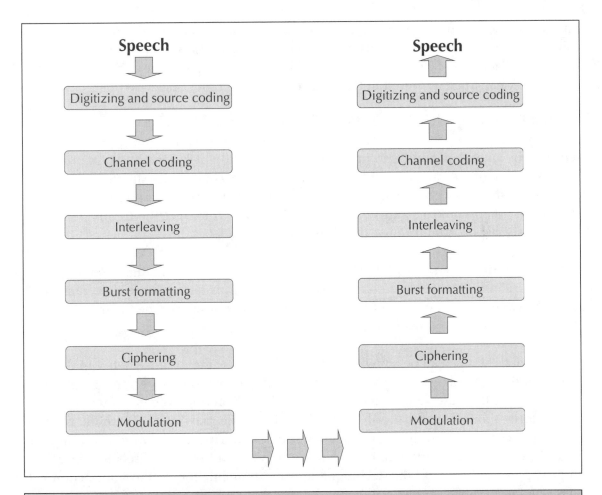

Figure 5.6 Sequence of operation from speech to radio wave

Interleaving: This step rearranges a group of bits in a particular way. This is to improve the performance of the error-correction mechanisms. The interleaving decreases the possibility of losing whole bursts during the transmission, by dispersing the errors.

Ciphering: Encrypts blocks of user data using a symmetric key shared by the mobile station and the BTS.

Burst formatting: Adds some binary information to the ciphered block. This additional information is used synchronization and equalization of the received data.

Modulation: The modulation technique chosen for the GSM system is the Gaussian Minimum Shift Keying (GMSK). Using this technique the binary data is converted back into analog signal to fit the frequency and time requirements for the multiple access rules. This signal is then radiated as radio wave over the air. Each time slot burst is 156.25 bits and contains two 57-bit blocks, and a 26-bit training sequence used for equalization (Fig 5.9). A burst is transmitted in 0.577 ms for a total bit rate of 270.8 kbps.

Multipath and equalization: At the GSM frequency bands, radio waves reflect from buildings, cars, hills, etc. So not only the 'right' signal (the output signal of the emitter) is received by an antenna, but also many reflected signals, which corrupt the information, with different phases are received. An equaliser is in charge of extracting the 'right' signal from the received signal. It estimates the channel impulse response of the GSM system and then constructs an inverse filter. In order to extract the 'right' signal, the received signal is passed through the inverse filter.

Synchronization: For successful operation of a mobile radio system, time and frequency synchronization are needed. Frequency synchronization is necessary so that transmitter and receiver frequency match (in FDMA). Time synchronization is necessary to identify the frame boundary and the bits within the frame (in TDMA).

The mobile station can be anywhere within a cell. Also, the distance between the base station and the mobile station vary. Due to mobility of the subscriber, the propagation time between the base station and the mobile keeps varying. When a mobile station moves further away, the burst transmitted by this mobile may overlap with the timeslot of the adjacent timeslot. To avoid such collisions, the **Timing Advance** technique is used. In this technique, the frame is advanced in time so that this offsets the delay due to greater distance. Using this technique and the triangulation of the intersection cell sites, the location of a mobile station can be determined from within the network.

5.4.1 An Example

In this section let us take an example of how and what happens within the GSM network when someone from a fixed network calls someone in a GSM network. Let us assume that the called party dialed a GSM directory number +919845052534. Figure 5.7 depicts the steps for this call processing.

The directory number dialed to reach a mobile subscriber is called the Mobile Subscriber ISDN (MSISDN), which is defined by the E.164 numbering plan. This number includes a country code and a National Destination Code, which identifies the subscriber's operator. The first few digits of the remaining subscriber number may identify the subscriber's HLR within the home PLMN. For example, the MSISDN number of a subscriber in Bangalore associated with Airtel network is +919845XYYYYY. This is a unique number and understood from anywhere in the world. In this example + means the prefix for international dialing like 00 in UK/India or 011 in USA. 91 is the country code for India (404 as defined in GSM). 45 is the network operator's code (Airtel in this case). X is the level number managed by the network operator ranging from 0 to 9. YYYYY is the subscriber code managed by the operator as well.

The call first goes to the local PSTN exchange. The PSTN exchange looks at the routing table and determines that it is a call to a mobile network. It forwards the call to the Gateway MSC (GMSC) of the mobile network. The MSC enquires the HLR to determine the status of the subscriber. If will decide whether the call is to be routed or not. If the user has not paid the bills, the call may not be routed. If the phone is powered off, a message may be played or forwarded to the voice mail. However, if MSC finds that the call can be processed, it will find out the address of the VLR where the mobile is expected to be present. It the VLR is that of a different PLMN, it will forward the call to the foreign PLMN through the Gateway MSC. If the VLR is in the home network,

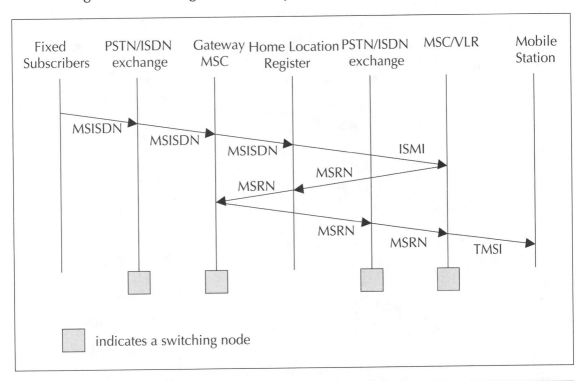

Figure 5.7 Call routing for a mobile terminating call

it will determine the Location Area (LA). Within the LA it will page and locate the phone and connect the call.

5.5 PLMN INTERFACES

The basic configuration of a GSM network contains a central HLR and a central VLR. HLR contains all security, provisioning and subscriber-related information. VLR stores the location information and other transient data. MSC needs subscriber parameter for successful call set-up. Figure 5.8 shows a basic configuration of a GSM mobile communication network.

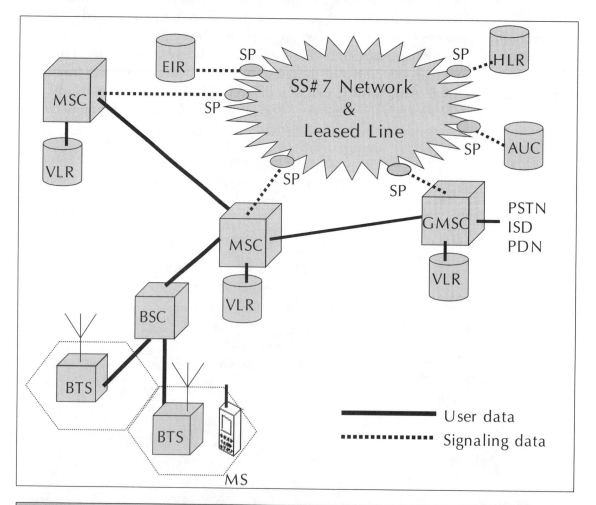

Figure 5.8 Configuration of a GSM PLMN

Within the switching and management system, the transmission rate is 2 Mbits/s. This 2 Mbits/sec interface is called E1 interface in India and in Europe. These are realized typically through microwave or leased lines. Any data related to user call (connection, teardown etc.) are processed with SS7 protocol for signaling using ISUP (ISDN User Part) stack between network nodes. For mobile specific signaling a protocol stack called MAP (Mobile Application Part) is used over the SS7 network. All database transactions (enquiries, updates etc.) and handover/roaming transactions between the MSC are performed with the help of MAP. For this purpose, each MSC uses registers known as SP (Signaling Point). These SPs are addressable through a unique code called Signaling Point Code (SPC). Signaling between MSC and BSS uses Base Station System Application Part (BSSAP) over SS7. Within BSS and at the air interface, signaling is GSM proprietary and does not use SS7.

5.6 GSM Addresses and Identifiers

GSM distinguishes explicitly between the user and the equipment. It also distinguishes between the subscriber identity and the telephone number. To manage all the complex functions, GSM deals with many addresses and identifiers. They are:

- **International Mobile Station Equipment Identity (IMEI):** Every mobile equipment in this world has a unique identifier. This identifier is called IMEI. The IMEI is allocated by the equipment manufacturer and registered by the network operator in the Equipment Identity Register (EIR). In your mobile handset you can type *#06# and see the IMEI.

- **International Mobile Subscriber Identity (IMSI):** When registered with a GSM operator, each subscriber is assigned a unique identifier. The IMSI is stored in the SIM card and secured by the operator. A mobile station can only be operated when it has a valid IMSI. The IMSI consists of several parts. These are:

  o 3 decimal digits of Mobile Country Code (MCC). For India MCC is 404.

  o 2 decimal digits of Mobile Network Code (MNC). This uniquely identifies a mobile operator within a country. For Airtel in Delhi this code is 10.

  o Maximum 10 decimal digits of Mobile Subscriber Identification Number (MSIN). This is a unique number of the subscriber within the home network.

- **Mobile Subscriber ISDN Number (MSISDN):** The MSISDN number is the real telephone number as is known to the external world. MSISDN number is public information, whereas IMSI is private to the operator. This is a number published and known to everybody. In GSM a mobile station can have multiple MSISDN number. When a subscriber opts for Fax and data, he is assigned a total of 3 numbers: one for voice call, one for fax call and another for data call. The MSISDN categories follow the international ISDN (Integrated Systems Data Network) numbering plan as the following:

o Country Code (CC): 1 to 3 decimal digits of country code

o National Destination Code (NDC): Typically 2 to 3 decimal digits

o Subscriber Number (SN): maximum 10 decimal digits.

The CC is standardized by the ITU-T through the E.164 standard. There are CCs with one, two, or three digits. For example the CC for USA is 1, for India it is 91, and for Finland it is 358.The national regulatory authority assigns the NDC. In India it is 94 for BSNL and 98 for all other operators. In India the subscriber number SN is 8 decimal digits. SN consists of 2 decimal digits of operator code, followed by one decimal digit level number with 5 decimal digit subscriber number. In India a MSISDN number looks like 919845062050. In this number 91 is the CC, 98 is the NDC, and 45062050 is the SN. In India SN is subdivided into operator code and subscriber code (45 is the operator code and 062050 is the subscriber code). Sometimes subscriber code is also subdivided into 1 digit level number (0 in this case) followed by 5 digit subscriber id (62050).

- **Location Area Identity:** Each LA in a PLMN has its own identifier. The Location Area Identifier (LAI) is structured hierarchically and unique. LAI consists of 3 digits of CC, 2 digits of Mobile Network Code and maximum 5 digits of Location Area Code.

- **Mobile Station Roaming Number (MSRN):** When a subscriber is roaming in another network a temporary ISDN number is assigned to the subscriber. This ISDN number is assigned by the local VLR in charge of the mobile station. The MSRN has the same structure as the MSISDN.

- **Temporary Mobile Subscriber Identity (TMSI):** This a temporary identifier assigned by the serving VLR. It is used in place of the IMSI for identification and addressing of the mobile station. TMSI is assigned during the presence of the mobile station in a VLR and can change (ID hopping). Thus, it is difficult to determine the identity of the subscriber by listening to the radio channel. The TMSI is never stored in the HLR. However, it is stored in the SIM card. Together with the current location area, a TMSI allows a subscriber to be identified uniquely. For an ongoing communication the IMSI is replaced by the 2-tuple LAI, TMSI code.

- **Local Mobile Subscriber Identity (LMSI):** This is assigned by the VLR and also stored in the HLR. This is used as a searching key for faster database access within the VLR.

- **Cell Identifier:** Within a LA, every cell has a unique Cell Identifier (CI). Together with a LAI a cell can be identified uniquely through Global Cell Identity (LAI+CI).

- **Identification of MSCs and Location Registeres:** MSCs, Location Registers (HLR, VLR), SCs are addressed with ISDN numbers. In addition, they may have a Signaling Point Code (SPC) within a PLMN. These point codes can be used to address these nodes uniquely within the Signaling System number 7 (SS#7) network.

5.7 NETWORK ASPECTS IN GSM

Transmission of voice and data over the radio link is only a part of the function of a cellular mobile network. A GSM mobile can seamlessly roam nationally and internationally. This requires that registration, authentication, call routing and location updating functions are standardized across GSM networks. The geographical area covered by

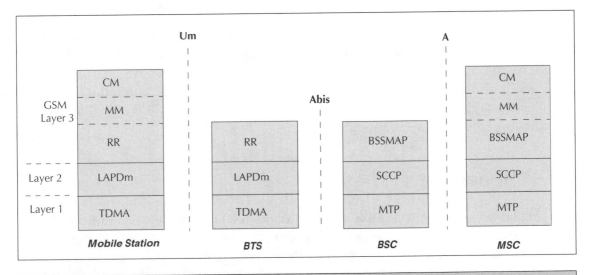

Figure 5.9 Signaling protocol structure in GSM

a network is divided into cells of small radius. When a call is in progress and the user is on the move, there will be a handover mechanism from one cell to another. This is like a relay race where one athlete passes on an object to another athlete. Though both roaming and handover functions are the basic characteristic of mobility, there is a difference between these functions. These functions are performed by the Network Subsystem, mainly using the Mobile Application Part (MAP) built on top of the Signalling System # 7 (SS7) protocol (Figure 5.9).

The signaling protocol in GSM is structured into three general layers, depending on the interface, as shown in Figure 5.9. Layer 1 is the physical layer, which uses the channel structures over the air interface. Layer 2 is the data link layer. Across the Um interface, the data link layer is a modified version of the LAPD protocol used in ISDN or X.25, called LAPDm. Across the A interface, the Message Transfer Part layer 2 of Signaling System Number 7 is used. Layer 3 of the GSM signaling protocol is itself divided into 3 sublayers.

- **Radio Resources Management:** Controls the set-up, maintenance, and termination of radio and fixed channels, including handovers.

- **Mobility Management:** Manages the location updating and registration procedures as well as security and authentication.

- **Connection Management:** Handles general call control, similar to CCITT Recommendation Q.931 and manages Supplementary Services and the Short Message Service.

Signaling between the different entities in the fixed part of the network, such as between the HLR and VLR, is accomplished through the Mobile Application Part (MAP). MAP is built on top of the Transaction Capabilities Application Part (TCAP, the top layer of SS7). SS7 is also used for many other Intelligent Network services within the GSM. The specification of the MAP is quite complex, and at over 500 pages, it is one of the longest documents in the GSM recommendations.

5.7.1 Handover

In a cellular network, while a call is in progress, the relationship between radio and fixed links is dynamic. The user movements may make a user move away or closer to a tower. When the user moves away from a tower, the radio signal strength or the power of the signal keeps reducing. This can result in change of the channel or cell. This procedure of changing the resources is called handover. This procedure is called 'handoff' in North

America. There are four different types of handover in the GSM system, which involve transferring a call between:

- Channels (time slots) in the same cell
- Cells (Base Transceiver Stations) under the control of the same Base Station Controller (BSC),
- Cells under the control of different BSCs, but belonging to the same Mobile services Switching Center (MSC)
- Cells under the control of different MSCs.

The first two types of handover, called internal handovers, involve only one Base Station Controller (BSC). To save signaling bandwidth, they are managed by the BSC without involving the Mobile services Switching Center (MSC), except to notify it at the completion of the handover. The last two types of handover, called external handovers, are handled by the MSC.

5.7.2 Mobility Management

The Mobility Management (MM) function handles the functions that arise from the mobility of the subscriber. MM is in charge of all the aspects related to the mobility of the user, especially the roaming, the location management, and the security/authentication of the subscriber. Location management is concerned with the procedures that enable the system to know the current location of a powered-on mobile station so that the incoming call routing can be completed.

When a mobile station is switched on in a new location area (for example, the user is roaming and has disembarked from an aircraft in a new city) or the subscriber moves to a new location area or a different operator's PLMN, the subscriber must register with the new network to indicate its current location. The first location update procedure is called the IMSI attach procedure where the MS indicates its IMSI to the network. When a mobile station is powered off, it performs an IMSI detach procedure in order to tell the network that it is no longer connected. Normally, a location update message is sent to the new MSC/VLR, which records the location area information, and then sends the location information to the subscriber's HLR. If the mobile station is authenticated and authorized in the new MSC/VLR, the subscriber's HLR cancels the registration of the mobile station with the old MSC/VLR. A location updating is also performed periodically. If after the updating time period, the mobile station has not registered, it is then deregistered.

Unlike routing in the fixed network, where a terminal is semi-permanently wired to a central office, a GSM user can roam nationally and even internationally. When there is an incoming call for a subscriber, the mobile phone needs to be located, a channel needs to be allocated and the call connected. A powered-on mobile is informed of an incoming call by a paging message sent over the paging channel of the cells within the current location area. The location updating procedures, and subsequent call routing, use the MSC and both HLR and the VLR. The information sent to the HLR is normally the SS7 address of the new VLR. If the subscriber is entitled to service, the HLR sends a subset of the subscriber information needed for call control to the new MSC/VLR, and sends a message to the old MSC/VLR to cancel the old registration.

An incoming mobile terminating call is directed to the Gateway MSC (GMSC) function. The GMSC is basically a switch, which is able to interrogate the subscriber's HLR to obtain routing information and thus contains a table linking MSISDNs to their corresponding HLR. A simplification is to have a GSMC handle one specific PLMN. Though the GMSC function is distinct from the MSC function, it is usually implemented within an MSC. The routing information that is returned to the GMSC is the Mobile Station Roaming Number (MSRN), which is also defined by the E.164 numbering plan. MSRNs are related to the geographical numbering plan, and not assigned to subscribers, nor are they visible to subscribers.

The most general routing procedure begins with the GMSC querying the called subscriber's HLR for an MSRN. The HLR typically stores only the SS7 address of the subscriber's current VLR. The VLR temporarily allocates an MSRN from its pool for the call. This MSRN is returned to the HLR and back to the GMSC, which can then route the call to the new MSC. At the new MSC, the IMSI corresponding to the MSRN is looked up, and the mobile is paged in its current location area (see Figure 5.7). As a rule of thumb, HLR is referred for incoming call; whereas VLR is referred for outgoing call.

5.7.3 Roaming Example

Let us assume that the user's mobile number is +919844012345. This is a number in Spice network in Bangalore. The mobile subscriber is roaming in Mumbai. Somebody from a fixed phone in Mumbai wants to talk to this Spice subscriber. The user (caller 'A' party) dials 09844012345 from Mumbai. This call will be switched at the PSTN network and will be routed to Spice network in Bangalore. The Spice MSC will look at the HLR and know that the subscriber (called 'B' party) is now within the coverage of a mobile operator (Orange) in Mumbai. The call will be routed to the Mumbai MSC at Orange.

The Orange MSC in Mumbai will look at its VLR to locate the Spice subscriber and route the call. However, in the process it will inform the Spice HLR about the MSRN. Also, when the call is over, the charging information will be forwarded to the Spice network. Please note that for the incoming call, the routing always happens via the home network resulting in the call routing from Mumbai to Bangalore to Mumbai. The calling party (person in Mumbai) pays long distance tariff for Mumbai to Bangalore; the called party (Spice subscriber) pays for Bangalore to Mumbai long distance tariff in addition to roaming airtime charges. For outgoing call, the home network is not referred (other than the first time authentication), resulting in the call being directly routed by the visiting network. Let us consider the opposite scenario; the Spice subscriber from Bangalore is still roaming in Mumbai and wants to call someone in Mumbai. The Spice subscriber dials the Mumbai number, the Orange MSC looks at the VLR and routes the call directly to the Mumbai number. In this case the Spice subscriber pays a local Mumbai-to-Mumbai call charge in addition to the airtime charges. Let us now look at a completely different scenario where both the caller and the called party are roaming in a foreign network. Let us assume that two subscribers 'A' and 'B' from Airtel Bangalore are visiting Kolkata. When 'A' calls 'B', 'A' dials the number of 'B' which is a Bangalore number. Therefore, the call will be routed to Airtel in Bangalore. In Bangalore it is found that 'B' is roaming in Kolkata, therefore the call will be routed back to Kolkata. If you notice, though both subscribers are in Kolkata, the call is routed through Bangalore and both of them pay the long distance charges. To avoid this, some network operators came up with something called Optimal Call Routing (OCR) or Direct Dialing. OCR will work only when the called party's VLR and the calling party's VLR are with the same network operator. Let us take the previous example and assume that both 'A' and 'B' have logged into Airtel network in Kolkata. While 'A' makes a call to 'B', he prefixes a # in front of the number like #09845062050. This being an outgoing call, the Airtel MSC in Kolkata will look at the Kolkata VLR first. As the number is prefixed with #, it assumes that the other number is roaming in the same network. Therefore it looks at its own VLR once again to see whether 'B' is available in its database. If yes, it routes internally without forwarding the call to the home network. In case of 'B', though it is an incoming call, it is routed directly through the VLR without referring to the HLR.

5.8 GSM FREQUENCY ALLOCATION

GSM in general uses 900 MHz band; out of this, 890-915 MHz are allocated for the uplink (mobile station to base station) and 935–960 MHz for the downlink (base station to mobile station). Each way the bandwidth for the GSM system is 25 MHz (Figure 5.10), which provides 125 carriers uplink/downlink each having a bandwidth of 200 kHz.

The ARFCN (Absolute radio frequency channel numbers) denotes a forward and reverse channel pair which is separated in frequency by 45 MHz.

$$\text{Mobile-to-base: } F_t(n) = 890.2 + 0.2(n-1) \text{ MHz}$$

$$\text{Base-to-mobile: } F_t(n) = F_r(n) + 45 \text{ MHz}$$

In practical implementation, a guard band of 100 kHz is provided at the upper and lower end of the GSM 900 MHz spectrum, and only 124 (duplex) channels are implemented. Since 1995, new bands have been added to the basic 900 MHz GSM. These bands are 1800 MHz and 1900 MHz. 1800 MHz band is licensed to the fourth GSM operators in India. The 1800 MHz band uses 1710–1785 MHz and 1805–1880 MHz (three times as much as primary 900 MHz) with a total of 374 duplex channels. GSM 900 uses the four-cell repeat pattern for the frequency reuse cell sets. In most cases, each cell is divided into 120-degree

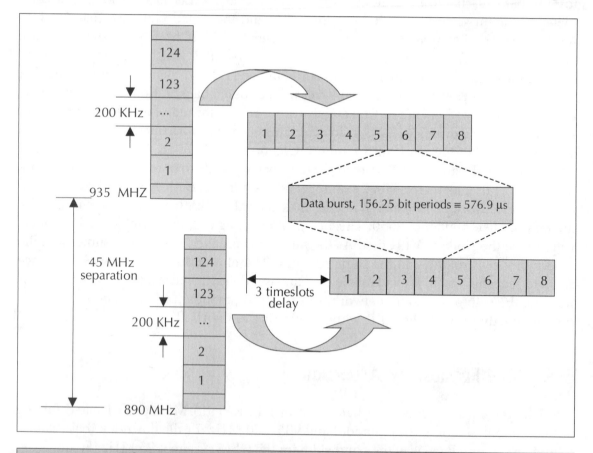

Figure 5.10 Carrier frequencies and TDMA frames

sectors, with three base transceiver subsystems in each cell. Each base transceiver has a 120-degree antenna. These 12 sectors (called cells in GSM system) share the 124 channels.

To share the bandwidth for multiple users, GSM uses a combination of Time-Division Multiple Access (TDMA) and Frequency-Division Multiple Access (FDMA) encoding. One or more carrier frequencies are assigned to each base station. Each of these carrier frequencies is then divided in time, using a TDMA scheme. The fundamental unit of time in this TDMA scheme is called a burst period and it lasts approximately 0.577 ms. Eight burst periods are grouped into a TDMA frame approximately 4.615 ms, which forms the basic unit for the definition of logical channels. One physical channel is one burst period per TDMA frame. Channels are defined by the number and position of their corresponding burst periods. A traffic channel (TCH) is used to carry speech and data traffic. Traffic channels are defined using a 26-frame multiframe, or group of 26 TDMA frames (see Figure 5.11). Out of the 26 frames, 24 are used for traffic, 1 is used for the Slow Associated Control Channel (SACCH) and 1 is currently unused.

5.9 Authentication and Security

The radio medium is open to everybody and anybody. Anybody who can get hold of a radio receiver can access GSM signal or data. Therefore, it is necessary and important that the communication over the wireless radio media is secured. The first step to the

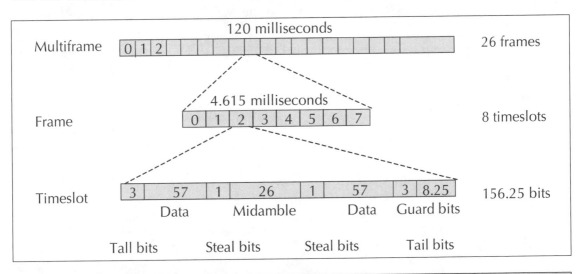

Figure 5.11 Organization of bursts, TDMA frames

GSM security is the authentication. The authentication of a user is done to ensure that the user is really the person who claims he is. Authentication involves two functional entities, the SIM card in the mobile phone, and the Authentication Center (AUC). Authentication is done by using an algorithm by name A3. Following the authentication, a key is generated for encryption. An algorithm by the name A8 is used to generate the key. A different algorithm called A5 is used for both ciphering and deciphering procedures. The ciphering is done on both signaling, voice and data. This in other words means that SS7 signal, voice, data, and SMS within GSM are ciphered over the wireless radio interface.

The GSM specifications for security were designed by the GSM Consortium in secrecy and are distributed only on a need-to-know basis to hardware and software manufacturers and to GSM network operators. The specifications were never exposed to the public. The GSM Consortium relied on Security by Obscurity, i.e. the algorithms would be harder to crack if they were not publicly available.

5.9.1 The MS Authentication Algorithm A3

During the authentication process the MSC challenges the MS with a random number (RAND). See Figure 5.12. The SIM card uses this RAND received from the MSC and a secret key Ki stored within the SIM as input. Both the RAND and the Ki secret are 128 bits long. Using the A3 algorithm with RAND and Ki as input a 32-bit output called signature response (SRES) is generated in the MS. This SRES is then sent back to

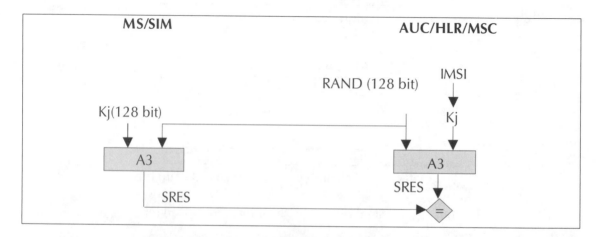

Figure 5.12 The workflow of authentication

the MSC as the response to the challenge. Using the same set of algorithms, the AUC also generates a SRES. The SRES from MS (SIM) and the SRES generated by the AUC are compared. If they are the same, the MS is authenticated. The idea is that no keys will be transacted over the air. However, if the SRES values calculated independently by the SIM and the AUC are the same, the Ki has to be same. If Ki is same, SIM card is genuine.

5.9.2 The Voice-Privacy Key Generation Algorithm A8

For any type of cipher, we need a key. If the key is random and difficult to guess, the cipher is relatively secured. In the GSM security model, A8 algorithm is the key generation algorithm (Figure 5.13). A8 generates a session key, Kc, from the random challenge, RAND, received from the MSC and from the secret key Ki. The inputs for A8 are the same set of 128-bit Ki and RAND as used in A3. The A8 algorithm takes these inputs and generates a 64-bit output. The keys are generated at both the MS (SIM) end and the network end. The BTS received the Kc from the MSC. The session key Kc, is used for ciphering, till the time the MSC decides to authenticate the MS once again. This might sometime take days.

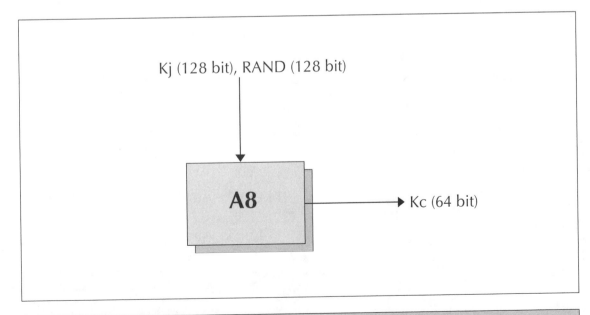

Figure 5.13 Session key (Kc) calculation

5.9.3 The Strong Over-the-Air Voice-Privacy Algorithm A5/1

In the GSM security model, A5 is the stream cipher algorithm used to encrypt over-the-air transmissions (Figure 5.14). The stream cipher is initialized all over again for every frame sent. The stream cipher is initialized with the session key, Kc, and the number of the frame being encrypted or decrypted. The same Kc is used throughout the call, but the 22-bit frame number changes during the call, thus generating a unique keystream for every frame.

Since the first GSM systems, many variations of A5 algorithms have been designed and implemented. The main motivation has been that the original A5 encryption algorithm was too strong. Strong and difficult encryption algorithm always attracts export restrictions. The original A5 algorithm was not allowed to be used outside Europe. Therefore, the first 'original' A5 algorithm was renamed A5/1. Other algorithms including A5/0, which means no encryption at all, and A5/2, a weaker over-the-air privacy algorithm were developed. The A5 algorithms after A5/1 have been named A5/x. Most of the A5/x algorithms are considerably weaker than the original A5/1.

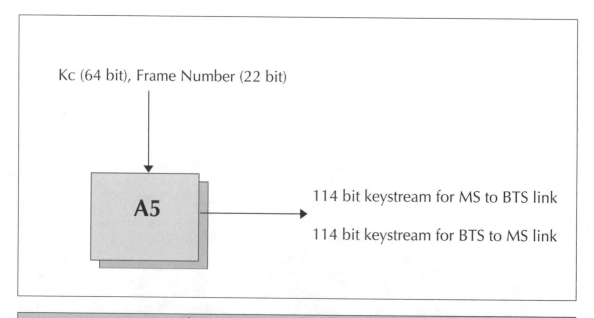

Figure 5.14 Keystream generation

References/Further Reading

1. GSM World: http://www.gsmworld.com.

2. Javier Gozálvez Sempere, An overview of the GSM system, http://www.telcor.-gob.ni/BCS/nd/gsm.html.

3. Siegmund M Redl, Matthias K Weber, Malcom W. Oliphant, GSM and Personal Communications Handbook, Artech House, 1998.

4. Gunnar Heine, GSM Networks: Protocols, Terminology, and Implementatio, Artech House, 1999.

5. Suthaharan Sivagnanasundaram, GSM Mobility Management Using an Intelligent Network Platform, Ph.D Thesis, University of London, 1997.

6. GSM Cloning: Reference: http://www.isaac.cs.berkeley.edu/isaac/gsm.html.

7. Vijay K Garg, Joseph E Wilkes, Principles & Applications of GSM Pearson Publication Asia, 2002.

8. Jorg Eberspacher, Hans-Jorg Vogel, GSM Switching, Services and Protocols, John Wiley & Sons, 1999.

9. William C. Y. Lee, Mobile Cellular Telecommunications Analogue and Digital Syatems, McGraw-Hill, 2000.

10. GSM Standard 03.01: Digital cellular telecommunications system (Phase 2+); Network functions, Release 1998, www.etsi.org.

11. GSM Standard 03.02: Digital cellular telecommunications system (Phase 2+); Network architecture, Release 1998, www.etsi.org.

12. GSM Standard 03.03: Digital cellular telecommunications system (Phase 2+); Numbering, addressing and identification, Release 1998, www.etsi.org.

13. GSM Standard 03.09: Digital cellular telecommunications system (Phase 2+); Handover procedures, Release 1998, www.etsi.org.

14. GSM Standard 03.20: Digital cellular telecommunications system (Phase 2+); Security related network functions, Release 1998, www.etsi.org.

REVIEW QUESTIONS

Q1: Describe the GSM architecture. Describe different elements in this architecture?

Q2: In GSM network, there are some databases used for various purposes. What are these databases? What are their functions?

Q3: What is handover/handoff? How is handoff different from roaming?

Q4: What is the role of AuC? How is authentication done in a GSM network? What are the different algorithms used for security in GSM?

Q5: What are HLR and VLR? Describe the functions of HLR and VLR in call routing and roaming?

Q6: What is a PLMN? How is a PLMN connected to PSTN and PDN?

CHAPTER 6

Short Message Service (SMS)

6.1 MOBILE COMPUTING OVER SMS

GSM supports data access over CSD (Circuit Switched Data). GSM is digitized but not packetized. In case of CSD, a circuit is established and the user is charged based on the time the circuit is active and not on the number of packets transacted. GPRS (General Packet Radio Service), also known as 2.5G, which is the next phase within the evolution of GSM, supports data over packets. WAP is a data service supported by GPRS and GSM to access Internet and remote data services. WAP has been covered in chapter 8. Other data services in GSM include Group 3 facsimile, which is supported by use of an appropriate fax adaptor. A unique data service of GSM, not found in older analog systems, is the Short Message Service (SMS). SMS enables sending and receiving text messages to, and from, GSM mobile phones. In this chapter we discuss SMS and developing applications using SMS bearer.

6.2 SHORT MESSAGE SERVICES (SMS)

Like many other eccentric technologies, SMS was also allegedly the right idea at the wrong time. On 3 December 1992, a scientist named Neil Papworth at Sema, a British technology company, sent the first text message 'Merry Christmas' to the GSM operator Vodafone. It was sent to Vodafone director Richard Jarvis in a room at Vodafone's HQ in Newbury in southern England. The message was an overly premature seasonal greeting, some three weeks ahead of the festivities. Vodafone offered this service as a text messaging service with a brand name TeleNotes service targeted for businesses community. The service was not at all popular in its early days. SMS was almost forgotten and became an unwanted child until seven years later in 1999 when other mobile phone operators started to allow customers to swap SMS. Today SMS is the most popular data bearer/service within GSM with an average of one billion SMS messages (at the end of

2002) transacted every day around the world, with a growth of on an average half a billion every month. The SS#7 signaling channels are always physically present but mostly
unused, be it during an active user connection or in the idle state. It is, therefore, quite
an attractive proposition to use these channels for transmission of used data. SMS uses
the free capacity of the signaling channel. Each short message is up to 160 characters in
length when 7-bit English characters are used. It is 140 octets when 8-bit characters (some
European alphabets) are used, and 70 characters in length when non-Latin alphabets
such as Arabic, Chinese or Hindi are used (70 characters of 16 bit Unicode).

6.2.1 Strengths of SMS

Following is a list of unique characteristics of SMS, which make this an attractive bearer
for mobile computing.

Omnibus nature of SMS: SMS uses SS7 signaling channel, which is available throughout the world. SMS is the only bearer that allows a subscriber to send a long distance
SMS without having long distance subscription. For example, you cannot make a voice
call to a mobile phone in UK unless you have an international calling facility. However,
you can send a SMS to a subscriber in UK, without having an international call facility.

Stateless: SMS is sessionless and stateless. Every SMS message is unidirectional and
independent of any context. This makes SMS the best bearer for notifications, alerts and
paging. SMS can be used for proactive information dissemination for 'unsolicited
response' and business triggers generated by applications (referred as 'Push' Figure 8.4).

Asynchronous: In http, for every command (e.g., GET or POST) there is a request and
a response pair making it synchronous at the transaction level. Unlike http, SMS is completely asynchronous. In case of SMS, even if the recipient is out of service, the transmission will not be abandoned. Therefore, SMS can be used as message queues. In
essence, SMS can be used as a transport bearer for both synchronous (transaction oriented) and asynchronous (message queue and notification) information exchange.

Self-configurable and last mile problem resistant: SMS is self-configurable. In the
case of Web or WAP, it is no trivial task to connect to a service from a foreign network
without any change in the configuration or preference setting. The device needs to be
configured interactively by the user or system administrator to access the network. This
makes the access dependent on the last mile. SMS has no such constraints. While in a
foreign network, one can access the SMS bearer without any change in the phone
settings. The subscriber is always connected to the SMS bearer irrespective of the home

and visiting network configurations. While roaming in a foreign network, even if the serving network does not have a SMSC (SMS Center) or SC (Service Center), SMS can be sent and received.

Non-repudiable: SMS message carries the SC and the source MSISDN as a part of the message header. Unlike an IP address it is not easy to handcraft a MSISDN address in the SMS. It is possible for an application connected to an SMS to handcraft a MSISDN address like "999" or even alphabetic addresses like "MYBANK". However, an application can not handcraft the SC address. Therefore, an SMS can prove beyond doubt the origin of the SMS.

Always connected: As SMS uses the SS7 signaling channel for its data traffic, the bearer media is always on. User cannot switch OFF, BAR or DIVERT any SMS message. When a phone is busy and a voice, data or FAX call is in progress, SMS message is delivered to the MS (Mobile Station) without any interruption to the call.

6.2.2 SMS Architecture

SMS are basically of two types, **SM MT** (Short Message Mobile Terminated Point-to-Point), and **SM MO** (Short Message Mobile Originated Point-to-Point). SM MT is an incoming short message from the network side and is terminated in the MS. SM MO is an outgoing message, originated in the user device (MS), and forwarded to the network for delivery. For outgoing message, the path is from MS to SC via the VLR and the IWMSC function of the serving MSC, whereas for incoming message the path is from SC to the MS via HLR and the GMSC function of the home MSC (Figure 6.1).

To use SMS as a bearer for Information exchange, the Origin server or the Enterprise server needs to be connected to the SC through a short message entity (SME) as in Figure 6.2. The SME in this case works as a SMS gateway, which interacts to the SC in one side, and the enterprise server on the other side.

6.2.3 Short Message Mobile Terminated (SM MT)

For a SM MT message, the message is sent from SC to the MS. This whole process is done in one transaction (Figure 6.3). For the delivery of MT or incoming SMS messages, the SC of the serving network is never used. This implies that a SMS message can be sent from any SC in any network to a GSM phone anywhere in the world. This makes any SM MT message mobile operator independent.

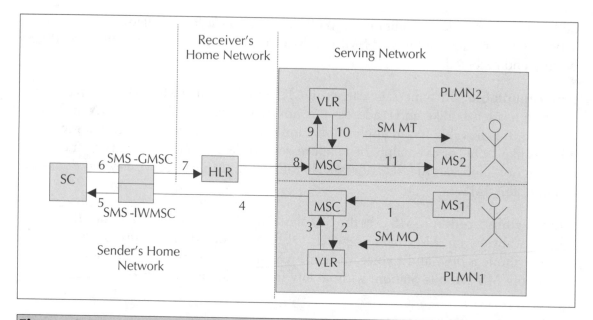

Figure 6.1 The main network structure serving as a basis for the short message transfer

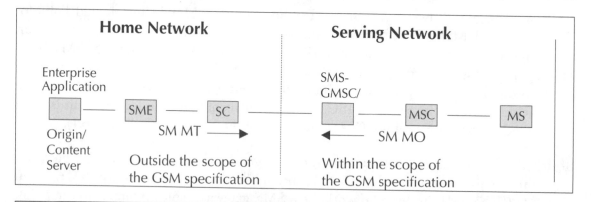

Figure 6.2 The basic network structure for SMS as information bearer

6.2.4 Short Message Mobile Originated (SM MO)

SM MO is an outgoing message originated in the MS where generally the user types in a message and sends it to a MSISDN number. For a MO message, the MSC forwards the message to the home SC. The SC is an independent computer in the network and works as a store and forward node. In SS7 terminology SC is a SCP (Service Control Point) within the SS7 cloud. MO message works in two asynchronous phases. In the first

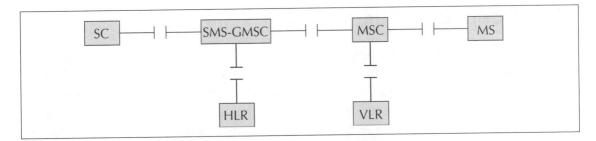

Figure 6.3 Interface involved in the short message mobile originated procedure

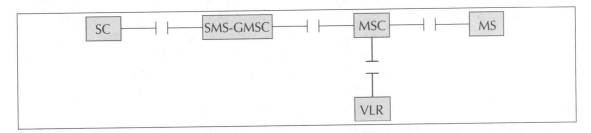

Figure 6.4 Interface involved in the short message mobile terminated procedure

phase, the message is sent from the MS to the home SC as a MO message (Figure 6.3). In the second phase, the message is sent from the home SC to the MS as a MT message (Figure 6.4). It is possible to attempt to send a SMS message to an invalid MSISDN number. In such a case, the message will be sent successfully from the MS to the SC. However, it will fail during the SC to the MS transfer.

6.2.5 SMS as an Information Bearer

SMS is a very popular bearer in the messaging domain. However, it is gaining popularity in other verticals like enterprise applications, services provided by independent service providers, and notification services.

To use SMS as a bearer for any information service, we need to connect the services running on the Enterprise Origin server to the SC through an SME (Short Message Entity) or ESME (External Short Message Entity). SME in any network is generally a SMS gateway. With respect to SMS, a GSM subscriber is always in control of the SC in the home network irrespective of the serving network. Thus, if there is any SMS-based data service in the home network, it will be available in any foreign network.

6.2.6 Operator-Centric Pull

For a SM MO to work it is mandatory that a SC is used. As a part of SMS value added services, operators offer different information on demand and entertainment services. These are done through connecting an Origin server to the SC via a SMS gateway. In different parts of the world a new industry vertical has emerged to address this market. These service providers are known as MVNO (Mobile Virtual Network Operators). Virtual operators develop different systems, services, and applications to offer data services using SMS. Many enterprises use these MVNOs to make their services available to mobile phone users. There are quite a few banks in India which offer balance enquiry and other low security banking services over SMS. For example, if a HDFC customer wants to use these services, he needs to register for the service. During the registration, the HDFC customer needs to mention the MSISDN of the phone which will be used for this service. Once a user is registered for the service, he enters 'HDFCBAL' and send the message to a service number (like 333 for example in the case of Escotel) as a MO message. SC delivers this MO message to the SMS gateway (technically known as SME–Short Message Entity) connected to this service number. SMS gateway then forwards this message to the enterprise application. The response from the enterprise application is delivered to the MS as a MT message from the SME. Even if the subscriber is in some remote region of a foreign network within GSM coverage, he can send the same SMS to the same service number in his home network. This makes the home services available in the foreign network. This also implies that an operator-centric SMS pull service is completely ubiquitous.

The connectivity between SC to SME and SME to Enterprise Origin server is not defined by GSM. However, there are a few de facto standard protocols for this communication. The most popular protocol is Short Message Peer to Peer (SMPP). There are certain other protocols like CIMD from Nokia as well. The connectivity between SME and Origin server could be anything like SOAP (Simple Object Access Protocol), or direct connection through TCP socket. However, common practice is through http. Http helps user to get information from the Internet via SMS. There is an open source for SMS gateway called Kannel, which supports a multitude of protocols and forwards the SMS enquiry as an http request and gets information from the Internet. This is how an SMS can be converted into a simple Internet access. Conventionally SMS queries are keywords driven like 'CRI' for live cricket score, or 'RSK 2627 3 03' to get the availability of sear/berth in Indian Railways train number 2627 (Karnataka Express) for March 3. There are applications where SMS is used in session oriented transactions. Applications like 'SMS chat' and 'SMS contests' need to remember the user context over multiple transactions.

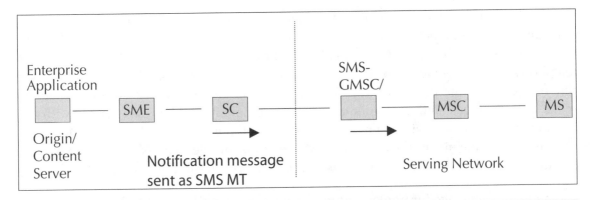

Figure 6.5 The basic network structure of the SMS Push

6.2.7 Operator Independent Push

We have seen that it is possible to send a SMS to any phone in any network. For example, a MT message can be delivered from a network in India to a MS of UK roaming in Germany (Figure 6.5). Which in other words means that any push, which may be an alert, notification or even response from a pull message generated by an application, can be serviced by any network and delivered to any GSM phone in any network without any difficulty.

Assuming that appropriate roaming tie-ups are in place, an enterprise can use SMS to send business alerts or proactive notifications to its customer anywhere, anytime on any GSM phone. With roaming tie-ups, operators reach an agreement on revenue share and call forwarding mechanism. Roaming tie-ups are a commercial issue rather than technical. Some credit card companies in India send SMS notifications to its cardholders in different networks using operator independent push.

6.2.8 Challenge for SMS as a Mobile Computing Bearer

When it comes to offering enterprise services using SMS, the scene becomes difficult to manage. Let us take the example of Indian Bank. In Delhi, a customer of this bank who is a subscriber of operator 'A' (Airtel) sends 'HDFCBAL' to 300 to know the balance in his account. In the same city of Delhi another customer of the same bank who happens to be a subscriber of a different operator 'B' (Essar) sends 'HDFCBAL' to 1234 to get the balance information. HDFC bank has a sizable population of customers in the Middle East. The same banking services, which are available in India, are not available in the Middle East. The reason being both cellular operators 'A' and 'B' connect to the bank's application through their private SC and SME, whereas the operators in Middle East do not have a SME to connect to the bank's application. This is like in the early days of telephony when an enterprise used to announce different customer care numbers for different cities. If the enterprise did not have an office in a city, the customers had to make long distance call to customer care in some other cities. All these changed with the introduction of 1-800 service. Enterprises need something similar to 1-800 in SMS. Also, this gives some identity to the enterprise. My Inc for example may like to publish a number like +9198375MYINC for any of its customer anywhere in the world.

The major challenge for implementing ubiquitous service through SMS requires operator independent SM MO messages or operator independent pull services. The SMS routing needs to work exactly in the same fashion as 1-800 services.

6.2.9 Operator Independent Pull

As the SME is always connected to the home network's SC, with the conventional framework, it is not possible to route mobile originated SMS messages to any application or any SME of choice. There are ways by which a SMS message can be routed to some enterprise SME connected to external SC. This is achieved through SAT, where the SAT application running on the SIM card changes the SC number during the transmission of the SMS and forces the SMS to recognize a different SC of a different network as its home SC. In this case also, technically the SMS is sent to the SME connected to the home SC. SMS has always been considered a revenue generating tool for cellular operators. Therefore, the current framework suits a cellular operator very well. If a SMS service is operator dependent, the cellular operator can use this to its advantage. In today's global scenario an enterprise or a MVNO has its customers around the world subscribing to different GSM networks. To make this possible, enterprises need operator-independent pull as well. Operator-independent pull services can be achieved using GSM modem technology described in the following sections. Also, the same can be done using Intelligent Network Technologies.

6.3 VALUE ADDED SERVICES THROUGH SMS

Value Added Services (VAS) can be defined as services, which share one or more of the following characteristics:

- Supplementary service (not a part of basic service) but adds value to total service offering.

- Stimulates incremental demand for core services offering.

- Stands alone in terms of profitability and revenue generation potential.

- Can sometimes stand-alone operationally.

- Does not cannibalize basic service unless clearly favorable.

- Can be an add-on to basic service, and as such, may be sold at a premium price.

- May provide operational and/or administrative synergy between or among other services and not merely for diversification.

A GSM operator's primary business goal is to offer the network infrastructure. Voice, SMS are basic services provided by a GSM operator. However, offering different other services using SMS as a bearer will be a VAS. There are various flavors and variations of VAS over SMS. We will give some examples and discuss how to develop them. The most popular VAS over SMS are entertainment and information on demand. Information on demand has three categories as described below.

1. **Static information.** This type of information does not change frequently. A good example is a restaurant guide. It is sufficient to update this type of information once in a fortnight, or even less frequently. These contents generally fall in mass market category.

2. **Dynamic information.** This type of information changes in days. For example the daily horoscope needs to be updated on daily basis. Mass market contents fall in this category as well.

3 **Real-time information.** This type of information changes continually. Third party contents fall in this category. For example, scores in a live cricket match or stock quote undergo continual change. All the enterprise contents will fall in this category. For enterprise content, the content will be obtained directly from the enterprise.

6.3.1 User Interface in SMS Value Added Services

We have already seen that SMS is sessionless. In chapter 1 we also have discussed Session oriented transaction and Short transaction (Section 1.4). Majority of services over SMS will use the Short transaction model. For a SMS-based service, the user interface is always keyword-based. This is something similar to the character-based command interfaces, where the first word is the keyword (command) and rest are the parameters for the command. For example, I want to know the latest news. For this I enter **News** and send it to the VAS service. If I want business news, I enter **News Biz**. **News** is the keyword.

Another example could be **RSA 2627 Bangalore New Delhi 20 01.** This example is for finding out the seat availability on the Indian Railways train number 2627 from Bangalore to New Delhi for 20 January. For Indian Railways, the tickets are available only for 60 days in advance. Therefore, we do not need the year. The response for this enquiry will be **Date: 20-1 Train: 2627 KARNATAKA EXP Class:2A Status: WL 31/WL 14 Class: 3A Status: WL 63/WL 51 Class: SL Status: WL 59/WL 29.** Please note that the response from a service can sometimes be more than 160 characters. If it is more than 160 characters, we need to split the response into multiple message responses. It is advised that while the message is broken into multiple messages, it is broken at the word boundary. It is also advised that a sequence number like ... **1/3,** ... **2/3,** and ... **3/3** is added in the first, second and third messages respectively.

6.3.2 VAS Examples

In this section we describe some of the popular value added services.

News/Stock Quotes Service

In a service like **News** or **Stock Quote**, we get the latest news or stock information. This will be a Short transaction. The keyword for news will be **News**, whereas the keyword for stock quote can be **BSE**. BSE Infosys, will give the stock price of Infosys at the Bombay Stock Exchange. These are examples for real-time information on demand. For services like **News** and **Stock Quote** we need to have a relationship with some content provider who will supply us the up-to-date information. For example, we could tie up with CNN for international news, Indian Express for general news, weather.com for weather news etc. For stock quote, we may need to tie up with a stock exchange like Bombay Stock Exchange or National Stock Exchange. We will receive live feed from these content providers and update the content database on a real-time basis. As and when a subscriber wants these information, we supply the latest information from the live database.

Session-based Chat Application

A chat service is essentially a session oriented transaction. In a chat service the user needs to log in. The user needs to explicitly log out or will be logged out implicitly following a period of inactivity. Every time the user sends a chat keyword, we need to know the previous transactions. Every SMS message carries the unique MSISDN number.

This unique MSISDN number of the phone can be used as the session key. In the chat software we remember the state of the transaction using this MSISDN.

Email through SMS

This is a very useful service and is a transaction-oriented dialogue. To send an email through SMS, the user message will be **mail roopa@iitb.ac.in we will meet tomorrow 6:00 pm**. This VAS will send a mail to Roopa with mail id roopa@iitb.ac.in. The body of the mail will be 'we will meet tomorrow 6:00 pm.' The mail will be sent to Roopa by a SMTP server.

Health Care Services

Health care applications need both Pull and Push. A typical health care application could be ICU (Intensive Care Unit) system. The system will include alerts to doctor. In status monitoring service, a doctor or a nurse can enquire the status of a patient in the ICU. A limited enquiry facility will be provided to one MSISDN outside the hospital staff. This could be for someone in the family. This enquiry will be a Short transaction. We will have alert services as well. In the alert service, nurses and doctors are notified periodically about the status of the patients in the ICU. The alerts can also be integrated with medical equipment.

Micro-Payment Services

Let us take an example of micro-payment for a vending machine. This will be a session-oriented dialogue. In this application there will be some id pasted on the vending machine. The user enters this number and sends to the VAS. The VAS will authenticate the user and check whether the user has sufficient money in credit. Based upon the credit the transaction will either be approved or rejected. If approved, an authorization message will be sent back to the vending machine. The vending machine will ask the user to select merchandise. The user selects the merchandise, a soft drink for example. Once the merchandise is dispensed, the vending machine will send back a message to the VAS indicating that the merchandise has been dispensed. The price for the merchandise is debited from the user account.

6.3.3 Alert Services

These are proactive alert services. For a stock quote the alert services can be of the following kind.

Time-based: In this service, proactive alerts are sent to the mobile phone at a pre-assigned time of the day. The alert contains the stock quote of different scripts of the portfolio.

Watermark based: In this service whenever the stock price goes up or falls down to a certain level, alerts are sent. This information will help the subscriber to decide whether to buy or sell some particular stock.

For other services, like cricket score, it can be a periodic alert (every 10 minutes) during the match. There can be other alerts like inform the live score whenever a player is being out etc.

6.3.4 Location-based software

Location-based services could be road direction, restaurant guide etc. Some location-aware VAS services provide shopping alerts as well. In location-based services only the information relevant to the current location of the mobile phone (or the subscriber) is provided. In a shopping service, the user will receive alerts on discount or sale information when they pass through close the proximity of the shopping malls. In the case of a restaurant guide, let us assume that the subscriber is in a office on M.G. Road in Bangalore and sends **Res** to the VAS. Only the restaurants in, and around, M.G. Road will be provided as response to this request. When the same user asks for the same information in Mumbai, restaurants in Mumbai will be given as response. For location-aware software, the precise location of the user needs to be determined. The location of a mobile phone can be determined either from the network or from the device. Using **Time Advancing** techniques within the BTS, the location of the mobile phone can be determined. This technique however requires the support of the network. The other option is to find out the location from the device. Device-specific location awareness requires either of the following technologies:

1. Cell ID (CID)-based system.
2. Global Positioning System (GPS)-based system.

In a CID-based system, the CID of the current BTS is determined. The CID-based system needs a mapping of the cell id to the geographical location. To handle the growing subscribers, new cell sites are added and the CIDs reconfigured. In such cases the mapping between locations versus CIDs need to be synchronized. For CID-based system, the signal strength from all the different CIDs are extracted from the device and sent to the server through a SMS. The location of the user is determined using the signal strength and triangulation algorithms. In a GPS-based system, the location is detertmined through a GPS receiver installed within the phone. GPS provides facility to compute position, velocity

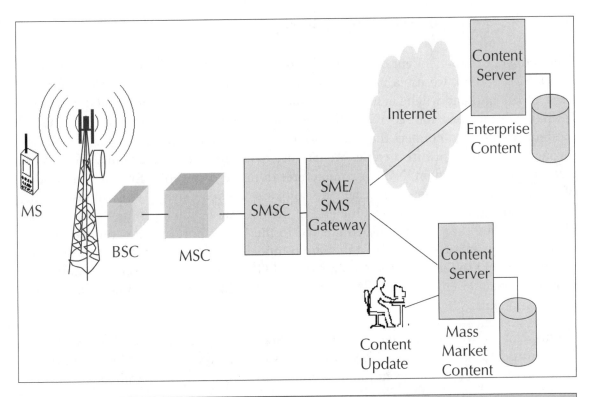

Figure 6.6 SMS Value Added Services Architecture

and time of a GPS receiver. To offer a travel direction through GPS, the GPS system will inform the application about the exact location of the phone. From the velocity it will also know the direction the user is moving. Based on the location and direction, the direction will be provided. Please note that sometimes it may not be a trivial task to take a U turn on a freeway or motorway. GPS-based system is not dependent on the network operator.

In Figure 6.6 we describe the basic value added service provisioning architecture for SMS. The reader should try to map an application scenario and get a feeling of how information travels across in the case of 'pull'/'push' .

6.4 ACCESSING THE SMS BEARER

There are two ways the SMS bearer can be accessed:

1. Use a mobile phone as a GSM modem and connect it to the computer.

2. Use the SMSC of an operator through SMPP or similar interface.

6.4.1 GSM Modem (Over the Air)

This is operator independent quick-fix solution. GSM modem does not have scalability. In a GSM modem, we use a normal cell phone as a data modem. The cell phone will have a SIM card and will be in a position to access the network as a normal GSM phone. It will be able to send and receive SMS messages. To convert this cell phone into a GSM modem, we need to connect the phone to a computer. This connectivity is established through either IrDa or direct cable. We can use data cables available from Nokia to use a Nokia phone as modem. Nokia manufactures different data cables for different models of Nokia phones. We need to install the software (device drivers) associated with the phone model into the computer. One end of the data cable will be connected to the cell phone and the other end will be connected to the COM port of the computer. This is similar to steps involved in using an external modem in a computer. Once all these are ready, one can use the cell phone as an external GSM modem and issue **AT** commands to transact data over the GSM/SMS bearer as done in case of any other Hayes compatible modems.

One can use the Hyper terminal (Go to **Start** -> **Programs** -> **Accessories** -> **Communications** -> **Hyper Terminal**) software and try sending SMS from a PC. Following is a very simple example of sending SMS from a hyper terminal.

Sent: AT

Recv: OK

Sent: AT+CMGF=1

Recv: OK

Sent: AT+CMGS="9810080856"

```
> This SMS message is being sent from a computer using
            hyper-terminal and my Nokia phone<ctrl-Z>
```

Recv: +CMGS: 122

```
OK
```

In this example, we use the standard Hayes Modem command sets. We send the AT (In Hayes terminology this is known as attention) command to the modem from the computer. The GSM modem responds by saying 'OK'. This means that the modem is ready and can take instructions. We then set the message format to text mode through CMGF command. In the next request we send AT+CMGS="9810080856". This is to send a SMS message to a mobile with MSISDN 9810080856. The GSM modem accepts the request and responds with a '>' sign. This is a prompt from the modem requesting

AT command can also be used for other functions of the phone. Most of the functions available as a part of MMI (Man Machine Interface), are available through AT command. Examples could be sending a SMS, read a SMS; check battery power or write a phone book entry. Following is a list of the AT commands supported for SMS.

SMS Text Mode

AT+CSMS	Select Message Service
AT+CPMS	Preferred Message Storage
AT+CMGF	Message Format
AT+CSCA	Service Center Address
AT+CSMP	Set Text Mode Parameters
AT+CSDH	Show Text Mode Parameters
AT+CSCB	Select Cell Broadcast Message Types
AT+CSAS	Save Settings
AT+CRES	Restore Settings
AT+CNMI	New Message Indications to TE
AT+CMGL	List Messages
AT+CMGR	Read Message
AT+CMGS	Send Message
AT+CMSS	Send Message from Storage
AT+CMGW	Write Message to Memory
AT+CMGD	Delete Message

SMS PDU Mode

AT+CMGL	List Messages
AT+CMGR	Read Message
AT+CMGS	Send Message
AT+CMGW	Write Message to Memory

for the user input. The user enters the data followed by a control 'Z'. '^Z' (control z, 0x1A) is used to indicate the end of message. When the message is sent, the GSM modem responds with a number 122. This number is the message id of the message successfully sent.

In a text mode the user sends the message as a text. In text mode the message is ASCII encoded. However, during transmission this is converted in TPDU (Transfer Protocol Data Unit) encoding. The message can alternatively be encoded in the TPDU format as well. For ringing tones or picture messages, we need to send binary data; it is therefore mandatory that these data be encoded in TPDU mode. In TPDU mode, a text will also be encoded in binary. For example if you want to send 'Mobile Computing' in text mode, you send a string of 8 bit unsigned integers with values 0x4D, 0x6F, 0x62, 0x69, 0x6C, 0x65, 0x20, 0x43, 0x6F, 0x6D, 0x70, 0x75, 0x74, 0x69, 0x6E, 0x67. This in bit streams will look like the following:

01001101-01101111-01100010-01101001-01101100-01101001-00100000-01000011-
01101111-01101011-01110000-01110101-01110100-01101001-01101110-01100111

It is sufficient to use 7 bits for any ASCII character. Therefore, in TPDU mode we use only 7 bits out of the 8 bits. In TPDU mode we remove the most significant bit of every character to convert the above bit stream into:

1001101-1101111-1100010-1101001-1101100-1101001-0100000-1000011-1101111-1101011-
1110000-1110101-1110100-1101001-1101110-1100111

Each byte now has only 7 bits of information. To make an 8 bit byte, we borrow bits from following characters. The least significant bit from the following 7 bits character is taken and made as the most significant bit of the preceding character. This will make the bit stream look like:

11001101-10110111-00111000-11001101-01001110-10000011-10000110-11101111-
00110110-10111100-01001110-01001111-10111011-11001111

Therefore, the same text 'Mobile Computing' in TPDU mode will look like 0xCD 0xB7 0x38 0xCD 0x2E 0x83 0x86 0xEF 0x36 0xBC 0x4E 0x4F 0xBB 0xCF. To send a message in text mode the message format needs to be set by using AT+CMGF command. The value of 1 for CMGF is for text mode whereas the value 0 is for TPDU mode. In case of text mode we enter ^Z as the end of string (like the null character or the ^@ in case of 'C' string). In case of TPDU mode the length of the string needs to be explicitly given along with the ^Z.

Let us take an example of sending **Mobile Computing** in TPDU format to a MSISDN +919845062050 using a SMSC whose service center number is +919810051914.

```
AT+CMGF=0            // set SMS PDU mode on
OK
AT+CMGS=28           // length of the SMS PDU,
                     // The RP layer SC address octets are not counted in the length.
>0791198901509141110 00C911989546002050000A711CDB738CD2E8386EF3
6BC4E4FBBCF<ctrl-z)
    +CMGS: 212       // message reference is shown
    OK
```

Table 6.1 explains the fields of the above TPDU.

Table 6.1 RP SC Address-Value field followed by a TPDU in hexadecimal **form**

RP SC address (optional)	Value	Description	Status
	07	Address length	Length of the address is 7. Including the type of numbering plan indication.
	91	Type of address	International address using ISDN telephone numbering plan.
	19 89 01 50 91 41	Short message service center address	The short message service center number. F.ex 198901509141 is encoded as 91 98 10 05 19 14. In this case the address takes 6 octets. +919810051914 (Airtel in Delhi) encoded as 198901509141
TPDU Octet 1 bits	Value (hex11)	Description	Status
7	0	TP-Reply-Path	Reply path no set
6	0	TP-User-Data-header-indicator	Indication that user data doesn't contain additional header.
5	0	TP-Status-Report-Request	Not requested
4	1	TP-Validity-Period-Format	Relative format (bits 4 and 3)

Table 6.1 (*continued*)

	3	0	TP-Validity-Period-Format	Relative format (bits 4 and 3)
	2	0	TP-Rejected-Duplicates	Do not reject duplicates in SC
	1	0	TP-Message-Type-Indicator	Type : SMS-SUBMIT (from phone to network), (bits 1 and 0)
	0	1	TP-Message-Type-Indicator	Type : SMS-SUBMIT (from phone to network), (bits 1 and 0)
TPDU Octet 2	Value	Description		Status
	00	TP-Message-Reference		Given by the phone, application/user does not need to fill this octet.
TPDU Octet 3	Value	Description		Status
	0C	Address length in semi-octets.		Length of the address is 12 in semi-octets. Length the type of numbering plan indication.
TPDU Octet 4	Value	Description		Status
	91	Type of address		International address using ISDN telephone numbering plan.
TPDU Octet 5–10	Value	Description		Status
	91 98 45 06 20 50	TP-Destination-Address		The destination telephone number. F.ex +919845062050 is encoded as 198954600205. The address can be 2 to 12 octets long. +919845062050 (Asoke's phone) encoded as 198954600205
TPDU Octet 11	Value	Description		Status
	00	TP-Protocol-Identifier, consist one octet. For the details, see GSM 03.40 specification, version 7.2.0, page 53.		Parameter identifying the above layer protocol, if any. Note that for the straightforward case of simple MS-to-SC short message transfer, the TP-Protocol-Identifier is set to the value 00.

Table 6.1 (*continued*)

TPDU Octet 12 bits	Value (hex00)	Description	Status
7	0	TP-Data -Coding-Scheme used in TP-User- Data, consist one octet. See GSM 3.38	Functionality (bits 7 and 6) related to usage of bits 4-0.
6	0		Functionality (bits 7 and 6) related to usage of bits 4-0.
5	0		Indicates that text is uncompressed.
4	0		Indicated that bits 1 and 0 have no message class meaning.
3	0	Alphabet being used (bits 3 and 2)	7bit message
2	0	Alphabet being used (bits 3 and 2)	7bit message
1	0	Reserved	No meaning, indicated by bit 4
0	0	Reserved	No meaning, indicated by bit 4
TPDU Octet 13	Value	Description	Status
	A7	TP-Validity -Period (Relative format). See GSM 03.40, version 7.2.0, page 55 for details	A7 -> 24 hours
TPDU Octet 14	Value	Description	Status
	11	TP-User -Data-Length	Parameter indicating the length of the TP-User-Data field to follow. Represented as amount of septets (integer). 11 hex -> 17 septets. This is because of 7-bit user data. User data is coded to seven databits, because SMS have to be sent to air in 7 bit format. Length includes the user data header (not included in this example) and data itself.

Table 6.1 (*continued*)

TPDU Octet 15–29	Value	Description	Status
	CDB73 8CD2E 8386EF 36BC4 E4FBB CF	TP-User-Data	The user data. Format of the user data depends, what kind of message is sent. This example includes text string **Mobile Computing**. 16 septets + fill bits = 14 octets.

Encode the short message for ringing tone (8 bit, User-data-header)

Nokia phones can be customized with personalized ringing tone. Ringing tones can be sent as SMS messages. These are also called EMS (Extended Message Service). Following is an example of sending a ringing tone to a Nokia phone with MSISDN +919845062050. Ringing tone is a 8-bit binary message in TPDU format. The name of the tone is 'test'. This name will be displayed in the menu.

AT+CMGF=0   // set SMS PDU mode on

OK

AT+CMGS=50   // length of the SMS PDU

                 // The RP layer SC address is not included in this example.

>0051000C91198954600205F515A72406050415811581024A3A51D195CDD00
8001B205505906105605585505485408208499000<ctrl-z>

+CMGS: 214   // message reference is shown

OK

Following is the note for Indian National Anthem Jana-gana-mana. If one wants to send Jana-gana-mana as a ringing tone, one will compose a SMS with the following code as the TP-User-Data. Also, we need to modify the MSISDN accordingly.

02 4A 3A 69 8C E9 71 B5 E4 81 91 BD 8D D4 04 00

```
3A D9 34 91 41 34 15 41 54 15 41 54 15 41 54 15
21 54 15 41 34 15 41 62 10 81 52 15 41 54 13 21
34 13 42 0E 21 24 D0 45 00 00
```

Sending a picture (bitmap) OTA (Over The Air)

A picture message can occupy the complete display area of the phone. For Nokia phones, the maximum size of the picture message is 72×28 pixels. The maximum size of the operator logo and the CLI logo is 72×14 pixels.

Each semi-octet in the OTA bitmap presents 4 pixels in the original bitmap. Because one row takes 18 semi-octets, the whole 72×14 size (operator logo and CLI logo) bitmap takes 18×14 = 252 semi-octets = 126 octets. In the case of the picture message the whole 72×28 size bitmap takes 18×28 = 504 semi-octets = 252 octets (as it must be = sent using concatenate message). For details please refer to Nokia site at www.nokia.com.

Reading a Message through GSM modem

In previous examples we discussed how we to send SMS or EMS messages through GSM modem. When you send a SMS from a computer, the GSM modem always acts as a pass-through. However, for input message it behaves differently. When a message is received by a GSM modem, by default it is routed to the phone inbox. Therefore, to read a SMS from the GSM modem, we need to ensure that the SMS is forwarded to the computer rather than the phone local store. For this we use the CNMI command. The value 2,1,0,0,0 signifies that whenever a SMS is received, an interrupt will be raised on the COM port followed by the messages flushed on the COM port. This transformation is described in detail in GSM 03.38 standard. Following is an example of reading a message 'hellohello'

```
AT+CMGF=0        // set SMS PDU mode on
AT+CNMI=2,1,0,0,0 // set the modem-computer interface
AT+CMGR              // read the message
```

The data we read from the port will be:

07917283010010F5040BC87238880900F10000993092516195800 3C16010

In the above octet sequence there are three parts: An initial octet indicating the length of the SMSC information ('07'), the SMSC information itself ('917283010010F5'), and the SMS_DELIVER part (specified by ETSI in GSM 03.40).

All the octets above are hexa-decimal 8-bit octets, except the Service center number, the sender number and the timestamp; they are decimal semi-octets. The message part in the end of the PDU string consists of hexadecimal 8-bit octets, but these octets represent 7-bit data (Table 6.2). Semi-octets are binary coded decimals, e.g. the sender number is obtained by performing internal swapping within the semi-octets from '72 38 88 09 00 F1' to '27 83 88 90 00 1F'. The length of the phone number in this example has 11 digits (odd). Note that a proper octet sequence cannot be formed by this number. This is the reason why a trailing F has been added. The time stamp, when parsed, equals '99 03 29 15 16 59 80', where the first 6 octets represent date in YYMMDD format; the following 6 octets represents time in HHMMSS format; and the last two represents timezone related to GMT. Timezone 1 signifies 15 minutes. For all operators in India timezone will be 22 (GMT+5.5).

Table 6.2 A received TPDU

RP SC address (optional)	Value	Description	Status
	07	Address length	Length of the SMSC information (in this case 7 octets).
TPDU Octet 1			
	91	Type-of-Address of the SMSC.	(91 means international format of the phone number)
TPDU Octet 2–7			
	72 83 01 00 10 F5	Service center number (in decimal semi-octets).	The length of the phone number is odd (11), so a trailing F has been added to form proper octets. The phone number of this service center is "+27381000015".

Table 6.2 (*continued*)

TPDU Octet 1 bits	Value (hex 51)	Description	Status
7	0	TP-Reply-Path	Parameter indicating that reply path exists.
6	0	TP-User-Data-Header-Indicator	Indication that user data contains an additional header.
5	0	TP-Status-Report-Request	This bit is set to 1 if a status report is going to be returned to the SME
4	0	TP-Validity-Period-Format	Relative format (bits 4 and 3)
3	0	TP-Validity-Period-Format	Relative format (bits 4 and 3)
2	1	TP-More-Message-to-Send	This bit is set to 0 if there are more messages to send
1	0	TP-Message-Type-Indicator	type:SMS-DELIVER (from network to phone), (bits 1 and 0 both set to 0)
0	0	TP-Message -Type - Indicator	type:SMS-DELIVER (from network to phone), (bits 1 and 0 both set to 0)
TPDU Octet 3	Value	Description	Status
	0B	Address length	Length of the address is 11 in decimal.
TPDU Octet 4	Value	Description	Status
	91	Type of address	International address using ISDN telephone number plan.
TPDU Octets 5–10	Value	Description	Status
	72 38 88 09 00 F1	TP-Destination-Address	The destination telephone number +27838890001 is encoded as 72 38 88 09 00 F1. In this case the address is 11 digit, therefore a F is added to make it occupiy 6 octets. The

Table 6.2 (*continued*)

			address field can be anywhere between 2 to 12 octets long.
TPDU Octet 11	Value	Description	Status
	00	TP-Protocol-Identifier, consist one octet.	Short Message Type 0. This means that the ME must acknowledge receipt of the short message but may discard its contents.
TPDU Octet 12 bits	Value (hex 15)	Description	Status
7	0	TP-Data -Coding -Scheme used in TP-User-Data, consist one octet. See GSM 3.38	Functionality (bits 7 and 6) related to usage of bits 4-0.
6	0		Functionality (bits 7 and 6) related to usage of bits 4-0.
5	0		Indicates that text is uncompressed
4	0		Indicated that bits 1 and 0 have message class meaning.
3	0	Alphabet being used (bits 3 and 2)	8 bit data
2	0	Alphabet being used (bits 3 and 2)	8 bit data
1	0	Message class (bits 1 and 0)	Class 1, Default meaning: ME-specific
0	0	Message class (bits 1 and 0)	Class 1, Default meaning: ME-specific
TPDU Octet 13	Value	Description	Status
	99 30 92 51 61 95 80	TP-Service-Center-Time-Stamp	Format Year, Month, Day; Hour, Minute, Second; Timezone relative to GMT with 1 unit as 15 minutes. 0x99 0x30 0x92 0x51 0x61 0x95 0x80 means 29 March 1999 15:16:59 GMT+2

Table 6.2 (*continued*)

TPDU Octet 14	Value	Description	Status
	0A	TP-User -Data-Length	Parameter indicating the length of the TP-User-Data field to follow. Represented as amount of octets (integer). 0A hex -> 10 octets. Length includes the user data header and data itself. User data length, length of message. The TP-DCS field indicated 7-bit data, so the length here is the number of septets (10). If the TP-DCS field were set to indicate 8-bit data or Unicode, the length would be the number of octets (9).
TPDU Octets 15–50	**Value**	**Description**	**Status**
	E8329BFD4697D 9EC37	TP-User -Data	The user data. Format of the user data depends, what kind of message is sent. This message includes user data header and the user data itself. The example is 8-bit octets representing 7-bit data. The user data is "hellohello" message.

6.4.2 Example code for GSM modem

Following is an example code for GSM modem. This is written in Visual Basic. The code uses Microsoftmscomm controls. The mscomm controls use the COM1 port for communication.

In the example line 13 is for setting of the communication port and the interface between the computer and the modem.

Line 14–24 is for initialization of the GSM phone as modem.

Line 28–46 is to send a SMS.

Line 50–52 is for reading SMS from the modem.

6.4.3 SMPP

```
1     ' (c) 2002
2     ' GSM Modem implementation using MSCOMM and Nokia phone
3     .
4     .
5     .
6     ' Set up the communications port
7     MSComm1.CommPort = 1 ' Set COM1 for MSCOMM
8     ' Set for 9600 baud, no parity, 8 data, and 1 stop bit.
9     MSComm1.Settings = "9600,N,8,1"
10    ' Tell the control to read entire buffer when Input is used
11    MSComm1.InputLen = 0
12    ' Open the port
13    MSComm1.PortOpen = True
14    ' AT commands are terminated by Carriage Return & Line feed
15    ' Send an initial 'AT' command to the phone
16    MSComm1.Output = "AT" ' Write AT on COM1
17    MSComm1.Output = Chr$(13) ' Write Carriage Return
18    MSComm1.Output = Chr$(10) ' Write Line Feed
19    ' The phone will respond with an 'OK'
20    ' Set the GSM modem so that all SMSs are forwarded to our program
21    MSComm1.Output = "AT+CNMI=1,2,0,1,0" ' Write AT on COM1
22    MSComm1.Output = Chr$(13) ' Write Carriage Return
23    MSComm1.Output = Chr$(10) ' Write Line Feed
24    ' The phone will respond with an 'OK'
25    .
26    .
27    .
28    ' Set up the phone for a text message
29    MSComm1.Output = "AT+CMGF=1" & Chr$(13) & Chr(10)
30    ' The phone will respond with an 'OK'
31    ' Prep for SMS, give destination type and destination address.
32    ' Enter the destination type & address to prep for SMS
33    ' e.g. AT+CMGS="+919845170882",^Z
34    MSComm1.Output = "AT+CMGS="
35    MSComm1.Output = Chr$(34) ' The start quote character
```

36	MSComm1.Output = "+919845170882" ' Mobile number with country code
37	MSComm1.Output = Chr$(34) ' The end quote character
38	MSComm1.Output = Chr$(13) ' Write Carriage Return
39	MSComm1.Output = Chr$(10) ' Write Line Feed
40	' The phone will return a'>' prompt, and await entry of the SMS message text.
41	' Now send the text to the phone and terminate with (Ctrl-Z)
42	MSComm1.Output = "This is a test message" ' Frame the message
43	MSComm1.Output = Chr$(26) ' Add the ^Z
44	' The phone will respond with a conformation containing the
45	' Close the port
46	MSComm1.PortOpen = False
47	.
48	.
49	.
50	' Read the input buffer
51	buffer = MSComm1.Input
52	InpStr = StrConv(buffer, vbUnicode)
53	.
54	.

This is an operator dependent solution with high scalability. The SMS data speed is about 300 bits/sec. Using a GSM modem, it takes about 4 seconds to read a message, and about 8 seconds to send a message, resulting in about 5 message pairs/minute. GSM modem can be used for operator independent SMS application. The only limitation of GSM modem is that it cannot scale. The theoretical limit is about 300 message pairs in an hour. If the transaction rate is low, GSM modem can work out to be a convenient and economical way of using SMS for mobile computing. However, if we need reliability and scalability, GSM modem technology is not the best answer. For such cases we need carrier grade solution. For this we need connection directly to the SMSC of the operator. This is achieved through SMPP (Short Message Peer to Peer) protocol.

The SMPP protocol is an open, industry standard protocol designed to provide a flexible data communications interface for transfer of short message data between a Message Center (SC or SMSC) and a VAS application, such as a WAP Proxy Server, Voice Mail server, EMail Gateway or other Messaging Gateway (Figure 6.7). An SMPP client is termed a External Short Message Entity (ESME), and is connected to the SC.

SMPP release v3.4 presently supports Digital Cellular Network technologies which include the following.

- GSM

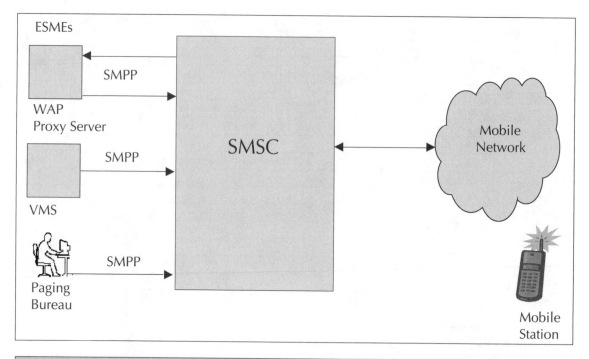

Figure 6.7 Communication between SMSC and SME

- IS-95 (CDMA)
- CDMA 1X/CDMA 2000
- ANSI-136 (TDMA)
- IDEN

Using the SMPP protocol, an SMS application system called the ESME may initiate an application layer connection with an SC over a TCP/IP or X.25 connection and send and receive short messages. The ESME may also query, cancel or replace short messages using SMPP.

SMPP supports a full featured set of two-way messaging functions such as the ones described below.

- Transmit messages from an ESME to single or multiple destinations via the SMSC.
- An ESME may receive messages via the SMSC from other SMEs (e.g. mobile stations).

- Query the status of a short message stored on the SMSC.

- Cancel or replace a short message stored on the SMSC.

- Send a registered short message (for which a 'delivery receipt' will be returned by the SMSC to the message originator).

- Schedule the message delivery date and time.

- Select the message mode, i.e. datagram or store and forward.

- Set the delivery priority of the short message.

- Define the data-coding type of the short message.

- Set the short message validity period.

- Associate a service type with each message e.g. voice mail notification.

SMPP Protocol Overview

SMPP protocol is an open message-transfer protocol that enables short message entities (SMEs) outside the mobile network to interface with an SC. Non-mobile entities that submit messages to, or receive messages from an SMSC are known as External Short Message Entities (ESMEs).

The SMPP protocol defines operations and data as described below.

- Set of operations for the exchange of short messages between an ESME and an SMSC.

- Data that an ESME application must exchange with an SMSC during SMPP operations.

Subscribers to an SMS-capable Cellular Network may receive short messages on a Mobile Station (MS) from one or more ESMEs. The examples of such ESME applications include the following.

- Voice mail alerts originating from a VMS (Voice Messaging System), indicating voice messages at a customer's mailbox.

- Numeric and alphanumeric paging services.

- Information services. For example, an application that enables mobile subscribers to query currency rates or share-price information from a database or the WWW and have it displayed as a short message on the handsets.

- Calls directly dialed or diverted to a message-bureau operator, who forwards the message to the SMSC, for onward delivery to a subscriber's handset.

- A fleet management application that enables a central station to use the SMSC to determine the location of its service vehicles and notify the closest vehicle of a service request in their area.

- Telemetry applications. For example, a household meter that transmits a short message to a utility company's billing system to automatically record customer usage.

- WAP Proxy Server. A WAP Proxy Server acts as the WAP gateway for wireless internet applications. A WAP Proxy Server may select an SMS or USSD bearer for sending WDP (Wireless Data Protocol) datagrams to and receiving WDP datagrams from a mobile station.

There is an open source SMS gateway available from www.kannel.org. Along with SMS, Kannel gateway supports WAP and MMS (Multi Media Messaging) interfaces. Kannel offers http interface for message transfer and administrating of the gateway.

Kannel divides its various functions into different kinds of processes (Figure 6.8), called boxes, mostly based on what kinds of external agents it needs to interact with.

- The **bearerbox** implements the bearer level of SMS. As part of this, it connects to the SMS centers. Definitions of different TCP/IP ports, usernames, passwords etc. are required to be defined for this connection.

- The **smsbox** implements the rest of the SMS gateway functionality. It receives textual SMS messages from the bearerbox, and interprets them as service requests, and responds to them in the appropriate way. All the services will be handled and managed by this box.

There can be only one bearerbox, but any number of smsboxes in a single Kannel instance. Duplicating the bearerbox is troublesome. Also, each SMS center can be connected only to one client. While it is possible to have each SMS center served by a different process, it has been deemed not to give enough extra reliability or scalability to warrant the complexity. Having multiple smsboxes can be beneficial when the load is very high. Although the processing requirements as such are fairly low per request, network bandwidth from a single machine, or at least operating system limits regarding the number of concurrent network connections are easier to work around with multiple processes, which can, if necessary, be spread over several hosts. Each box is internally multithreaded. For example, in the bearerbox, each SMS center connection is handled by a separate thread. The thread structures in each of the boxes are fairly static

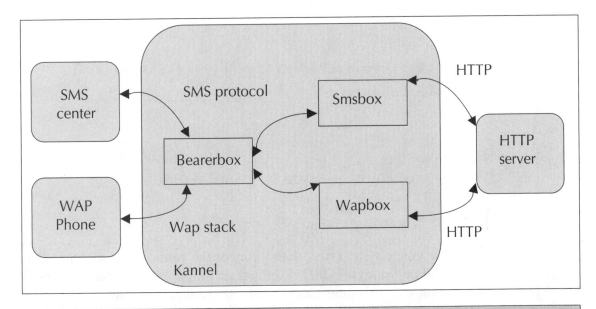

Figure 6.8 Boxes of Kannel

i.e. the threads are mostly spawned at startup, instead of spawning a new one for each message.

In Kannel we configure the SMS services like:

```
# SMSC Configuration
group = smsc
smsc = smpp
smsc-id = hssup
host = 203.168.14.69
port = 15009
smsc-username = smppabc
smsc-password = abcocse
address-range = "^333$"
# SMSBOX Configuration
group = smsbox
bearerbox-host = localhost
```

```
sendsms-port = 333
global-sender = 333
####################################################
# Service Configuration
# News
group = sms-service
keyword = news
url = "http://sms.apps.com/SMSService.jsp?keyvalue=
          %a&mobileno=%p"
max-messages = 4
```

We install Kannel on a Linux system and then connect the same to the SMSC. The connection to SMSC will be through TCP/IP. Kannel also support proprietary protocol like CIMD2 from Nokia SC. In either case, a part of configuration shall define the behavior of bearerbox; some will define the smsbox.

Pull messages

In case of pull transaction, the user enters a message with a keyword and then sends the same to a service number (333 in this example). Let us assume that the network offers a news service. To access the breaking news, the user needs to enter **News brk** and then send it to 333. During binding of the SMS gateway we intimate the SC that we are listening for 333. Therefore, all the messages sent to 333 will be routed to our SMPP gateway. In the Kannel configuration file, we mention that whenever there is a message with keyword **news**, it should be forwarded to a http URLhttp://sms.apps.com/ SMSService.jsp. To service the user with appropriate response, we need to know the request with all the parameters and the MSISDN number of the phone. These are transferred from Kannel gateway to the URL through %a and %p. The complete URL would appear as follows.

```
sms.apps.com/SMSService.jsp?keyvalue=news%20brk&mobileno=91
9845052534.
```

The response of the http request will be forwarded directly to the user (919845052534) by Kannel gateway. If the response from the content/origin server is more than 160 characters, Kannel splits the message into multiple messages. During splitting, Kannel ensures that the messages are split at word boundary. The max-messages parameter defines the limit of maximum number of messages as response. If we set the max-messages to 0, no response will be sent to the user, though there could be some response coming from the http request.

Push messages

In case of push messages, the message is sent through http interface as well. An application uses an http URL to communicate with the Kannel gateway and to send SMS messages. Kannel delivers these messages to the SC. Following are a few examples of sending a text, ring tone (binary), picture (binary), and Hindi Unicode messages to a mobile phone 919811557988. To offer certain level of security Kannel allows the user authentication through user-id and a password to access these URLs.

```
# TEXT
http://kannel.apps.com:31333/cgi-bin/sendsms?user=
    tester&pass=foobar&to=919811557988&text=hello

# BINARY (Ring Tone)
http://kannel.apps.com:31333/cgi-bin/sendsms?user=
    tester&pass=foobar&to=919811557988&udh=%06%05%04%15%81%00%
    00&text=%02%4A%3A%71%5D%85%B1%AC%81%BD%98%81%31%A5%99%94%0
    4%00%4F%20%CA%E8%38%93%89%20%82%0C%2E%C3%0C%38%83%0C%2C%C2
    %A9%2A%92%08%20%42%08%2C%C3%0C%38%93%89%20%82%0C%2C%C3%0C%
    38%83%0C%2C&coding=2
```

The significance of UDH in the above example is as follows:

06 – Length of the UDH,

05 – IEI, Information Element Identifier (Application port addressing scheme, 16 bit port address)

04 – IEDL, Information Element Data Length

1581 – Information Element Data (Destination Port)

0000 – Information Element Data (Originator Port)

BINARY (Picture - 3 Messages)

```
http://kannel.apps.com:31333/cgi-bin/sendsms?user=
tester&pass=foobar&to=919811557988&udh=%0B%05%04%15%8A%00%
00%00%03%72%03%01&text=%FF%FF%FF%FA%FE%03%FF%FF%FF%FF%FF%F
F%F7%FE%01%FF%FF%FF%FF%FF%FF%EB%FC%02%FF%FF%FF%FF%FF%FF%F1
%F0%01%FF%FF%FF%FF%FF%FF%F8%E0%00%FF%FF%FF%FF%FF%FF%F0%00%
01%FF%FF%FF%FF%FF%FF%E0%00%00%FF%FF%FF%FF%FF%FF%E0%00%00%7
F%FF%FF%FF%FF%FF%AB%80%28%EF%FF%FF%FF%FF%FF%45%F1%FC%5F%FF
%FF%FF%FF%FF%A2%EA%E8%CF%FF%FF%FF%FF%FF%F1%11%10%5F%FF%FF%F
F%FF%FF%80%88%A0%BF%FF%FF%FF%FF%FF%C0%00%01%1F%FF%FF%FF%FF"
```

The significance of UDH is as follows:

0B — Length of the UDH

05 — IEI, Information Element Identifier (Application port addressing scheme, 16 bit port address)

04 — IEDL, Information Element Data Length

158A — IED, Information Element Data (Destination Port)

0000 — IED (Originator Port)

00 — IEI (Concatenated Short Message, 8 bit reference number)

03 — IEDL (Information Element Data Length)

03 — IED (Total number of concatenated messages 0-255)

01 — IED (Sequence number of current short message)

```
# Hindi UNICODE

http://kannel.apps.com:31332/cgi-bin/sendsms?user=
tester&pass=foobar&to=919811557988&text=%09%50&coding=3
```

Text is a single character OM in Hindi. The range of Hindi/Devanagari, as decided by Unicode Consortium is within the range 0900–097F.

REFERENCES/FURTHER READING

1. Introduction to SMS: http://www.gsmworld.com/technology/sms/intro.shtml.

2. Nokia, AT Command Set for Nokia GSM Products, Nokia Mobile Phones 2000.

3. GSM Standard 03.40: Digital cellular telecommunications system (Phase 2+); Technical realization of the Short Message Service (SMS) Point-to-Point (PP), Release 1997, http://www.etsi.org.

4. GSM Standard 03.39: Digital cellular telecommunications system (Phase 2+); Interface protocols for the connection of Short Message Service Centres (SMSCs) to Short Message Entities (SMEs), Release 1998, http://www.etsi.org.

5. GSM Standard 03.19: Digital cellular telecommunications system (Phase 2+); Subscriber Identity Module Application Programming Interface (SIM API); SIM API for Java Card (TM); Stage 2, Release 1999, http://www.etsi.org.

6. GSM Standard 03.47: Digital cellular telecommunications system (Phase 2+); Example protocol stacks for interconnecting Service Centre(s) (SC) and Mobile-services Switching Centre(s) (MSC), Release 1998, http://www.etsi.org.

7. GSM Standard 03.48: Digital cellular telecommunications system (Phase 2+); Security mechanisms for SIM application toolkit; Stage 2, Release 1999, http://www.etsi.org.

8. Nokia, AT Command Set for Nokia GSM Products, Nokia Mobile Phones 2000.

9. GSM Standard 11.11: Digital cellular telecommunications system (Phase 2+); Specification of the Subscriber Identity Module–Mobile Equipment (SIM–ME) interface, Release 1999, http://www.etsi.org.

10. GSM Standard 11.14: Digital cellular telecommunications system (Phase 2+); Specification of the SIM Application Toolkit (SAT) for the Subscriber Identity Module–Mobile Equipment (SIM-ME) interface, Release 1999, http://www.etsi.org.

11. Short Message Peer to Peer Protocol Specification v3.4, Document Version:- 12-Oct-1999 Issue 1.2, http://smsforum.net/.

12. Kannel WAP and SMS Gateway: http://www.kannel.org.

13. Asoke K Talukder, Siddhartha Chhabra, Sudarsan T.S, Gowrishankar K.A, Debabrata Das, Ubiquitous Rescue Operation at Affordable Cost through Location-Aware SMS, Proceedings of National Conference on Communication, IIT-Kharagpur, 28–30 January 2005, pp 651–655.

14. Asoke K Talukder, Bijendra Singh, Debabrata Das, Ubiquitous-Trustworthy-Secure Data Communication in Hybrid Cellular-Internet Networks, Proceedings of IEEE INDICON 2004, December 20–22, 2004, IIT Kharagpur, pp 135–138.

15. Jesudoss Venkatraman, Vijay Raghavan, Debabrata Das, and Asoke K Talukder, Trust and Security Realization for Mobile Users in GSM Cellular Networks, Trust and Security Realization for Mobile Users in GSM Cellular Networks, Proceedings of Asian Applied Computer Conference, Kathmandu October 29–31, 2004; LNCS 3285 pp-302–309.

16. Rishi Pal, Asoke K Talukder, Global Service Portability and Virtual Home Environment through Short Message Service (SMS) and GSM, Proceedings of the 15th International Conference in Computer Communication (ICCC2002), pp 8-18, 12-14 August 2002, Mumbai, India.

17. Kurian John, Asoke K Talukder, Ubiquity and Virtual Home Environment for Legacy Applications and Services, Proceedings of the 15th International Conference in Computer Communication, pp 371–381, 12–14 August 2002, Mumbai, India.

REVIEW QUESTIONS

Q1: What are various strengths of SMS? Explain all of them. Also, state what are the applications areas where these strengths can be used?

Q2: Explain the difference between SM MT and SM MO?

Q3: How do you develop location based applications using SMS?

Q4: How do you use SMS in an application using GSM Modem (Over-The-Air)? Explain different phases in this technology?

Q5: What is SMPP protocol? Why and when is it used? What are the basic difference between SMPP based technology and GSM Modem technology?

CHAPTER 7

General Packet Radio Service (GPRS)

7.1 INTRODUCTION

People love freedom. In Hollywood movies, we see actors moving around the room and talking on the telephone holding the phone in the left hand and the handset in the right. Today all this has changed. We have cordless phones and the wireless mobile phones. People can move around (inside and outside home or even vehicles) and still talk. As the world is changing, people's expectations are also changing. People are looking for freedom from wire with respect to data. Pundits believe that the trend taking place in fixed networks whereby the growth of data traffic is overtaking that of voice traffic, will also influence the wireless networks. GSM started with voice in mind and offered whatever a wireless voice user wanted. The popularity of GSM, Internet, and digital communication forced GSM to look for wireless data with higher band-width. General Packet Radio Service (GPRS) is a step to efficiently transport high-speed data over the current GSM and TDMA-based wireless network infrastructures.

7.2 GPRS AND PACKET DATA NETWORK

Until 3G/UMTS/IMT-2000 becomes a reality, and perhaps long afterwards, GPRS will thrive in both vertical and horizontal markets where high-speed data transmission over wireless networks is necessary. The deployment of GPRS networks will allow a variety of new applications ranging from mobile e-commerce to mobile corporate VPN access. Deployments of GPRS networks have already taken place in several countries in Europe and the Far East. In Mumbai and Delhi GPRS was launched quite sometime ago.

7.2.1 Capacity and Other End-user Aspects

GPRS has the ability to offer data speeds of 14.4 KBps to 171.2 KBps, which allow for comfortable Internet access. It allows for short 'bursty' traffic, such as e-mail and web browsing, as well as large volumes of data. To support GPRS operations, new protocols and new network devices are required. By allowing information to be transmitted more quickly, immediately and efficiently across the mobile network, GPRS may well be a relatively less costly mobile data service compared to SMS and Circuit Switched Data. For GPRS, no dial-up modem connection is necessary. It offers fast connection set-up mechanism to offer a perception of being 'always on'. This is why GPRS users are sometimes referred to as being 'always connected'. This is like SMS, which is an always-on service. Immediacy is one of the advantages of GPRS compared to Circuit Switched Data.

7.2.2 Quality of Service (QoS)

The Quality of Service (QoS) requirements of typical mobile packet data applications are very diverse. For example the QoS for realtime multimedia content is different from web browsing or email transfer. GPRS allows definition of QoS profiles using the parameters of service precedence, reliability, delay and throughput.

- **Service precedence** is the priority of a service in relation to another service. There exist three levels of priority: high, normal, and low.

- **Reliability** indicates the transmission characteristics required by an application. Three reliability classes are defined, which guarantee certain maximum values for the probability of loss, duplication, mis-sequencing and corruption (an undetected error) of packets.

- **Delay** parameters define maximum values for the mean delay and the 95-percentile delay. The delay is defined as the end-to-end transfer time between two communicating mobile stations or between a mobile station and the signaling interface to an external packet data network.

- **Throughput** specifies the maximum/peak bit rate and the mean bit rate.

Using these QoS classes, QoS profiles can be negotiated between the mobile user and the network for each session.

7.2.3 Integral Part of the Future 3G Systems

The different approaches to third generation (3G) wireless systems (IMT-2000, UMTS, CDMA, WCDMA, 3GPP, 3GPP2 etc.) were intended to address the challenge of voice-to-data crossover and integration. The complexities of new and exciting wireless technologies have slowed down progress in their development and widespread deployment. To lessen the impact of the delay in implementing 3G wireless systems, GPRS is introduced as an intermediate step to efficiently transport high-speed data over the current GSM and TDMA-based wireless network infrastructures. GPRS is therefore called the 2.5G (two and half G or two and half generation) in the evolution process of wireless cellular networks.

7.3 GPRS NETWORK ARCHITECTURE

GPRS uses the GSM architecture for voice. In order to offer packet data services through GPRS, a new class of network nodes need to be introduced as an upgrade to existing GSM network. These network nodes are called GPRS support nodes (GSN). GPRS support nodes are responsible for the delivery and routing of data packets between the mobile stations and the external packet data networks (PDN). There are two types of support nodes, viz., SGSN (Serving GSN) and GGSN (Gateway GSN). Figure 7.1 depicts GPRS system components for data services.

Serving GPRS Support Node (SGSN): A serving GPRS support node (SGSN) is at the same hierarchical level as the MSC. Whatever MSC does for voice, SGSN does the same functions for packet data. SGSN's tasks include packet switching, routing and transfer, mobility management (attach/detach and location management), logical link management, and authentication and charging functions. SGSN processes registration of new mobile subscribers and keeps a record of their location inside a given service area. The location register of the SGSN stores location information (e.g., current cell, current VLR) and user profiles of all GPRS users registered with this SGSN. SGSN sends queries to Home Location Register (HLR) to obtain profile data of GPRS subscribers. The SGSN is connected to the base station system with Frame Relay.

Gateway GPRS Support Node (GGSN): A gateway GPRS support node (GGSN) acts as an interface between the GPRS backbone network and the external packet data networks. GGSN's function is similar to that of a router in a LAN. GGSN maintains routing information that is necessary to tunnel the Protocol Data Units (PDUs) to the SGSNs that service particular mobile stations. It converts the GPRS packets coming from the SGSN into the appropriate packet data protocol (PDP) format for the data networks like Internet or X.25. PDP sends these packets out on the corresponding packet data network. In the other direction, PDP receives incoming data packets from data networks and converts them to the GSM address of the destination user. The readdressed packets are sent to the responsible SGSN. For this purpose, the GGSN stores the current SGSN address of the user and his or her profile in its location register. The GGSN also performs authentication and charging functions related to data transfer.

7.3.1 GPRS Network Enhancements

In addition to the new GPRS components (SGSN and GGSN), some existing GSM network elements must also be enhanced in order to support packet data. These are:

Base Station System (BSS): BSS system needs enhancement to recognize and send packet data. This includes BTS upgrade to allow transportation of user data to the SGSN. Also, the BTS needs to be upgraded to support packet data transportation between the BTS and the MS (Mobile Station) over the radio.

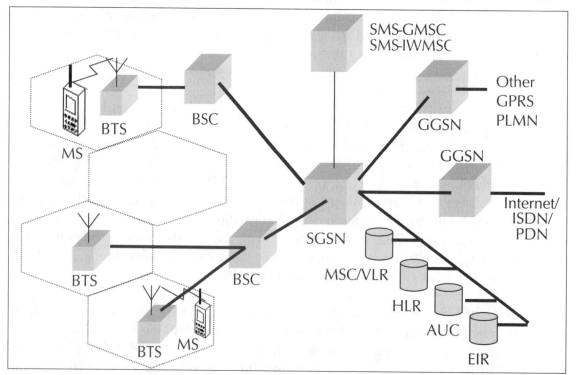

Figure 7.1 GPRS system architecture

Home Location Register (HLR): HLR needs enhancement to register GPRS user profiles and respond to queries originating from GSNs regarding these profiles.

Mobile Station (MS): The mobile station or the mobile phone for GPRS is different from that of GSM.

SMS nodes: SMS-GMSCs and SMS-IWMSCs are upgraded to support SMS transmission via the SGSN. Optionally, the MSC/VLR can be enhanced for more efficient co-ordination of GPRS and non-GPRS services and functionality.

7.3.2 Channel Coding

Channel coding is used to protect the transmitted data packets against errors. The channel coding technique in GPRS is quite similar to the one employed in conventional GSM. Under very bad channel conditions, reliable coding scheme is used. In reliable coding scheme many redundant bits are added to recover from burst errors. In this scheme a data rate of 9.05 KBps is achieved per time slot. Under good channel conditions, no encoding scheme is used resulting in a higher data rate of 21.4 KBps per time slot. With eight time slots, a maximum data rate of 171.2 KBps can be achieved.

7.3.3 Transmission Plane Protocol Architecture

Figure 7.2 illustrates the protocol architecture of the GPRS transmission plane, providing transmission of user data and its associated signaling.

Signaling Plane

The protocol architecture of the signaling plane comprises protocols for control and support of the functions of the transmission plane. This includes GPRS attach and detach, PDP context activation, control of routing paths, and allocation of network resources. The signaling architecture between SGSN and the registers like HLR, VLR, and EIR uses the same protocols as GSM. However, they are extended to support GPRS-specific functionality. Between SGSN and HLR as well as between SGSN and EIR, an enhanced MAP (Mobile Application Part) is employed. MAP is a mobile network-specific extension of the Signaling System SS#7 used in GSM. It transports the signaling information related to location updates, routing information, user profiles, and handovers. The

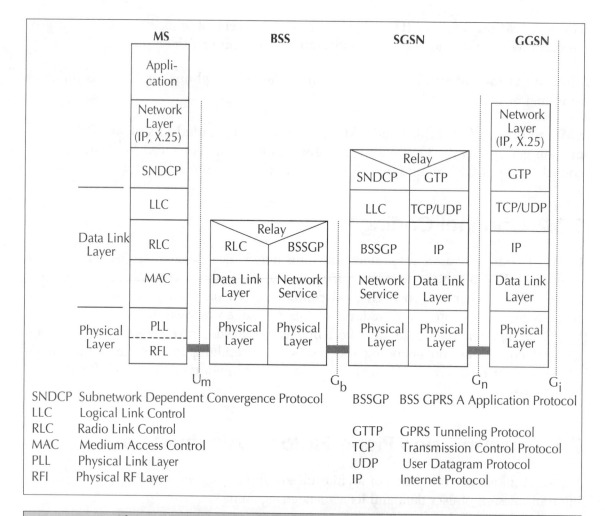

Figure 7.2 Transmission plane and GPRS protocol stack

exchange of MAP messages is accomplished over the transaction capabilities application part (TCAP) and the signaling connection control part (SCCP). The base station system application part (BSSAP+) is an enhancement of GSM's BSSAP. It is used to transfer signaling information between the SGSN and the VLR.

GPRS Backbone

GPRS backbone includes the transmission plane between SGSN and GGSN. User data packets and related signaling information within the GPRS network are encapsulated using the GPRS Tunneling Protocol (GTP). The GTP protocol is used in both intra-PLMN (between SGSN and GGSN within one PLMN) and inter-PLMN (between SGSN and GGSN of different PLMNs). In the transmission plane, GTP protocol tunnels the user data packets through the GPRS backbone by adding GPRSspecific routing information. GTP packets carry the user's data packets from both IP and X.25 data networks. Below GTP, the standard protocols TCP or UDP are used to transport the GTP packets within the backbone network. X.25 expects a reliable data link; therefore TCP is used for tunneling X.25 data. For IP based user data, UDP is used as it does not expect reliability in the network layer or below. Ethernet, ISDN, or ATM-based protocols may be used in the physical layer in the IP backbone. In essence, in the GPRS backbone we have an IP/X.25-over-GTP-over-UDP/TCP-over-IP transport architecture.

BSS–SGSN Interface

The BSS and SGSN interface is divided into the following layers:

Sub-Network Dependent Convergence Protocol (SNDCP): The SNDCP is used to transfer data packets between SGSN and MS. Its functionality includes:

- Multiplexing of several connections of the network layer onto one virtual logical connection of the underlying LLC layer.

- Segmentation, compression, and decompression of user data.

Logical Link Control (LLC): a data link layer protocol for GPRS which functions similar to Link Access Procedure–D (LAPD). This layer assures the reliable transfer of user data across a wireless network.

Base Station System GPRS Protocol (BSSGP): The BSSGP delivers routing and QoS-related information between BSS and SGSN.

Network Service: This layer manages the convergence sublayer that operates between BSSGP and the Frame Relay Q.922 Core by mapping BSSGP's service requests to the appropriate Frame Relay services.

Air Interface

The air interface of GPRS comprises the physical and data link layer.

Data Link Layer

The data link layer between the MS and the BSS is divided into three sublayers: the logical link control (LLC) layer, the radio link control (RLC) layer and the medium access control (MAC) layer.

Logical Link Control (LLC): This layer provides a reliable logical link between an MS and its assigned SGSN. Its functionality is based on HDLC (High-level Data Link Control) protocol and includes sequence control, in-order delivery, flow control, detection of transmission errors, and retransmission (automatic repeat request, ARQ). Encryption is used in this interface to ensure data confidentiality. Variable frame lengths are possible. Both acknowledged and unacknowledged data transmission modes are supported. This protocol is an improved version of the LAPDm protocol used in GSM.

Radio Link Control (RLC): The main purpose of the radio link control (RLC) layer is to establish a reliable link between the MS and the BSS. This includes the segmentation and reassembly of LLC frames into RLC data blocks and ARQ of uncorrectable data.

Medium Access Control (MAC): The medium access control (MAC) layer controls the access attempts of an MS on the radio channel shared by several MSs. It employs algorithms for contention resolution, multiuser multiplexing on a packet data traffic channel (PDTCH), and scheduling and prioritizing based on the negotiated QoS.

Physical Layer

The physical layer between MS and BSS is divided into two sublayers: the physical link layer (PLL) and the physical RF Layer (RFL).

Physical Link Layer (PLL): This layer provides services for information transfer over a physical channel between the MS and the network. These functions include data unit framing, data coding, and the detection and correction of physical medium transmission errors. The Physical Link layer uses the services of the Physical RF layer.

Physical RF Layer (RFL): This layer performs the modulation of the physical waveforms based on the sequence of bits received from the Physical Link layer above. The Physical RF layer also demodulates received wave forms into a sequence of bits that are transferred to the Physical Link layer for interpretation.

Multiple Access Radio Resource Management

On the radio interface, GPRS uses a combination of FDMA and TDMA. As in GSM (Figure 5.10), GPRS uses two frequency bands at 45 MHz apart; viz., 890–915 MHz for uplink (MS to BTS), and 935–960 MHz for downlink (BTS to MS). Each of these bands of 25 MHz width is divided into 124 single carrier channels of 200 kHz width. Each of these 200 kHz frequency channels is divided into eight time slots. Each time slot of a TDMA frame lasts for a duration of 156.25 bit times and contains a data burst.

On top of the physical channels, a series of logical channels are defined to perform functions like signaling, broadcast of general system information, synchronization, channel assignment, paging or payload transport. As with GSM, these channels can be divided into two categories: traffic channels and signaling channels. Traffic channel allocation in GPRS is different from that of GSM. In GSM, a traffic channel is permanently allocated for a particular user during the entire call period (whether any data is transmitted or not). In contrast, in GPRS traffic, channels are only allocated when data packets are sent or received. They are released after the transmission of data. GPRS allows a single mobile station to use multiple time slots of the same TDMA frame for data transmission. This is known as multislot operation and uses a very flexible channel allocation. One to eight time slots per TDMA frame can be allocated for one mobile station. Moreover, uplink and downlink are allocated separately, which efficiently supports asymmetric data traffic like Internet where the bandwidth requirements in uplink and downlink are different.

In GPRS, physical channels to transport user data packet is called data traffic channel (PDTCH). The PDTCHs are taken from a common pool of all channels available in a cell. Thus, the radio resources of a cell are shared by all GPRS and non-GPRS mobile stations located within the cell. The mapping of physical channels to either packet switched data (in GPRS mode) or circuit switched data (in GSM mode) services are performed dynamically depending on demand. This is done depending on the current traffic load, the priority of the service and the multislot class. A load supervision procedure monitors the load of the PDTCHs in the cell. According to the demand, the number of channels allocated for GPRS can be changed. Physical channels not currently in use by GSM can be allocated as PDTCHs to increase the bandwidth of GPRS.

7.3.4 Security

GPRS security functionality is similar to the existing GSM security. The SGSN performs authentication and cipher-setting procedures based on the same algorithms, keys and criteria as in GSM. GPRS uses a ciphering algorithm optimized for packet data transmission. Like its predecessor, a GPRS device also uses SIM card.

7.4 GPRS NETWORK OPERATIONS

Data transmission in a GPRS network requires several steps as described below in the context of the protocol layers described in the previous section. Once a GPRS mobile station is powered on, it 'introduces' itself to the network by sending a 'GPRS attach' request. Network access can be achieved from either the network side or the MS side of the GPRS network.

7.4.1 Attachment and Detachment Procedure

In order to access the GPRS services, an MS needs to make its presence known to the network. It must register itself with an SGSN of the network. This is done through a GPRS attach. This operation establishes a logical link between the MS and the SGSN. The network checks if the MS is authorized to use the services; if so, it copies the user profile from the HLR to the SGSN, and assigns a packet temporary mobile subscriber identity (P-TMSI) to the MS. In order to exchange data packets with external PDNs after a successful GPRS attach, a mobile station must apply for an addresse. If the PDN is an IP network, it will request for an IP address; for a X.25 network it will ask for a X.25 DTE (Data Terminal Equipment) address. This address is called PDP (Packet Data Protocol) address. For each session, a PDP context is created. It contains the PDP type (e.g., IPv4), the PDP address assigned to the mobile station (e.g., 129.187.222.10), the requested QoS, and the address of the GGSN that will function as the access point to the PDN. This context is stored in the MS, the SGSN and the GGSN. With an active PDP context, the MS is 'visible' to the external PDN. A user may have several simultaneous PDP contexts active at a given time. User data is transferred transparently between the MS and the external data networks trough GTP encapsulation and tunneling. User data can be compressed and encrypted for efficiency and reliability.

The allocation of the PDP address can be static or dynamic. In case of static address, the network operator permanently assigns a PDP address to the user. In the other case, a PDP address is assigned to the user upon activation of a PDP context. The PDP address can be assigned by the home network (dynamic home–PLMN PDP address) or by the visited network (dynamic visited–PLMN PDP address). In case of dynamic PDP address assignment, the GGSN is responsible for the allocation and the activation/deactivation of

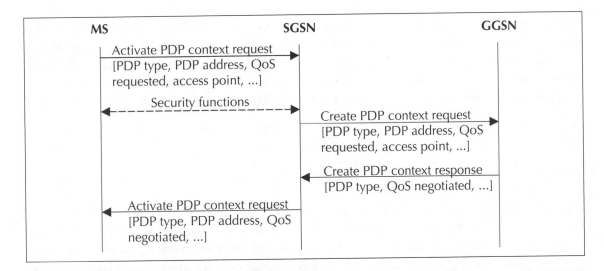

Figure 7.3 PDP Context activation

the PDP addresses. This function is similar to the DHCP (Dynamic Host Configuration Protocol) function.

Figure 7.3 shows the PDP context activation procedure. Using the message 'activate PDP context request,' the MS informs the SGSN about the requested PDP context. If the request is for dynamic PDP address assignment, the parameter PDP address will be left empty. In following steps security functions (e.g., authentication of the user) are performed. If authentication is successful, the SGSN will send a 'create PDP context request' message to the GGSN. The GGSN creates a new entry in its PDP context table, which enables the GGSN to route data packets between the SGSN and the external PDN. The GGSN returns a confirmation message 'create PDP context response' to the SGSN, which contains the PDP address. The SGSN updates its PDP context table and confirms the activation of the new PDP context to the MS ('activate PDP context accept').

The disconnection from the GPRS network is called GPRS detach. All the resources are released following a GPRS detach. Detach process can be initiated by the mobile station or by the network.

7.4.2 Mobility Management

As a mobile station moves from one area to another, mobility management functions are used to track its location within each PLMN. SGSNs communicate with each other to update the MS's location in the relevant registers. The mobile station's profiles are preserved in the VLRs that are accessible to SGSNs via the local MSC. A logical link is established and maintained between the mobile station and the SGSN at each PLMN. At the end of transmission or when a mobile station moves out of the area of a specific SGSN, the logical link is released and the resources associated with it can be reallocated.

7.4.3 Routing

Figure 7.4 depicts an example of how packets are routed in GPRS. The example assumes two intra-PLMN backbone networks of different PLMNs. Intra-PLMN backbone networks connect GSNs of the same PLMN or the same network operator. These are private packet-based networks of the GPRS network provider; for example, Airtel GSNs in Bangalore connecting to Airtel GSNs in Delhi through a private data network. In the diagram, these intra-PLMN networks are connected with an inter-PLMN backbone. An inter-PLMN backbone network connects GSNs of different PLMNs and operators. To install such a backbone, a roaming agreement is necessary between two GPRS network providers. For example, Airtel GSNs in Bangalore connect to Hutch GSNs in Delhi. The gateways between the PLMNs and the external inter-PLMN backbone are called border gateways. Among other things, they perform security functions to protect the private intra-PLMN backbones against unauthorized users and attacks.

We assume that the packet data network is an IP network. A GPRS mobile station located in PLMN1 sends IP packets to a host connected to the IP network, e.g., to a Web server connected to the Internet. The SGSN that the mobile station is registered with encapsulates the IP packets coming from the mobile station, examines the PDP context and routes them through the intra-PLMN GPRS backbone to the appropriate GGSN. The GGSN decapsulates the packets and sends them out on the IP network, where IP routing mechanisms are used to transfer the packets to the access router of the destination network. The latter delivers the IP packets to the host.

Let us assume the home-PLMN of the mobile station is PLMN2. An IP address has been assigned to the mobile by the GGSN of PLMN2. Thus, the MS's IP address has the same network prefix as the IP address of the GGSN in PLMN2. The correspondent host is now sending IP packets to the MS. The packets are sent out onto the IP network and are routed to the GGSN of PLMN2 (the home-GGSN of the MS). The latter queries the HLR and obtains the information that the MS is currently located in PLMN1. It

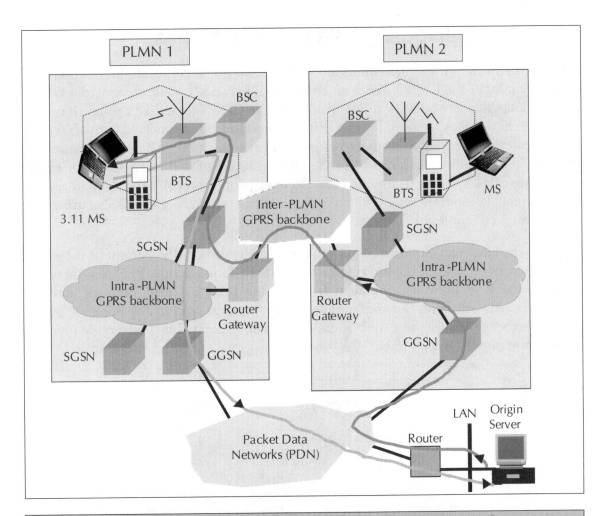

Figure 7.4 GPRS system architecture and routing example

encapsulates the incoming IP packets and tunnels them through the inter-PLMN GPRS backbone to the appropriate SGSN in PLMN1. The SGSN decapsulates the packets and delivers them to the MS.

The HLR stores the user profile, the current SGSN address, and the PDP addresses for every GPRS user in the PLMN. For example, the SGSN informs the HLR about the current location of the MS. When the MS registers with a new SGSN, the HLR will send the user profile to the new SGSN. The signaling path between GGSN and HLR may be used by the GGSN to query a user's location and profile in order to update its location register.

7.4.4 Communicating with the IP Networks

A GPRS network can be interconnected with Internet or a corporate intranet. GPRS supports both IPv4 and IPv6. From an external IP network's point of view, the GPRS network looks like any other IP sub-network, and the GGSN looks like a usual IP router. Figure 7.5 shows an example of how a GPRS network may be connected to the Internet. Each registered user who wants to exchange data packets with the IP network gets an IP address. The IP address is taken from the address space of the GPRS operator maintained by a DHCP server (Dynamic Host Configuration Protocol). The address resolution between IP address and GSM address is performed by the GGSN, using the appropriate PDP context.

Moreover, a domain name server (DNS) managed by the GPRS operator or the external IP network operator is used to resolve host names. To protect the PLMN from unauthorized access, a firewall is installed between the private GPRS network and the external IP network. With this configuration, GPRS can be seen as a wireless extension of the Internet all the way to a mobile station or mobile computer. The mobile user has direct connection to the Internet.

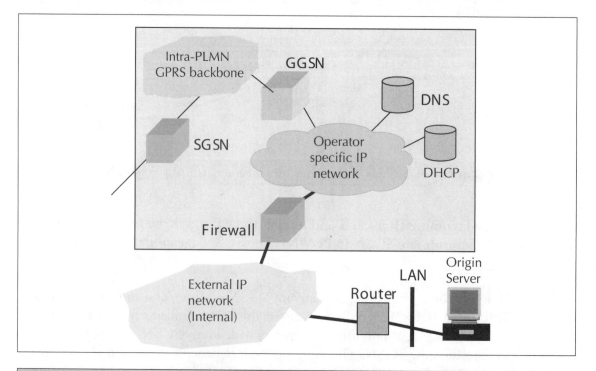

Figure 7.5 Example of a GPRS–Internet connection

7.5 DATA SERVICES IN **GPRS**

A wide range of corporate and consumer applications are enabled by GPRS services. A user is likely to use either of the two modes of the GPRS network. These are **Application mode** or **Tunneling mode**.

Application mode: In this mode the user will be using the GPRS mobile phone to access the applications running on the phone itself. The phone here acts as the end user device. All GPRS phones have WAP browser as an embedded application. This browser allows browsing of WAP sites. Some GPRS devices support mobile execution environment (MExE classmark 3). These devices support development of client application that can run on the device. The device operating execution environments supported are Symbian and J2ME. Applications can be developed in C/C++ or Java.

Tunneling mode: This mode is for mobile computing where the user will use the GPRS interface as an access to the network. The end user device will be a large footprint device like laptop computer or small footprint device like PDAs. The mobile phone will be connected to the device and used as a modem to access the wireless data network. For these devices, access can be gained via a PC Card (PCMCI) or via a serial cable to a GPRS-capable phone. These 'black-box' devices do not have display, keypad and voice accessories of a standard phone.

7.5.1 GPRS Handsets

A GPRS terminal can be one of three classes: A, B or C. A Class A terminal supports GPRS data and other GSM services such as SMS and voice simultaneously. This includes simultaneous attach, activation, monitor, and traffic. As such, a Class A terminal can make or receive calls on two services simultaneously. In the presence of circuit-switched services, GPRS virtual circuits will be held or placed on busy rather than being cleared. SMS is supported in Class A terminal. Like GSM, a SMS can be received while a voice or data call is in progress.

A Class B terminal can monitor GSM and GPRS channels simultaneously, but can support only one of these services at any time. Therefore, a Class B terminal can support simultaneous attach, activation, and monitor but not simultaneous traffic. As with Class A, the GPRS virtual circuits will not be closed down when circuit-switched traffic is present. Instead, they will be switched to busy or held mode. Thus, users can make or receive calls on either a packet or a switched call type sequentially but not simultaneously. SMS is supported in Class A terminal. Like GSM, a SMS can be received while a voice or data call is in progress.

A Class C terminal supports only nonsimultaneous attach. The user must select which service to connect to. Therefore, a Class C terminal can make or receive calls from only the manually selected network service. The service that is not selected is not reachable. The GPRS specifications state that support of SMS is optional for Class C terminals.

7.5.2 Device Types

In addition to the three types of terminals, each handset will have a unique form factor. Terminals will be available in the standard form factor with a numeric keypad and a relatively small display. Other types of phones with different form factors, color displays, with cameras are common. Smart phones with built-in voice, nonvoice and Web-browsing capabilities are common too. Smart phones have various form factors, which may include a keyboard or an icon drive screen.

7.5.3 Bearers in GPRS

The bearer services of GPRS offer end-to-end packet switched data transfer. GPRS is planned to support two different kinds of data transport services. These are the point-to-point (PTP) service and the point-to-multipoint (PTM) service. Out of these two, PTP is available now; PTM will be available in future releases of GPRS. The PTP service offers transfer of data packets between two users.

GPRS will support the following types of data services:

SMS: Short message service was originally designed for GSM network. GPRS will continue to support SMS as a bearer. Please refer to chapter 6 for details on SMS and application development using SMS.

WAP: WAP is Wireless Application Protocol. It is a data bearer service over HTTP protocol. WAP uses WML (Wireless Markup Language) and a WAP gateway. Please refer to Chapter 8 for details of WAP and application development using WAP.

MMS: MMS is Multimedia Messaging Service. This is the next generation messaging service. SMS supports text messages whereas MMS supports multimedia messages. MMS uses WAP and SMS as its lower layer transport. Video, audio pictures or clips can be sent through MMS. Please refer to chapter 8 for details of WAP and application development using WAP.

7.6 APPLICATIONS FOR **GPRS**

In this section we describe some applications which nedd higher data bandwidth and suitable for GPRS.

7.6.1 Generic Applications

There are many applications suitable for GPRS. Many of them are of generic type, some of them are specific to GPRS. Generic applications are applications like information services, Internet access, email, Web Browsing, which are very useful while mobile. These are generic mass market applications offering contents like sports scores, weather, flight information, news headlines, prayer reminders, lottery results, jokes, horoscopes, traffic information and so on. Using Circuit Switched Data (CSD as in GSM), user experience for using these applications have never been enduring. Due to higher bandwidth, mobile Internet browsing will be better suited to GPRS. Access to corporate Intranet can add new dimension to mobile workers. Mobile commerce is another generic applications people may like to use while mobile. Banking over wireless is another generic application. Some Indian banks are offering banking over GPRS/WAP.

7.6.2 GPRS-Specific Applications

Chat: Chat is a very popular service in Internet and GSM (over SMS). Groups of like-minded people use chat services as a means to communicate and discuss matters of common interest. Generally people use different chat services; one, through Internet (offered by Yahoo, ICQ etc.) and the other, using SMS (offered by mobile operator). GPRS will offer ubiquitous chat by integrating Internet chat and wireless chat using SMS and WAP.

Multimedia Service: Multimedia objects like photographs, pictures, postcards, greeting cards and presentations, static web pages can be sent and received over the mobile network. There are many phones available in the marketplace where a digital camera is integrated with the phone. These pictures can be sent as an electronic object or a printed one. Sending moving images in a mobile environment has several vertical market applications including monitoring parking lots or building sites for intruders or thieves. This can also be used by law enforcement agents, journalists, and insurance agents for sending images of accident site. Doctors can use these applications to send pictures of patients from a health center for expert help.

Virtual Private Network: GPRS network can be used to offer VPN services. Many Bank ATM machines use VSAT (Very Small Aperture Terminal) to connect the ATM system with the banks server. As the bandwidth in GPRS is higher, many banks in India are migrating from VSAT to GPRS-based networks. This is expected to reduce the transaction time by about 25%.

Personal Information Management: Personal diary, address book, appointments, engagements etc. are very useful for a mobile individual. Some of these are kept in the phone some in the organizer and some in the Intranet. Using J2ME and WTAI (Wireless Telephony Application Interface) the address book, the diary of the phone can be integrated with the diary at the home office. GPRS and other bearer technology will help achieve this.

Job Sheet Dispatch: GPRS can be used to assign and communicate job sheets from office-based staff to mobile field staff. Customers typically telephone a call center whose staff takes the call and categorize it. Those calls requiring a visit by field sales or service representative can then be escalated to those mobile workers. Job dispatch applications can optionally be combined with vehicle positioning applications so that the nearest available suitable personnel can be deployed to serve a customer.

Unified Messaging: Unified messaging uses a single mailbox for all messages, including voice mail, fax, e-mail, SMS, MMS, and pager messages. With the various mailboxes in one place, unified messaging systems then allow for a variety of access methods to recover messages of different types. Some will use text-to-voice systems to read e-mail and, less commonly, faxes over a normal phone line, while most will allow the interrogation of the contents of the various mailboxes through data access, such as the Internet. Others may be configured to alert the user on the terminal type of their choice when messages are received.

Vehicle Positioning: This application integrates GPS (Global Positioning System) that tell people where they are. GPS is a free-to-use global network of 24 satellites run by the US Department of Defense. Anyone with a GPS receiver can receive their satellite position and thereby find out where they are. Vehicle-positioning applications can be used to deliver several services including remote vehicle diagnostics, ad hoc stolen vehicle tracking and new rental car fleet tariffs. In India this application is becoming popular in logistics industry.

Location-based Services and Telematics: Location-based services provide the ability to link push or pull information services with a user's location. Examples include hotel and restaurant finders, roadside assistance, and city-specific news and information.

All systems developed for Intelligent Transportation System (ITS) are built around GPRS and GPS technology. Location can be determined either through GPS or cell id from the operator. This technology also has vertical applications such as workforce management and vehicle tracking.

7.7 LIMITATIONS OF GPRS

There are some limitations with GPRS, which can be summarized as:

Limited Cell Capacity for All Users: There are only limited radio resources that can be deployed for different uses. Both Voice and GPRS calls use the same network resources. Use for one data precludes simultaneous use for voice. If the tariffing and billing is not done properly, this may have impact on revenue.

Speed Lower in Reality: Achieving the theoretical maximum GPRS data transmission speed of 172.2 kbps would require a single user taking over all eight time slots without any error protection. It is unlikely that a network operator will allow all time slots to be used by a single GPRS user. Additionally, the initial GPRS terminals are expected to be supporting only one, two or maximum three time slots.

Support of GPRS Mobile Terminate Connection for a Mobile Server not Supported: As of date a GPRS terminal can only act as a client device. There are many services for which the server needs to be mobile. An example could be a mobile healthcare center for rural population. For such applications the server needs to be on the mobile network and user needs to connect to the server. Using GPRS network, such communication is not possible.

7.8 BILLING AND CHARGING IN GPRS

There is a saying in the wireless business community, 'Data sells, voice pays.' Tariffing of data in wireless network has always been a challenge. For voice networks tariffs are generally based on distance and time. This in other words means that user pays more for long distance calls. They also pay more if they keep the circuit busy by talking for a longer period of time. In a voice system, charging is the fundamental part of the architecture. On the other hand, data services have evolved from research and education without any concept of charging. In packet network keeping the circuit busy does not have any meaning. Also, charging a customer by the distance traversed by a packet does not make any sense. Many times due to congestions packets traverse much longer distance than the optimum distance.

7.8.1 Tariffing

The main challenge for a network operator is to integrate these two models and generate revenue. Decisions on charging for GPRS by packet or simply a flat monthly fee are contentious but need to be made. Charging different packets at different rates can make things complicated for the user, whilst flat rates favor heavy users more than occasional ones. It is believed that the optimal GPRS pricing model will be based on two variables, time and packet. Network operators will levy a nominal per packet charge during peak times plus a flat rate. There will be no per packet charge during non-peak times. Time and packet-related charging will encourage applications such as remote monitoring, meter reading and chat to use GPRS at night when spare network capacity is available. Simultaneously, a nominal per packet charge during the day will help to allocate scarce radio resources, and charge radio heavy applications such as file and image transfer more than applications with lower data intensity. It has the advantage of automatically adjusting customer charging according to their application usage.

7.8.2 Billing

GPRS is essentially a packet switching overlay on a circuit switching network. The GPRS specifications stipulate that the minimum charging information that must be collected are:

- Destination and source addresses
- Usage of radio interface
- Usage of external Packet Data Networks
- Usage of the packet data protocol addresses
- Usage of general GPRS resources and location of the Mobile Station

Since GPRS networks break the information to be communicated down into packets, at a minimum, a GPRS network needs to be able to count packets to charging customers for the volume of packets they send and receive. Today's billing systems have difficulties handling charging for today's data services. It is unlikely that circuit switched billing systems will be able to process a large number of new variables created by GPRS.

GPRS call records are generated in the GPRS Service Nodes. The incumbent billing systems are often not able to handle real time Call Detail Record flows. As such, an intermediary charging platform is a good idea to perform billing mediation by collecting the charging information from the GPRS nodes and preparing it for submission to the billing system. Packet counts are passed to a Charging Gateway that generates Call Detail Records that are sent to the billing system.

The billing of the services can be based on the transmitted data volume, the type of service, and the chosen QoS profile. It may well be the case that the cost of measuring packets is greater than their value. The implication is that there will not be a per packet charge since there may be too many packets to warrant counting and charging for. For example, a single traffic monitoring application can generate tens of thousands of packets per day. Thus the charging gateway function is more a policing function than a charging function since network operators are likely to tariff certain amounts of GPRS traffic at a flat rate and then need to monitor whether these allocations are far exceeded. The billing of roaming GPRS subscribers from one network to another is still a challenge.

REFERENCES/FURTHER READING

1. Introduction to GPRS: http://www.gsmworld.com/technology/gprs/intro.shtml.

2. Christoffer Andersson, GPRS and 3G Wireless Applications, John Wiley & Sons, 2001.

3. Kurian John, Asoke K Talukder, Ubiquity and Virtual Home Environment for Legacy Applications and Services, Proceedings of the 15th International Conference in Computer Communication, pp 371–381, 12-14 August 2002, Mumbai, India.

4. Christian Bettstetter, Hans-Jorg Vogel, Jorg Eberspacher, GSM Phase 2+ General Packet Radio Servie GPRS: Architecture, Protocols, and Air Interface, IEEE Communications Surveys, vol. 2 no. 3, Third Quarter 1999.

5. GSM Standard 03.01: Digital cellular telecommunications system (Phase 2+); Network functions, Release 1998, http://www.etsi.org.

6. GSM Standard 03.02: Digital cellular telecommunications system (Phase 2+); Network architecture, Release 1998, http://www.etsi.org.

7. GSM Standard 03.03: Digital cellular telecommunications system (Phase 2+); Numbering, addressing and identification, Release 1998, http://www.etsi.org.

8. GSM Standard 03.09: Digital cellular telecommunications system (Phase 2+); Handover procedures, Release 1998, http://www.etsi.org.

9. GSM Standard 03.20: Digital cellular telecommunications system (Phase 2+); Security related network functions, Release 1998, http://www.etsi.org.

10. GSM Standard 03.60: Digital cellular telecommunications system (Phase 2+); General Packet Radio Service (GPRS); Service description; Stage 2, Release 1998), http://www.etsi.org.

REVIEW QUESTIONS

Q1: What is the difference between GSM and GPRS? What are the network elements in GPRS that are different form GSM?

Q2: How is data handled in GPRS?

Q3: How is data routing done in GPRS? IN what respect is data routing different from voice routing?

Q4: Describe four applications suitable for GPRS?

Q5: Describe what are the limitations of GPRS?

CHAPTER 8

Wireless Application Protocol (WAP)

8.1 INTRODUCTION

We are moving towards a net-centric world, where Internet is becoming part of our environment. Along with the physical environment, we also acquire information and knowledge from the Internet. The appetite for data and information over communication networks are growing day by day. This appetite has made people look at cellular networks for data access as well. In 2G cellular networks, data access was possible over mobile phones using SMS (Short Message Service) (Chapter 6) and WAP (Wireless Application Protocol) over circuit (circuit switched data–CSD). In a circuit, the user pays for the circuit even during the idle period when there is no data transmission. Also, the data speed supported by CSD is in the range of 9.6K bits per second. For Internet access 9.6KBps speed is unlikely to offer a good user experience. GPRS (Chapter 7) is designed to overcome some of these constraints of GSM and offer a higher data rate.

In chapter 7 we have discussed how General Packet Radio Service (GPRS) is designed to efficiently transport high-speed data over the current GSM and TDMA-based cellular networks. GPRS is a packet network with higher bandwidth compared to CSD in GSM. Deployments of GPRS networks have already begun in several countries including India. GPRS allows for short 'bursty' traffic, such as e-mail and web browsing, including large volumes of data. GPRS has the ability to offer data speeds from 14.4KBps to 171.2KBps. Internet traffic is asymmetric in the sense that the traffic volume from network to the user agent (client device) is significantly higher compared to the traffic in reverse direction. GPRS manages asymmetric traffic quite well by dynamically configuring bandwidth. Being a packet network, GPRS may well be a relatively less expensive mobile data service compared to SMS or CSD.

Wireless application protocol (WAP) is designed for access to Internet and advanced telephony services from mobile phones. WAP pays proper sensitivity to the constraints of these devices like small display, limited keys on the keypad, no pointer device like mouse etc. Independent of their network, bearer and terminals, a user will be able to access Internet and corporate intranet services while mobile. Net–net using WAP, a mobile user will be able to access the same wealth of information from a pocket-sized device as they can from a desktop. Though WAP can be used from a variety of networks, GPRS and 3G networks are more suited for these applications. In this chapter we will discuss application development using the WAP and MMS (Multimedia Messaging Service).

8.1.1 Evolution of Wireless Data and WAP

In 1992, Nippon Telegraph and Telephone Corporation (NTT), a leading telephone company in Japan spun off a wireless division and named it DoCoMo. DoCoMo was derived combining two syllables Do and como. Do in Japanese mean 'everywhere' and como was for Communications. DoCoMo's success has been due to i-mode, its mobile Internet service. DoCoMo developed a language called cHTML (Compant Hyper Text Markup Language) and a gateway. Using these frameworks an i-mode user can use the Internet.

In 1994, in USA a company named Unwired Planet was founded to develop and market a platform for Internet access through wireless devices like PDA. Unwired Planet developed a comprehensive framework including browser, gateway and markup language. The markup language was called HDML (Handheld Device Markup Language). Unwired Planet also developed HDTP (Handheld Device Transport Protocol). Unwired Planet launched all these technologies in 1995.

In 1995, Ericsson another leading wireless company in Europe, began work on a protocol known as ITTP or Intelligent Terminal Transfer Protocol. ITTP was designed with the intent of making it easy for call control and add services to mobile telephony platforms.

In 1997, Nokia, the other major wireless company in Europe developed the TTML (Tagged Text Mark-up Language). TTML was designed to allow a mobile phone to communicate with a World Wide Web site via gateway. Following TTML, Nokia also introduced Narrowband Sockets (NBS) to create wireless messaging applications for PCs communicating with GSM 'smart phones'.

Out of all these different technologies, i-mode in Japan was the most successful service. It started growing like a bushfire. In USA and Europe, though there were different services and protocols offered by different companies, none of them could match the popularity of DoCoMo. Realizing that these competing wireless protocols could fragment and possibly destroy the potential market, in June of 1997, Ericsson, Motorola, Nokia, and Unwired Planet (now known as Phone.com) joined hands to launch the WAP Forum (www.wapforum.com). WAP Forum is now known as Open Mobile Alliance. The goal of this effort was to produce a refined, license-free protocol, which is independent of the underlying airlink standard. The WAP inherited its main characteristics and functionality from HDML and HDTP developed by Unwired Planet, the Smart Messaging specification based on TTML and NBS developed by Nokia, and the ITTP specification developed by Ericsson. The first release of the WAP 1.0 specifications was released in the spring of 1998.

8.1.2 Networks for WAP

As part of the WAP Forum's goals, WAP will be accessible from (but not limited to) the following networks:

- GSM-900, GSM-1800, GSM-1900
- GPRS
- CDMA IS-95, cdma2000
- TDMA IS-136
- i-mode
- 3G systems–IMT-2000, UMTS, W-CDMA, Wideband IS-95.

WAP can be used through 2G, 2.5G, and 3G networks. There is a perception that WAP or MMS requires GPRS networks. This is not correct. WAP and MMS can be accessed technically from a 2G network using CSD. However, high-speed data networks like GPRS are more suitable for WAP and MMS applications.

8.2 WAP

WAP forum develops standards for application deployment over wireless devices like PDAs and mobile phones. WAP is based on layered architecture. The WAP Protocol Stack is similar to the OSI network model (Figure 8.1). These layers consist (from top to bottom) of:

- Wireless Application Environment (WAE)
- Wireless Session Protocol (WSP)

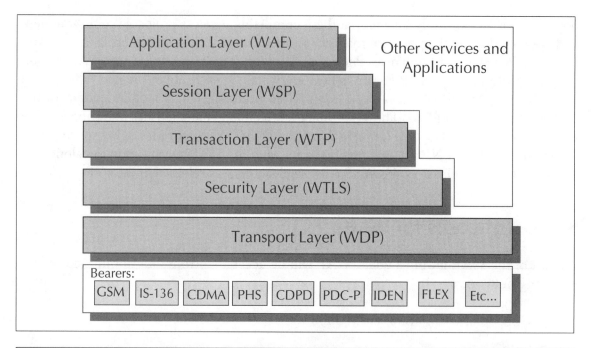

Figure 8.1 WAP layered Architecture and protocol stack

- Wireless Transaction Protocol (WTP)
- Wireless Transport Layer Security (WTLS)
- Wireless Datagram Protocol (WDP).

The application environment of WAE comprises multiple components to provide facilities like:

- User agent: the browser or a client program.

- Wireless Markup Language (WML): a lightweight markup language, similar to HTML, but optimized for use in wireless devices.

- WMLScript: a lightweight client side scripting language, similar to JavaScript in Web.

- Wireless Telephony Application: telephony services and programming interfaces.

- WAP Push Architecture: mechanisms to allow origin servers to deliver content to the terminal without the terminal requesting for it.

- Content Formats: a set of well-defined data formats, including images, phone book records and calendar information.

WAP supports different types of bearer networks. These are GSM, IS-136, CDMA, PHS (Personal Handyphone System), CDPD (Cellular Digital Packet Data), etc.

8.2.1 WAP Application Environment (WAE)

The primary objective of the WAP application environment (WAE) is to provide an interoperable environment to build services in wireless space. It covers system architecture relating to the user agents, networking schemes, content formats, programming languages and shared services based on World Wide Web (WWW) technologies. Content is transported using standard protocols in the WWW domain and an optimized HTTP-like protocol in the wireless domain. WAE architecture allows all content and services to be hosted on standard Web servers. All contents are located using WWW standard URLs. WAE enhances some of the WWW standards to reflect some of the telephony network characteristics.

A WAP request from the browser (user agent) is routed through a WAP gateway (Figure 8.2). The gateway acts as an intermediary between the client and network through a wireless last mile (GSM, GPRS, CDMA etc.). The gateway does encoding and decoding of data transferred from and to the mobile user agent. The purpose of encoding is to minimize the size of data transacted over-the-air. Reduced data size reduces the computational

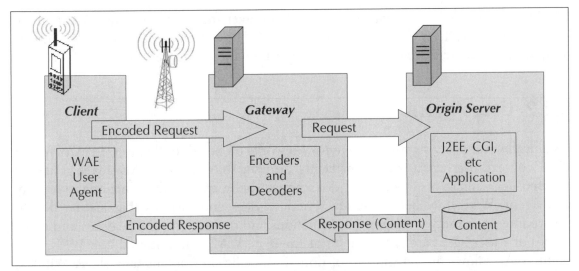

Figure 8.2 WAE Logical Model

power required by the client to process that data. In most cases the WAP gateway resides on TCP/IP network. The gateway processes the request, retrieves contents from the server using Java servlets, J2EE, CGI scripts, or some other dynamic mechanism. The data is formatted as WML (Wireless Markup Language) and returned to the client. The client device can employ logic via embedded WMLScript for client-side processing of WML.

The major elements of the WAE model include:

- *WAE User Agents:* User facing client software (browser). User agents are integrated into the WAP architecture. They interpret network content referenced by a URL. WAE includes user agents for two primary standard contents: encoded WML and compiled WMLScript.

- *Content Generators:* Applications on origin servers that extract standard content in response to requests from user agents. Content servers are typically HTTP servers as used in WWW.

- *Standard Content Encoding:* A set of well-defined content encoding, allowing a WAE user agent to navigate web content. Standard content encoding includes compressed encoding for WML, bytecode encoding for WMLScript, standard image formats, business, and calendar data formats.

- *Wireless Telephony Applications (WTA):* A collection of telephony specific extensions for call and telephony feature control.

WAE defines a set of user agent capabilities that is exchanged between the client and the server using WSP. These capabilities include global device characteristics as WML version, WMLScript version, floating-point support, image formats and so on.

8.2.2 User Agent

Technically user agent signifies an agent who works on behalf of the user. In WWW and WAE context, user agent is the user facing browser software. In WAE this is generally referred to as micro-browser. WAE does not formally specify the functionality of any user agent. WAE only defines fundamental services and formats that are needed to ensure interoperability among implementations and different layers. Features and capabilities of a user agent are left to the implementers. WAE is not limited to a WML user agent. WAE allows the integration of domain-specific user agents as well. A Wireless Telephony Application (WTA) user agent has been specified as an extension to the WAE specification for the mobile telephony environments. This covers features like call control as well as other applications in the telephones, such as phonebooks and calendar applications. Following sections provide details on WML, and WMLScript.

8.2.3 User Agent Profile (UAProf)

The User Agent Profile (UAProf) specification allows WAP to notify the content server about the device capability. UAProf is also referred to as ***Capability and Preference Information*** (CPI). CPI is passed from the WAP client to the origin server through intermediate network points. It is compatible with Composite Capability/Preference Profile (CC/PP) of the W3C (WWW Consortium). This CPI may include, hardware characteristics (screen size, color capabilities, image capabilities, manufacturer, etc.), software characteristics (operating system vendor and version, support for MExE, list of audio and video encoders, etc.), application/user preferences (browser manufacturer and version, markup languages and versions supported, scripting languages supported, etc.), WAP characteristics (WML script libraries, WAP version, WML deck size, etc.). In a WSP response, it transmits information about the client, user, and network that will be processing the content.

Devices that support UAProf architecture provide a URL in the WAP or HTTP session header. This URL points to a XML file that describes the profile of that device. Many vendors have their own public HTTP-servers where service providers can download device profiles as standardized XML documents. In case of MMS (Multimedia Message Service), the MMSC (MMS Controller) is able to pick the profile address from the protocol header and fetch the respective device profile. Device profile information is used by the MMSC to format the content to best suit the terminal's capabilities. Content can be adapted based on the device capabilities. For example, the scaling of a bitmap and adjusting its color map may be required to fit the display size or reduce the size of an image or a music file.

8.2.4 Wireless Markup Language (WML)

WML is a tag-based document manipulation language. It shares a heritage with HTML of W3C and HDML of Unwired Planet. WML is designed to specify presentation and user interaction on mobile phones and other wireless devices. These devices suffer from different constraints like small displays, limited user-input facilities, narrow band network connections, limited memory resources, limited computational resources, and absence of pointer devices like mouse.

WML implements a **deck** and **card** metaphor. A *deck* is a logical representation of a document. Decks are made up of multiple *cards*. Each WML card, in a deck, performs a specific task for a particular user interaction. To access a document, a user navigates to a card; reviews its contents, makes a choice or enters requested information, and then moves to another card. WML decks can be stored in 'static' files and fetched by CGI, JSP or ASP scripts. It can also be dynamically generated by a Java servlets running on an origin server.

WML has a wide variety of features, including:

- *Support for Text and Images:* WML provides the facility to render text and images to the user. WML provides a limited set of text mark-up elements. These include:

  - E*mphasis* elements like bold, italic, big, etc.

  - L*ine breaks* models like line wrapping, line wrapping suppression, etc.

  - T*ab columns* that support simple tabbing alignment.

- *Support for User Input:* WML supports several elements to solicit user input. WML includes a *text entry* control that supports text and password entry. Text entry fields can also be masked to prevent the user from entering unwanted character types. WML includes an *option selection* control that allows the author to present the user with a list of options that can set data, navigate among cards, or invoke scripts. WML supports client-side validation by allowing the author to invoke scripts to check the user's input.

- *Task invocation Controls:* These controls initiate a navigation or a history management task such as traversing a link to another card (or script) or popping the current card off of the history stack. WML also allows several navigation mechanisms using URLs. Navigation includes HTML-style hyperlinks, inter-card navigation elements, as well as history navigation elements.

- *International Support:* The character set for WML document is the Universal Character Set. Currently, this character set is identical to Unicode 2.0.

- *MMI Independence:* WML's specification of layout and presentation enables terminal and device vendors to control the MMI design for their particular products.

- *Narrow-band Optimization:* WML specification includes different technologies to optimize traffic on a narrow-band device. This includes the ability to specify multiple user interactions (cards) in one document transfer (a deck).

- *State and Context Management:* Each WML input control can introduce variables. The lifetime of a variable state can extend beyond a single deck. The state can be shared across multiple decks without having to use a server to save intermediate state between deck invocations.

WML is mostly about rendering text. Many tags that are available in HTML and demand a high resource are generally not supported in WML. The use of tables and images are restricted. Since WML is an XML application, all tags are case sensitive (<wml> is not same as <WML>). In WML all tags must be properly closed. Cards within a deck can be related to each other with links. A card element can contain text, input-fields, links, images etc. When a WML page is accessed from a mobile phone, all the cards in the page are downloaded from the WAP server. Navigation between the cards is done inside the phone without any extra access trips to the server. Following is an example of Hello World in WML.

```
<?xml version="1.0"?>
<!DOCTYPE wml PUBLIC "-//WAPFORUM//DTD WML 1.1//EN"
"http://www.wapforum.org/DTD/wml_1.1.xml">

<wml>
        <card id="no1" title="Card 1">
                <p>Hello World!</p>
        </card>

        <card id="no2" title="Card 2">
                <p>Welcome to WML!</p>
        </card>
</wml>
```

In the above example, the WML document is an XML document. The DOCTYPE is defined to be wml, and the DTD is accessed at www.wapforum.org/DTD/wml_1.1.xml.

The document content is inside the <wml>...</wml> tags. Each card in the document is inside <card>...</card> tags, and actual paragraphs are inside <p>...</p> tags. Each card element has an id and a title.

When the above WML deck is executed, the result will look like this (Figure 8.3):

Figure 8.3 Output from the example Hello world application

8.2.5 WMLScript

WMLScript is an extended subset of JavaScript and forms a standard means for adding procedural logic to WML decks. WMLScript is used to do client side processing. Therefore, it can be used very effectively to add intelligence to the client and enhance the user interface. Using WMLScript, it is possible to access the device resources. WMLScript provides the application programmer with a variety of interesting capabilities. These are as follows:

- The ability to do local validation of user input before it is sent to the content server.

- The ability to access device resources, functions, and peripherals.

- The ability to interact with the user without reference to the origin server.

Key WMLScript features include:

- *JavaScript-based scripting language:* WMLScript is based on industry standard JavaScript solution and adapts it to the narrow-band environment.

- *Procedural Logic:* WMLScript adds the power of procedural logic to WML.

- *Compiled implementation:* WMLScript can be compiled down to a more space efficient bytecode that is transported to the client.

- *Event-based:* WMLScript may be invoked in response to certain user or environmental events.

- *Integrated into WAE:* WMLScript is fully integrated with the WML browser. WMLScript has access to the WML state model and can set and get WML variables.

- *International Support:* WMLScript supports Unicode 2.0.

- *Efficient extensible library support:* WMLScript can be used to expose and extend device functionality without changes to the device software.

- *Data types:* Following basic data types are supported in WMLScript: *boolea*n, *integ*er, *floating-point, string and invali*d. WMLScript attempts to automatically convert between the different types as needed.

8.2.6 Wireless Telephony Application (WTA, WTAI)

WAP offers WTAI (Wireless Telephony Application Interface) functions to create Telephony Applications. This is achieved through a wireless telephony application (WTA) user-agent using the appropriate WTAI function. For example let us say that we want to book a table for a lunch meeting in a restaurant. From the WAP application, we go to the restaurant site and get the telephone number. In normal case we note down the telephone number on a piece of paper, exit from the browser session, and then make a voice call to book the table. In case of WTAI that is not required. We can display an action item **call** in the WAP screen and make a call straight from the WAP page. The WTAI function libraries are accessed from server side using URL's; or at the client side through WMLScript.

There are different library functions to do different telephony functions:

- Voice Call Control: This library handles call set-up and control of device during an ongoing Call. The call may be either outgoing or incoming.

- Network Text: Using this library, SMS text messages can be integrated with the WML, WMLScript functions.

- Phonebook: Using this library, the phonebook entries in the device can be manipulated.

- Call Logs: Using this library, call logs in the device can be accessed.

8.2.7 WAP Push Architecture

The WAP Push framework allows information to be sent to a client device without a previous user action. In a normal client/server model, a client requests for a service or information from a server. The server then responds to this request by transmitting information back to the client. This is referred to as **pull** technology (Figure 8.4), where the client pulls information from the server. In addition to this type of synchronized request response transaction, WAP offers **push** technology (Figure 8.5). Push is also based on the client/server model, but there is no explicit request from the client before the server transmits its content. This can be termed as unsolicited response. In other words, 'pull' transactions are always initiated from the client, whereas, 'push' transactions are server-initiated. Push technology is helpful to implement alerts and notification.

8.2.8 The Push Framework

The push content generally is originated in a server in the Internet that needs to be delivered to a mobile phone. The Push Initiator contacts the Push Proxy Gateway (PPG) from the Internet side, delivering content for the destination client (Figure 8.6). The PPG then

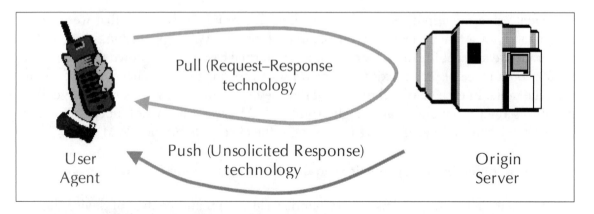

Figure 8.4 Pull versus Push technology

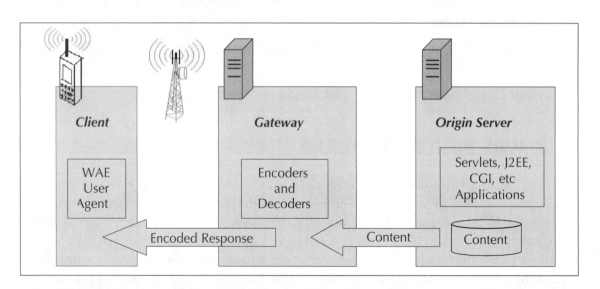

Figure 8.5 WAE Push-based Model

forwards the content to the mobile network to be delivered to the destination client over-the-air. In addition to providing simple proxy gateway services, the PPG is capable of notifying the Push Initiator about the final outcome of the push operation. It may even wait for the client to accept or reject the content in two-way mobile networks (MMS uses this function).

The Internet-side PPG access protocol is called the *Push Access Protocol*. The WAP-side (OTA) protocol is called the *Push Over-The-Air Protocol*.

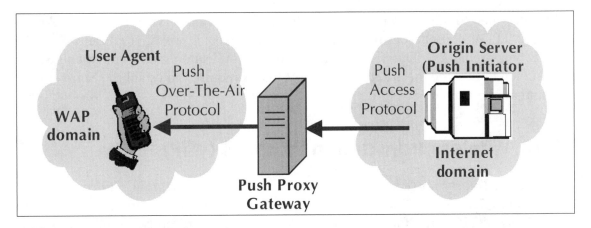

Figure 8.6 Push framework with PPG (Push Proxy gateway)

8.2.9 Wireless Session Protocol (WSP)

The Wireless Session Protocol (WSP) provides a consistent interface between two session services (client and server). It provides the cooperating client/server applications to:

(a) Establish a reliable session from client to server and close it in an orderly manner.

(b) Agree on a common level of protocol functionality using capability negotiation.

(c) Exchange content between client and server using compact encoding.

(d) Suspend and resume the session.

Currently the scope of WSP is suited mostly for browsing applications. It offers both connection-oriented and connectionless service. The connectionless service is most suitable, when applications do not need reliable delivery of data and do not care about confirmation. The connection-oriented session services are divided into following categories:

- Session Management facility
- Method Invocation facility
- Exception Reporting facility
- Push facility
- Confirmed Push facility
- Session Resume facility.

WSP is designed to function on the transaction and datagram services between WAE and the WTP. WSP itself does not require a security layer; however, applications that use WSP may require it. The transaction, session or application management entities are assumed to provide the additional support that is required to establish security contexts and secure connections.

8.2.10 Wireless Transaction Protocol (WTP)

The Wireless Transaction Protocol (WTP) runs on top of a datagram service and provides a lightweight transaction-oriented protocol that is suitable for implementation in 'thin' clients. WTP allows for interactive browsing (request/response) applications and supports three transaction classes: unreliable with no result message, reliable with no result message, and reliable with one reliable result message. WTP provides the following features:

- Three classes of transaction service:
    - o Unreliable one-way requests
    - o Reliable one-way requests
    - o Reliable two-way request-reply transactions
- Optional user-to-user reliability: WTP user triggers the confirmation of each received message
- Optional out-of-band data on acknowledgements
- PDU concatenation and delayed acknowledgement to reduce the number of messages sent
- Asynchronous transactions.

8.2.11 Wireless Transport Layer Security (WTLS)

WTLS is a security protocol based upon the Transport Layer Security (TLS) protocol. WTLS and TLS are derived from the Secure Sockets Layer (SSL) protocol. WTLS is intended for use with the WAP transport protocols and has been optimized for use over narrow-band communication channels. WTLS provides the following features:

- Data integrity: WTLS contains facilities to ensure that data sent between the terminal and an application server is unchanged and uncorrupted.

- Privacy: WTLS contains facilities to ensure that data transmitted between the terminal and an application server is private and cannot be seen by any intermediate parties that may have intercepted the data stream.

- Authentication: WTLS contains facilities to establish the authenticity of the terminal and application server.

- Denial-of-service protection: WTLS contains facilities for detecting and rejecting data that is replayed or not successfully verified. WTLS makes many typical denial-of-service attacks harder to accomplish and protects the upper protocol layers.

8.2.12 Wireless Data Protocol (WDP)

The Transport layer protocol in the WAP architecture is referred to as the Wireless Datagram Protocol. The WDP layer operates above the data capable bearer services supported by the various network type general transport service. WDP offers a consistent service to the upper layer protocols of WAP and communicates transparently over one of the available bearer services. While WDP uses IP as the routing protocol, unlike the Web, WAP does not use TCP. Instead, it uses UDP (User Datagram Protocol), which does not require messages to be split into multiple packets, and sent out only to be reassembled on the client. Due to the nature of wireless communications, the mobile application must be talking directly to a WAP gateway, which greatly reduces the overhead required by TCP.

Since the WDP protocols provide a common interface to the upper layer protocols the Security, Session, Application layers are able to function independently of the underlying wireless network. This is accomplished adapting the transport layer to specific features of the underlying bearer. By keeping the transport layer and the basic features consistent, global interoperability can be achieved using mediating gateways.

8.2.13 WAP Gateway

WAP gateway acts as a middleware which performs coding and encoding between cellular device and the web server. The WAP gateway can be located either in a telecom network or within a computer data network (an ISP). A user from a WAP device requests for a WAP page using a URL, the gateway establishes a connection to the target WAP site. It collects the document from the site. Then the WAP page is 'compiled' and converted into binary code. Binary code takes far less space compared to the WML source. This realizes quicker delivery. The code is then sent across to the phone or the wireless device over the air. When the phone receives the stream of octets, it 'de-compiles' it.

The client browser does the reverse operation of compilation by decompiling the binary code. This will allow the client to regenerate the normal WML page and then displays it on the device. We talked about Kannel SMS gateway in chapter 6. Kannel also has a free opensource WAP gateway. The WAP phones all have a maximum allowed size for a compiled WAP page. Basic functions of a WAP gateway (Figure 8.7) are:

- Implementing WAP protocol stack
- Protocol translation between phone and server
- Compress WML pages to save bandwidth
- User authentication and billing.

The WAP protocols are designed to operate over a variety of different bearer services, including short message, circuit-switched data, and packet data. The bearers offer different levels of quality of service with respect to throughput, error rate, and delays. The WAP protocols are designed in such a fashion that it can compensate for or tolerate this varying level of QoS (quality of service). External Interfaces of a gateway are:

- SMS center, using various protocols
- HTTP servers, to fetch WML pages
- WAP devices using WAP protocol stack.

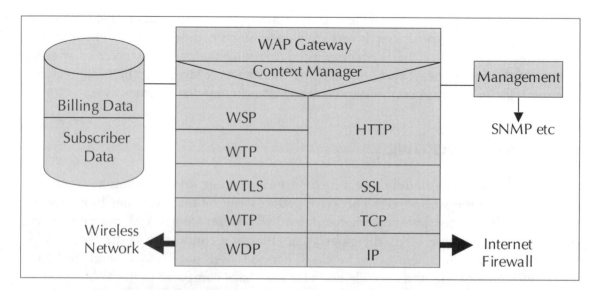

Figure 8.7 Architecture of a WAP gateway

Many WAP gateways include additional functions. These relate to user authentication and charging. For charging, it captures the usage data. The gateway does not actually include a billing system itself but it provides the user and the service provider the usage data. The usage data is given to the billing system of the operator. From the user's point of view, the gateway is also responsible for optimizing WAP usage as far as possible. The gateway keeps the number of packets small to keep costs down and make the best use of available bandwidth.

Generally Internet is not self-configurable. The same is true with WAP. This means that when we move from one network to another network, there may be a need to configure the client device to suit the network parameters of the serving network. This also may depend on the type of the network. The configuration of WAP will require an IP address of the WAP gateway. Though the WAP gateway can be from home network, due to security and charging reasons, service providers do not allow the usage of external WAP gateways. Therefore, at a minimum, two parameters need to be changed. These are the telephone number for the WAP dial-up connection and the IP address of the WAP gateway. The rest can be configured once only. There could also be some dependence of the settings on other security parameters.

8.3 MMS

In the GSM world, the popularity of SMS (Short Messaging Service) surprised everybody. SMS was launched in 1992 and has become the most successful wireless data service to date. SMS was originally designed to carry text message. Vendors started thinking adding more life to the text message. Result was enhanced SMS (EMS). EMS offered a combination of text and simple pixel-image (pictures) and melody (ringing tone). Technology for SMS and EMS is already discussed in Chapter 6. We can refer to SMS as the first generation of messaging, whereas EMS can be considered as the second generation messaging. SMS was person-to-person, whereas EMS was content-to-person. The popularity of the first and the second generation messaging made device vendors think of next generation messaging. In the third generation of messaging, the message content will be multimedia objects. This is called Multimedia Messaging Service or MMS in short. An MMS message can contain formatted text, graphics, data, animations, images, audio clips, voice transmissions and video sequences.

Though MMS is targeted for the 3G networks, it can work under a 2G or 2.5G network as well. All it needs is a MMS handset and the MMS infrastructure. There are two standards bodies producing specifications relating to MMS messages. These are WAP Forum

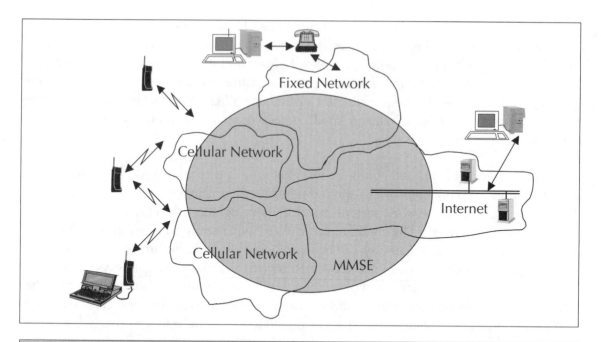

Figure 8.8 General view of MMS provision within the different networks

and the 3GPP (third Generation Partnership Project). The standards produced by these two bodies in turn use existing specifications from two Internet standards bodies: the W3C (World Wide Web Consortium) and the IETF (Internet Engineering Task Force). The standards from the WAP Forum specify how messages are composed and packaged whereas the standards from the 3GPP specify how messages are sent, routed, and received.

Figure 8.8 shows a generalized view of the Multimedia Message Service architecture for a 3G messaging system. It combines different networks and network types. It integrates existing messaging systems within these networks. A user takes a picture using his mobile phone and sends it to another person using the MMS functionality (person-to-person). The picture can be sent directly as a MMS message or can be sent as an attachment to an email message. MMS messages can also be automatically generated and sent through software (content-to-person). For example, a user could ask for the day's weather forecast to be sent to her phone each morning complete with animated maps and audio of the weatherman.

In the first phase of the MMS, users should be able to create presentation slides through software. The layout and ordering of the slides are specified through a language called SMIL (Synchronization Multimedia Integration Language). This may be a slide show with multiple or even a single slide. The slide show may combine a number of still pictures or animations into one MMS message. The display areas of these slides area divided into different sections. Currently, there are only two sections per slide–one for an image and one for the text. It is also acceptable to have either just an image region or just a text region. In the second phase, users should be able to record their own video content (10 second video clip) and send it via MMS. The contents of the slides–the actual images, text, and audio–are separate pieces that are sent along with the slides. The maximum size of the entire packaged message that first generation devices can support is 50 kB.

8.3.1 MMS Architecture

The connection between different networks in Figure 8.8 is provided by the Internet protocol and its associated set of messaging protocols. This approach enables messaging in wireless networks to be compatible with messaging systems found on the Internet. Multimedia Message Service Environment (MMSE) encompasses various elements required to deliver a MMS (Figure 8.9). This includes:

- MMS Client: This is the entity that interacts with the user. It is an application on the user's wireless device.

- MMS Relay: This is the system element that the MMS client interacts with. It provides access to the components that provide message storage services. It is responsible for messaging activities with other available messaging systems. The SMS relay along with the MMS content server is referred to as MMSC (MMS Controller).

- WAP Gateway: It provides standard WAP services needed to implement MMS.

- MMS Server: This is the content server, where the MMS content is generated

- Email Server: MMS can integrate seamlessly to the email system of Internet.

The messages that transit between the MMS Client and MMS Relay pass through WAP Gateway. Data is transferred between the MMS client and WAP gateway using WAP Session Protocol (WSP). Data is transferred between the WAP gateway and the MMS Relay using HTTP.

8.3.2 MMS Transaction Flows

As mentioned earlier, the MMS service is realized by the invocation of transactions between the MMS Client and the MMS Relay. The general transactions of sending and retrieving messages do not depend on what type of client the message is sent to or

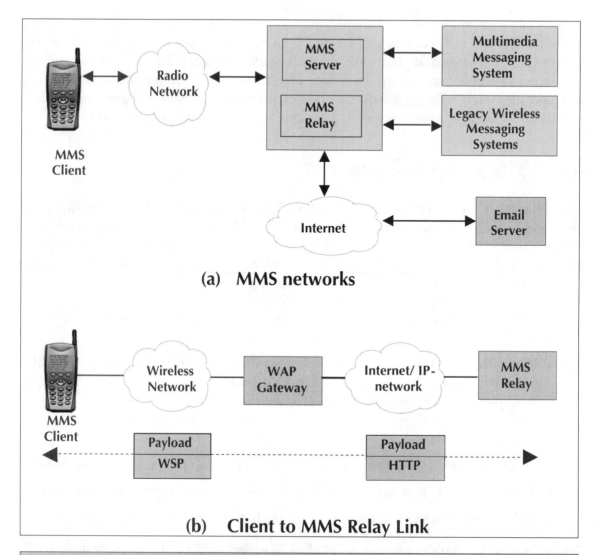

(a) MMS networks

(b) Client to MMS Relay Link

Figure 8.9 MMS Environment

received from. The other endpoint for the message may be another MMS Client or a client on a legacy wireless messaging system or it may even be an email server.

The above message exchanges can be considered to form the following logically separate transactions (Figure 8.10):

- MMS Client (sender) sends a message to MMS Relay

  (M-Send.req, M-Send.conf)

- MMS Relay notifies MMS Client (recipient) about a new message arrival (M-Notification.ind, M-NotifyResp.ind)
- MMS Client fetches (recipient) a message from MMS Relay (WSP GET.req, M-Retrieve.conf)
- MMS Client (recipient) sends a retrieval acknowledgement to MMS Relay (M-Acknowledge.req)
- MMS Relay sends a delivery report about a sent message to MMS Client (sender) (M-Delivery.ind)

From this list it is clear that MMS uses 8 types of messages to perform messaging transactions. The M-Notification.ind function is generally done through SMS. This is a special type of SMS. This SMS is not forwarded to the SMS inbox; it is forwarded to the MMS client. Client notifies the user about the arrival of a MMS message. The user then fetches the message from the relay. Some terminals allow the facility to configure the MMS client so that the message is fetched automatically. The multimedia messaging

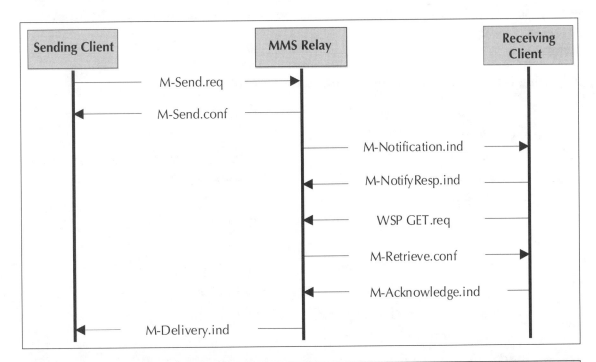

Figure 8.10 Example of MMS Transaction Flow–Delayed Retrieval

PDUs (Protocol Data Units) consist of MMS headers and a message body. The message body may contain any content type such as text, image, audio, and video. The message body is used only when the multimedia message is sent or retrieved. All other PDUs contain only the MMS-headers part.

8.3.3 SMIL (Synchronized Multimedia Integration Language)

Synchronized Multimedia Integration Language (SMIL) is an XML-based language specified by the W3C. It is a markup language for specifying how and when multimedia clips will play. SMIL integrates streaming audio and video with images, text or any other media type. MMS messages are sent using SMIL as the presentation language. The presentation part specifies how the various other parts of the message should be presented to the user—at what time and in which place in relation to the other parts. MMS adapted a limited subset of SMIL, often referred to as 'MMS SMIL'. Links can be to various places inside the current presentation to allow the user to jump around from place to place within the timeline of the presentation. They can also point to a website, for example, letting the user get more information, downloads, etc. using the terminal's browser.

Example

Following is a simple example of the SMIL for an MMS message.

```
<smil>
    <head>
            <meta name="title" content="vacation photos" />
            <meta name="author" content="Radha Krishna" />
            <layout>
                    <root-layout width="160" height="120"/>
                    <region id="Image" width="100%"
                    height="80" left="0" top="0" />
                    <region id="Text" width="100%"
                    height="40" left="0" top="80" />
            </layout>
    </head>
    <body>
```

```
<par dur="8s">
        <img src="FirstImage.jpg" region="Image" />
        <text src="FirstText.txt" region="Text" />
        <audio src="FirstSound.amr"/>
</par>
<par dur="7s">
        <img src="SecondImage.jpg" region="Image" />
        <text src="SecondText.txt" region="Text" />
        <audio src="SecondSound.amr" />
</par>
    </body>
</smil>
```

The example above is for a terminal whose screen will displays the slide in a portrait orientation, where the height is greater than the width. On a PC screen, the SMIL slides are all displayed exactly 160 pixels wide and 120 pixels tall. The total slide area is divided into two smaller areas. The image region will be 80 pixels tall and always appears above the 40 pixel tall text area. On an MMS client however this will be different. The screen may not be large enough to accommodate the layout. Each slide in turn contains at least two elements: one for the image region and one for the text region. Two of the slides also contain an audio element that will be played when the slide is viewed. In normal SMIL, the names of the layout regions (image and text in our MMS message) are just handy names for generic regions that can contain any type of content. In MMS SMIL, however, the image region must contain an image element and the text region, a text element.

As we can see, the SMIL markup is very similar to HTML or WML markup language. The entire message body is enclosed within <smil></smil> tags and the message (or document) itself has both head and body sections. The head section contains information that applies to the entire message. The title and author meta fields here correspond to the From and Subject fields of the message. These meta fields are optional. Under MMS implementations of SMIL, a client is free to re-format the layout in a way best suited to the client's display. Actual slides are within the body section of the message. These slides are denoted with the par–for parallel tag. Parallel denotes that all the elements within the tag are to be displayed simultaneously. The dur attribute for each slide is the duration of the slide in the slide show. Again, the receiving client is free to modify or ignore this, replacing duration with a button for the next slide, for example.

The following are the specific media formats that will be supported in the first generation of MMS systems.

- **Images:** Image formats supported are baseline JPEG with JFIF exchange format, GIF87a, GIF89a, and WBMP. The maximum guaranteed image resolution is 160 pixels wide by 120 pixels high. Larger images are supported, but need to be converted for the target device. The browser safe color palette (256 colors) is recommended for color image. JPEG is better suited for rendering photographs; whereas, GIF is a better choice for line drawings.

- **Text:** The text of the message may use us-ascii, utf-8, or utf-16 character encoding. The supported character sets on any client will always be at least all of ISO 8859-1.

- **Audio:** Audio should be encoded as AMR (Adaptive Multi Rate), a codec used for voice in GSM and 3G networks. Many clients will also support iMelody for ring tones.

8.3.4 MMS Interconnection, Interoperability and Roaming

Like any other service MMS also has to meet the challenges of interoperability and roaming. Interoperability of MMS means the ability of terminals to exchange mutually acceptable messages between terminals from different vendors, or network components like MMSCs, and with WAP gateways. This includes the end-to-end exchange of formats and protocols. MMS roaming means that a subscriber can send and receive MMS messages when roaming in another network.

The main method for GPRS roaming is **PLMN roaming** where the home PLMN GGSN is used. Other method is **ISP roaming** where the visited PLMN GGSN is used. When the user is roaming, MMS messages are sent via normal packet data traffic between the **home** operator network and the **roaming** operator network. In addition to this, the roaming customer must be able to receive SMS from the home SMSC. To achieve MMS roaming, both GPRS and SMS roamings are required. Participating operators need to have a packet data roaming agreement and SMS roaming agreement in place. A roaming agreement means the technical and commercial agreement between operators on interoperability and charging. Charging includes functions like exchange of charging data, billing the subscriber, and sharing the revenue. Operators must solve the problem of handling the interconnection charge, first within one country and then globally. They both collect statistics from traffic volumes, with clearing based on statistics and agreements. In practice, there are three ways for operators to arrange MMS

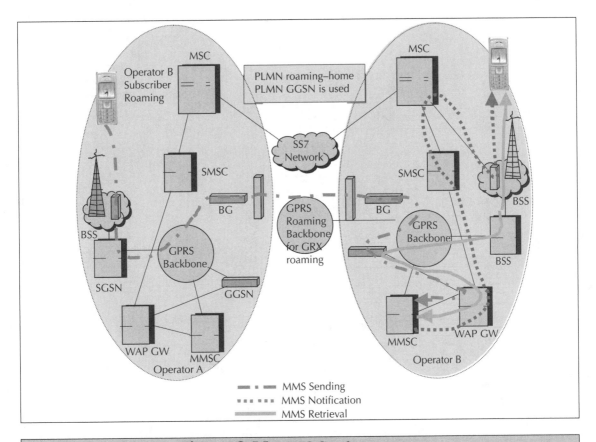

Figure 8.11 MMS Sender roaming

interconnection: using GRX (GPRS Roaming Exchange), VPN over Internet, or VPN over leased lines. Figure 8.11 depicts MMS sender roaming, whereas Figure 8.12 shows MMS receiver roaming.

OMA and 3GPP have defined three domains for multimedia messages. The first of these is the Core MM Content Domain where full interoperability is guaranteed. The second is the Standard MM Content Domain, where terminals and multimedia messages are still compliant with MMS standards but terminals have certain freedoms. The third domain is the unclassified MM Content Domain, giving full freedom to create multimedia messages.

8.3.5 MMS Device Management and Configuration

MMS services sometime require complex configuration. For example, the settings required for MMS include MMSC IP address, connection type and about ten other

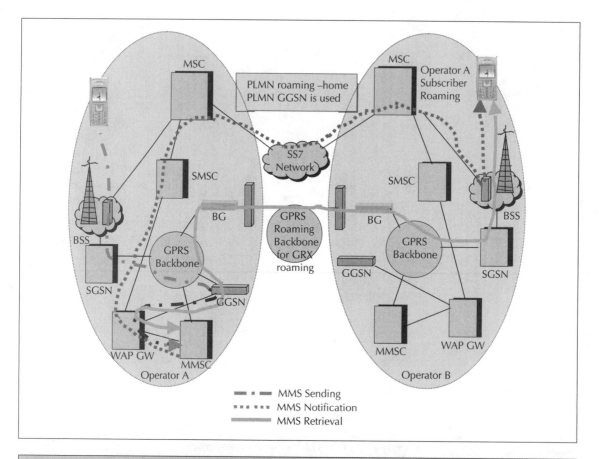

Figure 8.12 MMS sender roaming

parameters. Therefore, there is a need to be able to configure the users' devices, by providing device settings over the air. The OMA device management architecture consists of two components: OMA Client Provisioning and a continuous management technology that is based on the SyncML Device Management specification. Client Provisioning is a messaging based provisioning technology that sends settings over the air to the device and configure. All the user has to do is to accept the sent settings and the device will be correctly configured and ready for use.

8.4 GPRS Applications

For GPRS or WAP there are no specific services. These are the same services over the wireless media that can run either on GSM or 3G. However, as GPRS offers a higher bit rate, the user experience is better. Though, it is not mandatory, MMS based contents are better suited for GPRS networks. Some of the applications that are better suited for GPRS network are:

- Dating: This is an interactive dating system that exploits data bearer rather than the SMS bearer.

- Games: Online games, cartoons can be better suited for GPRS network.

- Sending/Receiving Fax/Email on the Handset: Different mobile office applications are quite well suited for GPRS network.

- Location aware applications: Location aware application combined with MAP and other type of travel related applications will have a better reach in GPRS network.

- VPN: It is possible to deploy wireless virtual private networks over GPRS network.

- Multimedia downloads: Various picture, music, video clips download will be part of this.

8.4.1 Digital Rights Management

MMS opens up lots of new possibilities for content providers to offer innovative premium mobile services, such as comic services, news services, sports updates, movie and music clips. As these services take off, it is crucial to take into account Digital Rights Management (DRM) for the distribution and consumption of mobile content. With DRM, the content owner or service provider can determine if and how his content can be distributed if at all from person to person. Of course, some content will be available even without DRM. All the content created by mobile users themselves such as capturing and sending of photo and video clips, or advertising content like sponsored movie trailers that promote new films in the cinema, will be free. Companies involved in mobile content services such as ring tones, wallpapers, or games where the business plan is based on being able to collect the rightful payment, will be interested in implementing DRM technology.

8.4.2 OMA Digital Rights Management

OMA has proposed Digital Rights Management and OMA Download standards. By implementing OMA DRM, service providers can allow end users to preview content before making a purchase decision. OMA DRM also allows end users to distribute content to other users via superdistribution. The OMA DRM standard will govern the use of mobile-centric content types. The first DRM standard, OMA DRM version 1.0, was officially approved in October 2002. The standard provides three DRM methods as following:

- **Forward-lock** is intended for the delivery of subscription-based services. The device is allowed to play, display or execute the MMS, but it cannot forward the MMS object. The content itself is hidden inside the DRM message that is delivered to the terminal. Examples will be news, sports, information and images that should not be sent on to others.

- **Combined Delivery** enables usage rules to be set for the media object. This method extends Forward-lock by adding a rights object to the DRM Message. Rights define how the device is allowed to render the content and can be limited using both time and count constraints. This method allows previews.

- **Separate Delivery** protects higher value media and enables superdistribution. This allows the device to forward the media, but not the rights. This is achieved by delivering the media and rights via separate channels, which is more secure than combined delivery. The media is encrypted into DRM Content Format (DCF) using symmetric encryption, while the rights hold the Content Encryption Key (CEK), which is used by the DRM User Agent in the device for decryption. Recipients of superdistributed content must contact the content retailer to obtain rights to either preview or purchase media.

REFERENCES/FURTHER READING

1. Charles Arehart, et al, Professional WAP, Wrox Press, 2000.

2. Wireless Application Protocol Architecture Specification, WAPForum, Apr-1998.

3. Wireless Application Protocol Wireless Markup Language Specification, Version 1.2, WAPForum, November-1999.

4. Wireless Application Protocol WMLScript Language Specification Version 1.1, WAPForum, Nov-1999.

5. Wireless Application Protocol Wireless Application Environment Specification Version 1.2, WAPForum, November-1999.

6. Wireless Application Protocol Wireless Application Environment Overview, WAPForum, November-1999.

7. Wireless Application Group User Agent Profile Specification, WAPForum, Nov-1999.

8. Wireless Application Protocol Push OTA Protocol Specification, WAPForum, Nov-1999.

9. Wireless Application Protocol Push Architectural Overview, WAPForum, Nov-1999.

10. Wireless Transport Layer Security Version 06-Apr-2001 Wireless Application Protocol WAP-261-WTLS-20010406-a, WAPForum.

11. Wireless Application Protocol Wireless Telephony Application Interface Specification, WAPForum, Nov-1999.

12. Wireless Application Protocol WAP 2.0, Technical White Paper, WAPForum, January 2002.

13. Multimedia Messaging Service: Service Aspects; Stage 1, 3GPP 3G TS 22.140 Release 1999.

14. Multimedia Messaging Service: Functional Description; Stage 2, 3GPP 3G TS 23.140 Release 1999.

15. WAP MMS Architecture Overview, WAP-205-MMSArchOverview.

16. WAP MMS Client Transactions, WAP-206-MMSCTR.

17. WAP MMS Encapsulation Protocol, WAP-209-MMSEncapsulation.

18. Nokia white paper on "MMS Entering into the Next Phase".

19. MMS Technology Tutorial: http://www.nokia.com/support/tutorials/MMS/en/.

20. MMS: http://www.gsmworld.com/technology/mms/index.shtml.

21. Synchronized Multimedia Integration Language: http://www.w3.org/TR/REC-smil/.

REVIEW QUESTIONS

Q1: Describe the WAP protocol stack. What are the functions of different layers in this protocol stack?

Q2: What is WTAI (Wireless Telephony Application Interface)? Why is it important to have such a function? Describe an application where WTAI can make the user experience better.

Q3: What is WAP Push? How is push different from pull?

Q4: What is a WAP Gateway? What are its functions?

Q5: What is MMS? How is it different from Short Message Service and Extended Message Service? Describe the MMS architecture?

Q6: What is SMIL? How is SMIL used in MMS?

CHAPTER 9

CDMA and 3G

9.1 INTRODUCTION

The popularity and growth of cellular phones is keeping technology and business people on their toes. Technologists are busy developing ever newer technologies to offer better user experience. Operators and service providers, on the other hand are, coming up with innovative applications and services to get a share of this market. Users today expect better quality of voice and data services while on the move. Not too long ago, hardly any one would have imagined mobile phone being used not only for voice communication, but also for watching a video clip or as a network interface for a laptop. CDMA and 3G expressly support such a versatile usage.

Many of these opportunities and challenges made the scientific and business community look at the spread spectrum technology as an option for wireless communication. Mobile phone technology had a reincarnation from first generation analogue (using FDMA) to second generation digital (using TDMA). The next incarnation is from second generation digital TDMA to third generation packet (using CDMA). CDMA is a specific modulation technique of Spread Spectrum technology. Third generation or 3G is more of a generic term to mean mobile networks with high bandwidth. Looking at the success of second generation GSM (using TDMA and roaming) and also the potential of second generation cdmaOne (IS-95 using CDMA), it was quite apparent that the next generation networks would have to be a combination of the best of these two technologies with amalgamation of some of the recent technology innovations.

9.1.1 How it started

Let us tell you an interesting story about the origin of the spread-spectrum technology. Have you heard the name of the famous Hollywood actress Hedy Lamarr? She was born in Vienna in 1914 as Hedwig Eva Maria Kiesler (note the full name). In 1933 Hedy Kiesler married the Austrian industrialist Fritz Mandl, CEO of the Hirtenberger Patronenfabrik, then one of the world's leading arms producers. Fritz was interested in control systems and conducted research in that field. Hedy was so beautiful that he was obsessed with keeping Hedy at his side all the time. He would take her even to business meetings and parties. Hedy received an education in munitions manufacturing from her husband and other Nazi officials through these meetings. Hedy escaped to London in 1937 and later traveled to USA to become one of the better known actresses in Hollywood. What many people may not know is that Hedy Lamarr helped the Allies win World War II, and she was the original patent holder of Spread-Spectrum Technology, which is at the foundation of today's CDMA (Code Division Multiple Access), Wireless LAN, IMT-2000 (International Mobile Telecommunication–2000), 3G (Third Generation), and GPS (Global Positioning System) technology.

With the help of an electrical engineering professor from the MIT, Hedy Lamarr and George Antheil, a film music composer patented 'Secret Communication System' in 1942. Like many other great technologies, the idea of 'Secret Communication System' was ahead of its time. Electronic technologies were beginning to develop and in the 1950s, engineers from Sylvania Electronic Systems Division began to experiment with the ideas in the Secret Communication System patent, using digital components. They developed an electronic spread-spectrum system that handled secure communications for the US during the Cuban Missile Crisis in 1962. It was in the early 1960s that the term 'spread spectrum' began to be used. Today it refers to digital communications that use a wide frequency spreading factor (much wider than typical voice telephone communications), and are not dependent on a particular type of tonality (such as a human voice) in the transmitting waveform.

In the mid-1980s, the US military declassified spread-spectrum technology. Immediately, the commercial sector began to develop it for consumer electronics. Qualcomm was the first to use this technology for commercial deployment of CDMA. 3G has been in gestation since 1992, when the International Telecommunications Union (ITU) began work on a standard called IMT-2000. IMT stands for International Mobile Telecommunications; the number 2000 initially had three meanings: the year that services should become available (year 2000), the frequency range in MHz that would be used (2000 MHz or 2 GHz), and the data rate in Kbits/sec (2000 KBps or 2 MBps).

9.2 SPREAD-SPECTRUM TECHNOLOGY

In a conventional transmission system, the information is modulated with a carrier signal and then transmitted through a medium. When transmitted, all the power of

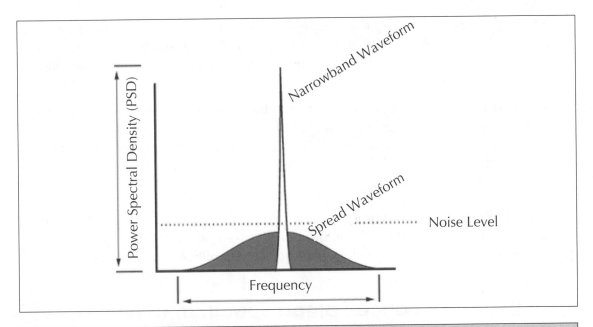

Figure 9.1 Narrow band and Spread Spectrum

the signal is transmitted centered around a particular frequency. This frequency represents a specific channel and generally has a very narrow band. In spread spectrum we spread the transmission power over the complete band as shown in Figure 9.1.

In spread spectrum the transmission signal bandwidth is much higher than the information bandwidth. There are numerous ways to cause a carrier to spread; however, all spread-spectrum systems can be viewed as two steps modulation processes. First, the data to be transmitted is modulated. Second, the carrier is modulated by the spreading code, causing it to spread out over a large bandwidth. Different spreading techniques are:

- **Direct Sequence (DS):** DS spread spectrum is typically used to transmit digital information. A common practice in DS systems is to mix the digital information stream with a pseudo random code.

- **Frequency Hopping (FH):** Frequency hopping is a form of spreading in which the center frequency of a conventional carrier is altered many times within a fixed time period (like one second) in accordance with a pseudo-random list of channels.

- **Chirp:** The third spreading method employs a carrier that is swept over a range of frequencies. This method is called chirp spread spectrum and finds its primary application in ranging and radar systems.

- **Time Hopping:** The last spreading method is called time hopping. In a time-hopped signal, the carrier is on–off keyed by the pseudo-noise (PN) sequence resulting in a very low duty cycle. The speed of keying determines the amount of signal spreading.

- **Hybrid System:** A hybrid system combines the best points of two or more spread-spectrum systems. The performance of a hybrid system is usually better than can be obtained with a single spread-spectrum technique for the same cost. The most common hybrids combine both frequency-hopping and direct-sequence techniques.

Amateurs and business community are currently authorized to use only two spreading techniques. These are frequency hopping and direct sequence techniques. Rest of the spread spectrum technologies are classified and used by military and space sciences.

9.2.1 Direct Sequence Spread Spectrum (DSSS)

Direct Sequence Spread Spectrum (DSSS) is often compared to a party, where many pairs are conversing, each in a different language. Each pair understands only one language and therefore, concentrates on his or her own conversation, ignoring the rest. A Hindi-speaking couple just homes on to Hindi, rejecting everything else as noise. Its analogous to DSSS is when pairs spread over the room conversing simultaneously, each pair in a different language. The key to DSSS is to be able to extract the desired signal while rejecting everything else as random noise. The analogy may not be exact, because a roomful of people all talking at once soon becomes very loud. In general, Spread-Spectrum communications is distinguished by three key elements:

1. The signal occupies a bandwidth much larger than what is necessary to send the information.

2. The bandwidth is spread by means of a code, which is independent of the data.

3. The receiver synchronizes to the code to recover the data. The use of an independent code and synchronous reception allows multiple users to access the same frequency band at the same time.

In order to protect the signal, the code used is pseudo-random, which makes it appear random while being actually deterministic, which enables the receivers to reconstruct the code for synchronous detection. This pseudo-random code is also called pseudo-noise (PN). DSSS allows each station to transmit over the entire frequency all the time. DSSS also relaxes the assumption that colliding frames are totally garbled. Instead, it assumes that multiple signals add linearly.

DSSS is commonly called Code Division Multiple Access or CDMA in short. Each station is assigned a unique *m*-bit code. This code is called the CDMA chip sequence. To transmit a 1 bit, the transmitting station sends its chip sequence, whereas to send 0, it sends the complement chip sequence. Thus if station *A* is assigned the chip sequence 00011011, it sends bit 1 by sending 00011011 and bit 0 by sending 11100100. Using bipolar notations, we define bit 0 as +1 and bit 1 as −1. The bit 0 for station *A* will now become (−1 −1 −1 +1 +1 −1 +1 +1) and 1 becomes (+1, +1, +1, -1, −1, +1, −1, −1). Figure 9.2 depicts this with 6 chips/bit (011010). For manipulation of bits, we XOR (addition with modulo2) the input bits, in bipolar notations we multiply to get the desired result:

```
0 XOR 0 = 0 ⟹ +1 × +1 = +1

1 XOR 1 = 0 ⟹ -1 × -1 = +1

1 XOR 0 = 1 ⟹ -1 × +1 = -1

0 XOR 1 = 1 ⟹ +1 × -1 = -1
```

Each station has its unique chip sequence. Let us use the symbol *S* to indicate the m-chip vector for station *S*, and $\bar{S}$ is for its negation. All chip sequences are pair-wise orthogonal, by which we mean that the normalized inner product of any two distinct chip sequences, *S* and *T* (written as *S·T*) is 0. In mathematical terms,

$$S.T = \frac{1}{m} \sum_{i=1}^{m} S_i . T_i = 0$$

$$S.S = \frac{1}{m} \sum_{i=1}^{m} S_i . S_i = 1$$

This orthogonality property is very crucial for mobile communication. Note that if $S·T = 0$ then $S.\bar{T}$ is also 0. The normalized inner product of any chip sequence with itself is 1. This follows because each of the m terms in the inner product is 1, so the sum is m. Also note that $S·\bar{S} = -1$.

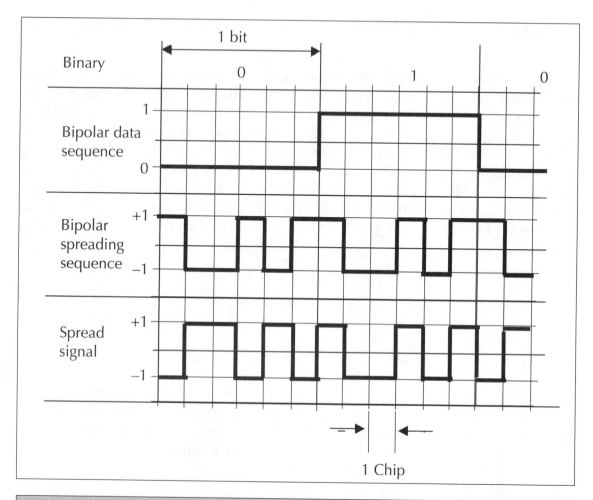

Figure 9.2 The CDMA Chip sequence

When two or more stations transmit simultaneously, their bipolar signals add linearly. For example, if in one chip period three stations output +1 and one station outputs −1, the result is +2. One can think of this as adding voltages: three stations outputting +1 volt and one station outputting −1 volts gives 2 volts.

In Figure 9.3 we see there are four stations A, B, C, and D with their chip sequences. In this example we have taken 8 chips. Figure 9.3(a) is the bit sequence of the chips whereas Figure 9.3(b) is the bipolar notations of the same. In Figure 9.3(c) we assume that there are six cases of four stations transmitting at the same time. In the first example, Figure 9.3(c), we assume that only C is transmitting bit 1. In the second example,

B transmits a bit 1, and C transmits a bit 1. Therefore, we get:

$$(-1 +1 -1 +1 +1 +1 -1 -1) = S_1$$

$$(-1 -1 +1 -1 +1 +1 +1 -1) + (-1 +1 -1 +1 +1 +1 -1 -1) = (-2\ \ 0\ \ 0\ \ 0 +2 +2\ \ 0 -2) = S_2$$

In the third example, station A transmits a 1 and station B transmits a 0, others are silent. In the fourth example, A and C transmit a 1 while B sends a 0. In the fifth example, all four stations transmit a 1. Finally, in the last example, A, B, and D transmit a 1, while C sends a 0. The result of these transmissions are different sequences S_1 through S_6 as given in Figure 9.3(d). All these examples represent only one bit time.

To recover the bit stream of any station, the receiver must know that station's chip sequences in advance. This is similar to the example of the party where different couples are conversing in different languages. We know someone is speaking in Hindi and may be someone else is speaking in French. The listener who knows Hindi can only

```
A:  00011011              A:  (-1,-1,-1,+1,+1,-1,+1,+1)
B:  00101110              B:  (-1,-1,+1,-1,+1,+1,+1,-1)
C:  01011100              C:  (-1,+1,-1,+1,+1,+1,-1,-1)
D:  01000010              D:  (-1,+1,-1,-1,-1,-1,+1,-1)

      9.3(a)                          9.3(b)


--1-    C                 S₁ = (-1,+1,-1,+1,+1,+1,-1,-1)
-11-    B + C             S₂ = (-2, 0, 0, 0,+2,+2, 0,-2)
10--    A + B̄            S₃ = ( 0, 0,-2,+2, 0,-2, 0,+2)
101-    A + B̄ + C        S₄ = (-1,+1,-3,+3,-1,-1,-1,+1)
1111    A + B + C + D     S₅ = (-4, 0,-2, 0,+2, 0,+2,-2)
1101    A + B + C̄ + D    S₆ = (-2,-2, 0,-2, 0,-2,+4, 0)

      9.3(c)                          9.3(d)

S₁.C = (+1+1+1+1+1+1+1+1)/8  =  1
S₂.C = (+2+0+0+0+2+2+0+2)/8  =  1
S₃.C = (+0+0+2+2+0-2+0-2)/8  =  0
S₄.C = (+1+1+3+3+1-1+1-1)/8  =  1
S₅.C = (+4+0+2+0+2+0-2+2)/8  =  1
S₆.C = (+2-2+0-2+0-2-4+0)/8  = -1

            9.3(e)
```

Figure 9.3 CDMA code arithmetic

understand the message from the partner speaking in Hindi. Someone knowing French can extract the French message.

DSSS does the recovery by computing the normalized inner product of the received chip sequence (the linear sum of all the stations that transmitted) and the chip sequence of the station whose bit stream it is trying to recover. Let us assume that we are interested in recovering the bit sequence of station C. If the received chip sequence is S (S_1, S_2, ... S_6) we compute the normalized inner product, S C. From each of the six sums S_1 through S_6, we calculate the bit by summing the pairwise products of the received S and the C vector of Figure 9.3(d) and then take 1/8 of the result. As shown in Figure 9.3(e), the product extracts the correct bit. Note that $S_3.C = 0$; this means that in the third example of 2(c) station C did not transmit. Also, note that $S_6 C = -1$; this means that in the sixth example station C transmitted a 0.

Walsh Function

The CDMA orthogonal codes are generated through Walsh function. Walsh functions are generated by code-word rows of special square matrices called Hadamard Matrices. These matrices contain one row of all 0s, with the remaining rows having an equal number of 1s and 0s. Walsh function can be constructed for block length $N = 2^j$, where j is an integer. The TIA IS-95 CDMA system uses a set of 64 orthogonal functions generated by using Walsh functions. The modulated symbols are numbered from 0 through 63.

The 64 x 64 matrix can be generated by using the following recursive procedure:

$$H_1 = [0]; \quad H_2 = \begin{pmatrix} 0 & 0 \\ 0 & 1 \end{pmatrix}; \quad H_2 = \begin{pmatrix} 0&0&0&0&0&0&0&0 \\ 0&1&0&1&0&1&0&1 \\ 0&0&1&1&0&0&1&1 \\ 0&1&1&0&0&1&1&0 \\ 0&0&0&0&1&1&1&1 \\ 0&1&0&1&1&0&1&0 \\ 0&0&1&1&1&1&0&0 \\ 0&1&1&0&1&0&0&1 \end{pmatrix} = \begin{bmatrix} \phi_1 \\ \phi_2 \\ \phi_3 \\ \phi_4 \\ \phi_5 \\ \phi_6 \\ \phi_7 \\ \phi_8 \end{bmatrix}; \quad H_{2N} = \begin{pmatrix} H_N & H_N \\ H_N & \overline{H_N} \end{pmatrix}$$

Where N is a power of 2 and $\overline{H}_N$ is the complement of H_N.

The period of time required to transmit a single modulation symbol is called a Walsh symbol interval and is equal to 1/4800 seconds (203.33 μs). The period of time associated with 1/64 of the modulation symbol is referred to as a Walsh chip and is equal to 1/307,200 seconds (3.255 μs). Within a Walsh symbol, Walsh chips are transmitted in the order 0, 1, 2, ... 63.

For the forward channel (base station to mobile station), Walsh functions are used to eliminate multiple access interference among users within the same cell. Steps followed are:

- The input user data of individual user is multiplied by orthogonal Walsh functions.

- All the data of all the users are combined.

- The combined data is then spread by the base station (BS) pilot pseudo-random (PN) code.

- This spread signal is then transmitted on a radio carrier.

- At the receiver, the mobile removes the coherent carrier and gets the spread signal.

- The mobile receiver multiplies the signal by the synchronized PN code associated with the base station to get the spread data.

- The multiplication by the synchronized Walsh function for the ith user will eliminate the interferences due to transmission from BS to other users.

In IS-95 or cdmaOne system different techniques are used for forward channel (BS to MS (Mobile Station)) and reverse channel (MS to BS) encoding. cdmaOne is the brand name of the service introduced by Qualcomm for digital mobile communication. The same technology was adopted by TIA (Telecommunication Industry Association) as IS-95 standard for second generation digital mobile communication in USA. Channelization in the forward link is accomplished through the use of orthogonal Walsh codes, while channelization in the reverse link is achieved using temporal offsets of the spreading sequence.

Different base stations are identified on the downlink based on unique time offsets utilized in the spreading process. Therefore, all base stations must be tightly coupled to a common time reference. In practice, this is accomplished through the use of the Global Positioning System (GPS), a satellite broadcast system that provides information on Greenwich Mean Time and can be used to extract location information about the receiver. This common time reference is known as system time.

There are two types of PN spreading sequences used in IS-95: the long code and the short codes. Both the PN sequences are clocked at 1.2288 MHz, which is the chipping rate. Two short code PN sequences are used since IS-95 employs quadrature spreading. These two codes are the in-phase sequence

$$P_I(x) = x^{15} + x^{13} + x^9 + x^8 + x^7 + x^5 + 1$$

and the quadrature sequence

$$P_Q(x) = x^{15} + x^{12} + x^{11} + x^{10} + x^6 + x^5 + x^4 + x^3 + 1.$$

These two sequences are generated using 15-bit shift register sequences; although they are nominally $2^{15} - 1 = 32767$ chips, a binary '0' is inserted in each sequence after a string of fourteen consecutive 0's appears in either sequence to make the final length of the spreading sequence an even 32768 chips.

The long code is given by the polynomial

$$P(x) = x^{42} + x^{35} + x^{33} + x^{31} + x^{27} + x^{26} + x^{25} + x^{22} + x^{21} + x^{19} + x^{18} + x^{17} + x^{16} + x^{10} + x^7 + x^6 + x^5 + x^3 + x^2 + x^1 + 1.$$

It is of length $2^{42} - 1$ chips as it is generated by a 42-bit shift register. It is primarily used for privacy, as each user of the mobile network may be assigned a unique temporal offset for the long code with reference to system time. Since the long code has a period of 41 1/2 days, it is nearly impossible to blindly detect a user's temporal offset. The offset is accomplished with the use of a long code mask, which is a 42-bit value that is combined with the shift.

BPSK and QPSK

The simplest form of a DSSS communications system employs coherent Binary Phase Shift Keying (BPSK) for both the data modulation and spreading modulation. But the most common form of DSSS uses BPSK for data modulation and QPSK (Quadrature Phase Shift Keyed) modulation for spreading modulation. QPSK modulation can be viewed as two independent BPSK modulations with 180 degree phase difference.

The input binary bit stream $\{d_k\}$, $d_k = 0, 1, 2, \ldots$ arrives at the modulator input at a rate $1/T$ bits/sec and is separated into two data streams $d_I(t)$ and $d_Q(t)$ containing odd and

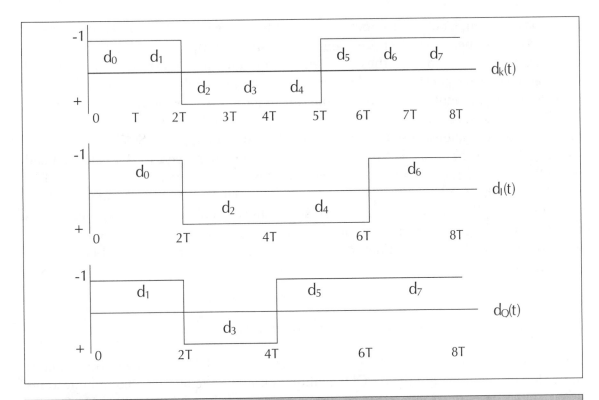

Figure 9.4 The QPSK modulation

even bits respectively like,

$d_I(t) = d_0, d_2, d_4 ,...$

$d_Q(t) = d_1, d_3, d_5 , ...$

QPSK can be viewed as two independent BPSK modulations. Figure 9.4 depicts an example of QPSK for a bit stream 00111000.

9.3 IS-95

Prof. George Cooper of Purdue University, USA did some work in the late 1970s in the field of spread-spectrum communications and its usage for mass commercial deployment. In a commercial cellular system we need to increase the transmission power when the mobile user moves further and reduce the power when the user comes closer to the base station. Cooper recognized the need for some type of power control system to overcome this near–far effect prevalent in CDMA systems, but could not achieve this. Qualcomm overcame some of these challenges and in the mid-1980s developed the first commercial spread-spectrum-based system for use in the cellular band. This system was considered an attractive alternative to the analogue FDMA technologies (AMPS, primarily) and TDMA systems (IS-54, IS-136, GSM). As mentioned earlier, this resulted in the Telecommunications Industry Association (TIA) developing the IS-95 standard. This standard formed the basis for the first CDMA systems deployed in the cellular band (from 800 to 900 MHz) in North America. This development eventually led to the TIA working with the T1P1 to develop the J-STD-008 standard for the PCS band (from 1800 to 1900 MHz). Since then, there has been some effort to enhance symmetric data rates for IS-95, resulting in the formation of a new standard in 1998, IS-95-B. The IS-95 family of standards is known as cdmaOne. It is a second generation digital mobile communication system.

9.3.1 Speech and Channel Coding

Normal audio range of human being is between 20 to 20 KHz. However, this range is normally used for high fidelity CD quality music. For telephonic communications where generally human voice is used, the frequency range of 300 to 3300 Hz is sufficient. For digitizing the speech, it is sufficient to sample at 8000 samples per second (assuming information bandwidth up to 4000 Hz). Therefore, to achieve telephone quality speech, 12 bits are sufficient to encode each sample. By using logarithmic sampling system 12 bits can be reduced to 8 bits per sample. This results in the PCM encoding of the speech and digitization of the voice at 64 KBps. This digitized voice is then passed through a coding scheme using Code-Excited Linear Prediction (CELP) algorithm. Linear Prediction Coding (LPC) is a combination of waveform coding and vocoder. Vocoder emulates the human vocal cord functions electronically and generates synthesized voice. In this process the analog voice is converted into 9.6 KBps digitized data.

In a mobile telecommunication environment, signal strength varies with location and movement of the mobile transmitter/receiver. Signal strength influences error rates, which in turn affect the quality of communication. Due to varying signal strengths, mobile telecommunication systems are susceptible to burst errors. Burst errors are groupings of errors in adjacent bits as compared to errors that are dispersed over the whole data block. IS-95 addresses the problem of burst errors by utilizing an error correction scheme based on encoding and interleaving. Generally, interleaving is used in conjunction with encoding (e.g., error-correcting codes) in order to lower the error rates. Interleaving is a technique in which encoded digital data is reordered before transmission in such a manner that any two successive digital data bits in the original data stream are separated by a predetermined distance in the transmitted data stream. Deinterleaving is the reverse of interleaving where data bits are reordered back to their original sequence.

9.3.2 IS-95 Architecture

The Key to the North American systems is the use of a common reference model from the cellular standards group TR-45. Different network entities within IS-95 are very similar to the network elements within a GSM network.

cdmaOne or IS-95 uses CDMA for its radio or last mile communication. Other than the radio interface, the rest of the network and especially the core is very similar to GSM.

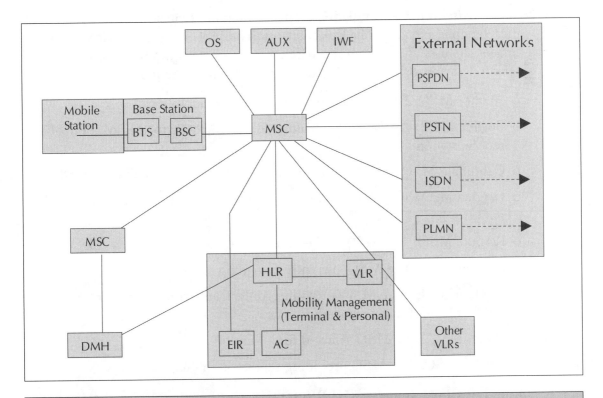

Figure 9.5 The IS-95 architecture model

The main elements of IS-95 (Figure 9.5) reference model are:

- **Mobile Station (MS):** This is the mobile phone unit with the user. The MS terminate the radio path on the user side and enables the user to gain access to services from the network. The MS can be a stand-alone device. It can have other devices (e.g., personal computers, fax machines) connected to it where it works as a pass through.

- **Base Station (BS):** The BS terminates the radio path and connects to the mobile switching center (MSC). BS is a system between the MS and the MSC. The BS is segmented into the BTS and BSC.

  - Base Transceiver Station (BTS): BTS consists of one or more transceivers placed at a single location and terminates the radio path on the network side.

  - Base Station Controller (BSC): The BSC is the control and management system for one or more BTSs. The BSC exchanges messages with both the BTS and the MSC. Some signaling messages may pass through BSC transparently.

- **Mobile Switching Center (MSC):** This is the main switching center equivalent to the telephone exchange in a fixed network. The MSC is an automatic system that interfaces the user traffic from the wireless network with the wireline network or other wireless networks. The MSC does one or more of the following functions:

  - Anchor MSC: First MSC providing radio contact to a call

  - Border MSC: An MSC controlling BTSs adjacent to the location of the mobile station

  - Candidate MSC: An MSC that could possibly accept a call or a handoff

  - Originating MSC: The MSC directing an incoming call towards a mobile station

  - Remote MSC: The MSC at the other end of an intersystem trunk

  - Serving MSC: The MSC currently providing service to a call

  - Tandem MSC: An MSC providing only trunk connections for a call in which a handoff has occurred

  - Target MSC: The MSC selected for a handoff

  - Visited MSC: The MSC providing service to the mobile station

- **Home Location Register (HLR):** HLR is the functional unit that manages mobile subscribers by maintaining all subscriber-related information. The HLR may be collocated with an MSC as an integral part of the MSC or may be independent of the MSC. One HLR can serve multiple MSCs or an HLR may be distributed over multiple locations.

- **Data Message Handler (DMH):** The DMH is responsible for collating the billing data.

- **Visited Location Register (VLR):** VLR is linked to one or more MSCs and is the functional unit that dynamically stores subscriber information obtained from the subscribers HLR data. When a roaming MS enters a new service area covered by the MSC, the MSC informs the associated VLR about the MS by querying the HLR after the MS goes through a registration procedure. VLR can be considered as cache whereas HLR is similar to a persistent storage.

- **Authentication Center (AC):** The AC manages the authentication associated with individual subscriber. The AC may be located within an HLR or MSC or may be located independent of both.

- **Equipment identity Register (EIR):** The EIR provides information about the mobile device for record purposes. The EIR may be located with the MSC or may be located independent of it.

- **Operations System (OS):** The OS is responsible for overall management of the wireless network.

- **Interworking Function (IWF):** The IWF enables the MSC to communicate with other networks.

- **External Networks:** These are other communication networks and can be a Public Switched Networks (PSTN), an Integrated Services Digital Network (ISDN), a Public Land Mobile Network (PLMN) or a Public Switched Packet Data Network (PSPDN).

9.3.3 IS-95 Channel Structure

IS-95 system operates on the same frequency band as the first generation AMPS (Advanced Mobile Phone System). It uses Frequency Division Duplex (FDD) with 25 MHz in each direction. It uses 824 to 849 MHz for forward link (base station to mobile station) and 869 to 894 MHz for reverse link (mobile station to base station). In digital communication, one data path maps onto one communication channel. In FDMA system one channel occupies a distinct frequency-band. In TDMA, it is a distinct timeslot within a frequency. In CDMA a channel is defined in terms of a code sequence and frequency. This results in offering a higher channel capacity, which translates into an overall higher bandwidth.

As mentioned earlier, IS-95 uses a different modulation and spreading technique for forward link and reverse link. In forward link 64 Walsh codes are used to map to 64 logical channels. The base station simultaneously transmits the user data for all mobiles in the cell by using different Walsh code for each mobile. This is then spread using a PN of length 2^{15} chips. The user data is spread to a channel chip rate of 1.2288Mchips. On the reverse link, channels are identified by long PN sequence.

For forward channels, base stations transmit information in four logical channel formats: pilot channels, sync channels, paging channels (PCH), and traffic channels (Code) (Figure 9.6). On the reverse link, all mobiles respond in an asynchronous fashion. The user data is encoded, interleaved, and then blocks of 6 bits are mapped to one of the 64 orthogonal Walsh functions. Finally, the data is spread by a user specific code of 42 bits (channel identifier). The reverse channel is organized in access channels and traffic channels.

At both the base station and the terminal, Rake receivers are used to resolve and combine multipath components, in order to improve the link quality. In IS-95, a three-finger Rake receiver is used at the base station.

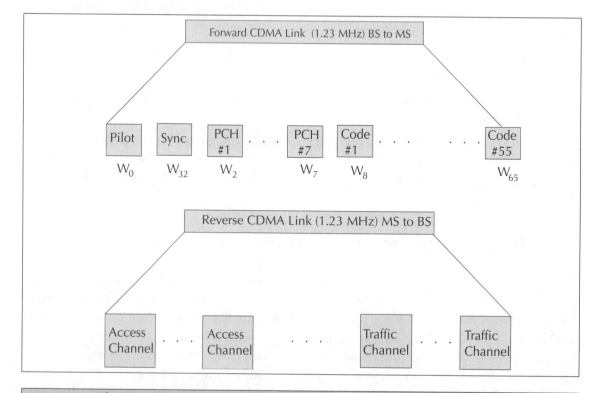

Figure 9.6 IS-95 Forward and Reverse link channel structures

In FDMA and TDMA multipath causes signal interference. In 1958 Price and Green proposed a method of resolving multipath problem in CDMA. In CDMA time shifted version of the same signal appears like a noise and almost uncorrelated. A signal that propagates from transmitter to receiver over multiple path can be resolved into separately fading signals by cross-correlating the received signal with multiple time shifted version of the same sequence. In the received three signals with maximum power are time shifted and added. The block diagram of this technique looks like a garden rake, hence the name rake receiver. A three-finger rake receiver takes three multipath signals.

Pilot Channel: The pilot CDMA signal transmitted by a base station provides a reference for all mobile stations. It is assigned the Walsh code W_0 (all 0). The pilot signal level for all base stations is kept about 4 to 6 dB higher than the traffic channel with a constant signal power. This is because the MS at the cell boundaries should be able to receive the pilot signal from other cells to decide when to perform handoff. The pilot signals from all base stations use the same PN sequences, but each base station is identified by a unique time offset. These offsets are in increments of 64 chips to provide 512 unique offsets.

Sync Channel: Sync channel is assigned the Walsh function W_{32} and is used with the pilot channel to acquire initial time synchronization. W_{32} has a pattern of 32 consecutive 0s and 32 consecutive 1s, which is ideal for synchronization. The Sync channel message parameters are: System Identification (SID), Network Identification (NID), Pilot short PN sequence offset index, Long-code state, System time, Offset of local time, Daylight saving time indicator and Paging Channel data rate (4.8 or 9.6kbps).

Paging Channel: There are up to seven paging channels that transmit control information to the terminals that do not have calls in progress. The paging channels are assigned the Walsh functions W_1 to W_7. Some of the messages carried by the paging channel include:

- System Parameter Message such as base station identifier, the number of paging channels and the page channel number.

- Neighbor List Message: information about neighbor base station parameters, such as the PN Offset.

- Access Parameters Message: parameters required by the mobile to transmit on an access channel.

- Page Message: provides a page to the mobile station.

- Channel Assignment Message: to inform the mobile station to tune to a new carrier frequency.

- Data Burst Message: data message sent by the base station to the mobile.

- Authentication Challenge: allows the base station to validate the mobile identity.

Access Channel: Access channel is used by a terminal without a call in progress to send messages to the base station for three principal purposes: to originate a call, to respond to a paging message, and to register its location. Each base station operates with up to 32 access channels. The messages carried by the access channel include:

- Order message: to transmit information such as base station challenge, mobile station acknowledgement, local control response and mobile station reject.

- Registration Message: sends to the base station information necessary to page the mobile such as location, status and identification.

- Data Burst message: user-generated data message sent by the mobile station to the base station.

- Origination message: allows the mobile station to place a call sending dialed in digits.

- Authentication Challenge Response message: contains necessary information to validate the mobile stations identity.

Forward Traffic Channel: channels not used for paging or sync can be used for traffic. Thus, the total number of traffic channels at the base station is 63 minus the number of paging and sync channels in operation at the base station. Information on the forward traffic channel includes the primary traffic (voice or data) secondary traffic (data) and signaling. When the forward link is used for signaling, some of the typical messages would be:

- Order message: similar to the order message in forward traffic channel.

- Authentication Challenge message: used to prove the identity of the mobile when the base station suspects its validity.

- Alert with Information message: allows the base station to validate the mobile identity.

- Handoff Direction message: provides the mobile with information needed to begin the handoff process.

- Analog Handoff Direction message: tells the mobile to switch to the analog mode and begin the handoff process.

- In-traffic System Parameters message: updates some of the parameters set by the System Parameters message in the paging channel.

- Neighbor List Update Message: updates the neighbor base station parameters set by the Neighbor List message in the paging channel.

- Data Burst message: data message sent by the base station to the mobile.

- Mobile Registration message: informs the mobile that it is registered and supplies the necessary system parameters.

- Extended Handoff Direction message: one of several handoff messages sent by the base station.

Reverse Traffic Channels: this channel can multiplex primary (voice) and secondary (data) or signaling traffic. Some of the typical messages that the reverse traffic channel carries are:

- Order messages: include base station challenge, parameter update confirmation, mobile station acknowledgement, service option request and response, release, connect, DTMF (Dual Tone Multi Frequency) tone etc.

- Authentication Challenge Response message: information to validate the mobile station.

- Pilot Strength Measurement message: information about the strength of other pilot signals that are not associated with the serving base station.

- Data Burst message: a user-generated data message sent by the mobile to the base station.

- Handoff Completion message: is the mobile response to a Handoff Direction message.

- Parameter Response message: is the mobile response to the base station to a Retrieve Parameters message.

9.3.4 IS-95 Call Processing

To set up a call or to transmit data, a data path needs to be established through a traffic channel. To establish a traffic channel, a mobile station in IS-95 goes through several states. They are:

- System initialization
- System idle state
- System access
- Traffic channel state

In the **system initialization** state the mobile acquires a pilot channel by searching all the PN offsets possibilities and selecting the strongest pilot (W_0) signal. Once the pilot is acquired, the sync channel is acquired using the W_{32} Walsh function and the detected pilot channel. Then the mobile obtains the system configuration and timing information.

Next the mobile enters the system **idle state** where it monitors the paging channel. If a call is being placed or received, the mobile enters the system **access state** where the necessary parameters are exchanged. The mobile transmits its response on the access channel and the base station transmits its response on the paging channel. When the access attempt is successful the mobile enters the **traffic state**. In the traffic state voice or data is transacted.

CDMA Registration

The registration process is used by the mobile device to notify its location, status, identification and other characteristics. Location information is required to page the mobile for an incoming mobile terminated call. When the MS does power on or power off it goes through the registration process as well. These functions are similar to GSM. Registration information is stored in HLR.

9.3.5 Authentication and Security

The Electronic Serial Number (ESN) of a IS-95 mobile station is a 32 bits binary number that identifies the mobile. It is factory-set and is not alterable in the field. All mobiles are assigned a unique ESN when manufactured. A mobile station also has a unique 15 digits number called Mobile Identification Number (MIN). This is the mobile's 10 digit directory number similar to the MSISDN number in GSM. The difference is that in IS-95, the mobile is assigned a number similar to the North American numbering scheme. For example, a fixed line number in Bangalore may have a directory number of 080-2593-2137 whereas a mobile phone may have a directory number of a GSM phone like 98450-62050. On the contrary in USA if a fixed line number is 1-630-858-7131, the mobile number can be 11-630-240-8900.

Whenever a mobile is turned on, it registers with the network. During the authentication process the network throws a global challenge to the mobile. The AC (Authentication Center) transmits a random number to the mobile station. The mobile station encrypts it using a key shared between the mobile station and the AC. The encrypted random message is sent back to the network. The AC checks this result with the result it calculates using the same random number and shared key. If these match, the mobile is authenticated. The global challenge in IS-95 is more frequent than in GSM. In IS-95, authentications even take place following successful handoffs. Following successful authentication, the VLR assigns a TMSI (Temporary International Mobile Subscriber Identity). The TMSI provides anonymity since it is a transient identity only the mobile and the network are aware of. The spreading PN (Pseudorandom Noise) sequence also play a role in security. For anybody to impersonate the CDMA traffic, the eavesdropper needs to know the PN sequence.

9.3.6 Handoff and Roaming

What is a handover in GSM is called a handoff in IS-95. When a subscriber moves away from a base station, the signal power reduces resulting in potential drop in connection. To ensure that the call does not break, some other base station closer to the

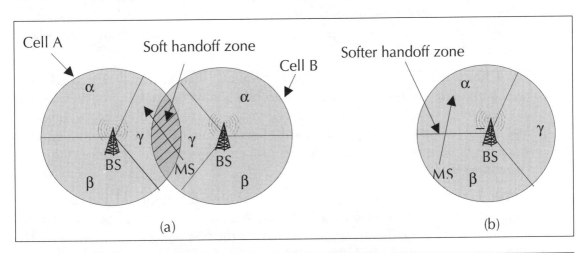

Figure 9.7 (a) Soft handoff; (b) Softer handoff

mobile station needs to attach the mobile to it and let the call continue without any interruption.

In CDMA, handoffs are handled differently compared to GSM. GSM handover is a hard handover. In a hard handover, the attachment with the current cell is broken first and then a new connection is set up with another cell. In GSM it is 'break before make'. In CDMA the spectrum is spread and everybody gets the same signal. Logically a mobile station in CDMA is always connected to different base stations at the same time. Therefore, handoff is managed by changing the attachment. There are three types of handoffs in CDMA. These are Soft handoff, Hard handoff, and Softer handoff.

> In CDMA a cell is divided into sectors. Like in GSM it is normally divide into 3 sectors each covering 120°. CDMA antennas are either Switched Beam System (SBS) or an Adaptive Antenna System (AAS). The SBS uses multiple fixed beams in a sector and a switch to select the best beam to receive a signal. In an AAS, the receiving signals by multiple antennas are weighted and combined to maximize the signal to noise ration (SNR).

- **Soft handoff:** This is the case of intercell handoffs (Figure 9.7(a)). Soft handoff is a process in which the control of a mobile station is assigned to an adjacent cell or an adjacent sector (in the same frequency) without dropping the original radio link. The mobile keeps two radio links during the soft hand-off process. Once the new communication link is well established, the original link is dropped. This process is also known as 'make before break,' which guarantees no loss of voice during hand-off. In Figure 9.7(a), as the user moves, a soft handoff takes place from Cell B to Cell A.

- **Hard handoff:** This is the case of interfrequency handoffs. CDMA to CDMA hard handoff is the process in which a mobile is directed to hand off to a different frequency assigned to an adjacent cell or a sector. The mobile drops the original link before establishing the new link. This is similar to a GSM handover. The voice is muted momentarily during this process. This handoff is completed very fast and cannot be noticed.

- **Softer Handoff:** A mobile communicates with two sectors of the same cell (Figure 9.7(b)). A rake receiver at the base station combines the best version of the voice frame from the diversity antennas of the two sectors into a single traffic frame. This is a logical handoff where signals from multiple sectors are combined instead of switching from one sector to another.

9.3.7 IS-95 Channel Capacity

In the first generation mobile networks the frequency channels were fixed and hence the capacity too. This is true with GSM as well. In GSM we multiply 125 frequencies with 8 timeslots to get 1000 channels. Therefore TDMA and FDMA capacities are bandwidth limited and hard-limited. The capacity of CDMA has a soft limit in the sense that we can add one additional user and tolerate a slight degradation of the signal quality. This is similar to a room full of people. Let us assume that people are talking to each other using a loudspeaker. In such a case not many people will get a chance to talk or to listen. However, more people can talk to each other if they converse in low voices. Another conclusion that can be drawn from this fact is that, any reduction in the multiple access interference converts directly and linearly into an increase in the capacity. The capacity of a CDMA system depends on the following criteria:

- **Voice Activity Detection (VAD).** The human voice activity cycle is 35 percent. This means that during a conversation people talk about 35% of the time. When users assigned to a cell are not talking, VAD will allow all other users to benefit due to reduced mutual interference. Thus interference is reduced by a factor of 65 percent. CDMA is the only technology that takes advantage of this phenomenon. It can be shown that the capacity of CDMA is increased by about 3 times due to VAD.

- **Sectorization for Capacity.** In FDMA and TDMA systems, sectoring is done to reduce the co-channel interference. In GSM there are total 1000 channels distributed between multiple operators, sectors and cells. The trunking efficiency of these systems decreases due to sectoring. This in turn reduces the capacity. On the other hand, sectorization increases the capacity of CDMA systems. Sectoring is done by simply introducing three (similar) radio equipments in three sectors. The reduction in mutual interference due to this arrangement translates into a 3-fold increase in capacity (in theory). In general, any spatial isolation through the use of multibeam or multisector antennas provides an increase in the CDMA capacity.

- **Frequency Reuse Considerations.** The previous comparisons of CDMA capacity with those of conventional systems primarily apply to mobile satellite (single-cell) systems. In the case of terrestrial cellular systems, the biggest advantage of CDMA over conventional systems is that it can reuse the entire spectrum over all the cells since there is no concept of frequency allocation in CDMA. This increases the capacity of the CDMA system by a large percentage (related to the increase in the frequency reuse factor).

As a rule of thumb the CDMA capacity is about four times that of TDMA and eight times that of FDMA.

9.4 CDMA versus GSM

GSM is a relatively mature technology, now several years in existence with a huge installation base. GSM has many experienced operators and equipment manufacturers. Interoperability within GSM is well proven. GSM is complete, open and has proven standards. GSM includes all the specifications from the handset other over the air, switch, interconnect it with switching, and every-aspect of mobile telecommunication. On the other hand, IS-95 is mainly a single vendor (Qualcomm cdmaOne) specification.

Table 9.1 GSM versus 3G

Functions	GSM	IS-95
Frequency	900MHz; 1800MHz (DCS180); 1900MHz (PCS 1900)	800MHz; 1900MHz
Channel bandwidth	Total 25 MHz bandwidth with 200KHz per channels, 8 timeslots per channel with frequency hopping	Total 12MHz with 1.25MHz for the spread spectrum
Voice codec	13Kbits/second	8 Kbits/sec or 13Kbps
Data bit rate	9.6Kbits/second expandable	9.6Kbits
Short message service	160 characters of text Supports	120 characters
SIM card	Yes	No
Multipath	Causes interference and destruction to service	Used as an advantage
Radio interface	TDMA	CDMA
Handoff	Hard Handover (handoff)	Soft Handoff (handover)
System Capacity	Fixed and limited	Flexible and higher than GSM
Economics	Expensive	Due to many technological advantages, dimension of investment per subscriber is expected to be lower than GSM

IS-95 only covers the air interface making it incomplete. Though there are many claims and counter claims, it is generally believed that CDMA has high potential to address some of the difficult challenges of the past quite effectively. These are described in Table 9.1

9.5 Wireless Data

Data transmission over wireless networks like CDMA or GSM is always a challenge. Typically raw channel data error rates for cellular transmission are 10^{-2}. This means that one in every 100 bits has an error. This is an error rate, which can be tolerated for voice transmission. This is because; our perception of hearing cannot detect it. Even if our ear is sometime able to detect it, our mind is able to correct it from the context. This error rate of 10^{-2} is too high for data transmission. An acceptable BER (Bit Error Rate) for data transmission is 10^{-6}. This means that one bit in a million can be tolerated as an error. In order to achieve this high level of reliability, it requires a design of effective error correction code and Automatic Repeat Request (ARQ). The CDMA protocol stack Figure 9.8 for data and facsimile has the following layers.

Application Interface Layer: This layer includes an application interface between the data source in the mobile station and the transport layer. The application interface provides functions like modem control, AT (Attention) command processing, data compression etc.

Transport Layer: The transport layer for CDMA asynchronous data and fax is based on TCP. TCP has been modified for IS-95.

Network layer: The network layer for CDMA asynchronous data and fax services is based on IP. The standard IP protocol has been enhanced for IS-95.

Sub-network Dependent Convergence Function: The SNDCF performs header compression on the header of the transport and network layers. Mobile station supports Van Jacobson TCP/IP header compression algorithm. Negotiation of the parameters for header compression is carried out using IPCP (Internet Protocal Control Protocal). The SNDCF sublayer accepts the network layer datagram packets from the network layer, performs header compression and passes that datagram to the PPP (Point to Point Protocol) layer. In the reverse operation, it receives network layer datagrams with compressed header from the PPP layer and passes it to the network layer.

Data Link layer: This layer uses PPP. The PPP Link Control Protocol (LCP) is used for initial link establishment and for the negotiation of optional link capabilities.

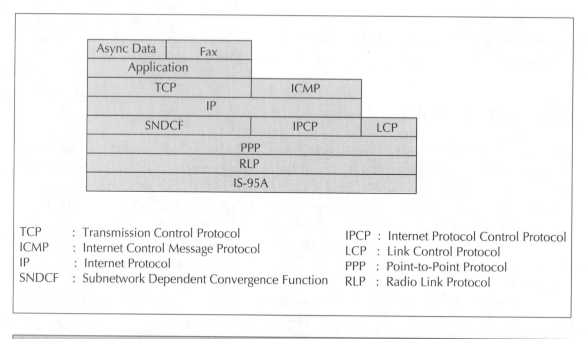

Figure 9.8 CDMA data protocol stack

Internet Protocol Control Protocol Sublayer: This sublayer supports negotiation of the IP address and IP compression protocol parameters. In general a mobile station does not have a permanent IP address. Therefore, the IP address needs to be negotiated and obtained from the network. IPCP does this job of leasing an IP address when the transport connection is established. The IP address is discarded when the connection is closed. This is similar to obtaining the IP address from a DHCP (Dynamic Host Configuration Protocol) server in a LAN environment.

Radio Link Protocol Layer: This layer provides octet stream service over the air. This service is responsible for reducing the error rate over the forward and reverse channels. There is no direct relationship between PPP packet and the traffic channel frame. A large packet may span multiple traffic channel frame. A single traffic channel frame may contain multiple PPP packets. RLP frames may be transported as traffic or signaling via data burst message.

9.5.1 Short Message Service

SMS in IS-95 is similar to SMS in GSM. Unlike GSM, the maximum size of a SMS in IS-95 is 120 octets. The SMS in IS-95 work the same way as in GSM. It supports

SMPP protocol and other features as in GSM. Like in GSM, the SMS in IS-95 uses the signaling channel for data transfer. SMS administration features include storage, profiling, verification of receipt and status enquiry capabilities.

9.6 Third Generation Networks

The telecommunications world is changing due to trends in media convergence and industry consolidation. The perception of mobile phone has changed significantly over the last few years. More changes predicted for the future are:

- The mobile devices will be used as an integral part of our lives.

- Data ('non-voice') usage of 3G will be important and different from the traditional voice business.

- A great deal of convergence will take place between information technology and communication technology.

- The look of the phone will be as important as its usage.

- Mobile communications will be similar in its social positioning. People will have only a mobile device.

To address these challenges and opportunities, the mobile telecommunication technology needs to adapt new techniques, facilities and services. The 3G system will offer a plethora of telecommunication services including voice, multimedia, video and high speed data. With 3G mobile Internet technology significant changes will be brought about in the day-to-day life of the people.

CDMA is the preferred approach for the third generation networks and systems. In North America cdma2000 is the version of 3G. cdma2000 standards are being driven by Telecommunication Industries Association (TIA). It uses the CDMA air interface, which is based on IS-95 and cdmaOne. In Japan 3G standard uses (Wideband Code Division Multiple Access) WCDMA (DoCoMo) version. This standard is being driven by ARIB. In Europe, Asia, Australia and many parts of the world 3G has been accepted as UMTS and WCDMA. UMTS/WCDMA is being driven by ETSI, and is the normal evolution from GSM/GPRS.

The main goal of UMTS (Universal Mobile Telecommunications System) is to offer a much more attractive and richer set of services to the users.

Beginning 1998 six partners–ARIB (Association of Radio Industries and Businesses), T1, TTA (Telecommunications Technology Association–Korea), ETSI in Europe, CWTS (China Wireless Telecommunication Standard group), TTC (Telecommunication Technology Committee–Japan) started discussions to cooperate for creating a standards for a third generation mobile system with a core network based on evolution for GSM and an access network based on all the radio access technologies supported by the different partners. This project was called the Third Generation Partnership Project (3GPP). About a year later ANSI decided to establish 3GPP2, a 3G partnership project for evolved ANSI/Telecommunications Industry Association (TIA)/Electronics Industry Association (EIA)-41 networks. There is also a strategic group called International Mobile Telecommunication Union-2000 (IMT-2000) within the International Telecommunication Union (ITU), which focuses its work on defining interface between 3G networks evolved from GSM on one hand and ANSI-41 on the other, in order to enable seamless roaming between 3GPP and 3GPP2 networks. 3GPP started referring 3G mobile system as Universal Mobile Telecommunication System (UMTS).

- **Universal Roaming:** Any user will be able to move across the world and access the network.

- **Higher Bit Rate:** over the air open the path toward multimedia applications.

- **Mobile-Fixed Convergence:** There is a need to offer users cross-domain services. An example is the tracking of a user's location in the mobile, fixed and Internet domain and automatically adapting the content of his incoming messages to SMS, voice message, fax or email. VHE (Virtual Home Environment) is the enabler to this service portability across networks and terminals in different domains.

- **Flexible Service Architecture:** By standardizing not the services themselves but the building blocks that make up services, UMTS shortens the time for marketing services from GSM and enhances creativity/flexibility when inventing new services.

9.6.1 IMT-2000

International Mobile Telecommunications-2000 (IMT-2000) is the global standard for third generation (3G) wireless communications defined by a set of interdependent ITU

Recommendations. IMT-2000 provides a framework for worldwide wireless access by linking the diverse systems of terrestrial and/or satellite based networks. It exploits the potential synergy between digital mobile telecommunications technologies and systems for fixed and mobile wireless access systems. IMT-2000 is working towards bringing all the different technologies, all the vendors, and all the standard making bodies to a common program to achieve the following vision of 3G networks. The underlying vision for IMT-2000 and the 3G capabilities includes:

- Common initial spectrum worldwide (1.8-2.2 GHz band)

- Multiple radio environments (cellular, cordless, satellite, LANs)

- Wide range of telecommunications services (voice, data, multimedia, internet)

- Flexible radio bearers for increased spectrum efficiency

- Data rates up to 2 Mb/s (phase 1) for indoor environments

- Maximum use of IN (Intelligent Networks) capabilities (for service provision and transport)

- Global seamless roaming and service delivery across IMT-2000 Family Member networks

- Support of VHE (Virtual Home Environment) and UPT (Universal Personal Telecommunication)

- Enhanced security and performance

- Integration of satellite and terrestrial systems to provide global coverage.

9.6.2 CDMA-2000

cdma2000 is the third generation version of cdmaOne or IS-95. The cdma2000 Radio Transmission Technology (RTT) is a spread spectrum, wideband radio interface. It uses CDMA technology as its underlying modulation technology. cdma2000 meets the specification for ITU (International Telecommunication Union) and IMT-2000. It addresses the specification for indoor, indoor-to-outdoor, pedestrian and vehicular environment. cdma2000 can operate in wide range of environments, viz.,

- Indoor/Outdoor picocell (<50 meter radius; e.g., one office floor)

- Indoor/Outdoor microcell (up to 1KM radius; e.g., a shopping mall)

- Outdoor macrocell (1–35KM radius)

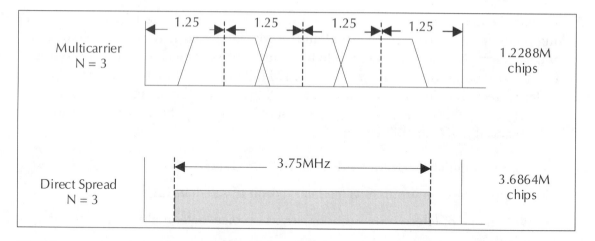

Figure 9.9 Multicarrier and Direct Spread in cdma2000

- Outdoor megacell (>35KM radius)

- Wireless in Local Loop (WiLL).

cdma2000 supports chip rates of N × 1.2288 Mcps (where N = 1, 3, 6, 9, 12). For N = 1, the spreading is similar to IS-95. However, for forward link QPSK modulation is used before the spread. There are two option for chips rate for N > 1. These are multicarrier and direct spread (Figure 9.9). In the multicarrier procedures for N > 1, the modulation symbols are demultiplexed onto N separate 1.25MHz carriers where N = 3, 6, 9, 12. Each of these carriers is then spread with 1.2288 M chips. For direct spread procedures for N > 1, the modulation symbols are spread on a single carrier with a chip rate of N × 1.2288M chips where N = 3, 6, 9, 12.

Two types of data services are currently under consideration for cdma2000. These are packet data and high speed circuit switched data. Packet data will be used for bursty traffic like Internet of mails. The circuit switched data can be used for delay sensitive real-time traffic. Video applications are potential candidates for circuit switch data as they need a dedicated channel for the duration of the call.

The cdma2000 will have phased development. The phase 1 of the cdma2000 effort, branded as CDMA 1x, employs 1.25 MHz of frequency bandwidth and delivers a peak data rate of 144 KBps for stationary or mobile applications. In India some of the WiLL operators (Tata Telecom and Reliance Infocomm) are using this technology for WiLL and mobile services. Reliance Infocomm in India is also offering data services with multimedia applications. The Phase 2 of cdma2000 development branded as CDMA

3x will use 5 MHz bandwidth. CDMA 3x is expected to support 144 KBps data for mobile and vehicular applications and up to 2 MBps data for fixed applications. The primary difference between second generation CDMA (cdmaOne or IS-95) and third generation CDMA (cdma2000) is bandwidth and peak data rate capability.

9.6.3 UMTS/WCDMA

The standards body for ETSI for 3G is called UMTS and 3GPP. Some of the CDMA encoding techniques are patented by Qualcomm. To avoid copyright issues, ETSI in Europe and ARIB in Japan have devised a different flavor of CDMA. This is branded as Wideband CDMA or WCDMA. WCDMA is also known as UTRAN (UMTS Terrestrial Radio Access Network) FDD (Frequency Division Duplex). Their responsibilities are similar with overlapping functions and responsibilities.

The physical layer of the universal mobile telecommunications system (UMTS) wideband code division multiple access (WCDMA) standard uses direct sequence spread spectrum (DSSS) modulation with a chip rate of 3.84 Mcps. The channel bandwidth is 5 MHz, this wider bandwidth has benefits such as higher data rates and improved

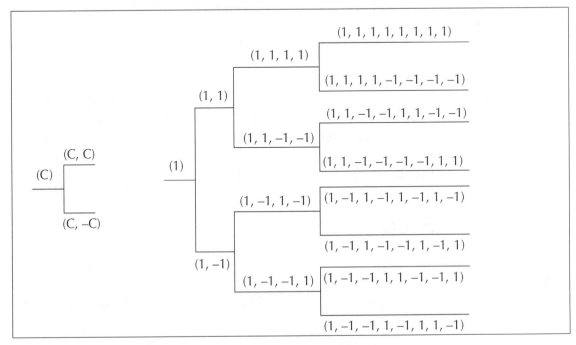

Figure 9.10 The Code Tree (Walsh Code) in WCDMA

multipath resolution. The data rates supported ranges from a few kb/s to 2 Mb/s. The physical layer supports two modes of operation: FDD (Frequency Division Duplex) and TDD (Time Division Duplex).

FDD and TDD Operational Modes

The frequency-division duplex (FDD) mode carries the uplink and downlink channels on separate frequency bands of 5 MHz each. This mode is typically used for large outdoor cells because it can support a larger number of users. The uplink and downlink transmissions in FDD mode are assigned fixed and equal frequency bands. This assignment works well when carrying voice traffic since such traffic tends to have uplink and downlink transmissions of approximately equal size. In time-division duplex (TDD) mode, the transmissions share the same frequency band by sending the uplink and downlink channels during different time slots. The TDD mode does not support as many users as the FDD mode, and hence the TDD mode is more suitable for smaller cells. In Internet the traffic pattern is asymmetric as the bandwidth requirement for download is more than the upload. Therefore, the TDD mode is more suited for carrying asymmetric data traffic like Internet. In TDD mode the uplink and downlink bandwidths can be modified by assigning more or fewer time slots to each link as and when necessary.

The Walsh codes in WCDMA are generated using the code tree as shown in Figure 9.10. If we look at these codes carefully, we will find that the WCDMA codes are same as IS-95. The spreading codes used in WCDMA are called orthogonal variable spreading factor (OVSF) codes, and the spreading factor can vary from SF=4 to SF=512. Since the 3.84 Mcps chip rate is held constant, higher data rates are obtained by using shorter spreading codes and lower data rates are obtained using longer spreading codes. Decreasing the spreading factor increases the data rate but reduces the number of users that can be supported because fewer codes are available at the shorter spreading factors.

9.6.4 Fixed Wireless

3G is commonly associated with mobile phones. However, the 3G specification includes the fixed wireless as well. Presently, we use separate links for data and voice. A fixed wireless will make it only one common link. The IMT-2000 specification makes specific provisions for 3G Fixed Wireless Access (FWA).

In most emerging economies, and developing countries, the wired infrastructure is inadequate. Fixed Wireless Access is expected to become the mainstream technology in such geographies. In developed countries, however, 3G residential wireless represents a new horizon for competitive access providers. Users can expect a wireless connection to provide somewhere between 1.5 Mbps and 2 Mbps data at home. In India some operators are using CDMA 1X technology for Wireless in Local Loop (WiLL). They are offering both mobile and fixed phones. These fixed phone lines are examples of fixed wireless.

Fixed wireless 3G is a converged, multimedia-driven technology. In fixed mode, 3G utilizes a point-to-multipoint network architecture that can transmit data and voice simultaneously at high speeds across core wireless infrastructure. Potential applications for 3G fixed services include business and home networking which creates a high-speed interface/gateway between an in-building 'network of networks' (e.g., wireless interworking of telephony, data, video, home energy monitoring, and security networks) and the outside world (e.g. the Internet and the PSTN).

9.7 Applications on 3G

Devices in 3G can works in multiple ways. It can run in a tunneling mode or in an application mode. In a tunneling mode the device works more as a pass through device or a modem. In this mode, the mobile phone is connected to another device like a laptop and functions as a wireless media interface. The intelligence of the phone is not used, only the communication interface of the phone is used.

In an application mode, applications run on the phone itself. A 3G mobile phone will support, SMS, WAP, Java etc. (MExE classmark 3). A MExE classmark 3 mobile device will have an execution environment that will allow application development for the client device. This application platform can be Java (through JavaPhone, PersonalJava, or J2ME, Java virtual machine), C/C++ (through Symbian, Brew or PalmOS) or Visual Basic (through Windows CE).

MExE classmark 3 devices will offer API to access device resource. These device resources will be SMS, messaging, diary, address book etc. In future, network related information will also be available to the MExE environment through API (Application Programming Interfaces). WTAI (Wireless Telephony Application Interface) can also be used in a WAP environment to access the telephone resource.

In 3G, there will be different types of client applications. These are:

1. Local

2. Occasionally connected

3. Online

4. Realtime.

Games, cartoons and similar applications are examples of local applications. These applications can be downloaded over the air and used offline. In an occasionally connected computing (OCC) environment, the user will connect to the network occasionally. Downloading and uploading of email are the best examples of OCC. Online applications will be the corporate applications. Examples of such applications will be online order booking or updating of inventory status. Realtime applications could be realtime stock update or applications for law-enforcement agents for realtime tracking or navigational systems.

9.7.1 3G Specific Applications

There will be different types of applications in 3G networks. These will be for both fixed wireless and mobile. Different types of applications are candidates for 3G. These include

- Personal Applications
- Content Applications
- Communication Applications
- Productivity Applications
- Business Applications.

Majority of these applications were discussed in previous chapters. We will discuss some of the applications, which are new and specific to 3G.

Virtual Home Environment

Conceptually, the Virtual Home Environment can be defined as a concept where an environment is created in a foreign network (or home network outside the home environment) so that the mobile users can experience the same computing experience as they have in their home or corporate computing environment while they are mobile and roaming. VHE is therefore aimed at roamers

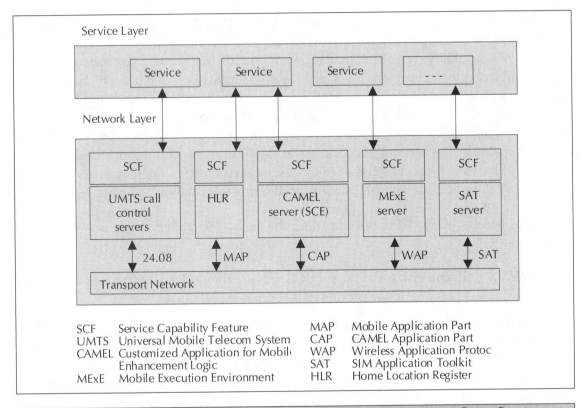

Service Layer

Network Layer

| SCF | SCF | SCF | SCF | SCF |

| UMTS call control servers | HLR | CAMEL server (SCE) | MExE server | SAT server |

24.08   MAP   CAP   WAP   SAT

Transport Network

SCF	Service Capability Feature	MAP	Mobile Application Part
UMTS	Universal Mobile Telecom System	CAP	CAMEL Application Part
CAMEL	Customized Application for Mobile Enhancement Logic	WAP	Wireless Application Protoc
MExE	Mobile Execution Environment	SAT	SIM Application Toolkit
		HLR	Home Location Register

Figure 9.11 The Virtual Home Environment architecture over Open Service Architecture

3GPP defined VHE as 'a system concept for personalisation service portability across network boundaries and between terminals'. The aim is to enable end users to access the services of their home network/service provider even when roaming in the domain of another network provider, thus making them feel 'virtually at home'. VHE will allow a user to personalize the set of services for which he/she has a subscription with his/her home network and provide these home services with the user's personalized 'look and feel' across different types of networks–mobile, public switched telephone network (PSTN), Internet. VHE will offer the same user experience over varied terminals–mobile, laptop, fixed phone, PDA, PC he/she might be using. Therefore in VHE environment device identification, location awareness and content adaptation are going to be major challenges. An example of one of the personal service settings could be 'from 9:00 AM to 7:00PM I want to be alerted for incoming messages from my boss.' The VHE will automatically adapt the type of messaging the user is using at that time. If the user is using WAP terminal but not roaming in a network that supports WAP, the VHE will convert the message into another format (may be SMS). Figure 9.11 depicts the VHE architecture.

The VHE specification introduces some new concepts related to open service architecture (OSA). This is an open interface between network layer and the service layer. There will be service capability servers (SCS), which will provide functionality to create new services in an object-oriented fashion. If we take the MSC as an example of an SCS, call control will be a class consisting of several call control related functions. For example, 'create a new call leg' or 'If time is between 9:00AM to 6:00PM connect call leg A to call leg B and charge B, else connect call leg A to call leg C and charge B'. To establish a call from point X to point Y, the telecom network has to establish many connections between networks; these are called legs. The classes of OSA are called service capability features (SCF). Examples of SCFs are call control, location/positioning information and notification.

As identified by 3GPP VHE specification, the SCSs and their roles in service provisioning are:

- **UMTS call control servers:** As SCS servers they offer mechanisms for applications to access basic bearer and call control capabilities.

- **Home Location Register (HLR) servers:** The HLR is an intelligent database that contains location and subscriber information including the tariff and service provisioning details. The MAP (Mobile Application Part) protocol allows the exchange of location and subscriber information between different networks and services.

- **Mobile Execution Environment (MExE) servers:** These servers will service MExE services with Java, WAP and WTAI.

- **SIM Application Toolkit:** Applications based on Smart card technologies. This will be STA (SIM Toolkit Application), Java card, or USIM applications.

- **Customized Application for Mobile Networks Enhancements Logic (CAMEL) servers:** Camel extends the scope of IN (Intelligent Networks) service provisioning to the mobile environment. International roaming on prepaid cards is implemented using CAMEL.

Personal Communication Networks

Personal Communication Networks (PCN) are digital telephone networking infrastructures, which supports personal numbering, individual service selection, moves towards unified billing and call anytime, anywhere through wireless digital telephony. Personal communication networks are centered around the twin concepts of deploying extensive digital cellular networks and the notion of a unique identification number called a Universal Personal Telecommunication (UPT) number.

At the core of the PCN concept is the idea that each subscriber is assigned a personal identification number. This number identifies the subscriber to the network and enables them to receive or initiate phone calls, regardless of their respective location. All universal personal telecommunication numbers are held in a networked database. As a PCN mobile phone moves from one micro-cell to another it uses the signalling network to notify the network that its location has changed. Alternately, as a call for a given number enters a switching exchange, the exchange triggers a signalling system request across the network to look up in the database how to handle the call. The database enquiry returns information on how the call should be handled. Let us assume that Mr. A's mobile telephone number at Delhi is +91-9811083712. He goes to London and rents a mobile phone for a week with number +44-77896-56872. PCN will offer a facility by which when someone calls Mr. A's Delhi number, the call will automatically be routed to the mobile phone in London.

USIM

USIM (Universal Subscriber Identity Module) is the smart card for third generation mobile phones. A SIM card in the mobile phone offers portability, security and individuality. Some standards such as the Personal Digital Cellular (PDC) standard in Japan that NTT DoCoMo uses for i-mode previously never had SIM cards. However, they will now have SIM cards with their 3G offerings. A SIM card helps to makes a device independent from the network. The USIM is the next generation of smart card based subscriber identity module (SIM). SIM card was initially designed as a simple security device for subscriber/network authentication including the ability to roam across networks. SIM card soon became a platform for storing SMS, user's phonebook and preferences. The USIM smart card will continue to perform basic subscriber/network authentication functions but in a more flexible way. For example, it will employ contextual mechanisms that are dependent on the type of network detected.

The USIM will also provide enhanced personalization in the form of comprehensive phonebooks. These are similar to palmtop organizers and include e-mail and Web addresses alongside phone numbers. In addition, more sophisticated USIMs with high-performance processors and cryptography capabilities are likely to be available by the time 3G networks start to roll out.

The USIM has the following features:

- 64 Kbytes memory.
- Card operating system based on either Java or MULTOS (a popular smart card OS standard).

- Backwards Compatibility with GSM. USIMs will not work in GSM phones, but existing GSM SIM cards will work in 3G/ UMTS devices.

- A number of security features from PKI (Public Key Infrastructure) to WIM (Wireless Information Module) to security algorithms will be incorporated into different vendor's USIMs.

- The 3GPP is committed to open interfaces for USIM cards with defined Application Programming Interfaces (APIs) making it possible for application developers and network operators to develop new services.

Audio/Video

Audio or video over the Internet will be either downloaded or streamed. In a downloaded environment the content is transferred, stored and played offline (local application). In a streamed environment the content is played as it is being downloaded, often in a burst, but not stored (online application). Downloaded content is generally of better quality. Audio/video contents are transferred using various different compression algorithms such as those from Microsoft or Real Networks or the MPEG-1 Audio Layer 3 (better known as MP3) protocol. MP3 is a compression/ decompression algorithm. MP3 was invented in 1987 in Germany and approved by the Moving Pictures Experts Group, in 1992.With 3G, MP3 files will be downloadable over the air directly to the phone.

Third generation applications will be used to download music, multimedia news etc. In India, Reliance Infocomm is offering services where consumers can download popular Hindi, Tamil, Kanada, Bengali and other regional language songs and plays. Many of these also include video clips from the film. One can also download news clips from popular TV channels like Star or CNN.

Voice over Internet Protocol (VoIP/Voice over Packet Network)

Another audio application for 3G is Voice over IP (VoIP). In 3G, VoIP is a data application where normal voice calls will use Internet or other packet networks.

Electronic Agents

Electronic agents will play an important role in the future. Electronic agents will be dispatched to carry out searches and tasks on the Internet and report back to their owners.

This is an efficient way to get things done on the move. In fact, one would be able to bid for items one may have wanted. The mobile agent will find out when, where and how the auction is proceeding.

Electronic agents are defined as 'mobile programs that go places in the network to carry out their owners' instructions. They can be thought of as extensions of the people who dispatch them.' Agents are 'self-contained programs that roam communication networks delivering and receiving messages or looking for information or services.' One example of agents could be in a manufacturing industry where an agent will move from one vendor's system to another and finally make the bill of material ordered in hours as opposed to weeks. This will help implement the just in time manufacturing system.

Downloading of Software and Content

As we move into future, more and more content will be digital. Today, software is increasingly downloaded electronically from the Internet rather than purchased as boxed product in stores. In future people will be able to borrow a book from a digital library sitting at home.

ENUM

ENUM is a protocol that is emerging from work of Internet Engineering Task Force's (IETF's) Telephone Number Mapping working group. The charter of this working group is to define a Domain Name System (DNS)-based architecture and protocols for mapping a standard telephone number to a Uniform Resource Identifier (URI). This URI can be used to contact a resource associated with that telephone number. The protocol is defined in RFC2916 'E.164 number and DNS'. E.164 defines the syntax for the international public telecommunication telephony numbering plan and URI defines the syntax for Uniform Resource Identifiers (URIs defined in RFC2396).

Using as an example the 10 digit phone number (and country code) +1-440-951-7997, the ENUM process for converting this phone number into a DNS address is as follows:

1. Remove all characters, save the +, to read: +14409517997.

2. All characters are removed and dots are placed between these digits: 1.4.4.0.9.5.1.7.9.9.7 (in DNS terms, each digit between the dots can then become a defined and distributed zone. For this example, delegation to North America

at the country code zone designation of '1'. The same can be accomplished at the area code zone.

3. The order of the digits is reversed: 7.9.9.7.1.5.9.0.4.4.1.

4. The ENUM domain e164.arpa is put at the end: 7.9.9.7.1.5.9.0.4.4.1.e164.arpa.

REFERENCES/FURTHER READING

1. Vijay K Garg, IS-95, CDMA and cdma2000, Pearson Education, 2003.

2. Flavio Muratone (Editor), UMTS Mobile Communications for the Future, John Wiley & Sons, 2000.

3. Christoffer Andersson, GPRS and 3G Wireless Applications, John Wiley & Sons, 2001.

4. 1xEV: 1x EVolution IS-856 TIA/EIA Standard, Airlink Overview, QUALCOMM, November 2001.

5. 3GPP TR 22.970: 3rd Generation Partnership Project; Technical Specification Group Services and System Aspects Service aspects; Virtual Home Environment (VHE), 1999.

6. Fundamentals of Wireless Communications & CDMA Qualcomm, Student Guide CDMA-050 80-13127-1 X6, January 24, 2000.

7. Andy Dornan, CDMA and 3G Cellular Networks, Network Magazine, http://www.networkmagazine.com/article/NMG20000831S0006.

8. Hedy Lamarr: http://www.inventions.org/culture/female/lamarr.html.

9. Ir. J. Meel, Spread Spectrum, Sirius Communications, October 1999.

10. Andrew S. Tanenbaum, Computer Networks, Prentice-Hall of India, 1999

11. ETSI ETS 300 779, Network Aspects (NA); Universal Personal Telecommunication (UPT); Phase 1 - Service description, 1997.

12. The Future Mobile Market Global trends and developments with a focus on Western Europe UMTS Forum Report # 8, March 1999.

13. Enabling UMTS Third Generation Services and Applications, UMTS Forum Report # 11, October 2000.

14. Loading Java into USIM: http://forum.java.sun.com/thread.jspa?threadID-=611101&messageID=3360952.

REVIEW QUESTIONS

Q1: What is Direct Sequence Spread Spectrum Technology? Explain how it works in the CDMA technology?

Q2: Describe the IS-95 architecture. Compare this architecture with the GSM architecture?

Q3: Describe different types of handoff. What are the differences between Hard handoff, Soft handoff, and Softer handoff?

Q4: Give 6 functions where CDMA is different from GSM?

Q5: Describe 3G networks. How is a 3G network different from a 2G CDMA network?

Q6: Describe Virtual Home Environment (VHE). Explain how VHE is realized in 3G networks?

CHAPTER 10

Wireless LAN

10.1 INTRODUCTION

Wireless Local Area Network (LAN) is a local area data network without wires. Wireless LAN is also known as WLAN in short. Mobile users can access information and network resources through wireless LAN as they attend meetings, collaborate with other users, or move to other locations in the premises. Wireless LAN is not a replacement for the wired infrastructure. It is implemented as an extension to a wired LAN within a building or campus.

10.2 WIRELESS LAN ADVANTAGES

Schools, campuses, manufacturing plants, hospitals and enterprises install wireless LAN systems for many reasons. Some of these are:

- **Mobility:** Productivity increases when people have access to data and information from any location. Decision-making capability based on real-time information can significantly improve work efficiency. Wireless LAN offers wire-free access to information within the operating range of the WLAN.

- **Low Implementation Costs:** WLANs are easy to set up, relocate, change and manage. Networks that frequently change, both physically and logically, can benefit from WLAN's ease of implementation. WLANs can operate in locations where installation of wiring may be impractical.

- **Installation Speed and Simplicity:** Installing a wireless LAN system can be fast and easy and can eliminate the need to install cable through walls and ceilings.

- **Network Expansion:** Wireless technology allows the network to reach where wires cannot reach.

- **Reduced Cost-of-Ownership:** While presently the initial investment required for Wireless LAN hardware is higher than the cost of wired LAN hardware, overall installation expenses and life-cycle costs is expected to be significantly lower. Long-term cost benefits are the greatest in dynamic environments requiring frequent moves, adds and changes.

- **Higher User to Install Base Ratio:** Wireless environment offers a higher user to capacity ratio. For example in a wired network like telephone, physical wire needs to be laid for each and every subscriber. Whereas, for a cellular network the ratio between subscribers and available channel is from 10 to 25 or even more. This means that if there is capacity for 100 channels, the network operator can safely have 2500 subscribers. Likewise in a wireless LAN, the network can offer a very high level of return on investment.

- **Reliability:** One of the common causes of failure in wired network is downtime due to cable fault. WLAN is resistant to different types of cable failures.

- **Scalability:** Wireless LANs can be configured in a variety of topologies to meet the needs of specific applications and installations. Configurations are easily changed and range from peer-to-peer networks suitable for a small number of users to full infrastructure networks of thousands of users that allow roaming over a broad area.

- **Usage of ISM band:** Wireless LAN operates in the unregulated **ISM** (Industrial Scientific and Medical) band (2.40 GHz to 2.484 GHz, 5.725 GHz to 5.850 GHz) available for use by anyone. A user need not go to the government to get a license to use the wireless LAN. In India 2.4 GHz band is made free for use in WLANs. The 5.7 GHz band is not yet unregulated as it may conflict with the C-band of satellite.

Wireless LAN is also commercially known as WiFi or **Wi-Fi**. Wi-Fi is an acronym for Wireless Fidelity. The **Wi-Fi™** logo is a registered trademark of the Wireless Ethernet Compatibility Alliance (**http://wi-fi.org**), a group founded by many companies that develop 802.11 based products.

10.2.1 Wireless LAN Evolution

Wireless LAN development started in an unstructured way. Some vendors started offering wireless communication between the corporate LAN and mobile devices (like laptop computers). This is like using a wireless keyboard or a wireless mouse. Protocols and interfaces were proprietary. However, within a short period of time many vendors started offering products in this space. These products were incompatible and soon interoperability became an issue. As IEEE is responsible for maintaining Ethernet LAN standards, IEEE assumed the responsibility of defining the wireless Ethernet LAN standards. The initial standard was published in June 1997. All these early 802.11 systems are first generation systems.

It was not until the introduction of the 11-Mbps 802.11b standard in September 1999 that the horizontal WLAN market achieved some semblance of legitimacy. Also, standards like 802.11a and 802.11g offered much higher bandwidth. All these are second generation WLANs. Second generation WLANs extended the security through 802.1x specifications and offered horizontal roaming. In a horizontal roaming, a user can move from one AP to another AP seamlessly.

In third generation WLANs, vertical roaming will be possible. Vertical roaming will provide seamless roaming between different networks. Third generation WLANs will integrate with third generation (3G) telecom networks. These WLANs will eliminate the boundaries between enterprise LAN (both wireline and wireless) systems and the public wireless systems for seamless roaming. It will extend the application of IP mobility standards. Also the security system is being extended. These will be achieved through standards like 802.11f and 802.11i.

10.2.2 Wireless LAN Applications

There are many areas and applications of wireless LAN. Wireless LAN is best suited for dynamic environments. Following are some of the examples.

Office/Campus Environment

WLAN is very useful in office environments and buildings with a big campus. In big buildings or in campuses people move between floors, rooms, indoors and outdoors. In an office environment, a person can move with his laptop to the meeting room and continue working. In a university campus, a student can move from the library to the cafeteria and continue working. In a hotel, a guest can move to the pool and work. In a hospital, a doctor can carry the patient information with him while on a regular round.

Factory Shop Floor

This includes environments like factory shop floor, warehouse, exhibition sites, retail shops, labs etc. These are very dynamic environments, where floor layouts change very frequently; objects within the building are constantly moving. Laying cables and setting up a wired LAN in these kinds of facilities are almost impossible. Wireless LAN can be very useful in such situations.

Homes

In homes WLAN can be used for convergence applications. These will include networking of different home devices like phones, computers and appliances.

Workgroup Environment

Any set-up where small workgroups or teams need to work together be it within a building or in the neighbourhood, WLAN can be very useful. This may include a survey team on top of a hill or rescue members after a natural disaster or an accident site. WLAN can be very useful in civil construction sites as well.

Heritage Buildings

There are many building of national heritage, where a data network needs to be set up. In a very old church for example, if we need to setup a virtual reality show, it is difficult to install a wired LAN. Wireless LAN can solve the problem.

Public Places

This includes airports, railway stations or places where many people assemble and need to access information.

War/Defense Sites

When there is a war or war game, access to networks help. There is some major research going on in US on mobile ad hoc networks for defense establishments.

10.3 IEEE 802.11 Standards

The IEEE 802 committee was set up in February 1980 (that is the origin of the name) to set the standard for local area networks. From time to time, IEEE came up with different standards in the LAN domain. This includes all the layers from physical, media access, and data link layer. When IEEE deliberated the standards, it was clear that wireless LAN will be different only at the physical and media access layer.

There are quite a few types of wireless LANs. However, IEEE 802.11 is gaining fast momentum in India and other parts of the world. When we refer to 802.11, we generally mean the generic 802.11 families of standards. There are many standards within this family with almost all the letters of the English alphabet starting from 'a' to 'x'. Different standards cover different aspects like bandwidths, modulation techniques, physical media, security etc. Table 10.1 is a list of all these approved standards, whereas table 10.2 refers to emerging standards.

Table 10.1 The IEEE Standards

Standard	Description	Status
IEEE 802.11	Standard for WLAN operations at data rates up to 2 Mbps in the 2.4-GHz ISM (Industrial, Scientific and Medical) band.	Approved in July 1997.
IEEE 802.11a	Standard for WLAN operations at data rates up to 54 Mbps in the 5-GHz Unlicensed National Information Infrastructure (UNII) band.	Approved in Sept 1999. End-user products began shipping in early 2002.
IEEE 802.11b	Standard for WLAN operations at data rates up to 11 Mbps in the 2.4-GHz ISM (Industrial, Scientific and Medical) band.	Approved in Sept 1999. End-user products began shipping in early 2000.
IEEE 802.11g	High-rate extension to 802.11b allowing for data rates up to 54 Mbps in the 2.4-GHz ISM band.	Draft standard adopted Nov 2001. Full ratification expected in 2003.
IEEE 802.15.1	Wireless Personal Area Network standard based on the Bluetooth specification, operating in the 2.4-GHz ISM band.	802.15.1-2002 conditionally approved on March 21, 2002.
IEEE802.1x	Port-based network access control defines infrastructures in order to provide a means of authenticating and authorizing devices attached to a LAN port that has point-to-point connection characteristics.	Approved June 2001

Table 10.2 IEEE Emerging Standards

Upcoming Standard	Description
IEEE 802.11e	Enhance the 802.11 Medium Access Control (MAC) to improve and manage Quality of Service, provide classes of service, and enhanced security and authentication mechanisms. These enhancements should provide the quality required for services such as IP telephony and video streaming.
IEEE 802.11f	Develop recommended practices for an Inter-Access Point Protocol (IAPP), which provides the necessary capabilities to achieve multi-vendor Access Point interoperability across a Distribution System supporting IEEE P802.11 Wireless LAN Links.
IEEE 802.11h	Enhance the 802.11 Medium Access Control (MAC) standard and 802.11a High Speed Physical Layer (PHY) in the 5GHz band. Objective is to make IEEE 802.11ah products compliant with European regulatory requirements.
IEEE 802.11i	Enhance the 802.11 Medium Access Control (MAC) to enhance security and authentication mechanisms.
IEEE 802.15 TG2	Developing Recommended Practices to facilitate coexistence of Wireless Personal Area Networks™ (802.15) and Wireless Local Area Networks (802.11).
IEEE 802.15 TG3	Draft and publish a new standard for high-rate (20Mbit/s or greater) WPANs™ .
IEEE 802.15 TG4	Investigate a low data rate WPAN solution with multi-month to multi-year battery life and very low complexity.

10.4 WIRELESS LAN ARCHITECTURE

10.4.1 Types of Wireless LAN

There are different types and flavors of wireless local area networks. Some of the most popular ones are:

- **802.11** In June 1997, the IEEE finalized the initial specification for wireless LANs: IEEE 802.11. This standard specifies a 2.4 GHz frequency band with data rate of 1 Mbps and 2 Mbps. This standard evolved into many variations of the specification like 802.11b, 802.11a, 802.11g, etc. using different encoding technologies. Today these standards offer a local area network of bandwidths going up to a maximum of 54Mbps.

- **HyperLAN** HyperLan began in Europe as a specification (EN 300 652) ratified in 1996 by the ETSI Broadband Radio Access Network group. HyperLAN/1, the current version works at the 5 GHz band and offers up to 24 MBps bandwidth.

Next version HyperLAN/2 (http://www.hyperlan2.com) will support a bandwidth of 54 Mbps with QoS support. This will be able to carry Ethernet frames, ATM cells, IP packets and support data, video, voice and image.

- **HomeRF** In 1998, the HomeRF Working Group (http://www.homerf.org) offered to provide an industry specification to offer Shared Wireless Access Protocol (SWAP). This standard will offer interoperability between PC and consumer electronic devices within the home. SWAP uses frequency hopping spread spectrum modulation and offers 1 Mbps and 2 Mbps at 2.4 GHz frequency band.

- **Bluetooth** Bluetooth was promoted by big industry leaders like IBM, Ericsson, Intel, Lucent, 3Com, Microsoft, Nokia, Motorola, and Toshiba. It was named after Harold Bluetooth, King of Denmark during 952 to 995 A.D., who had a vision of a world with cooperation and interoperability. Bluetooth is more of a wireless Personal Area Network (PAN) operating at 2.4 GHz band and offers 1Mbps data rate. Bluetooth uses frequency hopping spread-spectrum modulation with relatively low power and smaller range (about 10 meters).

- **MANET** Manet (http://www.ietf.org/html.charters/manet-charter.html) is a working group within the IETF to investigate and develop the standard for Mobile ad hoc NETworks.

10.4.2 Ad hoc versus Infrastructure Mode

Wireless Networks are of two types, infrastructure mode and ad hoc mode. In an infrastructure mode, the mobile stations (MS) are connected to a base station or Access Point (AP). This is similar to a star network where all the mobile stations are attached to the base station. Through a protocol the base station manages the dialogues between the AP and the MS. Figure 10.1 depicts a wireless LAN in infrastructure mode.

In an ad hoc mode, there is no access point or infrastructure. A number of mobile stations form a cluster communicate with each other. Figure 10.2 depicts wireless LAN in adhoc mode.

In an Infrastructure mode, 802.11 LAN is based on a cellular architecture where the system is subdivided into small clusters or cells (see Figure 10.1). Each cell is called Basic Service Set, or BSS. Depending on the topology one BSS is connected to other BSS or other infrastructure. In an ad hoc network, the BSS is completely independent. Therefore, technically an ad hoc network is termed as Independent BSS or IBSS. Whereas, in infrastructure mode the mobile stations form a cluster with an AP. Multiple

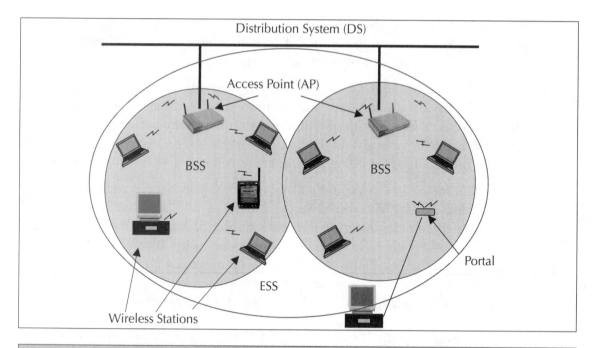

Figure 10.1 Wireless LAN in Infrastructure mode

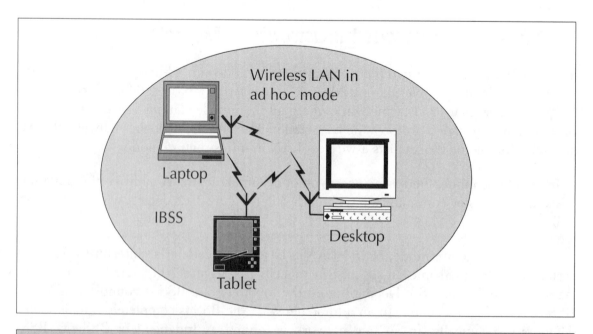

Figure 10.2 Wireless LAN in adhoc mode

such BSS form an Extended Service SET (ESS). The ESS is connected to the backbone LAN or the distribution system.

10.4.3 802.11 Architecture

In the 802.11 nomenclatures one cell or a BSS is controlled by one Base Station. This base station is called Access Point or AP in short. In some literature, access points are referred to as Hot Spots.

Although a wireless LAN may be formed by a single cell, with a single Access Point, most installations will be formed by several cells, where the access points are connected through some kind of backbone. This backbone is called Distribution System or DS (Figure 10.1). This backbone is typically Ethernet and, in some cases, is wireless itself (see Figure 10.3).

The whole interconnected Wireless LAN, including the different cells, their respective Access Points and the Distribution System, is seen as a single 802 network to the upper layers of the OSI model and is called as Extended Service Set (ESS).

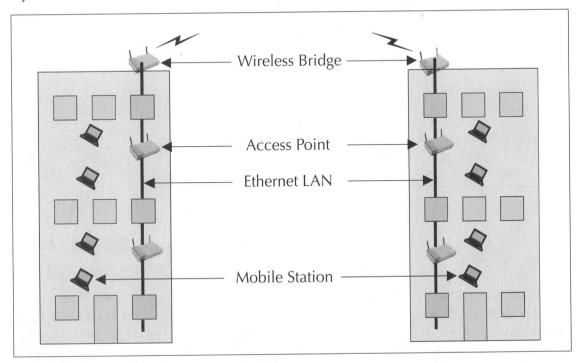

Figure 10.3 Two access points (Wireless Bridges) as part of the Distribution System

The 802.11 standard also defines the concept of a 'portal' (see Figure 10.1). A portal is a device that interconnects between an 802.11 and another 802 LAN. This concept is an abstract description of part of the functionality of a bridge.

Cells Design in Wireless LAN

For proper functioning of wireless LAN, neighboring cells (BSS) are set up on different frequencies, so that wireless LAN cards in each cell do not interfere with one another when they transmit signals. In order for these cells to work without interference, the DSSS standards define 13 different frequencies or channels (table 10.3). For Frequency Hopping Spread Spectrum (FHSS) there are 79 channels. These frequencies are typically 'non-overlapping'. This means that they operate in different sections of the radio spectrum or band.

Table 10.3 Channels within the 2.4GHz band

Channel No	Frequency (GHz)
1	2.412
2	2.417
3	2.422
4	2.427
5	2.432
6	2.437
7	2.442
8	2.447
9	2.452
10	2.457
11	2.462
12	2.467
13	2.472

In a design where there are many cells, effective use of these non-overlapping frequencies are very important.

In the following design (see Figure 10.4) there are four wireless cells (cells A through D). A cell is defined by the space and area the radio wave of a wireless LAN access point is able to cover. Cells A, B and C all use non-overlapping frequencies, while cell D uses the same frequency as cell A. The use of two frequencies by two cells will not have any effect on each other, as long as distance 'x' is great enough to ensure effective radio isolation from each other. Radio isolation or radio separation means that a device in cell A will

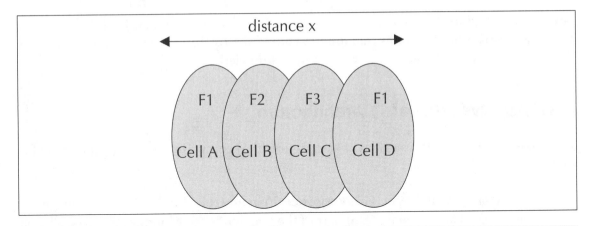

Figure 10.4 Cell design in a WLAN

not be able detect the signal transmitted by any device in cell D. This is because the air distance between two cells attenuates or weakens the radio signal so much that they cannot detect the radio signal of each other.

IEEE 802.11 Layers Description

The 802.11 standards cover definitions for both MAC (Medium Access Control) and Physical Layer. The standard currently defines a single MAC, which interacts with three PHYs (Figure 10.5) as follows:

- Frequency Hopping Spread Spectrum
- Direct Sequence Spread Spectrum, and
- InfraRed.

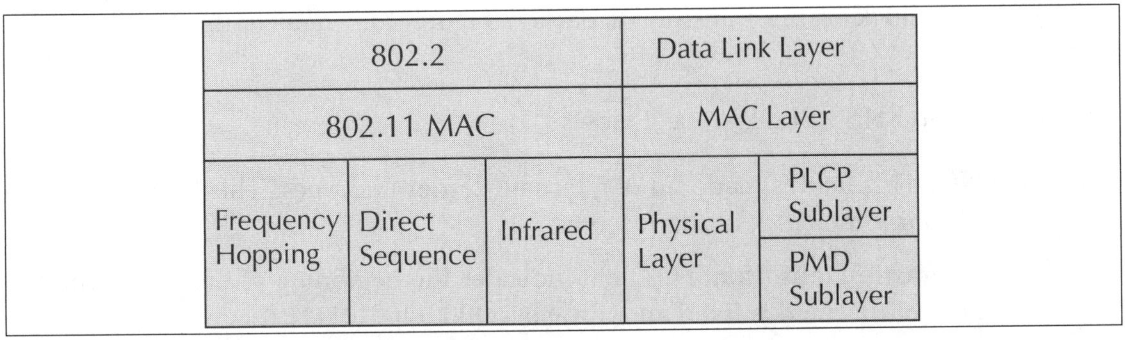

802.2			Data Link Layer	
802.11 MAC			MAC Layer	
Frequency Hopping	Direct Sequence	Infrared	Physical Layer	PLCP Sublayer
				PMD Sublayer

Figure 10.5 The 802.11 stack

Beyond the standard functionality usually performed by media access layers, the 802.11 MAC performs other functions that are typically done by upper layer protocols, such as Fragmentation, Packet Retransmissions, and Acknowledgements.

Physical Layer (Layer 1) Architecture

The architecture of the Physical layer (see Figure 10.5) comprises of the two sublayers for each station:

- **PLCP (Physical Layer Convergence Procedure):** PLCP sublayer is responsible for the Carrier Sense (CS) part of the Carrier Sense Multiple Access/Collision Avoidance (CSMA/CA) protocol. PLCP layer prepares the MAC Protocol Data Unit (MPDU) for transmission. The PLCP also delivers the incoming frames from the wireless medium to the MAC layer. PLCP appends fields to the MPDU that contains information needed by the physical layer transmitter and receiver. This frame is called PLCP Protocol Data Unit (PPDU). The structure of PLCP provides for asynchronous transfer of MPDU between stations. The PLCP header contains logical information that allows the receiving stations physical layer to synchronize with each individual incoming packet.

- **PMD (Physical Medium Dependent):** The PMD provides the actual transmission and reception of physical layer entities between stations through the wireless media. This sublayer provides the modulation/demodulation of the transmission.

FHSS (Frequency Hopping Spread Spectrum) Physical Layer

In FHSS mode, this layer carries the clocking information to synchronize the receiver clock with the clock of the transmitted packet. Figure 10.6 depicts the FHSS PPDU packet.

The fields in the FHSS PLCP are as follows:

1. **SYNC.** This field is made up of alternate zeroes and ones. This bit pattern is to synchronize the clock of the receiver.

2. **Start Frame Delimiter.** This field indicates the beginning of the frame and the content of this field is fixed and always is 0000110010111101.

3. **PSDU Length Word (PLW).** This field specifies the length of the PSDU in octets.

80 Bits	16 Bits	12 Bits	4 Bits	16 Bits	Variable PLCP Service Data Unit
SYNC	Start Frame Delimiter	PLW	PSF	Header Error Check	

Figure 10.6 Frequency Hopping Spread Spectrum PLCP

4. **PLCP Signaling (PSF).** This field contains information about the data rate of the fields from whitened PSDU. The PLCP preamble is always transmitted at 1Mbps irrespective of the data rate of the wireless LAN. This field contains information about the speed of the link. For example 0000 means 1 Mbps and 0111 signifies 4.5 Mbps bandwidth.

5. **Header Error Check.** This field contains the CRC (Cyclic Redundancy Check) according to CCITT CRC-16 algorithm.

FHSS PMD is responsible for converting the binary bit sequence into analog signal and transmit the PPDU frame into the air. FHSS PDM does this using the frequency hopping technique. The 802.11 standard defines a set of channels within the ISM band for frequency hopping. For US and Europe there are 79 1 MHz channels within 2.402 to 2.480 GHz band. The FHSS PMD transmits PPDU by hopping from channel to channel according to a particular pseudo-random hopping sequence. Once the hopping sequence is set in the access point, stations automatically synchronize to the correct hopping sequence.

Direct Sequence Spread Spectrum (DSSS) Physical Layer

DSSS PLCP is responsible for synchronizing and receiving the data bits correctly. Figure 10.7 depicts the DSSS PPDU packet.

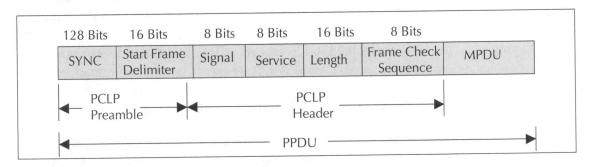

Figure 10.7 Direct Sequence Spread Spectrum PLCP Protocol Data Unit

The fields in the DSSS PLCP are as following:

1. **SYNC.** This field is made up of alternate zeroes and ones. This bit pattern is to synchronize the clock of the receiver with the received frame.

2. **Start Frame Delimiter.** This field indicates the beginning of the frame and the content of this field is fixed and is always 1111001110100000.

3. **Signal.** This field defines the type of modulation the receiver must use to demodulate the signal. When the value of this field is multiplied by 100 Kbps we get the bandwidth of the transmission. For 11 Mbps bandwidth this field will have a value of 01101110 (decimal 110). The PLCP preamble and the header are always transmitted at 1 Mbps. The bandwidth defined by this field applies to MPDU field.

4. **Service.** This field is not used and is usually 0.

5. **Length.** This field contains an unsigned 16-bit integer indicating the length of the frame. However, unlike the FHSS, this is not in octets. It is rather in microseconds. The receiver will use this to synchronize with the clock to determine the end of frame.

6. **Frame Check Sequence.** This is a 16-bit checksum based on CCITT CRC-16 algorithm.

DSSS PMD translates the binary digital sequence into analog radio signals and transmits the PPDU frame into the air. The DSSS physical layer operates within the ISM band. If we take the 2.4 GHz band, then it is between 2.4 GHz and 2.8435 GHz (802.11b and 802.11g) frequency band divided into multiple channels with 22 MHz width.

In DSSS the data is spread with a pseudo random noise (PN) code. This PN sequence is referred to as chip or spreading sequence. For 1MBps and 2MBps 802.11, the PN sequence is called the 11-bit Barker sequence. It is an 11 bit sequence of positive and negative ones like +1, −1, +1, +1, −1, +1, +1, +1, −1, −1, −1 (Figure 10.8). 5.5Mbps and 11 Mbps versions of 802.11b do not use the Barker sequence. They use Complementary Code Keying (CCK) technique instead. CCK is a set of 64 eight-bit code words used to encode data for 5.5 and 11 Mbps data rates. All these codes have unique mathematical properties that allow them to be correctly distinguished from one another by a receiver even in the presence of substantial noise and multipath interference.

Every bit in the data stream of the PPDU is modulated with this PN sequence. For example, in case of 802.11, a 1 in the data bit will be represented as +1,−1,+1,+1, −1,+1,+1,+1,−1,−1,−1; whereas a 0 in the data bit will be represented as −1,+1,−1,−1,+1, −1,−1,−1,+1,+1,+1 (Figure 10.8).

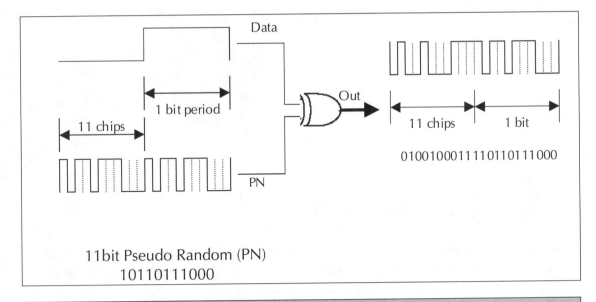

Figure 10.8 DSSS modulator

The DSSS used in wireless LAN and the DSSS used in the CDMA (IS-94 or CDMA-2000) for wireless MAN (Metropolitan Area Network) used in CDMA phones operate in similar fashion with some difference. In wireless LAN the chip used for each and every mobile station is the same. However, in case of wireless MAN the chip used for each different mobile station (for uplink or reverse path) are different.

The MAC Layer (Layer 2) Architecture

The MAC Layer defines two different access methods, the Distributed Coordination Function and the Point Coordination Function:

The Basic Access Method: CSMA/CA

The basic access mechanism, called the Distributed Coordination Function by IEEE standard, is Carrier Sense Multiple Access with Collision Avoidance mechanism (CSMA/CA). CSMA protocols are well known in the industry, the most popular being the Ethernet, which is a CSMA/CD protocol (CD stands for Collision Detection). In a wired environment (Ethernet for example) every station connected to the wire can sense the signal in the wire. In a wired LAN, if there is no activity or a collision of messages, every station connected to the LAN will be able to sense the collection almost instantly.

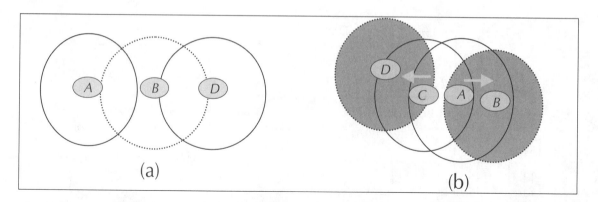

Figure 10.9 (a) Hidden terminal; (b) Exposed terminal

This is not true in the case of wireless media. In the case of wireless LANs, a Carrier Sense Multiple Access/Collision Avoidance (CSMA/CA) protocol is used, as it is not possible to detect a collision of data packets in mid air.

Consider the scenario with three mobile nodes as shown in Figure 10.9(a). The transmission of *A* reaches *B*, but not *C*. The transmission of *C* reaches *B*, but not *A*. However, the radio signal of *B* reaches both *A* and *C* making *A* and *C* both in the range of *B*. The net effect is *A* cannot detect *C* and vice versa.

A starts sending to *B*, *C* does not receive this transmission. *C* also wants to send to *B* and senses the medium. To *C* the medium appears to be free. Thus *C* starts sending causing collision at *B*. But now *A* cannot detect the collision and continues with its transmission. *A* is 'hidden' for *C* and vice versa.

Consider another case as shown in Figure 10.9(b). The radio transmission signal of *A* reaches *C* and *B*. The radio signal of *C* reaches both *A* and *D*. A wants to communicate to *B*, *A* starts sending signal to *B*. *C* wants to communicate to *D*, *C* senses the carrier and finds that *A* is talking to *B*. *C* has to wait till the time *A* finishes with *B*. However, *D* is outside the range of *A*, therefore waiting is not necessary. In fact *A*, *B* and *C*, *D* can communicate to each other in parallel without any collision, but according to the protocol that is not possible. *A* and *C* are 'exposed' terminals.

While Collision Detection mechanisms are a good idea on a wired LAN, they cannot be used on a Wireless LAN environment for two main reasons:

- Implementing a Collision Detection Mechanism requires the implementation of a Full Duplex radio capable of transmitting and receiving at the same time. This increases the cost significantly.

- In a Wireless environment we cannot assume that all stations will be able to receive radio signal from each other (which is the basic assumption of the Collision Detection scheme). The fact that a station wants to transmit and senses the medium as free (not able to sense signal from another station) does not necessarily mean that the medium is free (like the case of the hidden terminal) around the receiver area.

The mechanism behind CSMA/CA is as follows:

- When a wireless station (a wireless LAN device) wants to communicate, it first listens to its media (radio spectrum) to check if it can sense radio wave from any other wireless station.

- If the medium is free for a specified time then the station is allowed to transmit. This time interval is called Distributed Inter Frame Space (DIFS).

- If the current device senses carrier signal of another wireless device on the same frequency, as it wants to transmit on, it backs off (does not transmit) and initiates a random timeout.

- After the timeout has expired, the wireless station again listens to the radio spectrum and if it still senses another wireless station transmitting, continues to initiate random timeouts until it does not detect or senses another wireless station transmitting on the same frequency.

- When it does not sense another wireless station transmitting, the current wireless station starts transmitting its own carrier signal to communicate with the other wireless station, and once synchronized, transmits the data.

- The receiving station checks the CRC of the received packet and sends an acknowledgment packet (ACK). Receipt of the acknowledgment indicates to the transmitter that no collision occurred. If the sender does not receive the acknowledgment then it retransmits the fragment until it receives acknowledgment or is abandoned after a given number of retransmissions.

It can be seen from the above that the more times a wireless station has to back off or go into a random timeout, the less opportunity it has to transmit its data. This reduced opportunity for data transmission leads to less effective access to wireless bandwidth. This reduces the speed of the operation. In a worse case scenario the system would, after a number of retries, completely timeout and the wireless connection would be lost.

Virtual Carrier Sense

In order to reduce the probability of two stations colliding because they cannot sense each other's presence, the standard defines a Virtual Carrier Sense mechanism: A station wanting to transmit a packet first transmits a short control packet called **RTS** (Request To Send), which includes the source, destination, and the duration of the following transaction (the data packet and the respective ACK). The destination station after receiving this request packet responds with a response control packet called CTS (Clear to Send), which includes the same duration information.

All stations receiving either the RTS and/or the CTS, set their Virtual Carrier Sense indicator called Network Allocation Vector or NAV, for the given duration, and use this information together with the Physical Carrier Sense when sensing the medium. This mechanism reduces the probability of a collision on the receiver side by a station that is 'hidden' from the transmitter to the short duration of the RTS transmission because the station senses the CTS and 'reserves' the medium as busy until the end of the transaction. The duration information on the RTS also protects the transmitter area from collisions during the ACK (from stations that are out of range of the acknowledging station). It should also be noted that, due to the fact that the RTS and CTS are short frames, the mechanism also reduces the overhead of collisions, since these are recognized faster than if the whole packet was to be transmitted. The diagrams (Figure 10.10) show a transaction between stations A and B, and the NAV setting of their neighbors:

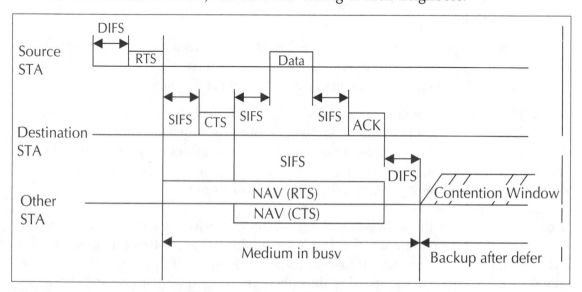

Figure 10.10 The CSMA/CA protocol

Fragmentation and Reassembly

Typical LAN protocols use packets several hundred bytes long (the longest Ethernet packet could be up to 1518 bytes long). There are several reasons why it is preferable to use smaller packets in a Wireless LAN environment:

- Due to the higher Bit Error Rate of a radio link, the probability of a packet getting corrupted increases with the packet size.

- In case of packet corruption (either due to collision or noise), the smaller the packet, the less overhead it causes to retransmit it.

- On a Frequency Hopping system, the medium is interrupted periodically for hopping, so, the smaller the packet, the smaller the chance that the transmission will be postponed after dwell time.

In a majority of cases, the wireless LAN uses standard Ethernet LAN as the backbone. Therefore, it is necessary that wireless LAN is able to handle Ethernet packets of 1518 bytes long. Also, any change in the protocol for wireless LAN may cause a major change in the protocol of the higher layers. Therefore, the IEEE committee decided to solve the problem by adding a simple fragmentation/reassembly mechanism at the MAC Layer of the wireless LAN. The mechanism is a simple Send-and-Wait algorithm, where the transmitting station is not allowed to transmit a new fragment until one of the following conditions happens:

1. Receives an ACK for the said fragment, or

2. Decides that the fragment was retransmitted too many times and drops the whole frame.

It should be noted that the standard does allow the station to transmit to a different address between retransmissions of a given fragment. This is particularly useful when

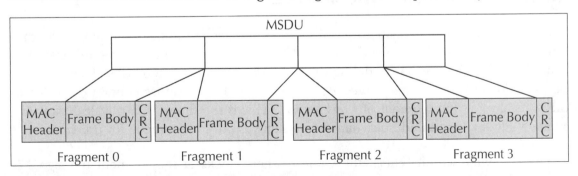

Figure 10.11 Frame Fragmentation

an AP has several outstanding packets to different destinations and one of them does not respond. Figure 10.11 shows a frame (MSDU) being divided to several fragments (MPDUs).

Inter Frame Spaces

The standard defines 4 types of spacing intervals. These are called Inter Frame Spaces (IFS). IFSs are used to defer a station's access to the medium and provide various levels of priorities:

- **SIFS (Short Inter Frame Space)**, is the shortest Inter Frame Space with the highest priority. RTS, CTS use SIFS intervals. SIFS value is a fixed value per PHY and is calculated in such a way that the transmitting station will be able to switch back to receive mode and be capable of decoding the incoming packet.

- **PIFS (Point Coordination IFS)**, is used by the Access Point (or Point Coordinator), to gain access to the medium before any other station. This value of PIFS is SIFS plus a Slot Time, i.e. 78 microseconds.

- **DIFS (Distributed IFS)**, is the Inter Frame Space used for a station willing to start a new transmission, which is calculated as PIFS plus one slot time, i.e. 128 microseconds.

- **EIFS (Extended IFS)**, is a longer IFS used by a station that has received a packet that it could not understand. This is needed to prevent the station (which could not understand the duration information for the Virtual Carrier Sense) from colliding with a future packet belonging to the current dialog.

Maintaining Stations Synchronized

Stations need to maintain synchronization. This is necessary to keep hopping and other functions like Power Saving synchronized. On an infrastructure BSS, synchronization is achieved by all the stations updating their clocks according to the AP's clock. The AP periodically transmits frames called Beacon Frames. These frames contain the value of the AP's clock at the moment of transmission. This is the time when physical transmission actually happens, and not when the packet was put in the queue for transmission.

The receiving stations check the value of their clocks at the moment the signal is received, and correct it to keep in synchronization with the AP's clock. This prevents clock drifting which could cause loss of synch after a few hours of operation.

Power Saving

Wireless LANs are typically related to mobile applications. In this type of application, battery power is a scare resource. That is why 802.11 standard directly addresses the issue of Power Saving. Power saving enables stations to go into sleep mode without losing information. The AP maintains a continually updated record of all stations currently in Power Saving mode. AP buffers the packets addressed to these stations until either the stations specifically request the packets by sending a polling request, or until the stations change their operation mode.

As part of Beacon Frames, the AP periodically transmits information about which power saving stations have frames buffered at the AP. If there is an indication that there is a frame stored at the AP waiting for delivery, then the station stays awake and sends a polling message to the AP to receive these frames.

10.5 MOBILITY IN WIRELESS LAN

When a station wants to access an existing BSS (either after power-up, sleep mode, or physically entering into the BSS area), the station needs to get synchronization information from the AP (or from the other stations when in ad hoc mode).

The station can get this information by one of two means:

- **Passive Scanning.** In this case the station just waits to receive a Beacon Frame from the AP, or
- **Active Scanning.** In this case the station tries to locate an Access Point by transmitting Probe Request Frames, and waits for Probe Response from the AP.

The Authentication Process

Once a wireless station has located an AP and decides to join its BSS, it goes through the authentication process. This is interchange of authentication information between the AP and the station, where the WLAN device proves its identity.

The Association Process

Once the station is authenticated, it then starts the association process which is the exchange of information about the stations and BSS capabilities, and which allows the DSS (the set of APs) to know about the current position of the station. A station is capable of transmitting and receiving data frames only after the association process is completed.

Roaming

Roaming is the process of moving from one cell (or BSS) to another without losing connection. This function is similar to the cellular phones' handover, with two main differences:

1. On a packet-based LAN system, the transition from cell to cell may be performed between packet transmissions, as opposed to telephony where the transition may occur during a phone conversation.

2. On a voice system, a temporary disconnection during handoff does not affect the conversation. However, in a packet-based environment it significantly reduces performance because retransmission is performed by the upper layer protocols.

The 802.11 standard does not define how roaming should be performed, but defines the basic tools. These include active/passive scanning, and a re-association process, where a station that is roaming from one AP to another becomes associated with the new AP. The Inter-Access Point Protocol (IAPP) specification addresses a common roaming protocol enabling wireless stations to move across multivendor access points. IAPP is the scope of IEEE standard 802.11f.

IAPP defines two basic protocols, viz., Announce protocol and Handover protocol. The announce protocol provides coordination information between access points. This information relates to network wide configuration information about active APs. The handover protocol allows APs to coordinate with each other and determine the status of a station. When a station associates with a different AP, the old AP forwards buffered frames for the station to the new AP. The new AP updates the necessary tables in the MAC layer to ensure that the MAC level filtering will forward frames appropriately. This type of roaming is called horizontal roaming.

Mobile IP is another protocol that is used to allow application layer roaming. Using Mobile IP, a mobile station can move from one type of network to another type of network. For example in an IMT-200 situation, the station moves from wireless LAN environment to a 3G wireless MAN environment. Mobile IP is described in Chapter 4.

10.6 DEPLOYING WIRELESS LAN

10.6.1 Network Design

The first step in designing a wireless network is to identifying the areas that need to be covered, the number of users and the types of devices they will use. From these requirements we need to determine how many access points (AP) are required and where they must be placed. The goal is to ensure adequate RF coverage to users. AP placement is typically determined using a combination of theoretical principles and a thorough site survey. Site survey is necessary to determine the required coverage; number, density, and location of APs. In office environments with walls (including cube walls) and other impediments, a typical range is 75 to 80 feet (23 to 24 meters).

In addition, the site survey can identify conditions that inhibit performance through path and multipath loss, as well as RF interference. Path loss refers to the loss of signal power experienced between the AP and the client system due to distance–walls, ceilings, and furniture–and the frequency of the transmission. Multipath loss occurs as an RF signal bounces off objects in the environment such as furniture and walls while en route to its destination. Use of APs with 'antenna diversity' help to correct for multipath loss. In 802.11b networks, antenna diversity is implemented using two antennas with supporting circuitry to improve signal reception.

RF interference is caused by other RF sources that also operate in the 2.4-GHz frequency band. These sources can include microwave ovens and cordless phones. In addition, emerging Bluetooth personal area network devices operating in this frequency band can interfere with 802.11 transmissions.

Scaling Capacity and Bandwidth

Figure 10.12 shows how 'aggregate bandwidth' in a localized coverage area helps to service a more dense population of wireless clients or to increase the bandwidth available to each wireless client in a coverage area. Channel 1, 6 and 11 are completely non-overlapping. To achieve 33-MHz, channels 1, 6 and 11 and be used as overlapping channels.

In the example shown figure 10.12(a), one AP provides up to 11 Mbps of bandwidth, which is shared by all wireless clients in the coverage area. As shown in figure 10.12(b), two more APs can be installed next to the original AP. Each provides an additional 11 Mbps of bandwidth to the same coverage area, for an aggregate bandwidth of up to

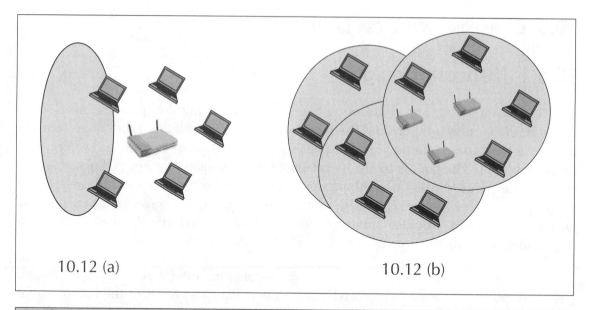

10.12 (a)   10.12 (b)

Figure 10.12 Scaling Aggregate Bandwidth (a) of 11 and (b) of 33 Mbps in a Localized Area by colocation

33 Mbps. This solution does not provide an individual wireless client with 33 Mbps of bandwidth. In 802.11 networks, each client associates with only one AP at a time and shares its bandwidth with other clients associated with the AP. Capacity and bandwidth can also be scaled by reducing the size of the coverage areas.

Channel Selection

Within the 2.4-GHz frequency band, the 802.11 standard defines 13 'center frequency channels' (see Table 10.3). Channel 1 (2.412 GHz), channel 6 (2.437 GHz), and channel 11 (2.462 GHz) are non-overlapping with large radio isolation band. Therefore, these channels are commonly used to minimize the complexity of configuring and managing channels.

Figure 10.13 shows a three-storey building serviced by nine APs configured with channels 1, 6 and 11. This arrangement minimizes interference between APs located on the same floor as well as APs in the neighboring floors. It also eliminates the bandwidth contention that occurs when two APs with overlapping coverage are configured with the same channel. When this happens, 802.11 wireless Ethernet carrier sense multiple access/collision avoidance (CSMA/CA) mechanisms ensure that users in both coverage areas can access the network. However, instead of providing two separate 11-Mbps

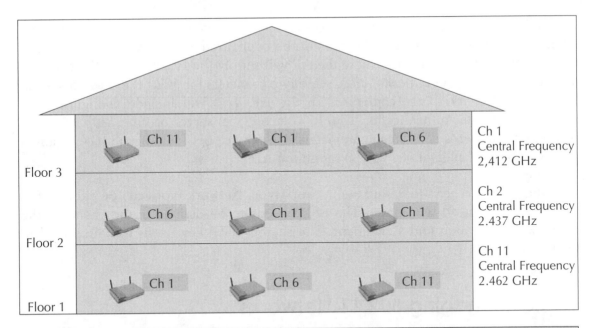

Figure 10.13 Sample Frequency Topology Using Channels 1, 6, and 11

channels and an aggregate bandwidth of 22 Mbps, the two APs provide only one 11-Mbps channel.

AP Transmission Power

The transmission power of most APs ranges from 1 mw up to 100 mw. Transmission power affects the effective range of the radio signal. The higher the transmission power, the longer the range of the signal (that is, the larger the coverage area). Higher power settings are appropriate in many large enterprise installations with cube-wall offices and a lot of open space. Lower settings are appropriate in environments such as test labs or small offices where the longer range is not required. Because lowering the transmission power reduces the range of an AP, lower power settings can also enable the wireless network to provide higher aggregate throughput. At lower power settings, more APs can be installed to serve a particular area than is possible at higher power levels.

10.6.2 Configuring the Wireless LAN

Configuration of a wireless LAN includes configuration of both the access point and the mobile station. The first level of configuration is to assign an IP address to the AP.

The WEP (Wired Equivalent Privacy) security, the shared key needs to be set both in the AP and the mobile station. The AP can also be configured as a DHCP (Dynamic Host Configuration Protocol) server where the AP will supply the IP address to the connecting client. Depending on the situation, security parameters for 802.1x (discussed later in this Chapter) or WEP are configured in the AP. This will include configuring the RADIUS (Remote Authentication Dial In User Service) server or other authentication servers like Kerberos etc. Other parameters like Service Set Identifier (SSID), channel selection, beacon interval etc. will be set on the AP.

In the client we need to define the network type. Network types can be either infrastructure mode or ad hoc mode. The SSID needs to be defined in the client for the network identification and attachment. The shared WEP key needs to be installed in the client.

10.6.3 Managing 802.11 Networks

Two key components to a successful wireless network deployment are good management and monitoring tools. Providing a stable and manageable network infrastructure with effective support, problem detection, and problem resolution is dependent upon a good foundation of network products and tools. For the 802.11 wireless network, this includes utilities on the client computer that allow the user to monitor the health of their radio connection, and the infrastructure tools used by IT to manage and monitor the wireless network. Most of the clients provide tools to check the health of the link.

Managing Access Points

The task of managing APs can be broken down into management and monitoring/reporting. Management tools are typically provided with the AP. Management tools allow IT staff to perform initial set-up and overall administration of an AP. Initial set-up includes tasks such as configuring the device name, channel selection, SSID settings, IP addressing, security settings, and Ethernet settings. Administration includes tasks such as changing IP addresses and WEP settings, upgrading firmware, performing AP remote re-boots, and analyzing AP network interfaces and AP client connections.

Monitoring and reporting tools can provide real-time monitoring and alerting as well as trend reporting for wireless network devices. These tools can allow IT staff to track network device health and receive alerts of critical events or outages.

Client Tools

For best results, choose client Network Interface Cards (NICs) with Wi-Fi certification of interoperability, as well as easy-to-use client utilities for diagnostics and determining the RF signal strength and quality. The user interface should provide pertinent information on link status, network statistics, configuration options such as SSIDs, WEP keys, and so forth. It should also allow users to easily maintain multiple profiles and to switch between them as required. Following are some free tools that can be used for this purpose.

Aida32 (http://www.aida32.hu/aida32.php) Aida32 provides direct and rapid access to a server's Event Viewer, User Manager, live lists (of Dynamic Link Libraries), open files, services in use and other hidden terminals. Developed, upgraded and maintained by Unlimited Possibilities in Budapest, Hungary, Aida32 works specifically on Win32 platforms and can be used to perform diagnostics and benchmarking.

Network Probe (http://www.objectplanet.com/probe/) is a protocol analyzer developed by ObjectPlanet of Oslo, Norway. This tool is designed to provide a real-time view of network traffic. The software can track and isolate traffic problems and congestion on network lines. It monitors conversations between hosts and applications, and shows network managers from and to where the network traffic is traveling.

Cflowd (http://www.objectplanet.com/probe/) and flowscan (http://www.caida.org/tools/utilities/flowscan/) Cflowd collects and correlates data from NetFlow, a part of Cisco's IOS that collects and measures data as it enters router or switch interfaces. The data can be used to monitor key applications, including accounting, billing and network planning, for corporate or service provider customers.

Kismet (http://www.kismetwireless.net/) is a network sniffer. It can spot unauthorized wireless access points. Unlike a standard sniffer, Kismet can identify and separate wireless use on the IP network. Kismet works with any 802.11b wireless card that can report raw wireless packets.

There are several excellent tools that are capable of mapping the access points in the area. They are also useful tools for installation/deployment and detecting rouge access points.

- NetStumbler (http://www.netstumbler.com) for Windows
- MiniStumbler (http://www.netstumbler.com) for Pocket PC
- PocketWarrior (http://www.pocketwarrior.org) for Pocket PC

- Kismet (http://www.kismetwireless.net) for Linux
- Dstumbler (http://www.dachb0den.com/projects/dstumbler.html) for NetBSD, FreeBSD, OpenBSD
- Wellenreiter (http://www.remote-exploit.org) for Linux, experimental BSD
- 802.11 Network Discovery Tools (http://sourceforge.net/projects/wavelan-tools/) for Linux
- iStumbler (http://homepage.mac.com/alfwatt/istumbler/) for Mac
- AirMagnet (http://www.airmagnet.com) for Windows, PDA
- THC-WarDrive (http://www.thc.org) for Linux
- PrismStumbler (http://prismstumbler.sourceforge.net) for Linux
- WaveStumbler (http://www.cqure.net/tools.jsp?id=08) for Linux
- ssidsniff (http://www.bastard.net/%7Ekos/wifi/) for Linux
- WaveMon (http://www.jm-music.de/projects.html) for Linux

Using such tools for driving around town discovering access points is called 'WarDriving' after 'WarDialing' where one used a modem to call a phone number at random or sequence in hope of finding an open or insecure server.

10.7 MOBILE AD HOC NETWORKS AND SENSOR NETWORKS

A mobile ad hoc network (MANET) is an autonomous system of mobile stations connected by wireless links to form a network. This network can be modeled in the form of an arbitrary graph. Ad hoc networks are peer-to-peer, multihop networks where data packets are transmitted from a source to a destination via intermediate nodes. Intermediate nodes serve as routers in this case. In an ad hoc network there will be situations when some of the nodes could be out of range with respect to some other nodes. When this happens, the network needs to reconfigure itself and ensure that the paths between two nodes are available. In an ad hoc network, communication links could be either symmetric (bidirectional) or asymmetric (unidirectional). Figure 10.14 shows a wireless ad hoc network.

To design a good wireless ad hoc network be it a sensor network or an information network, we need to account for various challenges. These are:

Dynamic topology. Nodes are free to move in an arbitrary fashion resulting in the topology changing arbitrarily. This characteristic demands dynamic configuration of the network.

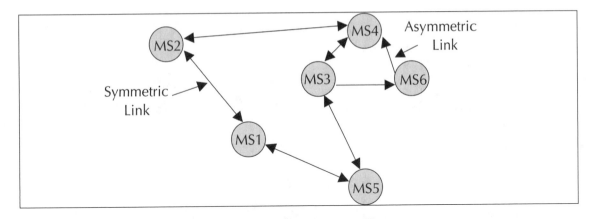

Figure 10.14 Mobile ad hoc network

Limited security. Wireless networks are vulnerable to attack. Mobile ad hoc networks are more vulnerable as by design any node should be able to join or leave the network any time. This requires flexible and higher openness.

Bandwidth limited. Wireless networks in general are bandwidth limited. In an ad hoc network more so because there is no backbone to handle or multiplex higher bandwidth.

Routing. Routing in a mobile ad hoc network is complex. This depends on many factors, including finding the routing path, selection of routers, topology, protocol etc.

10.7.1 Wireless Sensor Networks

Wireless sensor networks are a class of ad hoc networks. Sensor networks are very useful in unpredictable, unreliable environments. Sensor networks are primarily data collection points. They are widely used in defense, environmental, meteorological, and study of nature. A wireless sensor network is a collection of low-cost, low-power disposable devices. Each of these devices holds sensing, memory, and communication modules. Study of the movement of glaciers is done through wireless ad hoc networks. Sensor networks are generally unmanned. Sensors may not have any power source other than small batteries. Therefore power control is a major challenge in sensor networks to ensure long life of the network.

10.8 WIRELESS LAN SECURITY

In a wired network one has to be physically connected to transfer or receive data. This implies that it is possible to control the users in the network by controlling the physical

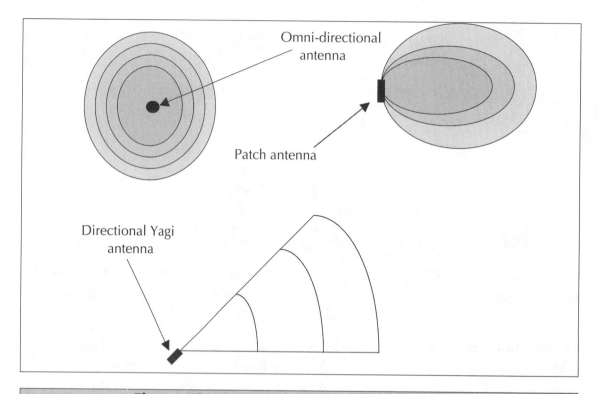

Omni-directional antenna

Patch antenna

Directional Yagi antenna

Figure 10.15 RF transmission pattern of antennas

access. Using a wireless network means using a radio transmitter and receiver. With varying degrees, radio signals will penetrate most building materials. Therefore, it is not possible to set up absolute physical boundary and expect that no outsider will be able to intrude into the network. With wireless networks we have no control of who might be receiving and listening to the transmissions. It could be someone in the building across the road, in a van parked in the parking lot or someone in the office above. Therefore, it is important that we understand the vulnerabilities of the wireless LAN and take necessary precautions.

As a part of the original specification, IEEE 802.11 included several security features, such as open system and shared key authentication modes; the Service Set Identifier (SSID); and Wired Equivalent Privacy (WEP). Each of these features provides varying degrees of security.

10.8.1 Limiting RF Transmission

It is important to consider controlling the range of RF transmission by an access point. It is possible to select proper transmitter/antenna combination that will help transmission of the wireless signal only to the intended coverage area. Antennas can be characterized by two features—directionality and gain. Omni-directional antennas have a 360-deg coverage area, while directional antennas limit coverage to better-defined areas (see Figure 10.15).

10.8.2 Service Set Identifier (SSID)

According to the 802.11 standard, a mobile station has to use the SSID of the access point for association between the NIC (Network Interface Card) in the client and the AP. The SSID is a network name (Id of the BSS or Cell) that identifies the area covered by an AP. The AP periodically broadcasts its SSID as a part of the management frame (beacon packet). As discussed earlier, the broadcast of beacon packet is necessary for clock synchronization. Unfortunately, as management frames of 802.11 are always sent in the clear, an attacker can easily listen on the wireless media for the management frames and discover the SSID to connect to the AP. The SSID can be used as a security measure by configuring the AP to broadcast the beacon packet without its SSID. The wireless station wishing to associate with the AP must have its SSID configured to that of the AP. If the SSID is not known, management frames sent to the AP from the wireless station will be rejected. It is also advised that the SSID of the AP is changed from the factory set defaults to some name, which is difficult to guess.

10.8.3 MAC Address Access Control

Many access points support MAC address filtering. This is similar to IPFiltering. The AP manages a list of MAC addresses that are allowed or disallowed in the wireless network. The idea is that the MAC address of the network card is unique and static. By controlling

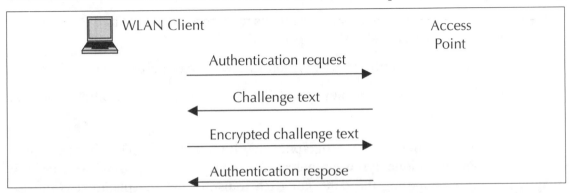

Figure 10.16 Shared key authentication

the access from known addresses, the administrator can allow or restrict the access of network only to known clients.

10.8.4 Authentication Modes

Two types of client authentication are defined in 802.11: Open System Authentication and Shared Key Authentication. Open system authentication is no authentication at all. Shared key authentication on the other hand (Figure 10.16) is based on the fact that both stations taking part in the authentication process have the same "shared" key.

It is assumed that this key has been transmitted to both stations through some secure channel other than the wireless media itself. In typical implementations, this is set manually on the client station and the AP. The authenticating station receives a challenge text packet (created using the WEP Pseudo Random Number Generator (PRNG)) from the AP. The station encrypts this PRNG using the shared key, and sends it back to the AP. If, after decryption, the challenge text matches, then one-way authentication is successful. To obtain mutual authentication, the process is repeated in the opposite direction.

10.8.5 WEP (Wired Equivalent Privacy)

WEP was designed to protect users of a WLAN from casual eavesdropping and was intended to offer following facilities:

- **Reasonably strong encryption.** It relies on the difficulty of recovering the secret key through a brute force attack. The difficulty grows with the key length.

- **Self-synchronizing.** Each packet contains the information required to decrypt it. There is no need to deal with lost packets.

- **Efficient.** It can be implemented in software with reasonable efficiency.

- **Exportable.** Limiting the key length leads to a greater possibility of export beyond US.

The WEP algorithm is the RC4 cryptographic algorithm from RSA Data Security. RC4 uses stream cipher technique. It is a symmetric algorithm and uses the same key for both enciphering and deciphering the data. For each transmission, the plaintext is bitwise XORed with a pseudorandom keystream to produce ciphertext. For decryption the process is reversed.

The algorithm operates as follows:

1. It is assumed that the secret key has been distributed to both the transmitting and receiving stations by some secure means.

2. On the transmitting station, the 40-bit secret key is concatenated with a 24-bit Initialization Vector (IV) to produce a seed for input into the WEP PRNG (Pseudo Random Number Generator).

3. The seed is passed into the PRNG to produce a stream (keystream) of pseudo-random octets.

4. The plaintext PDU is then XORed with the pseudo-random keystream to produce the ciphertext PDU.

5. This ciphertext PDU is then concatenated with the 24-bits IV and transmitted on the wireless media.

6. The receiving station reads the IV and concatenates it with the secret key, producing the seed that it passes to the PRNG.

7. The receiver's PRNG produces identical keystream used by the transmitting station. When this PRNG is XORed with the ciphertext, the original plaintext PDU is produced.

It is worth mentioning that the plaintext PDU is also protected with a CRC to prevent random tampering with the ciphertext in transit.

10.8.6 Possible Attacks

The possible security attacks on wireless LAN are:

- **Passive attacks** to decrypt traffic based on statistical analysis.

- **Active attack** to inject new traffic from unauthorized mobile stations, based on known plaintext.

- **Active attacks** to decrypt traffic, based on tricking the access point.

- **Dictionary-building attack** that, after analysis of about a day's worth of traffic, allows realtime automated decryption of all traffic.

- **Hijacking a session:** Following successful authentication, it is possible to hijack the session.

Analysis suggests that though these attacks are not common, it is possible to perform them using inexpensive off-the-shelf equipment.

10.8.7 802.1X Authentication

To prevent attacks on wireless LAN, the IEEE specification committee on 802.11 included the 802.1x authentication framework. The 802.1x framework provides the link layer with extensible authentication, normally seen in higher layers.

802.1x requires three entities (Figure 10.17):

- **The supplicant** Resides on the wireless LAN client
- **The authenticator** Resides on the access point
- **The authentication server** Resides on the server authenticating the client (e.g., RADIUS Kerberos, or other servers)

These are logical entities on different network elements. In a single network there could be many points of entry. These entries are through access points. Once the link between a supplicant (wireless station) and an authenticator (AP) is achieved, the connection is passed to the authentication server. The AP authenticates the supplicant through the authentication server. If the authentication is successful, the authentication server instructs the authenticator to allow the supplicant to access the network services. The authenticator works like a gatekeeper.

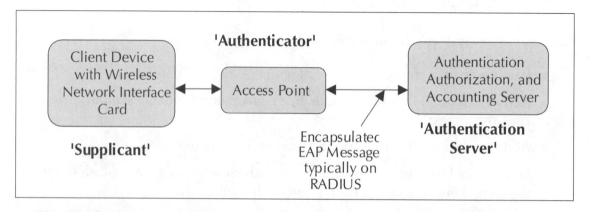

Figure 10.17 802.1x Setup

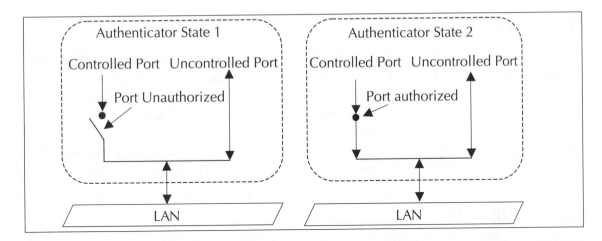

Figure 10.18 Effect of authorization state on controlled Ports

The authenticator creates one logical port per client, based on the client's association ID. This logical port has two data paths. The uncontrolled data path allows network traffic through to the network. The controlled data path requires successful authentication to allow network traffic through (Figure 10.18). In order to obtain network connectivity, a wireless client must associate with the AP. Complete association with an AP involves three states:

1. Unauthenticated and unassociated

2. Authenticated and unassociated

3. Authenticated and associated

IEEE 802.1x offers flexibility in authentication and possible encryption. After the link has been established, PPP (Point-to-Point Protocol) provides for an optional authentication phase before proceeding to the network layer protocol phase. This is called EAP (Extensible Authentication Protocol). Through the use of EAP, support for a number of authentication schemes may be added, including Smart cards, Kerberos, Public Key, One Time Passwords, CHAP (Challenge Handshake Authentication Protocol), or some other user-defined authentication systems.

There are still some vulnerabilities in the EAP. To overcome this, a new standard is being proposed in IETF to override the EAP proposal. This new standard is called PEAP (Protected EAP). PEAP uses an additional phase of security over and above EAP.

10.8.8 Wireless VPN

Virtual Private Network technology (VPN) has been used to secure communications among remote locations via the Internet since the 1990s. It is now being extended to wireless LAN. VPNs were traditionally used to provide point-to-point encryption for long Internet connections between remote users and the enterprise networks. VPNs have been deployed in wireless LANs as well. When a wireless LAN client uses a VPN tunnel, communications data remains encrypted until it reaches the VPN gateway, which sits behind the wireless AP. Thus, intruders are effectively blocked from intercepting all network communications.

10.8.9 802.11i

Task Group 'i' within IEEE 802.11, is developing a new standard for WLAN security. The proposed 802.11i standard is designed to embrace the authentication scheme of 802.1x and EAP while adding enhanced security features, including a new encryption scheme and dynamic key distribution. Not only does it fix WEP, it takes wireless LAN security to a higher level.

The proposed specification uses the Temporal Key Integrity Protocol (TKIP) to produce a 128-bit 'temporal key' that allows different stations to use different keys to encrypt data. TKIP introduces a sophisticated key generation function, which encrypts every data packet sent over the air with its own unique encryption key. Consequently, TKIP greatly increases the complexity and difficulty of decoding the keys. Intruders will not have enough time to collect sufficient data to decipher the key.

802.11i also endorses the Advanced Encryption Standard (AES) as a replacement for WEP encryption. AES has already been adopted as an official government standard by the US Department of Commerce. It uses a mathematical ciphering algorithm that employs variable key sizes of 128-, 192- or 256-bits, making it far more difficult to decipher than WEP. AES, however, is not readily compatible with today's Wi-Fi Certified WLAN devices. It requires new chipsets, which, for WLAN customers, means new investments in wireless devices. Those looking to build new WLANs will find it attractive. Those with previously installed wireless networks must justify whether AES security is worth the cost of replacing equipment.

10.9 WiFi versus 3G

3G offers a vertically integrated, top-down, service-provider approach for delivering wireless Internet access; while WiFi offers an end-user-centric, decentralized approach to service provisioning. Table 10.4 highlights characteristics of 3G versus WiFi:

Table 10.4 3G versus WiFi

Functions	3G	WiFi
Genesis	Evolved from voice network (realtime traffic) where QoS is a critical success factor	Evolved from data network (store and forward) where QoS is not a critical success factor.
Radio Interface	Use Spread Spectrum as the modulation technique.	Use Spread Spectrum as the modulation technique.
Access technologies	Access or edge-network facility. Offers alternatives to the last-mile wireline network. The wireless link is from the end-user device to the cell base station, which may be at a distance of up to a few kilometers.	Access or edge-network facility. Offers alternatives to the last-mile wireline network. The wireless link is a few hundred feet from the end-user device to the base station.
Bandwidth	3G supports broadband data service of up to 2Mbps. 3G will support 'always on' connectivity,	WiFi offers broadband data service of up to 54Mbps. WiFi offers 'always on' connectivity.
Business models/deployment are different	Service providers own and manage the infrastructure (including the spectrum). End customers typically have a monthly service contract with the 3G service provider to use the network	Users' organization owns the infrastructure. Following the initial investment, the usage of the network does not involve an access fee.

Spectrum policy and management	3G uses licensed spectrum. This has important implications for:
	  a. Cost of service
	  b. Quality of Service (QoS)
	  c. Congestion Management
	  d. Industry structure.
	The upfront cost of acquiring a spectrum license represents a substantial share of the capital costs of deploying 3G services. However, with licensed spectrum, the licensee is protected from interference from other service providers.
		WiFi uses unlicensed free, shared spectrum. Therefore, it does not involve any additional costs to acquire the spectrum.
		
		Unlicensed spectrum can be used by anybody and can be used for any purpose. This may cause interference in each other's right to use the spectrum. Therefore, WiFi imposes strict power limits on users (i.e., responsibility not to interfere with other users) and forces users to accept interference from others.

Contd. on next page

Table 10.4 (continued)

Status of Standards	The formal standards picture for 3G is perhaps clearer than for WiFi. For 3G, there is a relatively small family of internationally sanctioned standards, collectively referred to as IMT-2000.	WiFi protocol is one of the family of continuously evolving 802.11x wireless Ethernet standards, which itself is one of many Wireless LAN technologies that are under development.
Roaming	3G will offer well coordinated continuous and ubiquitous coverage. This offers a seamless roaming. To support this service, mobile operators maintain a network of interconnected and overlapping infrastructure that allows customers to roam around.	WiFi network growth is unorganized. Therefore seamless ubiquitous roaming over WiFi cannot be guaranteed. Also, WiFi technology has not been designed to support high-speed handoff associated with users moving between base station coverage areas.

REFERENCES/FURTHER READING

1. Introduction to Wireless LAN, The Wireless LAN Alliance, www.wlana.com.

2. IEEE 802.11 Working Group: http://grouper.ieee.org/groups/802/11/.

3. Eric Schreiber & Daniel Sigg, Clock Synchronization for Wireless LAN, Thesis DA-2002.18 Summer Term 2002.

4. Pabio Brenner, A Technical Tutorial on the IEEE 802.11 Protocol, Breezecom Wireless Communications, 1997.

5. Mustafa Ergen, IEEE 802.11 Tutorial, UC Berkeley, 2002.

6. William Lehr, Lee W. McKnight, Wireless Internet Access: 3G vs. WiFi, http://itc.mit.edu/itel/docs/2002/LehrMcKnight_WiFi_vs_3G.pdf.

7. Guoliang Li, Physical Layer Design for a Spread Spectrum Wireless LAN, MS Thesis, Virginia Polytechnic Institute and State University.

8. Wireless LAN protocol and its effect on cell design, Integrity Data Systems Pty Ltd White paper, A.B.N. 17 148 989 654, 2000.

9. Deploying 802.11B (Wi-Fi) In the Enterprise Network, Dell White paper, 2001.

REVIEW QUESTIONS

Q1: What are the advantages and disadvantages of wireless LAN? Under what situation is a wireless LAN desirable over LAN?

Q2: In an Ethernet LAN CDMA-CD is used, whereas in Wireless LAN CDMA-CA is used. What is CDMA-CA? Why is CDMA-CA used in Wireless LAN instead of CDMA-CD?

Q3: How are mobility and handoff managed in Wireless LAN?

Q4: What is WEP? Why is it considered unsafe? What are the mechanisms advised to ensure security in Wireless LAN?

Q5: How does 802.1x overcome the security vulnerabilities in WEP?

CHAPTER 11

Intelligent Networks and Interworking

11.1 INTRODUCTION

A communication network provides the service of transportation of payload of its subscribers. This payload can be voice, data or other types of payloads. Switches are at the heart of these networks routing the traffic from source to sink. Also, these work out procedures to charge the subscriber for the network service it provides. Switches used by the network operators are expensive, designed to do certain limited functions.

Intelligence network (IN) is a concept where intelligence is taken out of the central switch and distributed within the network. Through IN, intelligence is added into the network and placed in computer nodes that are distributed throughout the network. IN is an architecture, which separates the service logic from the switching service of the telephone exchanges. IN enables the establishment of an open platform for uniform service creation, implementation and management. Once introduced, services are easily customized to meet customer's need. It aims at rapid and economical service provisioning and facilitates customer control of network services. This makes telecom networks different from data networks. In a data network, endpoints are intelligent with no intelligence in the network, whereas, in telecom endpoints are dumb with all intelligence in the network.

11.2 FUNDAMENTALS OF CALL PROCESSING

To understand IN, we need to understand the basic function of a switch, i.e., the processing of a call. To connect a caller to a called number, the switch performs a series of functions and makes a series of decisions. Decisions are like:

- Is this subscriber allowed to place this call?
- Where is the call to be connected?

- How will the call be connected? What path will it take?
- What should be the cost of the call?
- Who should be charged for the call?

To make a successful call, there are many steps. Figure 11.1 depicts various steps in a telecommunication call set up and tear down. However, if they are simplified, it will look like:

Step 1 The A-subscriber lifts the handset. The moment a subscriber lifts the handset of a telephone, the switch in an exchange is able to detect that a subscriber has lifted the handset. At this moment the subscriber or the calling party (called the A-party) is welcome to enter the number of the destination or called party (called the B-party). The welcome message is a dial tone. The caller dials the telephone number of the called party.

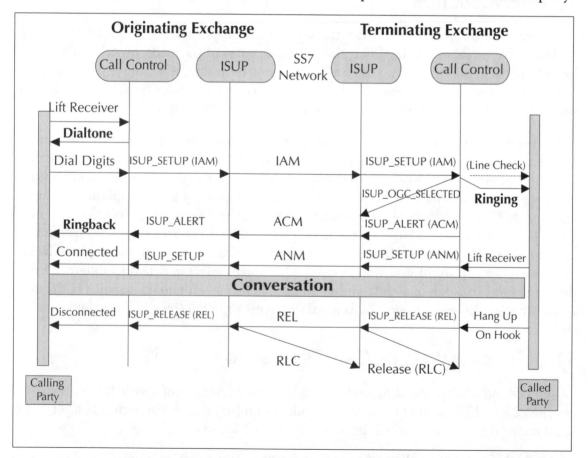

Figure 11.1 ISUP call establishment procedure

Step 2 The exchange receives B-subscriber's number. B's number is received by the switch. The switch analyses B's number to determine whether B is within the same exchange or a different exchange. If B is in a different exchange, it is called out-of-switch number.

Step 3 The exchange sets up the outgoing call. If B is within the same exchange, the line interface circuit of B is obtained and A's line is physically connected to that of B's. If B is from another exchange, a routing analysis is performed. The switch will make an attempt to connect to B using the routing path. At this stage ISUP messages move through the SS7 network from switching node to switching node. While the ISUP message is moving through the network, a voice path parallel to the ISUP path is being reserved for circuit establishment. If it is unable to set up a path due to the fact that B is busy, a busy tone is played for A. If B's phone is free, and the switch is able to establish a path between A and B, the circuit in the trunk is established. The telephone of B rings and A hears a ring tone. The ring tone of A is issued by A's exchange, whereas the ringing current (signal) to B's phone is sent by B's exchange.

The ISUP message sent from A's switch is called the IAM (Initial Address Message). This message contains all the information necessary for each switch to be able to consult its routing table and to select circuits that will result in connecting the circuit from end to end. A's switch receives a confirmation message called the ACM (Address Complete Message). When A's exchange receive this ACM message, it issues the ring tome to A. Once the phone is ringing, there is no further signalling being exchanged for a time.

A charging analysis is performed on the call request. Charging depends on subscriber's category, tariff plan, time of day, distance between the called and caller party etc. One of the registered tariffs is selected for the billing of the call.

Following the ring at B's telephone, B decides to answer to the call. B picks up the phone. When B's phone comes 'off hook' the switch at B's end sends an ANM (ANswer Message) backward into the SS7 network to A's exchange. Each switch is thus notified that the full circuit must now exist. The circuit reserved from A to B is now connected. The system switches from signal mode to traffic mode.

A and B enters into conversation mode. They talk as long as they want. The switches monitor the connection, primarily to enable the call to be charged.

Step 4 The subscribers conclude their conversation. The switch continues to scan the subscribers' lines even during the conversation. Eventually, of course, someone hangs up. The phone line once again goes 'on hook' and that is sensed at the subscriber interface of the switches serving the customer who hung up. Whichever office detects the hang up, that office sends a REL (release) message on to the previous switch in the circuit. Upon receiving the REL, each switch releases the circuit connection. At the same time, it returns an RLC (release complete) back to the switch that sent the REL. Switch by switch, the scenario continues until each switch in the circuit has released its circuitry and confirmed that action to the previous switch. Charging of the call stops, a CDR (Call Detail Record) is produced for charging.

11.3 INTELLIGENCE IN THE NETWORKS

In (Chapter 3) we have mentioned that we dial 1-600-111100 to talk to Microsoft customer service from anywhere (main cities) in India. We have also mentioned that this call is free. This is an example of IN. Let us now understand how it works. When a user dials this number from anywhere in India, the local exchange knows that this not a local number. It tries to find out the routing path for the number. It discovers that this is a virtual number and needs to refer to an IN node to obtain the routing path. The switch asks the IN node for the routing path. The IN node looks at its database and finds that this virtual number is mapped to a real number in Delhi (011-2629-2640). The IN node also informs the local switch (A's exchange) that A should not be billed; instead the B-party (Microsoft in this case) will be billed. The switch gets the routing path through another routing path enquiry on 011-2629-2640. Though the user has dialed 1-600-111100, the user gets connected to 011-2629-2640, and not charged for a STD call.

If we dial this number between 9:00 AM to 6:00 PM Monday through Friday, we can talk to Microsoft customer service representative for free. If it is any other time of the day, one will hear a recorded message, 'Welcome to Microsoft connect customer services. For product activation please dial 1, for all other services we work between 9:00 AM to 6:00 PM Monday through Friday. To leave a voice mail message please dial 2.' Let us make this example slightly more interesting. Let us assume that Microsoft has a few premium customers for whom the service is available 24 hours a day, 7 days a week. In this case the IN node will also check the telephone number of the A-party and the time of the call. If A's telephone number matches with the telephone number of the premium customers and the time is between 6:00 PM to 9:00 AM, the call will be diverted to a service engineer's home number. Also, please note that in this case the IN node has to make some even smarter decisions. It cannot blindly divert the call to a telephone of an engineer at Delhi. If the call has originated from the premium customer's Mumbai office, the IN node needs to divert the call to the service engineer at Mumbai. If the call is from

Bangalore, the call needs to be diverted to the service engineer at Bangalore. Depending on the conditions, the IN node gives the routing path. If the call is connected to a service engineer's number, the IN node tells A's exchange not to bill A, not even to bill B (service engineer in this case), but to bill a different entity (Microsoft).

If we did not have IN to implement this 600 service, we would need to add all the above complex logic into all the switches in all the exchanges in the country, which is an impossible task. However, it is easy to implement it in a separate node and just implement some logic in the exchanges that any number staring with 1-600 is a virtual number. The routing path for a virtual number is obtained from an IN node. IN services address following requirements:

- Remove the service data from the switching network and locate it in a centralized database. This database and service will be accessible from all the switching nodes.

- Separate the service logic from the switch and put it into an independent intelligent node. Whenever a new service is added to this node, it becomes available throughout the network. These intelligent nodes are called Service Control Point or SCP.

- A real time connection is needed between the switching nodes 'service switching points' (SSPs), and the 'service control points' (SCPs). This fast and standardized interconnection forms the basis for the IN architecture. Figure 11.2 shows the relationship between these network elements.

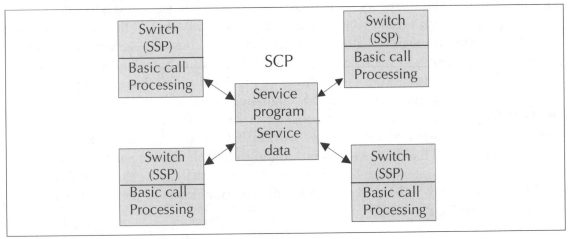

Figure 11.2 Intelligent Network approach

11.3.1 Standards For Intelligent Networks

In 1989, ITU and ETSI began work on IN standards. A phased approach of development was adapted to define the target IN architecture. Each phase of development intended to define a particular set of IN capabilities, known as a capability set (CS). Each capability set defines the requirements for one or more of the following areas:

- Service creation

- Service management

- Service interaction

- Service processing

- Network management

- Network interworking.

In March 1992 the first capability set (CS-1) of standard was approved. Work on CS-2 was started in 1994 that addressed basic aspects that were excluded from CS-1, such as IN management. Furthermore, work on CS-3 was started in 1995.

11.4 SS#7 SIGNALING

In a telecommunications network if the switch is like the heart, signaling is like the nerve. All the information related to command, control, and monitoring of the network activities are transmitted through signaling channels. Signaling channels carry data and information for the purposes of management of subscriber traffic. Signaling System number Seven or SS#7 or SS7 in short has been designed for signaling in telecommunication networks. SS7 is a digital packet network. SS7 signaling is also called common channel or out-of-band signaling. SS7 defines the procedures and protocol by which network elements in the public switched telephone networks (PSTN) exchange information to effect wireless (cellular) and fixed line call set-up, routing, control, charging as well as network management and maintenance. Switches are the 'glue' that holds the PSTN network together. Likewise, SS7 is held together by a digital counterpart of the switch known as a Signalling Transfer Point (STP).

SS7 comprises a series of interconnected network elements such as switches, databases, and routing nodes. Each of these elements is interconnected with links, each of which has a specific purpose. Main elements in SS7 network are STP, SSP, SCP, etc. These nodes are depicted in Figure 11.3 and has the following functions. The SS7 requires the use of continuously available transmission links. These links are 64K Bps channels. Channel 23 is the SS7 signal channel within a 1.544 MBps T1 line in USA and channel 16 within 2 MBps E1 lines in India and Europe. The job of the STP is to examine the destination address of messages it receives, consult a routing table, and send the messages on their way using the links that are selected from the routing tables. In a PSTN network, a telephone handset is an end point. Likewise, SEP (Signaling End Point) is an end point in the SS7 network. Similar to a telephone number, a Signaling End Points use an address known as a Signaling Point Code.

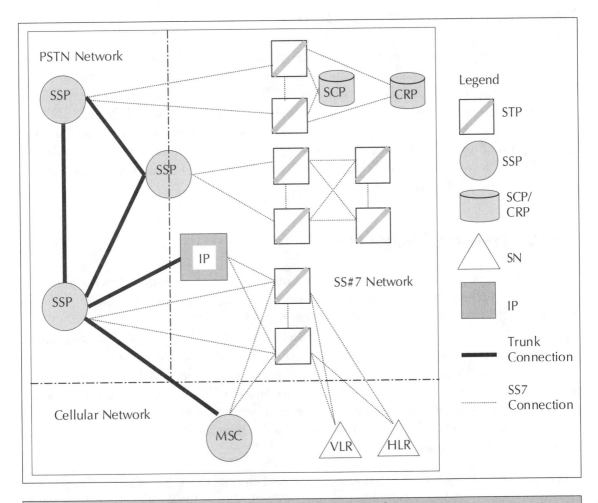

Figure 11.3 SS#7 Network Architecture

The STP (Signaling Transfer Point)

STP is like a switch or a node in the SS7 network doing the basic routing functions. For example, if an STP has links heading off toward the four compass points, it might be more 'appropriate' to direct a message addressed to Mumbai to a west-leading link than to an east-leading link. STP routing decisions are based on geography, distance, congestion, and least cost criteria. Once a SS7 message is delivered from a source to destination, a circuit on the same path is reserved for traffic. For fault tolerance, STPs are always installed in pairs with cross connections.

The SSP (Service Switching Point)

Service Switching Point is a switch in the SS7 network that can handle call set-up. The SSP has the ability to stop call processing, make queries to even unknown databases, and perform actions appropriate to the response. SSP is equipped with all the intelligence required to handle numerous feature capabilities. Example of a SSP will be a MSC in a cellular network. There is another switching point called a CCSSO (Common Channel Signaling Switching Office). These are end or tandem offices which have the capability to use the SS7 in what is referred to as a trunk signaling mode for call set-up. CCSSO is a limited version of the SSP.

The SCP (Service Control Point)

One of the first purely digital uses for the SS7 network was to provide a service to translate from one form of data to another. For example, switches need to maintain tables to translate dialed digits into routing information consistent with the international numbering system (for example iiitb number +91808410628). It is that plan that breaks India (country code 91) down into city code (80), exchange code (841), and finally to the line (0628) serving individual telephone. Let us take another example where a person has moved his home from one part of the town to another. When we dial the telephone number, it plays a recorded message like 'The number you have dialed is 28670203, the number has changed to 25320203. You may dial the new number or wait for a while to be connected automatically.'

When a virtual number like 1-600 in India or 1-800 in the US is dialed, there is no way for the switch to determine how to route this call. This is because such prefixes have no reference to the international numbering plan. In fact, a 1-600 number dialed in

Mangalore may be connected to a number in Bangalore, while the same number dialed in Pune may result in a connection to Mumbai. When that translation is returned to the switch, the number can be connected exactly as it would have been if it had been dialed in the first place. This database is located at an SS7 address called Signaling Point Code. SCPs are used for a variety of applications such as Calling Card verification, toll-free calls, tele-voting, premium tariff (1-900) calls etc. Such intelligent nodes make the network intelligent. This also frees the switch from trying to maintain ever larger routing tables, and enables the use of a broad range of services which depend on translations or digital data services of a variety of types.

SCP provides the access mechanism required for a service. These services may reside in the same location as the SCP or the SCP may serve as a 'front end' for services located elsewhere. In either case the SCP controls many services. To identify a service in a SS7 network, two parameters are required. These are SCP address and the service within the SCP.

CRP (Customer Routing Point)

The CRP provides on-premises control of the routing information requested by switches for translation of 800 type dialing. The operator of the CRP is a customer who requires rapid update and control of the translation of their own numbers.

Intelligent Peripheral (IP) is a peripheral process that deals with the requests made of it through the SCP by providing the services of a variety of equipments. IP includes database functionality of the SCP along with additional capabilities such as voice interaction and control of voice resources. Generally speaking, SCPs work well with requirements that call for voluminous data transactions. IPs, on the other hand, are best suited for special circumstance call processing involving voice resources and/or interaction. Example of an IP is a voice-activated system, where instead of dialing a number we can simply say 'call iiitb'. Here the call will be routed to an IP, which in turn will provide the iiitb telephone number.

Services Node (SN) A programmable IP is called Service Node (SN). Still, what one network calls an IP might be called a Services Node in another network.

Services Management System (SMS) This is a node in the SS7 network that provides a human interface to the database. This also provides the facility to update the database. SMS provides GUI/command line interfaces to update and manage services and the network. Service Operators configure the SMS to manage such mission-critical tasks as billing or access authorization.

11.4.1 SS#7 Protocol Stack

The SS7 protocol stack is depicted in Figure 11.4. In this model the upper 4 layers are called user parts whereas the lower are the message transfer parts.

Message Transfer Part

Message Transfer Part–Level 1 The Message Transfer Part Level 1 (MTP L1) is the 'physical layer'. It deals with hardware and electrical configuration. MTP level 1 is a part that deals with physical issues at the level of links, interface cards, multiplexors etc.

Message Transfer Part–Level 2 It is the last layer to handle messages being transmitted and the first layer to handle messages being received. It monitors the links and

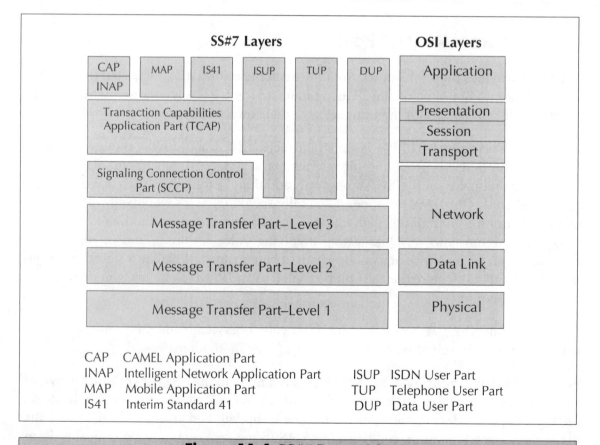

Figure 11.4 SS#7 Protocol Stack

reports on their status. It checks messages to ensure their integrity (both incoming and outgoing). It acknowledges good messages; it discards bad messages and requests copies of discarded messages. It provides sequence numbering for outgoing messages.

Message Transfer Part–Level 3 The MTP Level 3 provides the functions and procedures related to Message Routing (or Signaling Message Handling) and Signaling Network Management. MTP L3 handles these functions assuming that signaling points are connected with signaling links. The message routing provides message discrimination and distribution. Signaling Network Management provides traffic, link and routing management, as well as, congestion (flow) control.

SS7 User Parts

The layers above layer 3 of the SS7 stack consist of several different protocols. These protocols are called user parts and applications parts. They are the following.

ISDN User Part (ISUP)

The Integrated Services Digital Network User Part is used throughout the PSTN to provide the control information necessary for the set-up and tear-down of all circuits, both voice and data. Wireless networks also make use of ISUP to establish the necessary switch connections into the PSTN. ISUP offers two types of services, known as Basic and Supplementary services. Basic Services consist of services for setting up and tearing down of a call to the target number. Supplementary Services consist of services employed in passing all messages that may be necessary to maintain and/or modify the call.

Supplementary Services

The basic service provides the functionality for establishing circuits within the network. Supplementary services are all the other circuit-related services. These typically are those services where the messages are transported after the call path is established. Following are examples of supplementary services.

Call Forwarding This supplementary service enables incoming calls to be automatically redirected to some other telephone depending upon the following conditions. This facility is also known as 'call transfer or follow me' service. When the terminating switch receives the ISUP IAM message, it can forward the call depending upon conditions set in the SSP. Four different call forwarding services are available; all may be active at any one time.

- Unconditional: All calls will be forwarded to another telephone number unconditionally.

- On No reply: Call will be forwarded when the subscriber does not reply. This can also be qualified by how many ringing tomes will be supplied before forwarding. For example, if set to 5, then the telephone will ring 5 times and then the call will be forwarded to the number specified in the switch.

- On Busy: Call will be forwarded to the specified number when the subscriber line is busy.

- On Not Reachable: Call will be forwarded to the specified number when the subscriber is not reachable or out of coverage. This condition is valid only for mobile networks.

The target number can be any valid telephone number and can be a mobile phone (GSM, CDMA), a fixed phone (BSNL, Touchtel) or even a SCP offering some service like Voice Mail service. In many exchanges by default calls are diverted to voice prompt. For example if a mobile phone is powered off, exchange plays a message like 'number you have dialed is switched off': or from a fixed phone exchange 'the number you have dialed has been changed'. When a telecom company plays such voice prompts, the call is generally not charged. However, whenever a call is forwarded to a different number based on the conditions set by the subscriber, calls are always charged. In a GSM network, forwarding of calls can be set/reset from the subscriber's phone. For example I am traveling through a highway where I may not get GSM coverage all through. Therefore, I may like to forward a call to my office number when I am out of coverage. I enter * * 62 * 08025531234 # . Here 62 is the command for forwarding and 08025531234 is the number the call will be forwarded to. In GSM commands for unconditional forward is 21, forward for no reply is 61 and forward for busy is 67.

In a fixed line exchange, the follow me facility can be invoked by dialing different command from the phone. I can register for call transfer on no reply by dialing '1228 26632245'. In such a case when a call comes in my fixed line phone, and I do not reply the call within 30 seconds, the call will be diverted to 26632245.

Call Barring: This service allows a user to bar various categories of outgoing/or incoming calls based on following conditions.

- All Outgoing International Calls

- All Incoming Calls
- All Incoming Calls while Roaming

Like call forwarding, call barring can also be set from the subscriber's phone. If I want to bar all outgoing calls from my phone, I enter * 33 * 0000 #. To release the barring I need to enter # 33 * 0000 #. In India call barring is possible for STD and ISD lines. Also, passwords can be set for calls.

Voice mail: Voice mail is a supplementary service where the caller is forwarded to a voice mail service when the call does not mature. For mobile networks this can be due to one of the four conditions as explained in call forwarding. However, for fixed networks only the first three conditions are valid.

Multiparty call conferencing: In a multiparty service the served mobile subscriber is in control of one active call and one call on hold, both calls having been answered. In this situation the served mobile subscriber can request the network to begin the multiparty service. Notification will be sent towards the served mobile subscriber and all the remote parties in a multiparty call. A notification will always be sent to all remote parties every time a new party is added to the multiparty call. Notifications shall also be sent to remote parties when they are put on hold and when they are retrieved in accordance with normal Call Hold procedures.

Caller line ID: This supplementary service offers the facility by which the telephone number of the calling party is displayed on the phone. Mobile phones generally display the name of the caller. This is possible if the caller's telephone number is stored in the address book of the mobile phone. Some phone manufacturers also offer a facility to associate icons, ringing tones specific to caller, user groups. The ISUP IAM message contains the caller's telephone number. The same is passed on to the phone.

Alternate line service: Alternate Line Service offers a subscriber the convenience of two phone numbers for one mobile phone. These are useful when subscriber would like to have different numbers for different reasons, but would not like to carry multiple phones. For example, one number is for business calls and the other one is for personal calls. Each number may have its own associated ringing tone.

Closed user group: The Closed User Group (CUG) Supplementary Service enables subscribers, connected to a PLMN and possibly also other networks, to form closed user groups (CUGs) to and from which access is restricted. Members of a specific CUG can communicate among each other but not, in general, with users outside the group.

The network shall provide a subscription option in order to enable the user to specify a preferential CUG, for each basic service group included in at least one of the CUG(s).

Call Waiting: This feature allows a customer who is already in conversation with another to be informed by a Call Waiting tone that another call is waiting. The calling party will hear a ringing tone instead of an engage tone. In a fixed line exchange the same is registered by dialing 118 and deactivated by dialing 119.

Telephone User Part (TUP) Data User Part (DUP)

TUP handles analog circuits whereas digital circuits and data transmission capabilities are handled by the Data User Part. These services are no longer in use worldwide. In some countries (e.g., China, Brazil), the Telephone User Part (TUP) is used to support basic call setup and tear-down. In most parts of the world, ISUP is used for voice and data call management.

Signaling Connection Control Part (SCCP)

The SCCP provides connectionless (class 0) and connection-oriented (class 1) network services and extended functions including specialized routing (GTT-global title translation) and subsystem management capabilities above MTP Level 3. A global title is an address (e.g., a dialed 800 number, calling card number, or mobile subscriber identification number), which is translated by SCCP into a destination point code and subsystem number. A subsystem number uniquely identifies an application at the destination signaling point. SCCP is used as the transport layer for TCAP-based services.

Transaction Capabilities Application Part (TCAP)

The TCAP offers its services to user-designed applications as well as to OMAP (Operations, Maintenance and Administration Part) and to IS41-C (Interim Standard 41, revision C) and GSM MAP (Global Systems Mobile). TCAP supports the exchange of non-circuit related data between applications across the SS7 network using the SCCP connectionless service. Queries and responses sent between SSPs and SCPs are carried in TCAP messages. TCAP is used largely by switching locations to obtain data from databases (e.g. an SSP querying into an 600 number database to get routing and personal identification numbers) or to invoke features at another switch (like Automatic Callback or Automatic Recall). In mobile networks (IS-41 and GSM), TCAP carries Mobile

Application Part (MAP) messages sent between mobile switches and databases to support user authentication, equipment identification, and roaming.

Intelligent Network Application Protocol (INAP)

The INAP specifies the information flows between different entities of the IN functional model in terms of protocol data units (PDUs). The PDUs represent remote operations in the scope of TCAP.

Customized Applications for Mobile network Enhanced Logic

The CAMEL (Customized Applications for Mobile network Enhanced Logic) is a mechanism to help the network operator to provide the subscribers with the operator specific services when roaming. To support prepaid roaming services, CAMEL will be required.

CAMEL Application Part

CAMEL Application Part (CAP) is the application part based on CAMEL version 2. CAP is based on a sub-set of the Capability Set–1 (CS1) core INAP.

Mobile Application Part (MAP)

A protocol that enables real time communication between nodes in a mobile cellular network. This application part defines protocols to exchange subscriber related information from one cellular network to another cellular network. A typical usage of the MAP protocol would be for the transfer of location information from the VLR (Visitor Location Register) to the HLR (Home Location Register).

IS-41

Interim service, IS-41 is the counterpart of GSM-MAP for the US cellular networks. IS-41 is used to offer seamless roaming to the subscribers. One area in which GSM-MAP and ANSI-41 transport differ is in the area of roamer administration. GSM-MAP networks rely on an International Mobile Station Identifier (IMSI), as opposed to the Mobile ID Number (MIN) used in IS-41.

11.4.2 SS7 Signal Unit

Within a SS7 network signaling information is passed in the form of messages. These messages are called signal units. Signal units are continuously transmitted on any link in both directions. SS7 uses three different types of signal units:

- Fill-In Signal Units (FISUs)
- Link Status Signal Units (LSSUs)
- Message Signal Units (MSUs)

Figure 11.5 depicts different types Signal units.

Fill-In Signal Unit

FISUs is at the lowest level of service and does not carry any information. FISUs are transmitted when the link is idle and there is no payload (LSSUs or MSUs) to be transmitted.

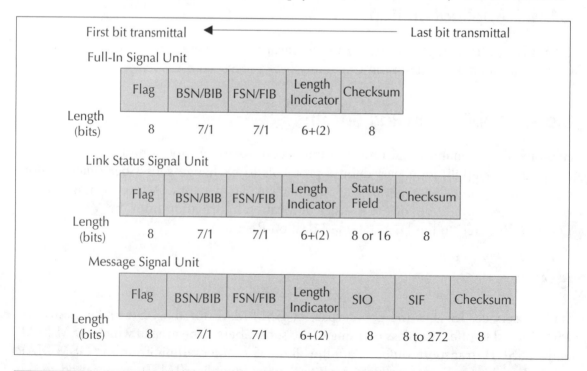

Figure 11.5 Different types of Signal units

Link Status Signal Unit

LSSUs transact information about the SS7 signaling link between nodes on either end of the link. When a link is identified to have failed, the signaling point that detects the error is responsible for alerting its neighboring signaling points. This information is contained in the status field of the signal unit. LSSUs signal the initiation of link alignment, quality of received traffic, and status of processors at either end of the link. LSSUs do not require any addressing information because they are only sent between signaling points. LSSU never get broadcast throughout the network.

Message Signal Units (MSU)

MSUs are the real workhorses within the SS7 network. All signaling related to call set-up call and tear-down, all signaling related to database query and response use MSU. Management functions of SS7 use MSUs. MSUs provide MTP protocol fields, service indicator octet (SIO) and service information field (SIF). The SIO identifies the type of protocol (ISUP, TCAP) and standard (ITU-TS, ANSI). The SIF transfers control information and routing label. The functionality of the MSU is defined through the contents of the service indicator octet (SIO) and the service information fields (SIF). The functionality of the MSU is defined through service indicator octet (SIO) and the service information fields (SIF). The SIO is an 8-bit field that contains three types of information: 4 bits to indicate service indicator (0–signaling network management; 1–signaling network testing and maintenance; 2–SCCP; 3–ISUP), 2 bits to indicate national (proprietary within a country) or international (ITU standard) message; remaining 2 bits to indicate priority with 3 the being highest priority. The service information field (SIF) defines the information necessary for routing and decoding of the message. SIF transfers control information and the routing label used in MTP Level 3. The routing label consists of the destination point code (DPC), originating point code (OPC) and signaling link selection (SLS) fields.

The common fields in all the Signals units are as following:

Flag: This contains a fixed pattern 01111110 and is used for clock synchronization

BSN: Backward Sequence number

BIB: Backward indicator bit

FSN: Forward Sequence number

FIB: Forward indicator bit

Length indicator: Out of 8 bits only 6 bits are used to indicate the length.

11.5 IN CONCEPTUAL MODEL (INCM)

INCM was developed to provide a framework for the design and description of each capability set and the target IN architecture. In an IN scenario there are four main actors. The first, the service user, is the end-user of the service. For example, this is the person who calls a free-phone or utilizes his calling card to call a friend while he is roaming. The second, the service subscriber, is the actor who subscribes to an IN feature. He can use it for himself or provide it to his customers. This could be an individual, a corporation, or a virtual service provider. The third, the service provider, who creates, deploys and supports IN services. This actor has contracts with service subscribers. These contracts specify the billing and the subscribed features. The fourth and last one is the network operator. This actor provides the infrastructure needed to support IN services. This actor has contracts with service providers. INCM captures the whole engineering process of the IN.

The INCM is structured into four planes (Figure 11.6) as follows:

- Service plane
- Global functional plane
- Distributed functional plane
- Physical plane.

The upper two planes focus on service creation and implementation, whereas the lower two planes addresses the network and physical needs.

Service Plane (SP)

This plane is of primary interest to service users and providers. It describes services and service features from a user perspective, and is not concerned with how the services are implemented within the network.

Global Functional Plane (GFP)

The GFP is of primary interest to the service designer. It describes units of functionality, known as service independent building blocks (SIBs) and it is not concerned with how

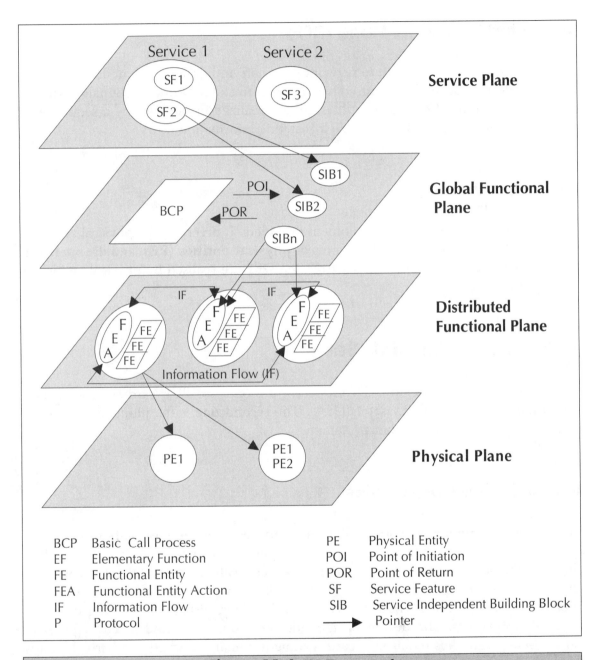

Figure 11.6 IN Framework

the functionality is distributed in the network. Services and service features can be realized in the service plane by combining SIBs in the GFP.

Distributed Functional Plane (DFP)

This plane is of primary interest to network providers and designers. It defines the functional architecture of an IN-structured network in terms of network functionality, known as functional entities (FEs). SIBs in the GFP are realized in the DFP by a sequence of functional entity actions (FEAs) and their resulting information flows.

Physical Plane (PP)

The PP is of primary interest to equipment providers. It describes the physical architecture for an IN-structured network in terms of **physical entities (PEs)** and the interfaces between them. The functional entities from the DFP are realised by physical entities in the physical plane.

11.5.1 Examples of IN Services

In this section we consider two IN services. First one is calling card service defined within the scope of capability set 1 (CS-1). The second one is the ultimate test for IN functionality called local number portability.

Virtual Calling Card Service

In a wireline telephone network, calling card services offer the possibility of making a call from any phone and charge the call to the user who makes the call, rather than charging the subscriber of the telephone line. The call needs to be set up without charging any of the lines involved in the call. This service is also known as automatic alternative billing service (AABS), account card calling, virtual card calling or credit card calling services. The distinction is made depending on whether a real physical card is being used (thus requiring a card reading terminal), or whether a 'virtual card is being used (requiring the user to dial an account/PIN number). In a cellular network the prepaid system also uses similar principles within the network; however, in the user interface the user need not enter the account information or PIN for every call they make. In a cellular network, authentication is done using the MSISDN number of the phone.

Consider a user who wants to make a call using a virtual calling card. The processing steps for this call is depicted in Figure 11.7 and will be as follows:

1. The user dials the service access code for the AABS (e.g., 1-600-234123). When connected to the service, there are voice prompts helping the user to navigate through the service menu. Then it asks the user to enter the destination phone number the user wants to talk to. User enters the number of the target telephone (e.g. 011-2549-9229).

2. The switch recognizes that the call is an IN call and the service switching point (SSP) sends an INAP query containing call information to the corresponding SCP. On receipt of the query the SCP starts the corresponding service logic program. The SCP determines an appropriate Intelligent Peripheral (IP) to query for the account code and PIN of the user for validation.

3. The SCP returns to the SSP a routing number of an appropriate IP and instructs the SSP to establish the connection to the IP.

4. The SSP routes the call to the IP and instructs the IP to start an appropriate dialogue with the user.

5. The IP asks the user for PIN.

6. The user enters the PIN, and the IP collects the response digits.

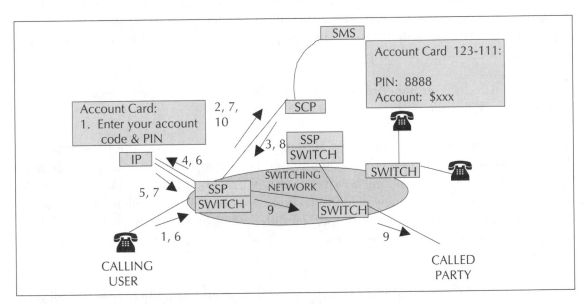

Figure 11.7 Automatic alternative billing service processing

7. The IP returns the information to the SSP. The SSP relays the account code and the PIN to the SCP. The SCP examines the PIN and checks that the account limit has not yet been exceeded.

8. The SCP instructs the SSP to disconnect the IP and to establish the connection to the destination number.

9. The SSP disconnects the IP and instructs the switch to establish a connection to the desired destination.

10. After the call is terminated, the SSP informs the SCP about the call charges. The SCP subtracts the charges of the call from the subscriber's balance.

Local Number Portability

The intent of Local Number Portability (LNP) is to open up local telephone service to competition. LNP was defined in the US in the Telecommunication Act of 1996 as the 'ability of users of telecommunication services to retain, at the same location, existing telecommunication number without impairment of quality, reliability or convenience when switching from one telecommunication carrier to another'. The regulators think that the biggest roadblock to competition in the telecom vertical is the ownership of the telephone numbers. Subscribers are reluctant to switch to a new service provider because they have to give up their existing telephone number when they switch to a new service provider. There are three phases of LNP.

Phase one is Service provider portability. This allows a subscriber to select a new service provider while keeping their existing telephone number.

Phase two is Service portability. This allows subscribers to change the type of service they have while keeping their telephone number. For example, if the subscriber changes from POTS (Plain Old Telephone Service) to ISDN (Integrated Service Digital Network) he or she has to obtain a new telephone number, because the switching equipment used to provide the IDNS service supports a different block of numbers. With LNP the subscriber does not have to give up the telephone number when changing the type of service.

Phase three is the third and most difficult phase, location portability. This will allow a subscriber to move from city to city or even state to state within USA while maintaining the same telephone number.

11.5.2 Wireless Intelligent Network

Wireless intelligent network (WIN) is a concept being developed by the Telecommunications Industry Association (TIA) Standards Committee TR45.2. The charter of this committee is to drive intelligent network capabilities, based on interim standard IS-41, into wireless networks.

Intelligent networks principles and functions are very much woven with almost all the services mobile networks offer. This starts from as simple as supplementary services like caller ID to complex functions like roaming. Many new IN services are proposed as part of the CS-1 service sets. One such service is voice-activated services. This is a hands-free service targeted for vehicular conditions. In a vehicle we need hands-free eyes-free service. Therefore, if we just say 'John Smith', the call will do speech recognition, look up a database, get the telephone number of John Smith and establish a call. Another IN example is selection of long distance carrier. Based upon the time of day, country etc. we can select the long distance carrier for routing the call.

11.6 Softswitch

The word Softswitch is derived from the combination of ***software and switch***. A softswitch is an API framework that is used to bridge a traditional telecommunication network and IP networks. A softswitch will manage traffic that contains a mixture of voice, fax, data and video. Softswitch can be considered as a telecommunication switch where the intelligence is outside the switch and driven by software. Softswitch will be used in future for VoIP and unified call control which will have voice, data, multimedia and instant message. This is conceptually an extension of intelligent network at a much broader sense. Softswitch is expected to address all the shortcomings of traditional local exchange switches. The various elements in a softswitch architecture network are:

- Call agent (media gateway controller, softswitch)
- Media gateway
- Signaling gateway
- Feature server
- Application server
- Media server
- Management, provisioning and billing interfaces.

Protocols that are supported by softswitches are MGCP, H.248 (Megaco), SIP, H.323, and Sigtran's suite of protocols and adaptation layers which include SCTP, IUA, M2UA, M3UA, SCUA, etc. In general, many of the telephony protocols and data protocols are supported by softswitches including CAS, SS7, TCAP, INAP, ISDN, TCP/IP, etc.

As viewed by the IP network, a media gateway is an endpoint or a collection of endpoints. Its primary role is to transform media from one transmission format to another, most often from circuit to packet formats, or from analog/ISDN circuit to packet as in a residential gateway. It is always controlled by a media gateway controller. Media server operates as a slave to a media gateway controller to perform media processing on media streams. The signaling gateway and media gateway must be deployed at the boundary between the PSTN and the Softswitch. All other components may be located anywhere within the network that makes sense with regard to latency of access, co-location of control, and other operational considerations.

11.7 Programmable Networks

In a world where everything is networked, there is a need to be able to program network components to adapt to application requirements. We need to have better control on the quality of service, security, application-dependent routing, intelligent caching, utilization of bandwidth, support mobility and sophisticated management functionality. It is therefore necessary to be able to dynamically program the resources within a network. These types of application-specific functions need to be dynamically programmed within the network components in order to support flexible and adaptive networks. Such networks are called programmable networks. Programmable networks will also address the need to 'open' the network up and accelerate its programmability in a controlled and secure manner for the deployment of new architectures, services and protocols. The separation of communications hardware (i.e., switching fabrics, routing engines) from control software is fundamental to making the network programmable. A programmable network is distinguished from other networking environments by the fact that it can be programmed from a minimal set of APIs to provide a wide array of higher level services.

11.8 Technologies and Interfaces for IN

We now know what IN is. We have gone into details of some IN applications. We have also learnt various layers and functions of user parts of SS7 stack. To develop an IN application we need to access one or more user parts in the SS7 stack. These applications will use native APIs supplied by the SS7 stack vendor. For some application, if we want to build the stack ourselves, we need to use APIs supplied by the SS7 hardware vendor. All this APIs are proprietary to the vendor. At a later time, if we want to use a stack from other vendor, or a different SS7, interface card, we need to change our application to suite new set of vendor specific APIs. This makes interoperability and development cycle complex and expensive. To address these challenges, some standardization is required. Using these standards, an application can use some universally supported APIs. Interfaces like class name, function name, function parameters for these APIs will be same and will be supported by all vendors across the board. This will make an IN application independent of the lower layer vendor. In the following sections we shall discuss some of these interfaces and standards.

11.8.1 PARLAY

The Parlay Group is an open multi-vendor consortium formed to develop open technology-independent APIs to access resources within a telecommunication network. Parlay integrates intelligent network (IN) services with IT applications via a secure, measured, and billable interface. Parlay is focused in defining umbrella architecture and API for Open Service Access (OSA). We have mentioned OSA in section 7.1 as a part of 3G (chapter 9). The OSA specifications define an architecture that enables service application developers to make use of network functionality through an open standardized interface.

Founded in 1998, The Parlay Group focused initial development of its APIs on functions such as call control, messaging and security. The Parlay Group was formed by a group of companies (BT, Microsoft, Nortel Networks, Siemens, and Ulticom, formerly DGM&S Telecom). The group first demonstrated a Parlay service in the UK and USA in December 1998.

In today's network, applications and services are part of the network operator's domain. This network-centric approach was good for specific applications. With the growth of Internet and new services, what is now needed is a solution that combines the benefits of the network-centric approach of economies of scale and reliability of telecommunication networks with the creativity and power of the IT industry. The Parlay APIs will allow services to be developed outside of the network that use the wide range of common functions at the center of the network. As networks of the future evolve, there will be a growing need to harmonize intelligence in the center of the network with intelligent devices at the network edge. This means that it should be possible to build applications, test and operate outside the network domain.

11.8.2 JAIN

JAIN is a community of companies led by Sun Microsystems under the Java Community Process that is developing Java APIs for next-generation systems consisting of integrated Internet Protocol (IP) or asynchronous transport mode (ATM), public switched telephone network (PSTN), wireless networks and intelligent networks (IN). These APIs include interfaces at the protocol level, for different protocols such as Media Gateway Control Protocol (MGCP), Session Initiation Protocol (SIP), and Transactional Capabilities Application Part (TCAP), as well as at higher layers of the telecommunications software stack. The JAIN APIs bring service portability, convergence, and secure network access to telephony and data networks. This is referred to as Integrated Networks. Furthermore, by allowing Java applications to have secure access to resources inside the network, the opportunity is created to deliver various services.

11.8.3 TINA

The IN is believed to be an essential revolutionary step towards the optimum restructuring of public networks. The Telecommunications Information Network Architecture Consortium (TINA-C) was founded to define a telecommunications information networking architecture (TINA) which would enable the efficient introduction, delivery and management of telecommunication services. Also, due to the rapid convergence of telecommunications and computing, the focus of attention moved away from the physical network to a software-based system. The kind of services to be supported by the TINA ranges from voice-based services to multimedia, multiparty services, the latter of which cannot be properly supported by current IN architectures. All of these software-based applications will run on a distributed hardware platform, which hides any distribution concerns from applications.

11.8.4 SS7 SECURITY

SS7 network is a data network used privately by the telecommunication operators. Though it is a global network, it can be considered to be private. Unlike the Internet, anybody cannot connect a device into this network. As it is private, it is considered secured. There has not been any record of attack on a SS7 network. However, this does not imply that the security in SS7 network is robust and impossible to break. With deregulation of telecommunication industry, network operators are obliged to allow private operators to install their IN nodes in the network. Also, with VoIP, Internet and other packet networks will be connected to the SS7 network. This makes it vulnerable to security attacks. However, as of date; it is quite secured.

REFERENCES/FURTHER READING

1. Travis Russell, Signalng System#7, McGraw-Hill, 2000.

2. Suthaharan Sivagnanasundaram, GSM Mobility Management Using an Intelligent Network Platform, Ph.D Thesis, University of London, 1997.

3. Intelligent Network (IN), The International Engineering Consortium.

4. Wireless Intelligent Network (WIN), The International Engineering Consortium.

5. Stéphane Nicoll, Overview of Intelligent Networks, 2001.

6. ETSI EG 201 781, Intelligent Networks (IN); Lawful Interception, 2000.

7. ETSI 300 374-1, Intelligent Network (IN); Intelligent Network Capability Set 1 (CS1); Core Intelligent Network Application Protocol (INAP); Part 1: Protocol specification, 1994.

8. American National Standard T1.111.8-2001 (T1.111a-2002), Numbering of Signalling Point Codes.

9. Standards Coordination Document No. 4, Intelligent Networks, Working Group Standards Coordination Permanent Consultative Committee I, 2000.

10. Prepaid Wireless Service Billed in Real Time, The International Engineering Consortium.

11. Olli Martikainen, Juha Lipiäinen, Kim Molin, TUTORIAL ON INTELLIGENT NETWORKS, IFIP IN Conference 1995.

12. Simply SS7, ADC NewNet Products, 2001.

13. VocalTec Softswitch Architecture Series 3000.

14. ETSI TS 100 518 V7.0.0 (1999-08): Digital cellular telecommunications system (Phase 2+); Closed User Group (CUG) Supplementary Services Stage 1 (GSM 02.85 version 7.0.0 Release 1998).

15. ETSI TS 100 517 V7.0.0 (1999-08): Digital cellular telecommunications system (Phase 2+); MultiParty (MPTY) Supplementary Services Stage 1 (GSM 02.84 version 7.0.0 Release 1998).

16. ETSI TS 100 907 V7.1.0 (1999-08): Digital cellular telecommunications system (Phase 2+); Man-Machine Interface (MMI) of the Mobile Station (MS) (GSM 02.03 version 7.1.0 Release 1998).

17. PARLAY: Draft ETSI ES 202 915-1 V0.0.4 (2003-03): Open Service Access (OSA); Application Programming Interface (API); Part 1: Overview.

18. GPS: http://www.aero.org/publications/GPSPRIMER/.

19. G. Lorenz, T. Moore, G. Manes, J. Hale and S. Shenoi, Securing SS7 Telecommunications Networks; Proceedings of the 2001 IEEE, Workshop on Information Assurance and Security United States Military Academy, West Point, NY, 5 6 June 2001, ISBN 0-7803-9814-9.

20. w3c CC/PP Working Group: http://www.w3.org/Mobile/CCPP/.

REVIEW QUESTIONS

Q1: Describe the steps involved in a telecom call setup?

Q2: What is intelligent network? When does a telecommunication network become Intelligent?

Q3: What are the elements in a SS#7 signalling network? If we compare a SS#7 signalling network with the IP network, which elements are unique and which elements are similar?

Q4: What is an SCP? What are its functions? Explain how do we use SCP to implement virtual calling card facility?

Q5: What is number portability? How is number portability different from telephone portability?

## CHAPTER 12

Client Programming

12.1 INTRODUCTION

We would like to begin by reminding ourselves that 'information is not knowledge, knowledge is not wisdom and wisdom is not foresight. Each grows out of the other and we need them all.' Arthur C. Clark (1997). In the same vein hardware, software and networks together make powerful applications possible. None is more or less important than the other.

12.2 MOVING BEYOND THE DESKTOP

This chapter aims to give us a quick overview of the current handheld computing landscape, its evolution and profitable usage. Mobile devices have traditionally been classified as phones, pagers, and personal data assistants (PDAs). Initially each had well-defined roles: cell phones provided communication capabilities similar to wired phones; pagers provided text messaging; PDAs provided portable data applications such as contacts, calendars and notes. What is interesting, however, is the way the technology is emerging at the present time. These seemingly independent streams are now merging and we have the device integration. Most cell phones now include address books and SMS (Short Message Services). Some pagers include e-mail access and almost all contemporary PDAs include communication capabilities. Devices today offer many permutations and combinations of these features leading to, 'one-size-fits-all' devices. A historical perspective of evolution helps to comprehend the growth and direction of technology. We shall, therefore, briefly review two independent streams of developments, one leading to mobile phone and the other to PDAs. The emphasis here will be on programming aspects but to understand the environment we will take a quick peek at the underlying OS.

Time	1920–1960	1960–1970	1970–1980	1980–1990	1990–
Devices	Marine radio and Vehicle mounted telephone	• Shoe phone • Briefcase cell phone • Bat mobile phone	• First hand held cell phone • NMT hand hand held • Car phone Tokyo	Nokia Ericsson Motorola NMT/PCS/GSM hand sets	Nokia Communicator
Carrier Technology	Analog radio	IMTS	Cellular Analog Systems.	PCS/GSM	GPRS/3G/UMTS/CDMA
Key features	Based on tubes large and Bulky basic voice only devices	Transistors allowed for miniaturization but still voice only	Cellular concepts were deployed but large heavy voice only systems.	GSM brought digital systems and limited data capability	Sleek light weight digital systems, capable of carrying multimedia data.

Figure 12.1 Evolution of Mobile Technology

Let us begin with the mobile phones. Briefly, wireless communications are enabled by packet radio, spread spectrum, cellular technology, satellites, infrared line of sight and microwave towers, and can be used for voice, data, video, and images. As we have already seen, a cell phone is an extremely sophisticated radio. To fully comprehend the present-day technology and features, it helps to first understand how these evolved over time.

As seen above in Figure 12.1, the radios and vehicle-mounted telephones were the first communication devices which continued to evolve over time spanning from the 20s to 60s of the twentieth century. The break through came with the invention of the transistors, which brought down the size, weight and subsequently the cost of the handsets. The real growth however started with the implementation of cellular concepts. The shift to digital technologies enabled data communications, leading to the devices of today. From huge monsters measuring 20" × 11" × 8.5" and weights close to 40 pounds that allowed only voice in the 1960s to 4.3" × 0.9" × 1.8" which allow multiple applications, multimedia, video conferencing facilities and even built-in cameras, today the cell phones have come a long way. Most current phones offer built-in phone directories, calculators and even games. Many of the phones incorporate some type of PDA or Web browser.

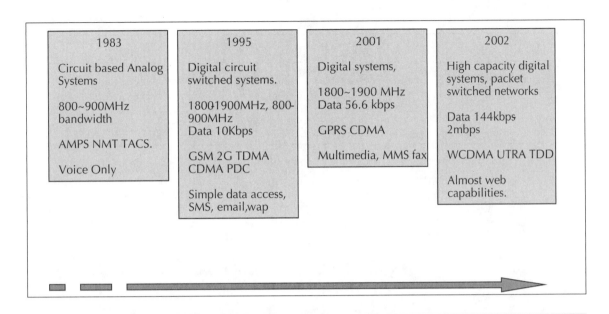

1983	1995	2001	2002
Circuit based Analog Systems	Digital circuit switched systems.	Digital systems,	High capacity digital systems, packet switched networks
800~900MHz bandwidth	18001900MHz, 800-900MHz	1800~1900 MHz Data 56.6 kbps	Data 144kbps
AMPS NMT TACS.	Data 10Kbps	GPRS CDMA	2mbps
Voice Only	GSM 2G TDMA CDMA PDC	Multimedia, MMS fax	WCDMA UTRA TDD
	Simple data access, SMS, email,wap		Almost web capabilities.

Figure 12.2 Evolution of cellular technology

It is but natural that the evolution of the handheld be closely mapped to the development of the networks. Thus today's mobile phones are a combination of a mobile and a data device rolled into one. The evolution of the cellular networks as we have already seen can be summarized as below. The move has been from voice communication to increasingly data-based communications. This is depicted in Figure 12.2.

Data over telecomm matured with the introduction of the world's first all-in-one communicator, the Nokia 9000 by Nokia in 1996. We will break at this time to see an alternate device, the PDAs that grew from a totally different requirement but have become an essential part of today's mobile computing environment.

A PDA as the name suggests is a personal digital assistant. Its major functionality is related to the storing, accessing and manipulating data. None other than the team of Star-trek conceived the PDAs in the early 1960s. But till the later half of the 1980s this remained mostly a concept. Apple demonstrated the first device. This device was the Newton. Figure 12.3 shows the evolution of the PDA.

From a stand-alone organizer the PDA has come a long way. Most of this significant evolution is due to the progress in hardware capabilities and development of underlying bearer networks. We now take a look at the current offerings. Most handheld devices

Time	1987	1988–1992	1993	1994–1996	1997–1999	2000....
Devices	Pison Organizer	Grid-pad, Atari, sharp	Message Pad, Zoomer, PenPad, Envoy	Palm, Marco, MagicLinc	Sharp, Palm VII	Nokia 9210, iPaq
Key features	Stand alone data organizer.	Handwriting recognition	Telephony application	Synchronization	Wireless link	Personal organizer come wireless data communicator.

Figure 12.3 Evolution of PDA

Mobile phones	PDA	Communicator
• Mostly voice using telephone network (PCS/CDMA/GSM/GPRS/UMTS etc)	• PIM (Personal information manager)	• A combination of the two.
• Provides phone book, CLI, messaging (SMS/EMS/MMS), limited data services.	• Network access using some external modern.	• Allows capabilities of both.
• Simple applications like email, ring tones, picture messages etc.	• Synchronization with PC using serial cable or IR.	• Can support powerful enterprise grade applications.

Figure 12.4 Comparison of capabilities of different devices.

can be classified as smart phones, PocketPCs or smart communicators (a combination of the two). The Figure 12.4 gives the features based on which this classification is made.

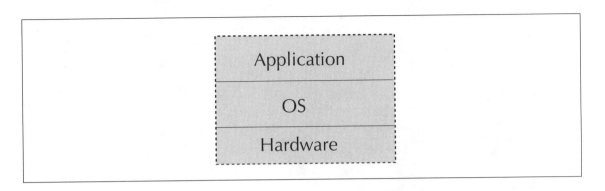

Figure 12.5 Structure of a mobile device.

Having seen where we came from, we will now see where we are going i.e., the applications. All applications have some basic requirement. At the minimum they need:

- Some input and provide some useful output.

- Memory for runtime and persistent storage.

- Some communication capabilities.

Most of these are device capabilities but programming at a device level is cumbersome and generally not advisable. So we need some level of abstraction on the bare hardware. This layer of abstraction is the device operating system. Most devices provide some kind of programming environment. Palm programming is mostly done in C/C++, symbian in C++/Java and WinCE or PocketPC in embedded VB or embedded VC. Roughly the architecture is shown in Figure 12.5. We will explore each of these layers one by one.

12.3 A PEEK UNDER THE HOOD: HARDWARE OVERVIEW

As we have said before the cell phones/ PDAs are among the most intricately built devices. Modern digital devices have a MIPS (million instructions per second) capability to process information (voice/data) stream. Both classes of devices essentially consist of:

- A microprocessor.

- A power source.

- A signal converter.

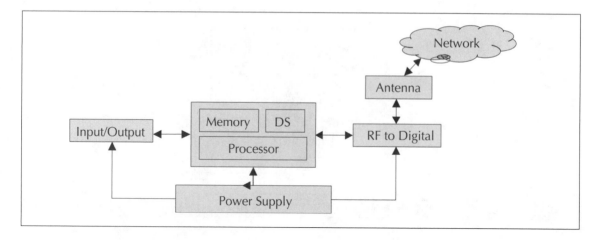

Figure 12.6 Overview of a digital communication device

- An I/O unit.

- Some memory (both persistent and volatile).

Figure 12.6 shows a simplified view. As we can see, here the power supply unit (generally lithium battery) provides the required power to all the components. The processor is the brain of the device; it handles all the processing in conjecture with the memory and the Digital Signal Processor (DSP), it interfaces to the I/O unit and also the external signaling system. There is also a RF to digital converter, which is the interface to the communication channel/network. It is responsible for converting the RF (Radio Frequency) signals to Digital and vice-versa. We will now take the cell phone and PDA separately.

12.4 Mobile Phones

As shown in Figure 12.7, internally a cell phone consists of the following parts:

- Antenna: This is the signal reception unit.

- Circuit board: This is the control unit of the system. It has several important chips mounted on it. The prominent ones amongst these are the following.

  1. The analog-to-digital and digital-to-analog conversion chips represented by the RF to Digital block in the figure above. This is the entry and exit point to

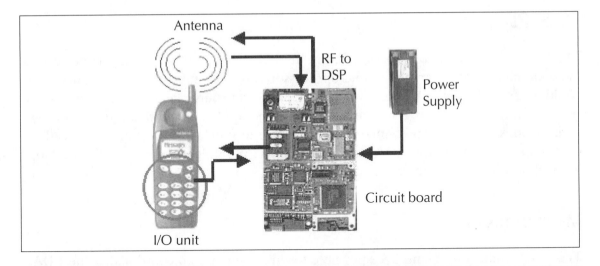

Figure 12.7 Components of a mobile phone

the phone. As the name implies, this chip is designed to convert analog network signals to digital for the phone to process and reverse.

2. The microprocessor controls and co-ordinates the handset functions. The most important of these are User Input/Output and network interfaces that include communication with the base station.

3. The ROM and Flash memory chips provide storage for the phone's storage requirements, which include system memory and application memory. Additional memory is also provided through external detachable memory cards.

4. The DSP (Digital Signal Processor) is a highly sophisticated chip that manages the signal manipulations.

5. The radio frequency (RF) chip manages the signal channels while the power section is responsible for power management and recharging.

- Display unit is the output unit, generally a Liquid Crystal Display (LCD) panel.
- Keyboard (qwerty or T): This is the input unit.
- Microphone: To facilitate speech transmission
- Speaker: The microphone's listener counter part
- Battery: The source of electrical energy.

12.5 PDA

The PDAs unlike the phones evolved from the PCs. They have a similar architecture as most desktops have namely a microprocessor, an I/O mechanism, memory, and additionally a wireless or IR port and of course a power source generally a battery. They typically have an operating system and mimic the applications on the desk tops e.g. address-book e-mail etc and hence need to be able to synchronize with the applications on the PC. As we see the PDAs are considered a portable extension to the desktop. We will now look at the internals of a typical PDA

Microprocessor

This is the control unit, and is responsible for initiating and co-coordinating the PDA's functions. Owing to intrinsic device limitations, PDAs use cheap low-end processors such as the Motorola Dragonball. These processors though very basic (about 16-75 MHz

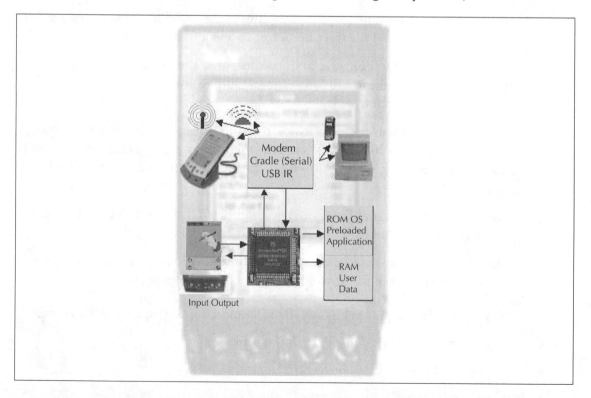

Figure 12.8 Components of PDA

capacities) are sufficient for the requirements of a PDA. However, this is an important consideration for application developers who consistently have to ensure that they do not overload the processor leading to a performance degradation and poor user experience. As we will see later, this is one of the factors which decides which part of processing should be performed on the device and what should be shifted to a backend server.

Operating System

The OS has a multitude of responsibilities very similar to those in the desk-top environment. The PDA operating systems however are fairly simple having fewer instructions and smaller footprints. The major market players are, Palm OS, Symbian and PocketPC. We will speak of these in more details in subsequent chapters.

Memory

The PDA has no concept of a hard drive. Instead, it has solid-state memory like Static RAM or Flash memory and these days we get removable memory cards and detachable memory sticks. It is common to have about 2 MB of memory. It is used to store the OS and application programs; these remain intact even when the machine shuts down. The user data however is stored in RAM. We will refer to it again when we explore the details of the OS.

Batteries

Most PDAs today use rechargeable lithium, nickel-cadmium or nickel-metal hydride batteries that last for about 2 hours on an average. Major culprits that drain batteries are:

- Additional memory
- Color LCD display
- High end features like Voice recording, camera, MP3 player etc

PDAs use advanced power management systems to extend the battery life. It is important for application developers not to start infinite loops or polling as this can potentially interfere with the system's power management. As a last word of caution it is essential to back up data to the PC whenever possible so that even if all the data on the device gets wiped out it can be restored from the PC.

LCD Display

The Liquid Crystal Display screen of a PDA is its I/O interface and is used for both input and output. The display may be grayscale (16) or color (65,536) has resolution of $160 \times 160, 240 \times 320$ etc. While backlit affords better reading in the dark, they consume more power, and hence, drain the battery faster.

Input Device—Buttons in Combination with Touch-Screen or Keyboard

PDAs mostly use a stylus and touch screen in combination with a handwriting recognition program. Most also have buttons to bring up frequently used applications. This is similar to using keyboard short-cuts. Some high-end devices may actually boast of a miniature 'qwerty' keyboard but these are exceptions and not the rule.

> The touch screen is an interesting piece. The set-up consists of multiple layers; on top is a thin plastic or glass sheet with a resistive coating on its bottom. This layer floats on a thin layer of nonconductive oil, which rests on a layer of glass coated with a similar resistive finish. Thin bars of silver ink line the horizontal and vertical edges of the glass by sending current first through the vertical bars and then the horizontal ones, the touch screen obtains the X and Y coordinates of the touchdown point. When the stylus touches the screen, the plastic pushes down through the gel to meet the glass (called a 'touchdown'). This causes a change in the voltage field, which is recorded by the touch screen's driver software, which determines the point of contact. The driver scans the touch screen thousands of times each second and sends this data to the application that is listening for it.

> How does **handwriting recognition** work? Using a **stylus**, we draw on the device's touch screen. Software inside the PDA converts the characters to letters and numbers. However, these machines don't really recognize handwriting. Instead, we must print letters and numbers one at a time. On Palm devices, the software that recognizes these letters is called **Graffiti**. In case one finds the graffiti difficult to use, it is possible to use an onscreen keyboard. It looks just like a regular keyboard, except that the letters are tapped with the stylus.

Input/Output Ports

A PDA must be able to communicate with a PC. This communication is called data synchronization or syncing and is typically done through a serial or USB port. This can be through a cable or the cradle. These days most come with an infrared port and offer telephone modem accessories to transfer files to and from a network.

12.6 DESIGN CONSTRAINTS IN APPLICATIONS FOR HANDHELD DEVICES

We have looked at both cells phones and PDAs the communicators are a cross between the two. These vary greatly across vendors and models from the same vendor. We shall not go into the hardware for communicators.

Another breed of device that is growing very fast is the java-enabled phone. These phones have Sun's Java virtual machine (j2me/Personal java) embedded in them. The biggest advantage that these devices enjoy is device, OS independence and application portability. This has caused an exponential increase in the number of applications that can run on these phones.

However, what is in all these devices for application developers? Common across these devices are the following characteristics:

- Low-end processors
- Small Screen Size
- Cumbersome input device
- Limited battery power
- Memory Limitations

Processing Power: The processing speeds start at about 16 MHz. Heavy-duty computation like encryption key generation is a heavy drain on the device's resources. Most devices that support security provide a special processor designed for the purpose. Computation on the device should be done judiciously. Offloading computation to a backend server is always a good idea. While designing a device resident client, architects and developers always have to walk the tight rope. Too many computations on the client while allowing for faster response eat into valuable memory. Shifting everything to the server leads to poor response, which is a frustrating experience for the user. So extreme caution has to be exercised. Based on the functionality and specificity of the software, clients are of three types

1. Thin clients: These are generic in nature and cater to a wide range of sources. They are similar to the web browsers. Communication from the server is mainly based on some flavor of Markup Language (ML). Note however that this requires a local parser. The server needs to send large amount of display information to the client, to adequately represent data, chocking up the transmission channel. Thus the thin client offers generality at the cost of bandwidth. This is a major consideration especially in networks where the users pay for the data and not call time.

2. Thick clients: The intelligence resides in the device and a call is made to the server only for data. Computation is done locally. While this approach resolves the bandwidth problem it introduces two more problems, namely, the size of the application on the device and distribution of the application when an update or fix happens.

3. Thin plus or semi thick clients: These lie somewhere in between. How thick or thin is decided based on the functionality of the application, the bandwidth availability and cost constraints.

Screen Size: Most handhelds have a small screen, limiting the information that we can display at one time. Hence, screens should be designed very carefully taking care to remove all extraneous information. It is a good practice to limit scrolling to two screens down. Navigation should be easy using single click as far as possible. A general rule of thumb is that the depth of navigation should be at most four clicks.

Cumbersome Input Devices: Most devices sport a stylus or T keypads or a very small 'qwerty' keyboard where keying in long strings is a pain. Hence, inputs should be kept to the minimum. As far as possible we should try to device inputs as a single click kind of option. (Yes or No can be substituted by radio buttons as scroll and click in most devices is easier than keying.)

Application Load Time: Generally, users switch on their handhelds for short durations. For Example, often it is to retrieve a contact number. The load time for applications should be low, as the users would not like to wait for the application to load each time.

Battery: Batteries are a scarce resource. One of the keys to the success of the application is long battery life. Activities like serial or IR communications, sound extended animation, and other tasks that use the CPU for long periods tend to consume large amounts of power.

Memory: All handheld devices have limited storage space, from 512K to 8MB, and a dynamic heap in the range of 32K to 256K (newer devices tend to have 8MB and 256K, respectively). Under such circumstances optimization is crucial. The Optimization mantra is heap first, speed second, and code size third.

Data Storage: All devices provide some amount of persistent storage. However, this is very limited. Different devices organize this in different ways. PalmOS based devices, for example, treat all storage as database blocks. We shall see more about it in the subsequent chapters. We will also discuss the techniques to optimize storage under different OSs.

Backward Compatibility: Backward compatibility is a key issue as the devices have evolved rather rapidly in a very short span of time and users are not expected to upgrade their system every time an advanced version becomes available. Hence, to gain wider acceptance we have to write applications in such a manner that they will run on all versions of the device and at the same time give the users of a later version better functionality.

Application Size: Applications need to be stored on the devices. Most devices will have an upper limitation on the size of executables. This needs to be taken into account during the design phase itself. Each feature should be visited multiple times with questions like 'Is it necessary?', ' Can it be done differently?', ' Can it be clubbed with some other feature?' 'Should it be done here or offloaded to a backend server?'. We should try to ensure that only absolutely necessary stuff resides on the device.

Finally, there is no ideal way to design a device resident client. All clients, which are done well, evolve over a period of time involving changes and undergo several iterations before a satisfactory mix is obtained.

To conclude, in the past few years, wireless technology has seen a phenomenal and dynamic growth. This has mirrored in the growth of the device segment as well. The new technologies are enabling mobile phones to be a combined camera, video camera, computer, stereo, radio and a host of other things. This has lead to an enormous potential for applications. The development of software applications and other key technologies has enabled subscribers to use the handset to connect to the Web to receive stock quotes, check e-mail, transmit data, and send faxes and much more. Some phones support personal digital assistant software that offers the convenience of a calendar, address book, calculator, and voice recorder. We saw the evolution of these devices and looked beneath the hood to get an overview of how things work. We also saw the major concerns in writing applications for these devices. In the next chapter we will explore programming for the Palm OS. We will briefly look at the operating system to the extent required by us to write good applications. Covering all the nuances of the system may require more exploration on part of the reader.

REFERENCES/FURTHER READING

1. Information on the cell phone internals and transmission technologies http://biz.howstuffworks. com/cell-phone.htm.

2. Information on the PDA internals and other details http://electronics.howstuff-works.com/pda.htm.

3. A detailed technical discussion on the birth and growth of mobile telephony http://www.international-phone-card.info/mobile_telephone_history.htm.

4. Good introduction and bird's eye view on many topics covered in this book http://www.peterindia.net/MobileComputing.html.

5. An excellent collection of articles and resources for mobile computing is available at http://www.bitpipe.com.

REVIEW QUESTIONS

Q1: What are the important differences between a desktop computer and a portable computer like PDA?

Q2: What are the challenges one need to keep in mind while designing a small footprint wireless device?

Q3: What are the design constraints for applications targeted for handheld devices?

CHAPTER 13

Programming for the Palm OS

13.1 INTRODUCTION

It is interesting to note that the origin of the concept of the PDA can be traced back to 1960 when Gene Roddenberry, the late creator of Star Trek, decreed that paper and pencils were debarred from the sets of starship enterprises. All communication and data collection would be through tricoders and communicators (the name is today retained to refer to smart communication and data devices). The term PDA however was coined by John Sculley, then the CEO of Apple Computer during his speech at the Winter Consumer Electronics Show. Apple was the first to ship a PDA under its Newton product line. Through the second half of the 1990s various companies including IBM, Sony, Samsung, NEC and others entered the market with their own variations of a PDA, but none achieved the spectacular success of Palm.

Various comparisons and reviews both commercial and technical, favor one over the other depending on the evaluation criteria and times when these were compiled. What everyone however accepts is that it was Palm that heralded the age of PDAs with the introduction of its PalmPilot 1000. This chapter introduces us to Palm OS and building applications for the Palm OS. The approach here as it is elsewhere is to give our readers an overview of the Operating System concepts and a guide to important APIs. Detailed discussions of the OS and programming intricacies are beyond the scope of this book. We begin by tracing Palm from its birth in 1996 and follow its turbulent journey spanning glory like none other down to its present; follow it up with a discussion about the Operating System architecture, the basics of programming and conclude with a look at the future directions planned for Palm OS.

13.2 History of Palm OS

The travails of Palm OS closely follow the fortunes of Palm Computing, the company under whose aegis Palm OS was developed. The story of Palm's conception is legendary. Right from when its founding father Jeff Hawkins carried a block of wood to every meeting to taking a practical approach that the user should learn the hieroglyphics of graffiti rather than putting together software that understands all nuances of human handwriting. These were based on learnings from costly mistakes made earlier and feedback from customers of an earlier product Zoomer. The Palm was designed with three commandments

- Handwriting recognition to be limited to simplified hieroglyphics
- Size should be small enough to fit into the pocket.
- A cradle to synchronize data with a PC

The result was a blueprint of what the PDA should be. But by 1994, successive debacles in the PDA market ensured that financers for the project were scarce. The lucky break came in the form of U S Robotics which lent its might to palm. The prototype, till then known as touchdown was christened PalmPilot 1000 and officially released in February 1996. Within the year they sold an unprecedented 350 thousand units. The volumes were growing and the tempo was kept up. At the same time certain management changes were happening, U S Robotics was acquired by 3Com in 1997. The next milestone was March 1998 when the fruits of investment and innovation saw PalmPilot III being released by 3Com. The device had double the RAM, supported infrared connections a better character recognition algorithm, more fonts, a handy cradle, an improved Palm OS 3.0 and stylish design. But much of it was only filling up the gaps in the earlier products. By then competition was also born in the form of WinCE still in its infancy.

At the dawn of the fateful year of 1999, everything was great; sales was at its peak, employees were motivated and competition non-existent. Trouble however was afoot; foresight was slow in coming. Two new models the PalmV and PalmVII that were minor variations of the PalmIII were released but failed to make a mark. The founding parents of Palm Computing, Jeff Hawkins and Donna Dubinsky, parted ways with 3Com to set-up a competitor, namely, Handspring. By September they came out with Visor. 2000 was quite uneventful except for the release of WinCE 3.0. The curtains however had come down the golden age of palm. 2001 saw both palm and handspring slowly rolling downhill. A global market recession leading to very low demand, fierce competition both within and without, saw the blue-eyed boy of the industry getting into red. Some good things too happened: New high-end models m500, m505 and m515, were released, these support for Secure Digital flash cards, a new version of operating system (Palm OS 4.0), new batteries with longer life and even a color display. The company was split giving birth to two divisions, one working on and licensing the Palm OS called PalmSource and the other manufacturing Palm devices called palmOne. The acquisition of Be, the move from dragonball processor to a more advanced ARM core based one. On the whole, though things were not rosy; the pda market at the close of 2001 was clearly in favor for Palm Computing and its allies.

2002 saw the effort to counter this downslide with the release of wireless capable devices like i705 and treo series from handspring. A series of belt-tightening measures both technically and financially in the previous year yielded some bright moments. They also had to counter a patent infringement suite from Xerox on the graffiti. This led to the licensing and adoption of jot. But now there was a need for a keyboard for applications like e-mail leading to more licensing from blackberry Rim. On the downslide was the release of PocketPC 2002 which already enjoyed substantial support from the elite corporate and of course the global reduction of the PDA market.

2003 was a continuation of the innovative efforts with lots of releases from the newly formed Palm one and palm source and handspring in the form of tungstone and zire. The best of course was the two versions of upgraded Operating System. The Chinese version of Palm OS gave it a presence in the huge Chinese market. The shift was now towards communication. So, later versions of Palm are tailored towards connectivity and communications. Some of the features supported include, WiFi, a 400MHz XScale processor, 64MB of RAM, an integrated keyboard, smaller size, built-in camera, longer battery life and more. As of this writing Palm OS 6 with all its glorious enhancements is looming large on the horizon. The focus of Palm OS 6 is mostly on wireless capabilities, security and more. We will look at some details in a later section.

At the dawn of 2004 Palm has its Dream Team in place; the founding fathers, whose creativity and experience is unparalleled, and the management who dragged Palm out of the abyss of 2001. In fact it looks like Palm is well poised to finally move from a defensive stance to that of an offencive take on the mammoth Microsoft. However another threat to Palm comes from the now mature and more advanced Symbian, with the likes of Nokia and Ericsson backing it.

What however decides the success of any operating system is the applications it supports. Here there is no beating the palm. The success of palm is reflected in the large number of application for the environment. Thousands of applications are available for the Palm OS serving various purposes from spread sheets, documents and presentations, database managers to trade information, e-books, business tools for CRM, order processing, surveying, records management, data collection, and inventory management, games, messaging applications like e-mail, fax, SMS, EMS, instant messengers, web and wap browsers, enhanced PIM tools, photo editors, audio and video tools and lots more.

Currently the field is all set for the battle of the giants, all placed favorably.

13.3 PALM OS ARCHITECTURE

We will now briefly look into the OS. Like most PDAs the high level view of the Palm OS device consists of three layers (Figure 13.1): Application, Operating System, and Hardware. Palm OS occupies approximately 300K bytes and can run in 32K bytes of RAM.

As shown in Figure 13.1, the heart of the OS is the kernel. Essentially the kernel handles all low-level communication with the processor, interrupts, multitasking facilities and messaging to the OS atop it. The kernel interfaces to the hardware via the hardware

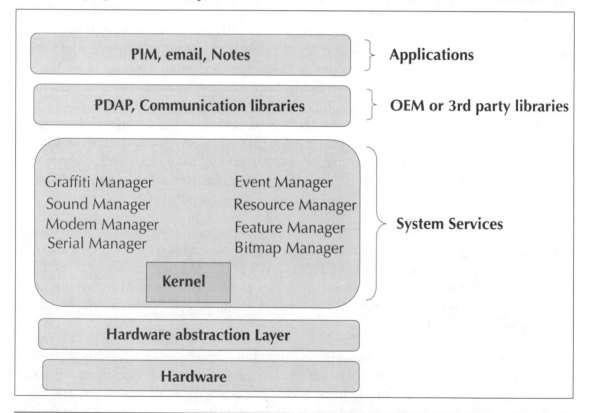

Figure 13.1 Architecture of Palm OS

abstraction layer. On top of the kernel there are the system services. Each service has a manager. E.g., event manager, graffiti manager, resource manager, feature manager, sound manager etc. To achieve faster execution these are generally mapped ROM commands. Then there the system libraries and independent third party libraries. Later versions of the OS also contain a PASE (Palm Application Compatibility Environment) which is an emulator for the older application ensuring backward compatibility. This essentially ensures backward compatibility. The topmost layer is the applications that use the underlying library to perform certain tasks. We will now see these components in some detail. Note a complete compendium of the OS intricacies is available at the palm source site. What follows is mostly a summary of the information provided by palm.

The kernel used in Palm OS is the AMX real-time, multitasking kernel, a product from KADAK Products Ltd. The kernel itself supports a lot of features but not all of these are available to the applications. Also since the devices features available in various models are different, we urge the readers to explore the features supported by the model on which the application is to be deployed of the users are advice. Some important features supported by the kernel are listed below.

13.3.1 Kernal Features

Multitasking: The kernel itself supports pretty advanced multitasking, including semaphores. But certain licensing limitations cause these features to be available only to the system functions and not the applications. So, for our purpose the OS is essentially single-tasked.

Interrupts: The kernel supports both maskable and non maskable interrupts in normal and nested modes. The handling is done through an interrupt specially written for it. It supports a mechanism to trap errors and is able to handle hardware interrupts. Interrupts can also initiate other tasks.

Time slicing and scheduling: This essentially allows the execution of several tasks according to their priority thereby supporting timers and time procedure. There are three types of triggers for task switching:

- Context switching: An application task requesting an implicit context switching

- Hardware interrupt: There is an interrupt controller inside Palm hardware system.

- Timer expiration: Each networking function has a timeout value to prevent the system from being idle in waiting state forever.

13.3.2 Memory

A Palm device has ROM and RAM Note: there is no hard disk. The ROM contains the OS and some other static data. Most new models use a flash ROM and hence allow OS updates. Since the onboard memory is pretty restricted, different versions of the OS support various types of extended memory in the form of memory cards. Theoretically there can be 256 such cards.

Internally the memory contains an identifier, a list of heaps and a database directory. Palm OS 3.x and 4.x used 16-bit addresses. The newer emersion 5 uses 32-bit bus. The main memory has an ID of 0. As shown in Figure 13.2 the memory is divided into three logical heaps namely Dynamic heap, Storage heap and ROM heap. As depicted in Figure 13.3, each heap has

- a header containing the unique heap ID, status flags, and the heap size,
- a master pointer table that is functionally similar to a page table and holds pointers to the beginning of each chunk
- a variable size chunk
- a terminator indicating the end of each chunk.
- additional reserved space for global variables.

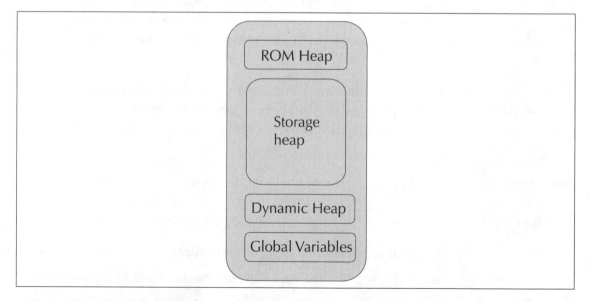

Figure 13.2 Memory Architecture

Dynamic Heap contains the operating system's global variables and data objects, user interface components, buffers, application data and an application stack, which also services system functions. The size varies between 32 kb to 256 kb depending on the Operating System version.

Storage Heap holds all the applications, user data, system patches, and any other persistent data in the system. The point to be noted is that data is not stored in files but rather in databases.

ROM Heap holds the operating system kernel. Interestingly its physical address in memory is at the higher end of the address space and helps to make compaction of the dynamic heap faster.

Database: There are two types of databases—record and resource databases. As the names suggest Record databases store user data while Resource databases store applications and free-form data. Record databases can have 64K records where resource databases can hold over 200 trillion records. Palm OS supports segmentation to overcome size restrictions on storage. It is the responsibility of the memory manager to handle logical and physical segmentation. Palm OS maps the databases to physical addresses and is analogous to the paging. Indexing is used to locate and retrieve data. Figure 13.4 shows a typical database architecture.

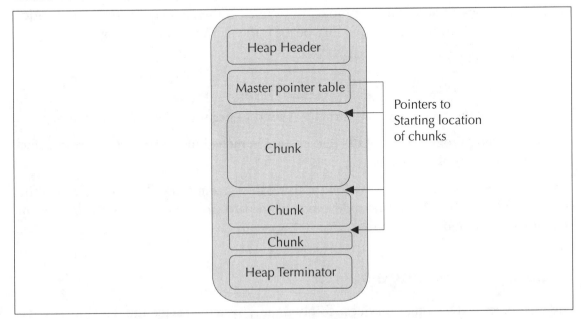

Figure 13.3 Heap Architecture

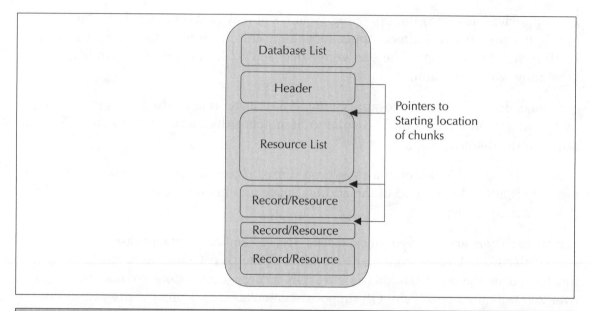

Figure 13.4 Database Memory Architecture

Cross card support is implemented since version 4 onwards in the form of VFS or virtual file system. All the application data is held in the RAM and updated in place. This leads to an important consideration 'Memory integrity' Which is the responsibility of the memory manager. Presiding over all three memory heaps is the Palm memory manager.

Memory Manager is responsible for

- Memory allocation,
- Manipulating the main data structures used to organize data within memory
- Providing Programming APIs that ensures standard memory access across different versions of Palm OS.

For faster response there is provision for using the system stack. This is also called the Feature Memory. But Application developers need to note that this taxes the system heap for very large data structures.

13.3.3 System Managers

The next layer is the system services. To obtain any system service the developers have to use the respective managers. All OS functionality the programmer wants to access must be accessed via the managers. The main language for programming Palm is C.

- Event Manager: Responsible for casting events handy for performing a global find or if the system should enter sleep mode

- Attention Manager: This is the manager responsible for alarms. Available from OS 4.0 onwards. it serves as a central modification point for all alarms set in the system

- Data & Resource Manager: Responsible for record and resource creation, modification and deletion.

- Exchange Manager: Responsible for all data transfer between several Palm OS devices. This includes IR, TCP/IP and Bluetooth.

- Feature Manager: Responsible for feature memory mentioned above in the section on memory.

- Graffiti Manager: All Graffiti is handled via the graffiti manager

- Memory Manager: This manager is responsible for all memory allocation and handling. It also ensures the integrity of the system's memory by validating every write call.

- Sound Manager: The sound manager allows synchronous and asynchronous sound (one of the few threads additionally available to the OS) and MIDI playback

- Telephony Manager: This manager was added in OS version 4.0, and as the name signifies allows access to telephony API.

- VFS Manager: This manager also was added in OS 4.0.

13.4 APPLICATION DEVELOPMENT

Palm operating system is event-based. Typical events are pen taps, button presses, menu selections and so on. Like Windows Programming, there is an event-loop. Users' actions generate a series of events. The events are FIFO queued in event queue. We query the OS for these events and perform actions based on these events.

13.4.1 EventLoop

The event loop, as the name suggests is, control excuting continuously in a loop through the eventQ of the current application. This means that each running application has an event queue which receives related events from the Operating System. The application fetches the event from this loop and processes it appropriately. The first step is to get an event using the `EvtGetEvent` function, it is then dispatch to its respective event handler. An event handler returns true if the event was successfully handled. Starting at the system level each event is successively sent to the next lower level handler for processing. The loop continues until the `appStopEvent` is received and the application closes. The steps are summed up below. A typical event loop control flow is shown in Figure 13.6.

1. Fetch an event from the event queue `EvtGetEvent()`.

2. An optional PreprocessEvent

3. The system is given the first chance to handle the event SysHandleEvent.

4. If step 3 above fails, then call MenuHandleEvent.

5. If step 4 above failed call applicationHandleEvent, a function provided by the application itself.

6. If step 5 failed call FrmDispatchEvent.

Figure 13.5 outlines the flow chart for a typical plam eventloop.

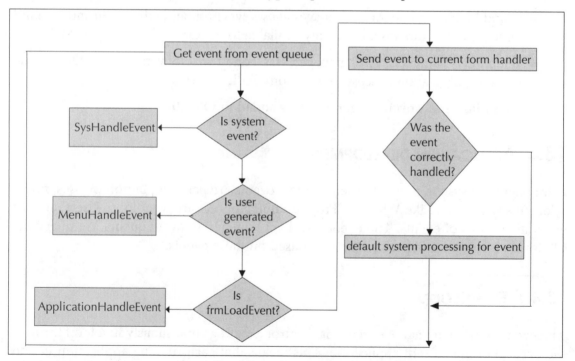

Figure 13.5 Event handling

Events data structures are defined in Event.h, SysEvent.h, and INetMgr.h. The system passes this to the application when the user interacts with the graphical user interface.

Some of the possible events are AppStopEvent,ctlEnterEvent, ctlExitEvent, ctlRepeatEvent, ctlSelectEvent, daySelectEvent, fldChangedEvent, fldEnterEvent etc

Some of the data associated with an event are

- type Specifies the type of the event.
- screenX xco-ordinate or the distance from the left margin of the window.
- screenY yco-ordinate or the distance from top left margin of the window.
- data Event specific data.

The system handles events like power on/power off, Graffiti® or Graffiti® 2 input, tapping input area icons, or pressing buttons. During the call to SysHandleEvent, the user may also be informed about low-battery warnings or may find and search another application.

MenuHandleEvent handles two types of events:

- brings up the menu.
- puts the events that result from the command onto the event queue.

FrmDispatchEvent first sends the event to the application's event handler for the active form

Thus the application gets to process events that pertain to the current form

If successful it returns true

Else call FrmHandleEvent to provide the system's default processing for the event.

ApplicationHandleEvent handles the frmLoadEvent it loads and activates application form resources and sets the event handler for the active form.

A sample event loop is given below.

```
static void AppEventLoop(void)
```

```
{
            UInt16 error;
            EventType event;
            do
            {
                    EvtGetEvent(&event, evtWaitForever);
                    // the handler for Palm system events
                    if (SysHandleEvent(&e))
                        continue;
                    // the menu event handler
                    if (MenuHandleEvent((void *)0, &e, &err));
                        continue
                    if (ApplicationHandleEvent(&event))
                        continue;
                    else
                      FrmDispatchEvent(&event);
            } while (event.eType != appStopEvent);
}
```

We will now proceed to see some aspects of programming for Palm.

Get:

Palm programming tools that we have used for our development purposes

- CodeWarrior Development Environment

It contains an IDE (Interactive Development Environment) for managing and building projects. Its main components include:

- Building project (compiling and linking)
- Debugging
- User interface resource design

- Palm Emulator

Palm Emulator is a desktop software product to emulate the execution of real Palm devices. It is normally used to test applications downloading them into a real Palm device.

- ROM image

If you own a device you can extract the ROM image from there or download one from the palm site. Instructions for both are available at www.palmos.com.

Set:

Now we need to set up our environments first. Details of setting up the environment are available at PalmSource site and are not duplicated here.

Go..

In most cases the beginning is the toughest part. Here is where we provide the user some insights and tips. My first application 'Hello Palm'.

13.4.2 First Palm

Each application has a PilotMain function equivalent to Main function in C. As shown in Figure 13.6, applications start at PilotMain function. Each application also has startApplication and stopApplication functions to load and save user preference settings, such as font size. The main body is the eventLoop whose function is to detect events and to perform corresponding actions. The eventLoop function loops forever until the user exits the application. We will see more details in a little while.

The first step is to launch an application. An application can be launched from the main application menu or another application. The operating system sends a **launch code** to PilotMain function. Launch code specifies how an application responds to the event.

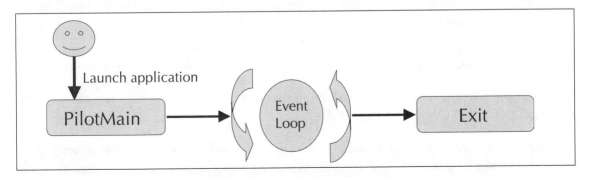

Figure 13.6 Application life cycle

During the launch, the application does the following:

- Initialize global variables from the values contained in the system and application preferences (like date and time formats).

- Find the application database by creator type or create and initialize one if it does not exist.

As in any C application, the first entry are the "include files" that contain the application definitions system files and related header definition files.

```
#include "firstPalm.h"
#include <System/SystemPublic.h>
#include <UI/UIPublic.h>
```

In our example we will look at launch codes, the main event loop and event handlers. When a user performs an action, for example, taps an application icon on the application launcher screen, the system generates the launch code *sysAppLaunchCmdNormalLaunch*, which tells the application to perform a normal launch and display its user interface. An application launches when it receives a launch code. Launch codes are a means of communication between the Palm OS and the application (or between two applications). Launch codes are declared in the header file SystemMgr.h. Different launch codes specify different actions. The launch code parameter is a 16-bit word value. All launch codes with values 0 are reserved for use by the system and for future enhancements. Launch codes 32768 are available for private use by applications. User-defined launch codes are also possible. Each launch code may be accompanied by two types of information:

- A parameter block, a pointer to a structure that contains several parameters. These parameters contain information necessary to handle the associated launch code. Typical Parameter blocks are declared in AppLaunchCmd.h, AlarmMgr.h, ExgMgr.h, and Find.h.

- Launch flags indicate how the application should behave. For example, a flag could be used to specify whether the application should display UI or not.

For a complete listing and details of all launch codes refer to *'Palm OS Programmer's API Reference'* available at the palm source site.

If an application can't handle a launch code, it exits without failure. Otherwise, it performs the action immediately and returns. When an application receives the launch code *sysAppLaunchCmdNormalLaunch*, it begins with a startup routine, then goes into an event loop, and finally exits with a stop routine.

As mentioned above, the entry point to every palm application is the PilotMain function.

UInt32 PilotMain(UInt16 cmd, MemPtr cmdPBP, UInt16 launchFlags)

cmd is the launch code for the application *sysAppLaunchCmdNormalLaunch*. In this case

cmdPBP is a pointer to a structure containing any launch-command-specific parameters, or NULL if the launch code has none.

launchFlags indicates the availability of the application's global variables, the application's state ready active etc. Launch flag values could be

- *SysAppLaunchFlagNewGlobals*. The system has created and initialized a new global values.

- *SysAppLaunchFlagUIApp*. Launch the UI.

- *SysAppLaunchFlagSubCall* . The application is calling itself indicating that the application is actually the current application.

Refer to the Figure 13.6 for application control flow. A skeletal pilot main could look like

```
UInt32 PilotMain(UInt16 cmd, MemPtr cmdPBP, UInt16 launchFlags)

{
        switch (cmd)
          {
                case sysAppLaunchCmdNormalLaunch:
                EventLoop();
                break;

                default:
                break;
          }
        return 0;
}
```

The PiltoMain function is essentially an event loop that continually handles the events. For convenience of understanding we have a separate function to illustrate the event loop. In this function, we continually process events for our application and the system.

Most applications will be using the *ApplicationHandleEvent()* and *FrmDispatchEvent()* to process user's instructions. What we do in the last two we will see in a little while. But first, let us see a working sample. This application just displays a message 'Hello Palm' and has an OK button. Clicking OK dismisses the application.

We need three files:

A **header file**: *firstPalm.h*

#define Form1   100

#define Ok   99

A **source file**: *firstPalm.c*

```
UInt32 PilotMain(UInt16 cmd, MemPtr cmdPBP, UInt16 launchFlags)
{
        unsigned short err;
        EventType e;
        FormType *pfrm;
        if (cmd == sysAppLaunchCmdNormalLaunch) // We will only
handle Normal Launch.
        {
                FrmGotoForm(Form1);   //Form1 has a code 100 and is
                defined in firstPalm.h .This will send a frmCloseEvent
                to the current form; send a frmLoadEvent and a
                frmOpenEvent to the specified form.

           do
        {
        EvtGetEvent(&e, 100);//poll for events every 100 millisecs
        if (SysHandleEvent(&e))
                        continue;
                if (MenuHandleEvent((void *)0, &e, &err))
                        continue;

                        switch (e.eType) //Which type of event is it?
```

```
    {
                case ctlSelectEvent: //A control object
        on the form has been selected.
                    if (e.data.ctlSelect.controlID ==
            Ok) // Is it the OK button.?
                    FrmCloseAllForms(); //Close all
            forms and return to main screen.
                      break;

                case frmLoadEvent: //A new form
FrmSetActiveForm(FrmInitForm(e.data.
frmLoad.formID));
                    break;
case frmOpenEvent://Set current focus to this form
                    pfrm = FrmGetActiveForm();
                    FrmDrawForm(pfrm);
                    break;

        default:
                if (FrmGetActiveForm())//If form has cur-
            rent focus handle //event
                FrmHandleEvent(FrmGetActiveForm(), &e);
            }
        } while (event.eType != appStopEvent);
    }
    return 0;
}
```

A **resource file**: *FirstPalm.rcp* Defines the resources used by the application.

Note it is always easier to use a resource editor to create/edit complex UI. But for our first application we will manually create it.

#include "hello.h"

FORM ID Form1 AT (0 0 140 140) //**Where on the screen to display the form.**

USABLE //**The form can respond to events.**

MODAL //Is it modal?

BEGIN

    *TITLE "Hello Palm" //***Title of the form**

    *LABEL "WOW My First Palm!" ID 200 AT (CENTER PREVBOTTOM+1) FONT 1//***A label to be displayed**

    *BUTTON "Ok" ID Ok AT (CENTER 100 AUTO AUTO) //***There is a button too**

END

VERSION 1 "1.0.0"

LAUNCHERCATEGORY ID 200 "Examples"

> You might like to insert the following before END and see what happens
>
> CHECKBOX "Unchecked" ID 2021 AT (CENTER PREVBOTTOM+2 AUTO AUTO)

13.4.3 Form

We will now see more details on forms. Forms are the palm's equivalent of windows in Windows operating system. They essentially act as containers for user-interface elements/widgets (e.g. buttons, lists, fields, checkboxes).

- Each form has a form event handler function, which contains the code to handle response for UI elements within the form. A form event handler routine is of the form: *Boolean FormEventHandlerType (EventType *eventP).*
- The *FrmDispatchEven()* routine provides indirect form-specific event handling by calling the form's event handler (*formEventHandler()* below).

The *ApplicationHandleEvent()* function is where we set the event handlers for all of our forms using

*void FrmSetEventHandler (FormType *formP, FormEventHandlerType *handler).*

The skeleton of the sample above then becomes as shown below.

```
static Boolean formEventHandler(EventPtr eventP)
```

```
    {
        Boolean handled = false;
        FormPtr frmP = FrmGetActiveForm(); //Get the currently
active form.
        switch (eventP->eType) //What kind of an event is it?
        {
            case frmOpenEvent: //Create a form defined by frmP
            case frmLoadEvent: //Load the form
            case ctlSelectEvent: //An object on the form has been
clicked.
            default:
            break;
        }
        return handled;
    }
```

The resulting app event handler would now look like

```
static Boolean AppHandleEvent(EventPtr eventP)
{
        UInt16 formId;
        FormPtr frmP;
        Boolean bRetVal = false;
        if (eventP->eType == frmLoadEvent) // is a form loading?
        {
            // Load the form resource.
            formId = eventP->data.frmLoad.formID;
            frmP = FrmInitForm(formId);
            FrmSetActiveForm(frmP); // make this the active form

            // Set the event handler for the form. The handler of the currently
            // active form is called by FrmHandleEvent each time is receives an
            // event.
            switch (formId)
```

```
        {
                case From1: //Is this our form? Form1
                        FrmSetEventHandler(frmP, MainFormHandleEvent);
                        break;
                default:
                        break;
        }
        bRetVal = true;
        }
        return bRetVal;
}
```

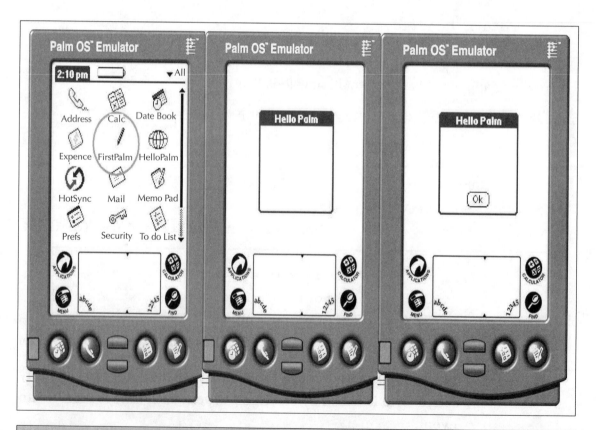

Figure 13.7a Output of sample code

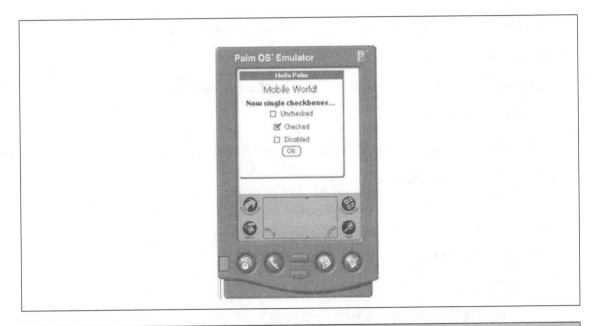

Figure 13.7b Output of sample code

We have seen three components 'form' 'label' and 'button'. The resulting output for the sample code above is shown in Fig 13.7a and Fig 13.7b. Palm OS provides us a comprehensive set of widgets for creating rich UI. A listing of the most commonly used ones is given below. The list is in no way complete or detailed. The *Palm OS Programmer's Companion* and the *Palm OS Programmer's API Reference* give a detailed description of all the APIs along with the sample code to use. These are freely available at the PalmSource site. Depending on the requirements you can use one or more of the following

13.4.4 GUI in Palm

To Display a collection of objects: **Form**

To display a menu for the user to choose from: **Menu**

To display a series of items: **List**

To have sub-choice in List-based on the first option similar to a drop down: **Pop-up list**

To display a non-editable string: **Label**

To display one or more lines of editable strings: **Text field**

To pop-up messages to the user: **Alert** (warning, error, or confirmation)

To use built-in keyboard item: **Keyboard Dialogue**

A submit or Execute command action: **Command button**

Selection / choice widgets

Select deselect options: **Check box**

Select a value: **Push button**

Movement items

To move control up or down: **Shift Indicator**

A progress control: **Slider**

Scrolling control: **Scroll bar**

Increment/decrement values: **Repeating button**

To display structured data: **Table**

And a custom control: **Gadget**

To handle user interfaces Palm OS provides certain low-level dedicated managers:

- The Graffiti Manager provides an API to the Palm OS Graffiti.

- The Key Manager manages the device buttons and key events.

- The Pen manager handles pen or stylus inputs.

Most applications do not need to access these managers directly; instead, applications receive and respond to events from the Event Manager. However if the developers so desire they can use these managers.

Creating and Handling Custom Events At times we may want to have custom events for our applications. Custom events are similar to system events. To post a custom event we use EvtAddEventToQueue() or EvtAddUniqueEventToQueue(), and retrieve them by calling EvtGetEvent(). However, due to the event handling mechanism of palm os, custom events have the drawback of not reaching the application. Refer to the Palm OS programer's manual for details.

13.5 Communication in Palm OS

Any discussion on handhelds is incomplete without describing its communications capability. Palm OS has a rich set of communication features that include TCP/IP, Bluetooth, serial connections, infrared beaming facilities, telephone interface etc. Of course not all devices will support all the features. The onus is on the developer to ensure that the device supports the required connection feature and or provide an alternate path for the same. As mentioned before, the palm source site gives excellent detailed references for all palm programming requirements. What follows is a brief introduction.

All forms of communication begin with establishing a connection between the sender and recipient. The Connection Manager manages at a high level all connections from the Palm handheld to external devices. In the Palm OS architecture individual connection protocols are described by profiles. A profile describes the sequence of plug-ins required to implement the protocol specified. A plug-in is the counterpart of a driver in the PC world. It is essentially a piece of code responsible for configuring, establish-ing, maintaining, and terminating connections at the low level. For example, a Profile for TCP/IP could contain PPP (point to point protocol) and a dialer plug-in. We will see details later. The main functions of the Connection Manager are listed below.

- Managing communications plug-ins.
- Managing communications profiles.
- Establishing and managing connections using profiles.

The process of using these profiles and configurations is simplified by the user interface framework, allowing editing, progress information and error display.

Figure 13.8 shows a top level view of the Connection Manager and its interaction with the plug-ins and profiles.

As shown in the figure, the connection manager interacts with two subsystems, the Connection Manager database and the plug-ins.

The Connection Manager database stores profiles, and interfaces. All objects in the database are identified by unique IDs. We will now see some details about these objects.

13.5.1 Profiles

As described earlier, the profile is a sequential listing of the plug-ins required to make a connection. Profiles can be created by the user via the Connection manager's

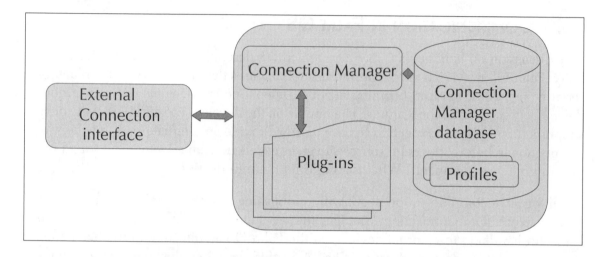

Figure 13.8 Connection Manager architecture

UI, programmatically, or during software installation. The Connection Manager also allows modification and management of existing profiles both through the Connection application or programmatically. When the application requests a connection, the Connection manager will search through its database for the corresponding profile. All profiles are associated with a priority and status. The manager returns the profile with the highest priority and status equal to "available". A profile can reference other profiles. These are called sub-profiles. Sub-profiles are expanded at runtime to produce the complete sequence. A special kind of sub-profile called macro profile can be used if inline expansion is required. Each entry in a profile is called a node. The analogy is to a graph where plug-ins represent nodes. Each node has an associated list of configuration parameters. Some of these parameters are mandatory and need to be set at the time of creating the profile. Profiles can be automatic or manual. In an automatic mode, the connection manager makes a best-effort choice based on availability and priority while in a manual mode the user is asked to select a profile of their choice. For example if we consider an internet connection the user may have multiple ISP accounts. The same ISP in a different city may have different dialing codes. So when the users are in their home town they might set it to automatic. While accessing from outstation they may want to choose the profile to be used. Or if a particular plan has better off-peak rates they may want to shift to that. kCncManualModeOption flag is used to set the connection mode. Aliases can be created using link objects. It is possible to represent a profile as a string. The general format is 'node no: properties'; successive node entries are separated by '/'. For more information the reader is referred to the programmers Manual from PalmSource.

Here is an example profile string for PPP over infrared:
'NetOut/IPIF/PPP:User='foo', Pass='bar'/IRIF'

> The application generally interacts only with the topmost component, which defines the connection type TCP/IP, in the example above and hence need-not know about the lower-level connections.

13.5.2 Interfaces

An interface is conceptually similar to its java counterpart. It is essentially an abstraction of similar plug-ins that logically group together. For example, the IrComm and serial plug-ins both provide the same RS-232 connection, so they can be abstracted by a general RS-232 interface. Note an interface has no associated code. It is simply an object in the Connection Manager database that relates to other plug-in objects. Typically, interfaces are used by plug-in developers. Applications normally don't need to create them.

13.5.3 Plug-ins

These are the actual workhorses. These are the code modules that implement the respective protocols. Plug-ins are built on the IOS (Input/Output Subsystem) framework and use lower-level IOS drivers and modules to interact with the hardware. Most plug-ins are configurable and provide adequate user interface to set these configuration parameters. Plug-ins are categorized based on the functionality that they implement. A general connection implementation consists of a top-level interface with the actual plug-ins lying below it.

Network Plug-ins

As the name suggests these provide connection to the network. The most commonly supported protocol, the TCP/IP, falls in this category. All network profiles start with the NetOut interface. The plug-ins lie below this in the hierarchy. Some of the plug-ins included are as below:

- **IPIF Plug-in**

    Manages IP configuration, domain name resolution, networking interfaces, network routes, DHCP and related configuration entries.

- **ILL Plug-in** (IP Link Layer)

  Implements a Data Link Provider Interface (DLPI).

- **PPP Plug-in**

  Implements a Point-to-Point (PPP) link and performs PPP negotiation. Configuration parameters to the PPP plug-in include user name, password, time-out, Maximum Receive Unit (MRU) size, authentication type, etc. The Script Plug-in works in association with PPP and provides login script capability.

- **DLE Plug-in**

  The DLE plug-in resides directly above the network hardware and provides the Ethernet framing interface.

Serial Plug-ins

The Serial plug-ins are an interface to serial communication hardware and is managed by the Connection Manager. Configuration parameters are device name, baud rate, number of data and stop bits, parity, etc. The serial plugins are essentially two components–a serial interface the SerialMgr interface and the various plug-ins as given below:

- USB Plug-in is an interface to USB hardware

- Infrared Plug-in handles infrared (IR) hardware.

- Bluetooth Plug-in manages Bluetooth connections.

- Telephony Plug-ins: Most of the current Palm OS devices provide telephony access both for voice and data. Two telephony plug-ins abstract the phone hardware,

- the phone Plug-in which is the the phone driver.

- and the DataCall for handling data calls.

13.5.4 Using the Connection Manager

We will now see how to use the Connection Manager. For a developer the Connection Manager is a shared system library managed by the OS. Essentially, we use the connection manager to perform the following:

- **Creating a Profile**

  All connections need a profile to connect to. We can either use an existing profile or create a new one. We can create an empty profile with CncProfileNew(). Next, we use CncProfileInsertItem() to insert items into the previously created functions. Items can be plug-ins, interfaces, sub-profiles, macros, links etc. Once the complete profile is ready, it can be submitted to the connection manager by a call to CncProfileSubmit().

  Another way to create a profile would be by passing the profile string to CncProfileDecode().

- **Changing a Profile**

  Changes to a profile happen when new component(s) are added, existing node(s) are deleted or configuration parameters are changed.

  Components can be added using CncProfileInsertItem().

  To delete we use CncProfileDeleteItem(). Note all deletes are cascaded meaning that all referencing profiles will also be removed.

  > To delete a profile from the Connection Manager database, call CncObjectDelete().

- **Finding Profiles**

  To find an existing profile we use CncObjectGetIndex() which is a name-based search or CncObjectFindAll() to retrieve all associated profiles and then iterate through the array returned above using CncProfileFindNext().

- **Managing Profiles**

  Profile management involves querying, addition, deletion and updating of the items/node. Locking and unlocking also fall in its preview.

  The process typically begins with CncObjectGetIndex() that returns the ProfileID

  This is followed by CncProfileGetItemIndex() to retrieve the index of the item/node to on which we want to operate, CncProfileGetItemId(). The number of items in a profile can be obtained using CncProfileGetLength(). Finally to read the iteminfo we use CncObjectGetInfo() and CncObjectSetInfo() to modify the values.

- **Configuring Components**

  As mentioned above, most plug-ins and nodes of a profile have configurable parameters. These can be accessed using CncProfileGetParameters() and CncProfileSetParameters(). We can also call the managers UI.

- **Invoking a Function in a Profile Plug-In**

  A plug-in can define requests that it will respond to. From within the application we can then send a request using a request parameter. This parameter tells the plug-in which parameter block to execute. While a plug-in may not respond to any user request it is mandatory for all plug-ins to implement kCncControl Availability() request, which returns the availability information about a particular plug-in.

- **Making a Connection**

  To make a connection from an existing stored profile, call CncProfileConnect(). It takes the ID of a stored profile and attempts to make the connection. On successful completion it returns an IOS file descriptor for the connection. This descriptor can now be used for read and write and close operations. An application may also choose to make a new profile dynamically and connecting from it. Note however that the developer is advised to make judicious choice here.

- Canceling or Disconnecting a Profile

  To cancel a connection we use CncProfileDisconnect().

  The connection Manager is also responsible for any clean-ups required in the process.

> The connection manager only provides a connection. Data I/O is the responsibility of the Exchange Manager, which we will see a little later.

13.5.7 Security Considerations

Handhelds are personal devices and are likely to hold sensitive information. When these devices are exposed to the external world, security concerns are paramount. Some notable steps to ensure security are listed below.

- The Connection Manager server, which is responsible for all task runs in the system process.

- Access to the system process is restricted, through plug-ins.

- Sensitive parameters, can be designated as write-only by plug-ins disabling read ensuring that malicious applications residing on the device have no access to them.

- The users have the choice not to store the password (they must enter it each time).

- Plug-ins are system processes; hence, they must be signed to guarantee authenticity. The installation manager will not allow users to install unsigned plug-ins.

13.5.8 Object Exchange

As mentioned above the Connection manager can only establish and manage a connection. But communication is all about sending and receiving data. Connection to the transport media is only one part of it. The actual reading and writing of data is the domain of the exchange manager. Palm OS supports read and write operations on default objects. A typed data object contains a stream of bytes plus some information about its contents. The content information includes any of these: a creator ID, a MIME data type, or a filename. An Address Book vCard object is a good example. It is identified by text/x-vCard MIME type.

The Exchange manager is essentially an interface that provides APIs to send and receive typed data objects. The Exchange manager is independent of the transport mechanism, providing transparent connectivity through the use of an exchange library.

Each protocol has its own specific exchange library that performs the actual communication with the remote device. The Exchange maintains a registry of libraries along with the information regarding the protocol that it implements and the data object it supports. When an application makes a call to the Exchange manager, in looks through its registry and forwards the request to the appropriate exchange library. Figure **13.9** shows the interactions between the application, the exchanger manager and the libraries. The library available on a device is dependent on the hardware capabilities. Some typically available libraries include: IR Library (IrDA), Local Exchange Library, SMS (Short Messaging System) Library, Bluetooth Library, and HotSync Exchange Library.

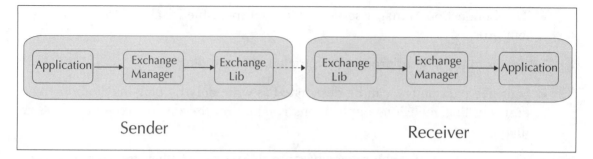

Figure 13.9 Communication using Exchange Manager

Now we will see how to send/receive data.

1. Create and initialize the exchange Socket Structure. The most important parameters to this structure are the library to be used and the data type to respond to.

2. At a high level, to send data we use:

the ExgPut() function to establish the connection with the exchange library, and

the ExgSend() function to send the actual data. (Can be called in a loop to send large chunks of data or multiple objects)

3. Finally, we use the ExgDisconnect() function to end the connection.

To be able to receive data, the application should further register for the corresponding MIME type.

As explained in the sections above, each application receives launch codes that it may choose to respond to. In this case the first one received is sysAppLaunchCmdExgAskUser which allows the developer to handle user inputs if any. Now launch the application with sysAppLaunchCmdExgReceiveData. This will actually receive the data.

1. Call ExgAccept() to accept the connection.

2. Call ExgReceive() receive the data (To receive multiple objects call within a loop).

4. Call ExgDisconnect() to end the connection.

A zero (0) return value indicates a successful transmission.

The only difference when sending and receiving databases is the use of ExgDBWrite() and ExgDBRead() with callbacks to ExgSend() and ExgReceive() respectively.

To implement a two-way communication, ExgGet() and ExgPut() can be used in combination with the ExgConnect().

Handling attachments and custom data types are beyond the scope of this book.

The last item in this section deals with the hot sync feature of Palm OS. Hot sync is the mechanism to synchronize values on the PC and the handheld. The advantage of using HotSync Exchange comes from its use of native format. This eliminates the need for custom conduits.

The HotSync exchange library supports two mode of operation:

- **The desktop scheme:** This is a direct exchange of a file to a HotSync desktop.

- **The send scheme:** Uses an exchange library that supports this scheme to send data.

The steps to communicate are the same as above namely:

1. Initialize the ExgSocketType
2. Call ExgPut().
3. Call ExgSend().
4. Call ExgDisconnect().

A newly installed application can use the steps in the previous section to receive data.

Another feature of the Exchange Manager is the PDI or Personal Data interchange which is essentially the exchange of personal information like business cards using a communication medium like IRDA. The Palm OS provides a PDI Library to facilitate this exchange. For more information about the PDI standards refer to the PDI consortium's web site at http://www.imc.org/pdi

More information regarding the Palm OS PDI library and APIs is available at palm source site.

13.6 MULTIMEDIA

Multimedia, as the word implies, is the capability to handle multiple modes of presentation, the various modes being text graphics sound video etc. A combination of these can be used to produce powerful user applications. But theses features require high-end computing power that the dragonball processor did not have. With the move to an ARM core things have changed drastically for Palm OS. The ARM infrastructure allows Palm OS to support multimedia features such as games, streaming video and MP3. A popular use of this capability is to take pictures and e-mail them. The new OS also lets mobile professionals handle multiple applications at once, such as Microsoft's Excel spreadsheets and Word documents, office e-mail and peer-to-peer wireless applications for data conferencing.

For the purpose of the application, multimedia is a data stream to/from the device. While recording/storing this data is encoded into an appropriate format like MIDI/MP3. While plying/retrieving the data it is decoded and sent to the hardware. This process of encoding and decoding is the job of a component called codec. Functionally, a codec translates media data from one format to another. Individual algorithms have specific codecs. Palm OS has several built-in codecs. A stream works within the purview of a session and represents data in a particular multimedia format. A session comprises a transaction which begins with opening a connection, transmission of data streams and finally closing the connection. Some of the parameters associated with a session are the source (where the data resides it could even be a camera or microphone) and the destination (the output might be a device like a speaker or screen, a file, or even a network stream). Each source and destination is connected by at least one stream representing the format of the data it carries. The stream further connects to a track. A track implements a codec to translate the data. Depending on the format there could be multiple tracks; for example, a movie session could contain two tracks, one audio and one video.

As with other features, multimedia has its own managers. Here we will cover the sound manager and multimedia library. The sound manager is an easy-to-use set of APIs for simple requirements like a system buzzer or alarm. For higher capability we need to use the multimedia API.

As seen above, the essence of multimedia programming is to create/read/write streams. Sound Manager can handle two types of sounds.

Simple sounds that can be played using

- SndPlaySystemSound() for a pre-defined system sound representing values like info alert, warning etc. and

- SndDoCmd()to play a single tone, the parameters to this function being pitch, amplitude, and duration.

- To play a standard MIDI File (SMF) Level 0 we can use SndPlaySmf()

[or information on MIDI and the SMF format, go to the official MIDI website, http://www.midi.org].

The following APIs can be used to play sound streams.

- SndStreamCreate()/SndStreamCreateExtended() to open and configure a new sound 'stream' from/into which we can record/playback buffers of 'raw' data; the second method is used if we need buffers of variable length. quantization, sampling rate, channel count etc which are some of the parameters to this function. A pointer to a callback function (SndStreamBufferCallback() or SndStream Variable BufferCallback()) is the key to the stream being created. This callback function is where the application implements its logic. The stream starts running with a call to SndStreamStart(), the callback function is called automatically, once per buffer of data. Timing is very important in this scheme of call back mechanisms for recording and playing.

- SndPlayResource() plays sound data that's read from a (formatted) sound file. The function configures the playback stream, based on the format information in the sound file header. Currently, only uncompressed WAV and IMA ADPCM WAV formats are recognized.

- The Sound Manager also provides functions to set the volume, namely, SndStreamSetVolume(). This function takes sound preference constant as its parameter. The constant can take the following values

  - prefSysSoundVolume default system volume:

  - prefGameSoundVolume for game sounds.

  - prefAlarmSoundVolume for alarms.

To play MIDI Files: MIDI data is typically stored in a MIDI database.

- The database type sysFileTMi1di identifies MIDI recorddatabases.

- The system MIDI database is further identified by the creatorsysFileCSystem.

The database by default holds system alarm sounds. Application MIDI records can either be stored in the same or in a separate database. The MIDI record header, the MIDI name, and the MIDI data are concatenated to form a MIDI record.

13.6.1 Multimedia Applications

We will now see the Multimedia Library. This is what covers the meat of the rich features provided by the device. Applications include the playback and recording of audio-visual media. Playback or recording sessions may be configured to run in the back-ground while the user uses other applications. Multiple components may interact with a session simultaneously. Note: the Multimedia Library does not provide a means for developers to write file format handlers or codecs.

As shown in the Figure 13.10 a typical multimedia application consists of

- A client application, the Media Player for example.
- The Multimedia Library, which provides the public APIs needed by the client applications to access multimedia features.
- A Server that runs as system process and spawns sessions that usually run as back-ground process.

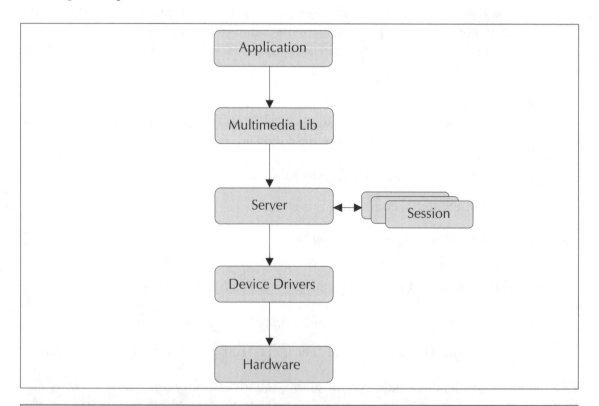

Figure 13.10 Multimedia manager architecture

- Device drivers that do the actual data communication with hardware components.

Multimedia Library

The Multimedia Library is a shared library that the system automatically loads and initializes. This section lists the Multimedia Library API used to play multimedia content. A detailed description can be found in the Multimedia documentation of the Exploring Palm OS series. The series also documents an example recording session.

1. The first step is to create a session MMSessionCreate().

2. Now to provide source and destination for the session MMSessionAddSource() MMSessionAddDest().

3. Then finalize the sources and destinations by calling MMSourceFinalize() and MMDestFinalize()

4. Next we need to associate the session with a track MMSessionAddTrack().

5. Finalize the Session MMSessionFinalize().

6. Now we can control the operations including start, stop, pause, capture, refresh etc MMSessionControl().

As always the last step before close is clean-up.

Telephony

The last item on the agenda is the Telephony API. The telephony API support is handled by the Telephony Manager. Application can open connections and make calls to it. As of now it supports only GSM/GPRS.

Testing the Telephony Environment

Before any application uses a device facility it is a good practice to verify if the given Environment supports those facilities. The Telephony Manager API TelIsServiceAvailable(), allows us to check for the availability of the required service. The manager also provides macros that do the same work. e.g, (TelIs*ServiceName*ServiceAvailable()).

Opening the Telephony Manager Library

After verifying that the service is available, we need to initiate a connection to the Telephony Manager library; we can use TelOpen()to use a system-selected phone profile or TelOpenPhoneProfile()to use a specific profile. The manager allows the application to verify support for function calls with TelIsFunctionSupported(), the corresponding macro being (TelIs*FunctionName*Supported()). The actual interface to the phone drivers is through the telephony server. The open call to the Telephony Manager library causes this server to be opened. Telephony Server uses the Connection Manager profile to retrieve the ID of the relevant drivers.

Making Synchronous and Asynchronous Calls

Calls are either synchronous (that will block the system till successful completion or an error causes the process to exit) or asynchronous (runs in the background and notifies the application when events such as an incoming SMS message, a call connection, battery status change, etc occurs).

Closing the Telephony Manager Library

All sessions with the manager should end with a call to TelClose() function, which does the cleanup and frees up any associated resources.

There are lots of other managers and libraries that support the Palm OS features. But a detailed discussion of all those is beyond the scope of this book. For more information and details about the API we urge the readers to refer to the PalmSource site.

13.7 ENHANCEMENTS IN THE CURRENT RELEASE

As of this writing the main excitement in the Palm world is its latest version **Palm OS® Cobalt**. We summarize the major enhancements of this version below. For details please refer to the documentation at the Palm source site.

13.7.1 System level

- A Frameworks-based architecture, allowing for component modularity, leading to plug and play opportunities, higher scalability and easy customization. Multi-tasking and multi-threading features allow for concurrent applications.

- A Protected Memory scheme allowing for greater stability and protection. Increase Memory of 256MB each for ROM and RAM allowing for rich multimedia and enterprise-grade applications.

- A Compatibility layer, PACE for backward compatibility.

13.7.2 Security

- The new OS comes with a CPM (Cryptographic Provider Manager) that features 128-bit encryption Pluggable default algorithms like RC4, SHA1, 3DES, MD5, SSL/TLS etc. Provision is also available for customized algorithms if needed. This has given a good boost for building end-to-end enterprise applications.

- Two new security modules the Authorization Manager for access restrictions and Authentication Manager for additional authentication mechanisms added by third parties.

13.7.3 Multimedia

- Support both audio and video for various standards like MP3, MPEG.

13.7.4 Database

- The updated Data Manager allows a standard database abstraction for 3rd party application developers.

- A new Database Access Control to create and control access to secure databases.

13.7.5 Communication

- Cobalt comes with a Modular, flexible, industry-standard STREAMS based Communications Framework. A generally improved look and feel and higher integration for Telephony and Bluetooth based applications have been provided.

- The Multi-tasking feature supports concurrent communication applications.

13.7.6 Display and UI

- Cobalt provides support for 320×320 pixel high-density displays, Graphics Rendering, Scalable Font APIs fonts for Latin locales and Multi process UI support.

13.7.7 Synchronization

- A new Sync Architecture with a new schema and secure databases, additional sync clients supporting more protocols.

- HotSync Exchange Provides access support for, many standard file formats. Drag and drop facility for sync has also been provided.

We approached the Palm OS from view of its past. We have climbed up the tech tree until we reached the version 5.0 and 6.0, along with their added possibilities and support. We have looked at the things that have made the Palm OS so strong compared to the other Operating Systems in the handheld market. After the discussion on the technical design of the operating system, starting at the Kernel, we investigated the way the Palm handles memory, how it stores and modifies data and applications, and again we had a short look at the limitations the memory of the Palm OS devices have. After a short overview over the system's managers, we moved on to the design of an application. How it can be started and how it handles events. We had a brief look at the different types of resources. The last stop was the new features in Cobalt, the official name for Palm OS 6. With this we conclude this chapter. In the next chapter we will look at another heavy weight Operating System the SymbianOS.

REFERENCES/FURTHER READING

1. The RTOS information from Kadak can be found at http://www.kadak.com/html/prls_x86.htm.

2. A collection of all documents from PalmSource is available for download at http://www.palmos.com/dev/support/docs/protein_books.html#devsuite.

3. Interesting articles on development using Palm OS can be found at http://www.asptechinc.com/posdevelop.asp.

4. Beaming "Using The Palm OS Exchange Manager" by Alex Gusev can be found at http://www.developer.com/ws/palm/article.php/3088941.

5. Download and install instructions for ROM is available at http://www.palmos.com/dev/dl/dl_tools/dl_emulator/generic_roms.html.

6. Sunit katkar has a very good introduction to programming for Palm OS at http://www.vidyut.com/sunit/palmpage.asp.

7. Robert Mykland, Palm OS Programming from Ground Up, Tata McGraw-Hill, 2000.

8. Lonnon R. Foster, Palm OS (Wireless + Mobile) Programming Bible, IDG Books, 2000.

REVIEW QUESTIONS

Q1: Descibe the architecture of Palm OS?

Q2: What are the basic considerations one has to keep in mind while developing applications for Palm OS?

Q3: What are the different types of communication mechanism available on a Palm OS?

Q4: What are the different security considerations in Palm OS?

Q5: Describe the telephony interfaces available in Palm OS. Why are these interfaces important?

CHAPTER 14

Wireless Devices with Symbian OS

Next in the series of environments for the handhelds is the Symbian OS. We will begin with a short introduction to this OS that evolved out of the original SIBO (**SIBO**) and later known as EPOC (**EPOC**). We will follow it up with a brief discussion of the OS architecture. We will then introduce ourselves to application development. Developing for Symbian OS is primarily of two kinds, one being OS programming which includes porting the OS to various target machines, device drivers, OS specific OEM enhancements etc. The second is application development that we will see a little later. There are excellent books written to cover different topics, some of these are listed in our reference section. Covering all nuances of programming is beyond the scope of this chapter. What we attempt here is to give a beginner's view. We assume an understanding of C/C++ programming languages and OOPS concepts. As before, a fore note to our readers: the OS has evolved through various versions, and OEMs will provide enhancements for their devices, hence it is advisable to check out the features available for their specific target devices before embarking on the journey of application building.

14.1 INTRODUCTION TO SYMBIAN OS

This OS traces its journey back to 1981, profits from Psion's flight simulator for the Sinclar ZX Spectrum bankrolled the creation of a database orientated pocket computer, the Organiser, (1984). It boasted of 32k of combined ROM and RAM and applications included a diary, database, clock, alarm, calculator and a simple programming language called OPL (Organiser Programming Language).This device, a success for Psion, formed the basis for the growth and evolution of EPOC.

The Organiser originally ran on an OS called the SIBO. It had an extremely small footprint of 384 Kb but supported a large number of applications including a spreadsheet.

Its hardware requirements were minimal, at 128Kb of ROM and it provided full multitasking, a feature even some of the more advanced PalmOS doesn't support. Psion shifted their focus to a smaller device resulting in the Series 3. We give here a brief time-line of the growth of the series 3.

- 1993, the 3a introduced the missing spreadsheet application;
- During 1995, PC synchronisation software PsiWin was introduced allowing data to be exchanged with PC applications.
- In 1996, the 3c introduced a built-in IR port; the Siena introduced a calculator pad along side a smaller screen.

The USP for the OS was robustness, its low power usage rich set of user friendly applications reliable with a functional hardware. There were limitations, amongst the most obvious ones being a 16 bit architecture. This prompted Psion to build a new 32 bit Os. The effort started in 1994 and was named Protea. The Protea had on its plate GUI, Communication and PcSync and a very pertinent switch to the ARM processor.

In 1997, for a cost of a little over £6 million, Psion completed the project and launched the Series 5. In the mean time EPOC was evolving, fax, Internet and e-mail were added. Finally, in 1998, the 3mx was introduced with a significantly faster processor.

Nokia had by this time launched the 9000 Communicator. This device packed in a mobile phone and PDA, and ran on the GEOS 16 bit.

A viable hardware and a superior OS indicated a predestined alliance. A Joint Venture emerged and Symbian was announced on 24 June 1998. The current shareholder of this enterprise are Ericsson, Nokia, Motorola, Panasonic, Sony Ericsson, Siemens, Samsung.

EPOC efforts now continued under Symbian and in June 1999 the first Symbian release of EPOC took place in parallel with the release of the Psion Series 5mx. The new and much improved OS, included new architecture for messaging, telephony, application access to contact and calendar data and more. The next logical step was a full WAP and Bluetooth stack support.

Symbian learned from Palm's experiences and tried to avoid infighting where all licensees created Psion clones and fought for market share on price considerations alone. This was made possible by the creation of DFRDs (Device Family Reference Designs). These provide the implementation blueprints, which ensure compatibility, while allowing oems to customize their products. The DFRDs also allow for the OS to be used across a large range of devices with varying capabilities. Currently there are three DFRDs:

Quartz, for Data and Voice devices where telephony and data are tightly integrated

Pearl, for Voice with Data devices

Crystal, for Data with Voice devices, similar to the Psion Series 5 or 7, that provides full wireless access to the Internet

And finally GT (Generic Technology) is the set of core technologies common to all the DFRDs. telephony components for example.

The year 2000 saw Symbian consolidating its position, expanding its licensee base and moving ahead with support for UMTS and GPRS.

The year 2001 saw the introduction of the version 6.x. Officially OS versions 6.x + are referred to as Symbian OS.

Two versions and many more licensees later, embedded in devices spread world over, the latest Symbian OS is Version 8.0.

The OS provides a rich set of facilities for smart mobile devices that combine the power of a PDA with mobile telephony and networked data services. Built, around an open and flexible architecture, it supports applications developed in a range of Programming Languages and environments, the core set of APIs as defined by GT are exposed to all devices. The Os further expands its usability by supporting most key standards including CLDC (MIDP) wap Bluetooth IPv4/v6 EDGE EGPRS IS-95, cdma2000 1x, and WCDMA SyncML DM 1.1.2 Unicode Standard version 3.0 etc. The current version supports the specific requirements of 2G, 2.5G and 3G mobile phones. Supported features include but are not limited to

Hardware support includes a full qwerty keyboard, a 0-9*# a T-Key pad, voice, hand-writing recognition and predictive text input.

OS Features include a hard realtime, multithreaded kernel working with state of art CPUs, peripherals and memories. Software support is provided through dedicated application engines for contacts, schedule, messaging, browsing, OBEX etc. Support for multimedia hardware acceleration. Direct access to screen and keyboard for high performance;

Application Development Environments include C++, Java (J2ME) MIDP 2.0, WAP. Application engines for audio and video recording, playback ,streaming, graphics, Unicode

Standard version 3.0 for international support etc. The system provides comprehensive security through encryption and certificates, secure protocols (HTTPS, SSL and TLS), WIM framework and authentication for installation of third party applications.

A rich set **Communication applications** in the form of SMS multimedia messaging (MMS), enhanced messaging (EMS), e-mail and SyncML DM 1.1.2. OTA and PC-based synchronization support is also provided. Supported protocols include TCP/IP, WAP, infrared (IrDA), Bluetooth and USB. Supported networks include GSM, GPRS/UMTS 3G networks CDMA (IS-95, cdma2000 1x, and WCDMA).

The Table 14.1 below shows the support for various features as the OS developed through the versions

Table 14.1 Symbian Versions

Version	Supported Features
Symbian OS v3.	Eikon UI, Agenda and office applications
Symbian OS v5.	Contacts application, Media Server, Messaging, Java (JDK 1.1.4), Telephony
Symbian OS v5.1	Uikon UI layer, Unicode
Symbian OS v6.0	Advanced GSM telephony, PersonalJava and JavaPhone, Quartz and Crystal UIs, WAP
Symbian OS v6.1	Bluetooth, GPRS
Symbian OS v7.0	IPv6, Metrowerks CodeWarrior support, MIDP Java, Multimedia Messaging, Multimode Telephony, Opera web browser, SyncML

14.2 SYMBIAN OS ARCHITECTURE

We will now briefly discuss the OS architecture. The strength of Symbian OS lies in its small footprint (the kernel is less than 200kb), adaptability to limited memory devices, a powerful power management model, a robust software layer confirming to industry standards, and support for integration with a plethora of peripheral hardware. The foundation for this is a fast, low power, low cost CPU core. The Symbian OS works atop the ARM architecture RISC processors (with V4 instruction set or higher).Supported processors including ARMv4T, ARMv5T, ARMv5TJ and Intel x86 (for the emulator).

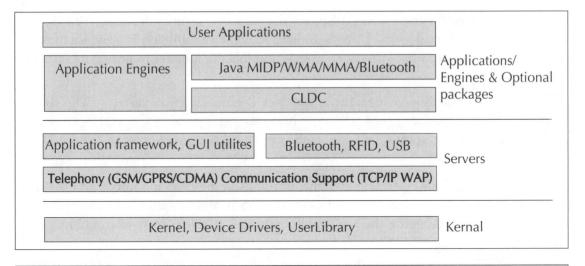

Figure 14.1 Symbian OS Architecture

The CPU is expected to be equipped with an integrated memory management unit (MMU) and a cache.

As in any other OS, the main objective of the OS is to provide hardware abstraction and manage system resources. A Symbian system can be divided into the 3 layers (Figure 14.1) where the bottom most interacts with the underlying harware/hardware abstaction layer as the case maybe. This layer includes the kernal, memory, device drivers and file services. On top of this, are the network and security support components. Also included are multimedia and communication protocol implimentations. The third layer is the application framework and applications support mechanism for PC synchronization, bluetooth and USB support. The topmost layer of course is the development environment and the appliacations themselves.

Symbian OS supports pre-emptive multitasking. All system services run in privileged mode, while user applications run in a user context. When an application requests a system service it is temporarily given privileged access through a context switch. This mechanism is conceptually similar to that in UNIX/WindowsNT. The MMU is designed for interrupt handling and privileged access modes. The CPU, MMU and cache along with timers and hardware drivers, all reside on the chip.

The kernel encapsulates the system services like multitasking, file services, power management, memory management and the various device drivers etc. It includes support for the telephony services for GSM,GPRS,CDMA and the security features. Version 8.0 boasts of powerful kernel architecture, with hard realtime capabilities. It provides the programming framework in the form of an abstraction making it easier to port Symbian OS.

> Symbian OS v8.0 is provided in application compatible two variants. The first variant, v8.0a uses the legacy kernel (EKA1) as per Symbian OS v6.1, v7.0 and v7.0s. The second variant v8.0b adopts the new hard realtime kernel (EKA2).

As most other mobile operating systems, it resides in the flash memory and executes-in-place. In the true spirit of plug and play, the OS is almost entirely implemented as DLLs. This helps in keeping the size small as there is a single copy of the library and everybody links to it.

> Symbian also makes extensive use of a special type of DLL called polymorphic DLL, conceptually these are similar to factory classes providing a public interface or function that applications can use. A notable example is the application DLL, which exports NewApplication(), to create an instance of an application

14.2.1 Hardware interfaces

All applications need access to hardware for functions like I/O media control etc. Optional hardware can be plugged in and the OS is expected to recognize and service it. As for other things the hardware support is also implemented as DLLS. All hardware access is restricted to the privileged mode i.e., all calls have to pass through the kernel.

Hardware is differentiated by the kernel's dependence on it. For example, timers, UART, DMA etc are essential for the OS. These are packaged along with the OS code and have direct kernel access.

Support for additional hardware is provided through a separate DLL, referred to as a kernel extension. Examples for kernel extension include keyboard, media devices and a lot more. Kernel extensions are detected and initialized at boot. Applications use the User library API to access kernel extensions.

And finally we have device drivers for optional or non essential hardware. Device drivers are true plug and play entities. Each device driver has two parts: a user side library for applications to link to and a kernel-side counterpart, for the actual hardware access. Note however that the screen buffer is an exception, here information is directly copied to the LCD display. The avoidance of frequent context switches allows for higher speeds.

Interestingly file system is also considered a device and access is provided through a device driver. The two components of this driver are a file system (generally FAT) and a media driver. The media driver is the kernel side library performing the actual operations.

14.2.2 Memory Management

Resource allocation and process life cycle are the prerogative of the kernel. It keeps track of event like thread death so that resources can be freed. As mentioned in the Palm OS chapter, memory is required for three purposes: To store the OS itself, persistent application and user data and runtime requirements. The volatile memory is provided through a RAM. The persistent memory is provided through Flash Memory. A flash memory though more expensive than ROM is the preferred option as it can be reprogrammed.

Symbian OS uses page memory architecture. It implements a two-level page table using 4 KB pages. This allows for efficient memory usage.

Owing to the multi-tasking capability of the OS, security becomes an important consideration. Symbian OS addresses security concerns in two ways. First, by using privileged and non privileged mode execution it takes care of restricting access to kernel and hardware access. Secondly it requires all applications to run in a virtual machine (VM) thereby protecting the applications from each other. Each application executes as a single process. (A single process is likely to have multiple threads). At launch or initialization, the application is allocated memory for its data, the outer page table stores a reference to this. When a context switch causes this process to be activated, all the pages are moved to a pre-defined location in the virtual memory map hence execution continues in the appropriate thread. Applications cannot make direct calls to the call to hardware drivers, they have to use User library APIs which in turn use system services through the kernel.

As mentioned earlier the primary concern of the MMU is to provide a protected mode system. Other functions of the MMU include

- Restriction on access to process data.

- Protection of application and OS code.
- Isolation of the peripheral hardware.

More information on the MMU and CPU architecture can be found at the Symbian site.

All handsets have limited runtime memory. Out of memory exceptions are quite likely. One way that Symbian uses to counter this is by having a clean-up stack, all partially constructed objects are placed here until their construction has been completed. If the phone does not have sufficient memory to complete object creation then it simply deletes the contents of this stack. By not allowing partially constructed object it avoids memory leaks as well as protects applications from potential data loss.

14.2.3 System software

All applications require system services of one type or the other. The Symbian OS System services framework operates in a client server mode where in most of the system services are provided as servers for example a file server, font and bitmap server, a media server etc. An application is a client that connects to these servers and requests their services. The client connects to the server using the kernel interfaces and uses a message passing mechanism for interaction. The server however runs in an unprivileged mode and will use other backend device drivers or kernel extensions to perform its tasks. From an application perspective we need to concentrate on the user library. The user library provides APIs to application framework and controlled access to the kernel. We will see the different frameworks and the applicable APIs as we proceed.

14.3 APPLICATIONS FOR SYMBIAN

The open architecture of Symbian enables Independent Software Vendors (ISVs) to focus on developing new applications for mobile phones. Third-party vendors provide software in the form of an installation (SIS) file. This file contains the libraries and resources of the application, secured by a certification system. This mechanism ensures a secure application where the vendor is identified as a trusted source. The installer is responsible for updating the file system with the files from the SIS file. Once installed, the user can launch the application.

Applications are generally divided into two parts the engine that implements the functionality and UI. This mechanism provides maximum opportunity for innovation to OEMs as the actual user interface is left to the manufacturers.

14.3.1 Development Environment

Symbian offers two development environments C++ and Java. (Another language called opl used to be available for EPOC). The SDKs, instruction to install and develop applications for each of these are available from the Symbian site. As mentioned above, the output of the development effort is an installable .sis file.

14.3.2 Java

As always it is advisable to check the supported versions on the target device. More so in case of Symbian java environment because at various times in its journey from EPOC to Symbian V8.0 Symbian has supported various flavors of Java including Personal Java, Java Phone APIs and MIDP. The latest one V8.0 supports J2ME MIDP 2.0 and CLDC 1.1.Additional libraries provided include Bluetooth 1.0 (JSR082), FileGCF and PIM (JSR075), Wireless Messaging 1.0(WMA) (JSR120), Mobile media (JSR 135), 3D graphics (JSR184). We will see MIDP 2.0 in our next chapter, Chapter 15 J2ME. Here we will concentrate on C++.

Note: V8.0 does not support PersonalJava and JavaPhone. External support for CDC is likely in future.

14.3.3 C++

Developing using C++ involves a Windows emulator that running on a PC that maps Symbian OS calls to Win32 APIs. To develop in C++, users will also require VC++. Details are provided at the Symbian site.

All environments have their own limitations and benefits. It is ultimately a design issue to choose one or the other. Most SDKs have an emulation environment that can mimic the exact target environment in terms of stack and heap sizes, run time memory utilization etc. We encourage our readers to explore and utilize this feature where-ever available.

14.3.4 HelloSymbian

First we will take a look at the structure and life cycle of a Symbian program. For our readers wishing to move from PalmOs to SymbianOS programming there is an interesting article that compares and contrasts the two.

OPL

Organiser Programming Language (OPL) used to be quite popular during the EPOC times. It is a BASIC like language originally meant for the PISON organizer. OPL has its pros and cones namely while it facilitated a compile and run and even sometimes a build on device kind of facility it has no direct access to Symbian services. (Some vendors provided OPXs to overcome this limitation). Later versions of Symbian Os support only C++ and java environments.

Owing largely to its long presence we are likely to come across a lot applications ranging from games and graphics packages to database applications developed using OPL.

Symbian categorizes user applications into "Applications" that have application logic as well as UI for the end user, and 'executables" or exes that perform certain tasks or services for others and require no user involvement. Generally servers and engines fall into the second category. Our mandatory hello program also falls into this, as it does not have any UI and simply prints "Symbian Hello" on the console. As we will see Symbian programming is a lot more complex and involved then the others that we have seen so far. The code below lists out "Symbian hello"

```
#include <e32base.h>
#include <e32cons.h>
LOCAL_D CConsoleBase* gConsole;
GLDEF_C int E32Main()
{
//Obtain a console
gConsole  =  Console::NewL(_L("SYMBIAN"),TSize(KConsFullScreen,
KConsFullScreen));

//Create a descriptor for the message
_LIT(SymbianHello,"Hello Symbian\n");

//print to the console
gConsole->Printf(SymbianHello);

//pause
User::After(1000000);
```

```
//exit
return 0;
}
```

The kernel and UserLibrary together constitute the E32. As is quite evident e32Cons.h contains the header information for the console. The e32base.h contains a few basic classes used by most Symbian applications e.g., CBase class which is the base for all objects. The CConsoleBase also inherits from this. E32Main() is our main() equivalent of 'c'_LIT that we see converts a 'c' string into a Symbian descriptor. Descriptors are the Symbian's way of handling Strings and binary data. Though a detailed discussion of descriptors is not possible in this text we encourage our readers to explore more about them. printf() becomes console->Printf because in Symbian prinf() is a method of CBase class from which the CConsoleBase inherits.

We saw the source file but a Symbian project has other files too. We will now proceed to examine these in details.

The first is the project specification file. Symbian applications can be built for different targets, the information for each target is a different file. For our purpose here we will only see the emulator or wins sample. This file has an .mmp extension and contains the following information:

TARGET HelloSymbian.exe // the name of the application to be generated

TARGETTYPE exe // Type of the target. Two possible types are app and exe

SOURCEPATH . //Location of the source files.

UID 0 // A unique identifier for the application. Used by the system to check for application integrity

SOURCE hellotext.cpp //name of the source files to included in the application
USERINCLUDE .

SYSTEMINCLUDE \epoc32\include

 // the locations of project-specific and system headers files are specified by the USERINCLUDE and SYSTEMINCLUDE statements respectively

LIBRARY euser.lib //libraries required for the project

Another very important file is the component definition file, this file is always called the bld.inf. This file contains information about the project specification files. In our case this is very simple but in a live application this file should list all specification files.

Applications can be built through various IDEs that you maybe using or from the command line. What ever the case the respective build commands should be documented in the product. The Symbian OS build environment is designed to minimise the complexity to developers of working with multiple program types. Every project is fully specified by its project file, and makefiles are generated from project files by the toolchain. Correctly declaring the target type in the project file will ensure the correct build process and generate an appropriate target. Thus, GUI applications are built as app type targets, ECom plug-ins as ecomiic type targets, and static interface DLLs as dll type targets.

In our case we are using the Nokia series 9200 communicator sdk, you could also be using the UIQ. The output of the program above is shown in Figure 14.2

Applications of the first category, i.e., those that have both an UI component and business logic, are structured differently. Symbian OS is an event-based system that responds to events generated either by the system or through user interaction, it implements its functionality through event handlers. UI based **Symbian programs follow an MVC model**. Since all applications must follow this, a brief discussion of the MVC pattern is included here.

The MVC pattern provides for a separation of the data and its presentation. An MVC object has three components **models** for data, **views** for display, and **controllers** that manage the flow of control between various views in response to events affecting them. The MVC abstraction can be graphically represented as shown in Figure 14.3

The system generates an event that is delivered to the controller. The controller is responsible for redirecting it to the appropriate view. The Model encapsulates the application functionality; it handles the data persistence and algorithms. The view interacts

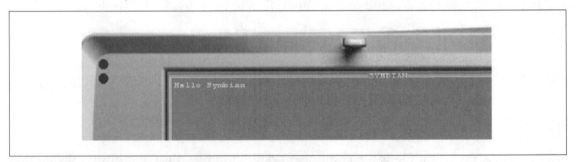

Figure 14.2 Hello Symbian

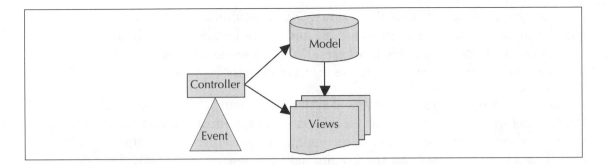

Figure 14.3 MVC overview

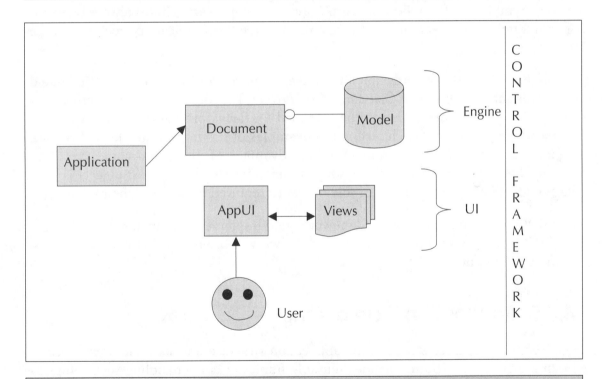

Figure 14.4 MVC in Symbian

with the model to obtain the relevant data. Any change or update to the data is returned back to the model for appropriate action.

The application is also structured as above. The various stages of application activation are shown in Figure 14.4. An application creates a document which contains the data model. All business logic including file and database operations are included here. This is the engine that we mentioned previously, it communicates with the system via the application and control frameworks (CF) and event-handlers. The other component is the user interface or UI this is where all the presentation logic resides. The event handlers are also implemented in the user interface class.

In the true spirit of an object-oriented system, Symbian OS provides most services as frameworks. A framework can be viewed as a collection of abstract base classes and some concrete classes. To use the framework, a programmer extends the abstract base classes, and provides new behavior. Traditionally frameworks were provided using polymorphicdlls. A polymorphic dll exports a single function. This creates a new instance of the newly derived framework class. We will see a practical usage of this when we discuss the HelloSymbian GUI application. Application developers will almost always be working with frameworks. Examples of frameworks include the Application architecture, the UIKON etc.

Symbian OS v7.0, onwards an alternative to polymorphic dlls (dynamic linked libraries) is supplied by the *ECom plug-in architecture*. This defines a generic framework that specifies plug-in interfaces, and be extended for developing new plug-ins. SyncML framework and Transport Architecture are examples of newer frameworks that require plug-ins to be written using ECom. Another advantage of the ECom comes from the fact that the responsibility of finding and instantiating suitable plug-in objects is now delegated to ECom, where polymorphic dlls were required to do these themselves.

Symbian OS includes a host of frameworks that implement its functionality. While it is beyond the scope of this book to discuss all, we will briefly look at two of the most commonly used ones.

14.3.5 Application Framework

The *Application Framework* defines the application structure and its basic user interface handling Figure 14.5. It also includes reusable frameworks for handling such things as text layout, user interface controls, and front end processors. It provides base classes for these and many of the key application concepts. This enables a licensee product to add its own specialist components that provide user interface elements suitable for its particular screen and input mechanisms. E.g. the UIQ. A key component of the application framework is Uikon, a standard framework common to all Symbian OS platforms. It not only provides the framework for launching applications, but it also provides a rich array of standard control components (for example, dialog boxes, number editors, and date editors) that, applications can make use of, at run time.

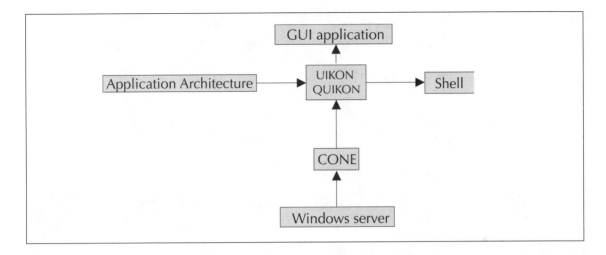

Figure 14.5 Overview of a Symbian application framework

CONE

Once the application is running, "events" are channeled to it via another part of the Symbian OS framework–the Control Environment (CONE). Events could be triggered by users, things such as key presses, or system events like the machine being turned off, the application coming to the foreground, etc.

UIKON/QUIKON

Cone itself doesn't provide any concrete controls or widget, that's the job of the system GUI like UIKONE or QUIKONE. Support for UI starts with the windows server, which manages the screen, pointer ad other navigation device on behalf of the all GUI programs within the system. It is a single server process that provides a basic API for client applications to use. CONE the control environment runs in the each application process and works with the windows server client side API to allow different parts of an application to share windows, and pointer events. A fundamental class delivered by CONE is CCoeControl, a control which is a unit of user interaction that uses some combination of screen, keyboard, and pointer. Many controls can share a single window. Other controls are derived from CColeControl. Together they specify a standard look and feel, and provide reusable controls and other classes that implement the look n feel.

All SymbianOS applications are required to have the components shown in Figure 14.6 alongside. The four main application framework classes are described below.

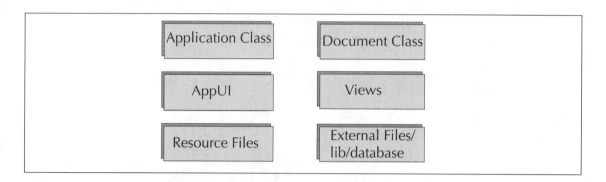

Figure 14.6 Components of a Symbian application

Readers should refer to Figure 14.7 as we go along.

- The Application class is a startup object and defines the application's properties. The base class for the application class is CEikApplication that inturn uses the CONE CApaApplication class. This class bootstraps the application and returns its system unique application ID. It is the responsibility of the application framework to create the Application. The Application constructs the Document, which in turn creates the AppUi. The AppUi constructs and owns the View(s).

- The Document object is the engine of the application and is used to store the application's persistent state. An application must have an instance of the Document class even if it has no persistent data, as it is required to launch the AppUi. The base class for documents is CApaDocument, the UIKone CEikDocument uses this class. In addition, the Document Class also contains and initializes the application's data model. The ways by which the application will save data and close down are coded in the Document Class.

- The AppUi, is responsible for creating the application's user interface controls, views, and handles UI commands and events such as Options menu commands, opening/closing files, and the application losing focus. It typically has no screen presence, i.e., it provides the foundation for the UI, but does not actually implement it. Instead drawing and screen-based interaction is delegated to the View (s) owned by the AppUI. It is also responsible for switching between Views. The base class for AppUI is CCoeAppUi.

- The View is a control, which displays data on the screen that the user can interact with. Typically, Views are notified of updates in the Model's state by an observer mechanism; they also pass user commands back to the AppUi. A View

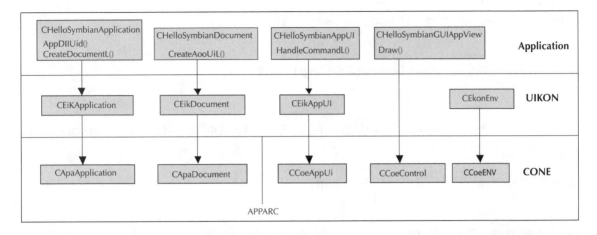

Figure 14.7 Application classes

is derived from CCoeControl or CEikDialog. This, in turn, springs from the Control framework base class, CCoeControl and from the view architecture class, MCoeView.

- Resource files are handled in almost the same way as they are in Palm OS. Symbian OS Resource files usually use the same names as their applications, with the extension .rss. When an application starts, it reads in the resource file. Further on the views are responsible for rendering it.

The application structure will become clearer once we walk through a sample program.

14.4.6 Source code

The example is a simple application containing a single view with the text "Hello Symbian!" drawn on it. This source file contains the single exported function required by all UI applications and the E32Dll function which is also required but is not used here.

NewApplication() is the starting point of any UI based application. This is the main purpose of the HelloSymbian_Main.cpp. The example is very simple containing a single view with the text "Hello World !" drawn on it.

```
#include "HelloSymbian.h"

// this source file contains the single exported function
required by all UI applications and the E32Dll function which
is also required but is not used here.
```

```
EXPORT_C CApaApplication* NewApplication()

    {

    return new CHelloSymbianApplication;

    }
```

This function is required by all EPOC32 DLLs. In this example, it does nothing.

```
GLDEF_C TInt E32Dll(TDllReason)

    {

    return KErrNone;

    }
```

Except in the rare case of insufficient memory the function NewApplication() will return an object of CHelloSymbian. The next step is the AppDllUid() this is a system level security check to ensure that identity of the application DLL. The code for HelloSymbian_Application.cpp is given below.

```
#include "HelloSymbian.h"

TUid CHelloSymbianApplication::AppDllUid() const

    {

    return KUidHelloSymbian;

    }
```

```
//This function is called by the UI framework at application
start-up. It creates an instance of the document class.
```

```
CApaDocument* CHelloSymbianApplication::CreateDocumentL()

    {

    return new (ELeave) CHelloSymbianDocument(*this);

    }
```

Notice that we have used KUidHelloSymbian, but where is this value defined? In the HelloSymbian_Application.h as shown below.

```
const TUid KUidHelloSymbian = { 0x01000000 };
```

Note this value must match the second value defined in the project definition file. This is only a sample Uid that we are using for illustration purpose. For a commercial application you will need to obtain a Uid from Symbian

```
class CHelloSymbianApplication: public CEikApplication
    {
private:
    // inherited from class CApaApplication
    CApaDocument* CreateDocumentL();
    TUid AppDllUid() const;
    };
```

So now our application is created. As we saw above the main purpose of the application is to Covey some information about the capabilities of the application Also acts as a factory for a default document.

Now we come to the Document class. In a file based application, this class handles all data operations. It also creates the appUI. The appUI is relevant when the application deals with modifiable data, contacts and notes for example. But irrespective of the need all application have to define an appUI. The CHelloSymbianDocument.cpp source is given below

```
#include "HelloSymbian.h"

CHelloSymbianDocument::CHelloSymbianDocument(CEikApplication&
aApp)

        : CEikDocument(aApp)
    {
    }
```

This is called by the UI framework as soon as the document has been created. It creates an instance of the ApplicationUI. The Application UI class is an instance of a CEikAppUi derived class.

```
CEikAppUi* CHelloSymbianDocument::CreateAppUiL()

    {
  return new(ELeave) CHelloSymbianAppUi;
    }
```

The corresponding header file containing the definitions for the constructor and destructors.

```
class CHelloSymbianDocument : public CEikDocument

    {
public:
    static CHelloSymbianDocument* NewL(CEikApplication& aApp);
    CHelloSymbianDocument(CEikApplication& aApp);
    void ConstructL();
private:
    // Inherited from CEikDocument
    CEikAppUi* CreateAppUiL();
    };
```

In a nutshell the document class

- Provides functionality for persistant data handling.
- Creation of the appUI.

Note Symbian has a two phase construction that includes definition and initialization. So far we have only seen the first phase.

The next step is the application UI which we created in the last phase. The applicationUI has two main functions:

Symbian OS and Series 90 Developer Platform 2.0 applications are required to handle events, such as key presses, generated by the system. The CONE environment provides the event-handling framework to an application. Events can be key presses, menu commands, screen-redraw events, or events from other controls. UI Controls and Application Views need to handle them in a manner consistent with the Series 90 UI Style Guide.

To capture commands to the application

To redirect user actions (e.g. keystrokes, pointer tap etc.) to the controls. We will come to controls later.

The concept of commands is similar in almost all environments. These are system commands and user defined commands. In Symbian OS command is a 32bit integer ID, which is defined in the resource file. Commands normally originate from user actions like menu selection button/key press etc. Irrespective of the source the application just needs to trap the commands that it is interested in and execute appropriate code. We will now proceed to examine the CSymbianHelloAppUI.cpp code to understand the command handler loop.

```
#include "HelloSymbian.h"
```

This is also the second phase constructor of the application UI class. The application UI creates and owns the one and only view for this example. This reads the resource file and constructs the menu and shortcut keys for the application.

```
void CHelloSymbianAppUi::ConstructL()
    {
        // BaseConstructL() completes the UI framework's con-
struction of the App UI.
        BaseConstructL();
```

Create the single application view in which to draw the text "Hello Symbian!", passing into it the rectangle available to it.

```
        iAppView = CHelloSymbianAppView::NewL(ClientRect());
    }
```

Since the appUi owns the view, destruction too is its responsibility

```
CHelloSymbianAppUi::~CHelloSymbianAppUi()
    {
    delete iAppView;
    }
```

The most interesting method here is the command handler. Conceptually this similar to the MIDP command handler. This method is called by UI framework when a command has been issued.

The command Ids are defined in the HelloSymbian.hrh file, however for this example we are using only the EEikCmdExit this is defined by the UI framework and is pulled in by including eikon.hrh. In further examples we will add a user defined command then this process will become clearer. For the sake of simplicity we have just added a button for exit. The code for this is in the view class which we will see next.

```
void CHelloSymbianAppUi::HandleCommandL(TInt aCommand)
{
        // UI environment
        CEikonEnv* eikonEnv=CEikonEnv::Static();
        switch (aCommand)
{
        case ECbaButton0:
        case EEikCmdExit:
                Exit();
                break;
                }
}
```

The ui header looks as below.

```
class CHelloSymbianAppUi : public CEikAppUi
  {
public:
  void ConstructL();
        ~CHelloSymbianAppUi();
private:
  // Inherited from class CEikAppUi
        void HandleCommandL(TInt aCommand);
private:
        CCoeControl* iAppView;
        };
```

So far we have created an application but now we need to associate a view with it and show it to the user. This is the equivalent of our Display in MIDP. The appUI owns one or more controls. Controls are the equivalents of MIDP displayable. Controls can draw to the screen and handle events. HelloSymbian is a very simple application that only draws a text on the screen and accepts the exit command.

Source file for the implementation of the application view class - CHelloSymbianAppView

```
#include "HelloSymbian.h"

CHelloSymbianAppView::CHelloSymbianAppView()
    {
    }
```

Static NewL() function to start the standard two phase construction.

```
//
CHelloSymbianAppView* CHelloSymbianAppView::NewL(const TRect&
aRect)
    {
    CHelloSymbianAppView* self = new(ELeave) CHelloSymbianApp
    View();
    CleanupStack::PushL(self);
    self->ConstructL(aRect);
    CleanupStack::Pop();
    return self;
    }

// Destructor for the view.
CHelloSymbianAppView::~CHelloSymbianAppView()
    {
    delete iHelloSymbianText;
    }

// Second phase construction.
void CHelloSymbianAppView::ConstructL(const TRect& aRect)
  {
      // UI environment
    CEikonEnv* eikonEnv=CEikonEnv::Static();
    // Fetch the text from the resource file.
    iHelloSymbianText = eikonEnv- >AllocReadResourceL(R_HEL-
    LOSYMBIAN_TEXT_HELLO);
```

```
// Control is a window owning control
CreateWindowL();
// this is the whole rectangle available to application.
The rectangle is passed to us from the application UI.
SetRect(aRect);
```

```
// At this stage, the control is ready to draw so we tell the
UI framework by activating it.
    ActivateL();
    }
```

//Drawing the view–in this example, consists of drawing a simple outline rectangle and then drawing the text in the middle. Those familiar with windows programming will see the similarities.

It involves obtaining the graphics context, setting it to the rect supplied from the header and drawing the text. Draw is actually an abstract class in CoeControl, we have to over-ride this and implement our drawing here. This is similar to the low level UI or canvas in MIDP. Most of the code is quite straight forward.

```
void CHelloSymbianAppView::Draw(const TRect& /*aRect*/) const
    {
 // Window graphics context
    CWindowGc& gc = SystemGc();
     // Area in which we shall draw
    TRect drawRect = Rect();
                // Font used for drawing text
    const CFont* fontUsed;
     // UI environment
    CEikonEnv* eikonEnv=CEikonEnv::Static();

     // Start with a clear screen
    gc.Clear();
     // Draw an outline rectangle (the default pen
     // and brush styles ensure this) slightly
```

```
    // smaller than the drawing area.
    drawRect.Shrink(10,10);
    gc.DrawRect(drawRect);
// Use the title font supplied by the UI
    fontUsed = eikonEnv->TitleFont();
    gc.UseFont(fontUsed);
    // Draw the text in the middle of the rectangle.
    TInt baselineOffset=(drawRect.Height() - fontUsed->Height
    InPixels())/2;
    gc.DrawText(*iHelloSymbianText,drawRect,baselineOffset,CG
    raphicsContext::
    ECenter, 0);
// Finished using the font
    gc.DiscardFont();
    }
```

The corresponding header file definition is:

```
class CHelloSymbianAppView : public CCoeControl
  {
public:
    static CHelloSymbianAppView* NewL(const TRect& aRect);
    CHelloSymbianAppView();
    ~CHelloSymbianAppView();
void ConstructL(const TRect& aRect);
private:
    // Inherited from CCoeControl
    void Draw(const TRect& /*aRect*/) const;
private:
    HBufC* iHelloSymbianText;
  };
```

Before we close the discussion, we will see two more important files the resource file and the project definition file.

The resource file as in palmOs contains the definitions for the resources. IDE would most likely provide a WYSIWYG resource editor which will provide a drag and drop facility for common controls and internally generate the resource file. However for our purpose we will manually edit this resource file. HelloSymbian resource file contains definitions for our text, the button and the exit hotkey. The button is actually defined in the .hrh file which is then included here.

```
RESOURCE RSS_SIGNATURE { }

RESOURCE TBUF { buf=""; }

RESOURCE EIK_APP_INFO

     {
     hotkeys=r_HelloSymbian_hotkeys;
     cba=r_HelloSymbian_cba;
     }


RESOURCE CBA r_HelloSymbian_cba

  {
  breadth=80;
  buttons=
      {

      CBA_BUTTON
      {
      id=ECbaButton0;
      txt="bye";
      bmpfile="";
      bmpid=0xffff;
      }
    };
  }
```

```
RESOURCE HOTKEYS r_HelloSymbian_hotkeys

  {
    control=
     {
     HOTKEY { command=EEikCmdExit; key='e'; }
     };
  }
```

```
RESOURCE TBUF r_HelloSymbian_text_hello { buf="Hello Symbian!"; }
```

Finally the application specification or mmp file. Except the Unique Uid everything else is the same as before in the text hello.

```
TARGET HelloSymbian.app
TARGETTYPE app
UID 0x100039CE 0x01000000

TARGETPATH \system\apps\HelloSymbian

SOURCEPATH .
SOURCE HelloSymbian_Main.cpp
SOURCE HelloSymbian_Application.cpp
SOURCE HelloSymbian_Document.cpp
SOURCE HelloSymbian_AppUi.cpp
SOURCE HelloSymbian_AppView.cpp

USERINCLUDE .
SYSTEMINCLUDE \epoc32\include

RESOURCE HelloSymbian.rss
LIBRARY euser.lib apparc.lib cone.lib eikcore.lib
```

Now it time to build and execute this application. The output on the 9200 emulator looks as below Figure 14.8. In a different emulator such as the UIQ, It will look different.

Figure 14.8 Output of Hello Symbian GUI

To summarize:

We first create the Application using NewApplication(), This then creates the Document class which further creates the AppUi. The AppUI creates and owns the views. The views are responsible for creating controls and rendering them on the screen. SymbianOS works in an event driven mode, where user interaction generates commands or creates events. The appUI implements the Command Handler code. The sequence of events is given in Figure 14.9.

Adding Menu and other simple controls is simple. We have seen how a button is implemented in the previous example, we will see how to define controls and handle a command and attach a control. Other examples will be available as samples from the SDK vendor. We encourage our readers to explore them.

14.7 CONTROLS AND COMPOUND CONTROLS

A control is an area of the screen that responds to user input events. It is similar to the MIDP Form. Application developers can also implement their own concrete controls by deriving from CCoeControl; OEMs provide device specific implementations of controls for example the Ckon (Series 9000 UI layer) user interface library from Nokia. This provides the OEMs with a facility to implement a standard look and feel across all the applications for their devices.

Controls can be of two types Window-owning and non-window owning controls. Window owning controls are akin to the top level UI elements like form in J2ME. They contain other controls and typically own the entire window which they occupy. The

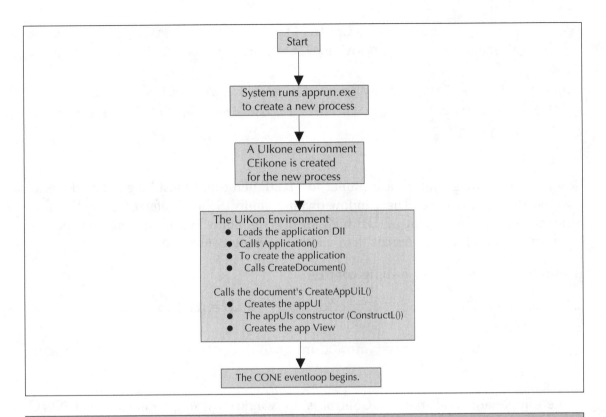

Figure 14.9 UIKone life cycle

following example shows the construction of a window-owning control:

```
CWOControl::ConstructL()
{
// Create the control
CreateWindowL();
// Set size
SetRectL(ClientRect());
// When created the control is blank we can then add things to it.
SetBlank();
// Defin borders
SetBorder(TGulBorder::EFlatContainer);
```

```
// Set appView note the similarity to MIDP display.setCurrent
   (displayable); ESkinAppViewWithCbaNoToolband is the default
   option.
CknEnv::Skin().SetAppViewType(ESkinAppViewWithCbaNoToolband);
// Finally activate the window. It is now ready to be drawn.
ActivateL();
}
```

Non-window-owning controls are similar to MIDP items and must be contained within a window-owning control. The window-owning control is also referred to as the container in this context. As in MIDP high-level items non-window owning controls are faster and require fewer resources than window-owning controls.

To construct a non-window-owning control:

```
CNWOControl::ConstructL(CCoeControl* aParentWOControl)
{
SetContainerWindowL(aParentWOControl);
}
```

In the sample above aParentWOControl is the window-owning container for CNWO Control.

14.7.1 Compound controls

A compound control is a group of associated controls. These are helpful in cases where the application requires the same set of controls, OK and Exit is a simple example of such a combination. A compound control will provide information about the contained controls via two virtual methods: CountComponentControls(), the number of component controls, and ComponentControl(TInt aIndex), which returns each of the controls by index (zero based). Figure 14.10 shows a comparison between simple and compound controls. The API reference on the Symbian site has more details on the same.

14.7.2 The Control Stack

Application UI contains controls for interacting with the users. Controls receive user commands through events like key-press navigation text in control area etc. A Control has to register itself to receive these events. The control stack is the structure that holds all controls

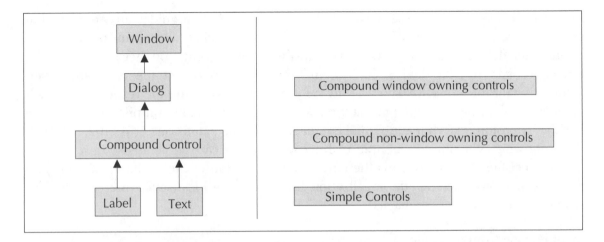

Figure 14.10 Simple and Compound controls

registered to receive events. Typically only a window owning control the container will add itself to the control stack. The container then distributes events to its component controls. Controls on the control stack are notified of events by the control framework calling the respective event listener. For example the key press events are notified via the OfferKeyEventL(). Details of constructing Dialogs are beyond the scope of the current text.

14.8 ACTIVE OBJECTS

As mentioned above Symbian OS is a preemptive multitasking operating system and hence supports multithreading. However multithreading is expensive in terms of resource usage and is generally not recommended. Active Objects are Symbian OS's way to enable a single thread to handle multiple asynchronous requests. Each Active Object implements two pure virtual functions: RunL() to handle request completion, and DoCancel() for cancellation of outstanding requests. Note: all user input events are handled through the same mechanisms, hence developers should ensure that the handler methods return fast and do not block. Long-running handler tasks should be broken up to run within the Active Object structure.

14.9 LOCALIZATION

In today's ubiquitous environment it is difficult to envisage an application that is restricted to a specific geographic location. Internationalization and localization are two important considerations in application design. Most UI style guides advise a flexible attitude towards item/control/widget layouts. This is especially important in custom or application defined controls, as the responsibility of presentation here lies with the developer and not he device. In System/OME defined object the risks are lesser. Internationalization or I18N should only be a matter of changing the resource file Each language will define its own compiled resource file. The default compiled resource file has the extension .rsc while individual languages have the extensions: .r01, .r02, .r03, etc. The following is a simple example of how localization can be achieved in a project. The project file (.mmp) contains the language information in the attribute LANG. The resource file (.rss) includes the localization file (.loc). The localization file contains the strings for the different languages.

Sample code from a loc file:

```
//Encoding information

CHARACTER_SET UTF8

//Default language specification



#ifdef LANG_01
#define str_sampleapp_string1 "String1 in default language"
#endif

// A different language resource

#ifdef LANG_02
#define str_sampleapp_string1 "String1 in language 2"
#endif

sample source code from resource file:

#include "loc file"
RESOURCE CBA r_softkeys_string1
{
  buttons =
```

```
{
    CBA_BUTTON
    {
        id = ESampleAppFunction; //note: should be defined in
        sampleapp.hrh
        txt = str_sampleapp_string1;
    }
};
}
```

To use the language 2 resources, we just change the mmp file's LANG attribute to LANG 01 02 03, and then build the resource files. This will build the resource files for default (01), (02), (03) where 01, 02, 03 could be English German Italian etc.

Resource files take care of strings but also to be considered in I18N are dates, time zones currency conversion and other locale sensitive data. These are handled through Tlocale and related classes. These are defined in E32STD.H;

14.10 SECURITY ON THE SYMBIAN OS

As mentioned in chapter 12 security consider-ations are paramount in handheld devices. Symbian OS provides a robust security mechanism that enables confidentiality, integrity and authentication. Similar to the J2ME provisioning mechanism Symbian OS also provides a mechanism for secure installation. Security in Symbian OS include APIs for standard cryptography algo-rithms, hash key generation, random number generation and certificate management, Public Key Cryptography, Certificates and Digital Signatures. Secure communication using SSL/TLS, WTLS etc. Detailed explanations of the algorithms are beyond the scope of this book

The Symbian OS security architecture fundamentally consists of two high level components:

- Certificate management (certman)

- Cryptography (cryptalg)

We do not see these directly, however all security related applications make use of these important among them are:

- Certificate management control panel item this allows the user to access the certificate management

- Software installation (authentication/digital signatures)

- Secure comms (SSL/TLS, WTLS, IPSec, etc.)

Certman as the name suggests handles the certificate management issues including checking validity of signed applications, key stores and PKI for secure comm. While CryptAlg contains the algorithms for hash functions, random number generation and many more. Certman depends on crypt for its algorithm requirements. The Figure 14.11 shows the security architecture.

With this we conclude this chapter. We began with a background on the birth and growth of this OS, followed it with an overview of the operating system architecture. We then looked at programming for Symbian OS with a couple of very simple example. Programming for Symbian is a long and interesting journey. Our attempt here has been to give an overview of what is involved. We encourage our readers to explore further.

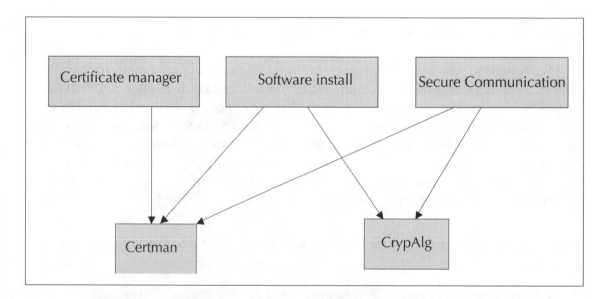

Figure 14.11 Security architecture

References/Further Reading

1. Ericsson in Symbian venture http://www.symbian.com/partners/ericsson.html.

2. Nokia in Symbian venture http://www.symbian.com/partners/nokia.html.

3. Motorola in Symbian venture http://www.symbian.com/partners/motorola.html.

4. Panasonic in Symbian venture http://www.symbian.com/partners/panasonic.html.

5. Sony in Symbian venture http://www.symbian.com/partners/sony-eric.html.

6. Siemens in Symbian venture http://www.symbian.com/partners/siemens.html.

7. Samsung in Symbian venture http://www.symbian.com/partners/samsung.html.

8. A detailed architecture of Symbian architecture is available at http://www.symbian.com/technology/create-symb-OS-phones.html.

9. The official Symbian OS developer zone for c++ is here. http://www.symbian.com/developer/development/cppdev.html.

10. The Nokia forum is a great place for Nokia handsets http://forum.nokia.com/Forum.

11. An excellent article for developers moving from Palm to Symbian http://www.symbian.com/developer/techlib/papers/SymbianOS_for_Palm/Symbian_OS_for_Palm_Developers.html.

12. A good resource for java on Symbian http://www.symbian.com/books/wjsd/support/wjsd-links-downloads.html.

13. Symbian tech Library is here http://www.symbian.com/developer/techlib.

14. A good book for beginners is Symbian OS C++ for Mobile Phones, From Symbian Press by Richard Harrison 2004. The Indian edition is published by Wiley Dream Tech India.

Review Questions

Q1: Describe the Symbian OS architecture. What are the functions of different layers in this architecture?

Q2: What are the different application environments available for Symbian OS?

Q3: What are the different types of control available in Symbian OS?

Q4: What are the different security procedures and protocols available on Symbian SO?

CHAPTER 15

J2ME

JAVA IN THE HANDSET

In this chapter we take a look at the java technologies available for mobile applications. Before we move into the specifics and language constructs it would be helpful to familiarize ourselves with the java kingdom and its governors. The chapter begins by briefly tracing the birth and growth of java. We will take a quick peek into the specifications, the authors, approvers and the implementations. With that background we shall look at the various programming aspects in mobile computing. We assume that the reader is familiar with java programming and OO (Object Oriented) concepts. There are a plethora of java books available for developers at all levels. The sun site "http://java.sun.com" is an excellent point of reference all information related to java.

15.1 WHY JAVA?

The contemporary languages both procedural and Object Oriented (C/C++ for example) had, inherent drawback of platform dependency; for example, a program compiled for windows could not run on UNIX. Overcoming this barrier has been one of the most touted reasons for the success of java. This was made possible because Java is a cross between a compiled and an interpreted language. The process is divided into two steps—first, the java compiles generates an intermediate code called the byte codes which are in the second step interpreted by the Virtual machine. Of course this requires that the VM be ported to all the environments on which the program has to run. As of this writing, most commercial OS (Operating Systems) have a jvm (java virtual machine) port for their environment, and hence this requirement is no longer a deterrent. Java, however, went a step ahead and propagated the idea that the same software should run on many different kinds of computers, platforms, consumer gadgets, and other devices.

What this means is that, at least in theory Java programs are device independent, i.e., a Java program will run on any Java virtual machine irrespective of the computer or operating system the virtual machine is running on. For example, a program written and compiled (to byte code) on a Windows 95 PC will run on a Symbian mobile phone having a jvm. The ground reality, however, is far from this, the limitations being, not those of java but of the underlying hardware that comprise a wide variety of configurations and capabilities. We have already visited these disparities in the chapter 12 *'Introduction to Client Programming'*. As we will see later in this chapter, this particular problem has been overcome by Sun using different flavors of java for machines with similar capacities.

The philosophy of JAVA can be very neatly summed up in the famous statement by Scott McNealy 'the network is the computer'. Which means it should be possible to 'pull' services whenever needed and 'push or offer' services wherever required. Essentially it means one single comprehensive language that can be used to write programs for all devices. This ties in neatly with Sun's new mantra 'Everyone, Everything, Everywhere, Every Time'. *'http://www.sys-con.com/Java/article.cfm?id=2070'*

Let's step back a few years and trace the birth and growth of java. For the coinsurers here is the Sun's official record 'http://web2.java.sun.com/features/1998/05/birthday.html'.

It was originally 'Green Project' aimed at developing an operating environment for ubiquitous consumer appliance and devices. James Gosling felt that the contemporary standard C++ was inappropriate, as it lacked the standard interfaces and reliability, essential to its environment. In 1991 he had the first avatar of java called 'oak' ready. Oak incorporated some cool new features in terms of UI, security, encryption etc. But it took the World Wide Web or internet to make oak successful. Oak saw commercial implementation in 1995 and was renamed java. Under the efforts of Bill Joy and Patrick Naughton, the first java 'killer app' saw daylight in the form of HOT-Java an interpreter for Netscape's applets. Sun and Netscape made history by giving away the software as well as the source to the developers online. A host of new applications that sent pieces of executable code to run on the client saw java firmly established in the world of software.

'We've got 1.2 billion JVMs running in the world in an extraordinary diversity of devices ... and who's driving it? You and three million other Java developers across the planet.' Jonathan Schwartz. EVP of Sun's Software Group, San Francisco, at the eighth JavaOne. To read the entire excerpt go to:

'http://www.sys-con.com/Java/article.cfm?id=2070'

As the use of the language grew, so did the clamor for new features and capabilities. Though just eight years old its popularity is overwhelming and still speeding at full throttle. Today, millions of developers worldwide are writing Java applications ranging from embedded mobile systems, to desktop computers, to large servers. Figure 15.1 depicts a brief time-line. To read blow by blow account refer to 'http://www.wired.com/wired/archive/3.12/java.saga.html'.

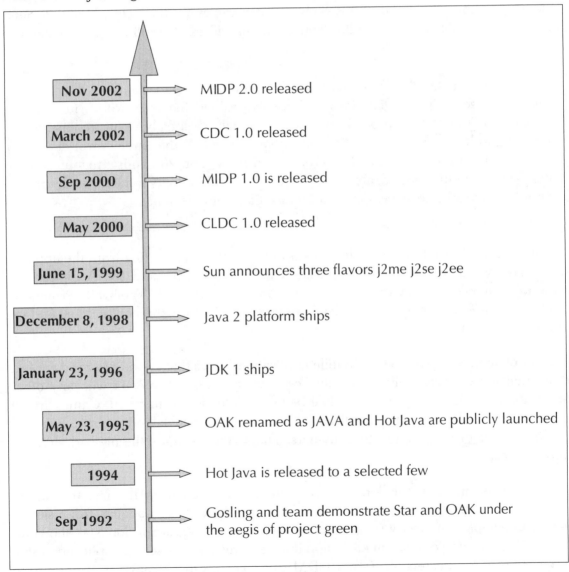

Figure 15.1 Timeline for Java and J2ME

15.2 THE THREE-PRONG APPROACH TO JAVA EVERYWHERE

The mobile market worldwide estimated at a whopping $80 billion ('http://wireless.java. sun.com/developers/business/articles/opportunities') has numerous stakeholders. Starting from the device manufacturers going down to the end users it includes a host of others like network operators, application and service providers and others, all held together by the underlying technologies and the tools. So we see companies like the lucent, Motorola, Sun, Nokia and a host of third-party vendors and service providers, all wanting a piece of the pie.

'According to researcher John Jackson, 'The Yankee Group estimates a current installed base of 97.7 million Java technology-enabled handsets worldwide, and expects continued rapid growth for J2ME penetration globally.' With researchers referring to Java technology as 'the common language of the wireless world,' and deployment of it by 53 network operators and more than 20 handset manufacturers, Java technology is cemented into the international mobility landscape.' 'http://wireless.java.sun.com/developers/business/articles/opportunities'

But all these depend on a seamless integration of all pieces starting from the application server which is hosting the application to the end user hand set. A typical top view of a mobile application will look as shown Figure 15.2. The white papers at 'http://java.sun.com/blueprints/wireless' are a wealth of information on building end-to-end mobile applications.

In brief, there are two types of mobile applications: ones that are device resident, i.e., those that utilize exclusively the device resources and do not interact with any other applications outside except maybe while being downloaded or installed. Games are an excellent example. But such applications are very restrictive and do not generate much in terms of revenue for the providers. The second type are the network enabled applications.

The real potential of this kind of application is derived from the fact that it can connect to other devices across networks and perform tasks ranging from simple e-mail to complex monetary transaction. A typical j2me transaction looks as shown in Figure 15.2. The device resident client app initiates a request to a server application using any of the various bearers like GSM, CDMA etc. The protocols may vary from http to WAP. The request received by the server is processed and the response is returned to the application. A session may be maintained over a series of transactions if required.

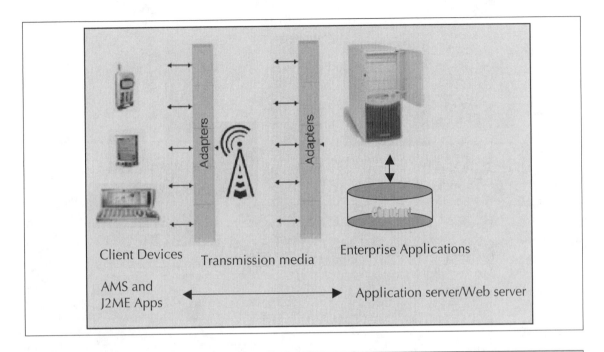

Client Devices

Transmission media

Enterprise Applications

AMS and
J2ME Apps

Application server/Web server

Figure 15.2 A typical mobile application architecture

As Figure 15.1 shows, initially everything was java, but sun soon realized that one size does not fit all. So with the release of java2 they identified java for three different platforms the low end or limited devices, the desktops and the high-end servers. The Figure 15.3 below shows a detailed diagram of the java spectrum.

Thus we have the Hotspot-powered Java 2 Enterprise Edition (J2EE) for the server side applications, the JVM-powered J2SE desktop environments and the Kilobyte Virtual Machine (KVM)/CVM powered J2ME for handheld devices and set-top boxes. A contemporary of the J2ME is the Java Card Technology for Smart cards. It is powered by a card-VM and is characterized by an extremely small footprint.

Again in a significant departure from traditional product, Sun evolved a process called "Sun Community Process" or JCP to define the functionality of Java. It is responsible for the development and approval of Java technical specifications. It's an open community where everybody is free to join and be involved in its process. Since most stake holders are actively involved, the JCP procedures ensure Java technology's versatility, stability and cross-platform compatibility. The open approach of JCP makes industry adoption very fast in fact as fast as 6 months from the release of

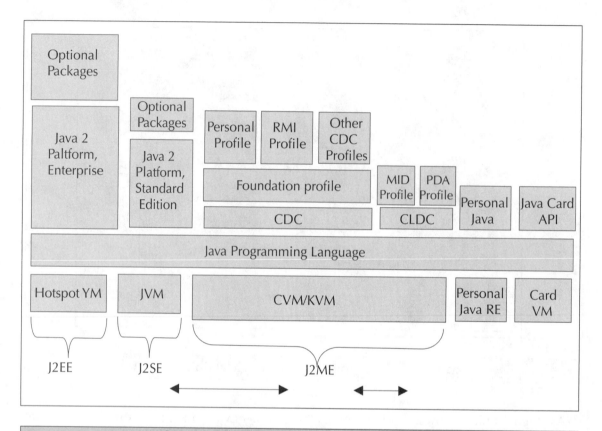

Figure 15.3 A three prong approach to Java

It is interesting to note that as early as 1998 just three years from inception, Java very nearly fractured into different flavors. Read more about how it was resolved at 'http://cgi.cnn.com/TECH/computing/9901/18/javasplit.idg'

the final specs.) For more details on how the process works, you can visit "*http://www.jcp.org*".

The fact that there are 55 JCPs for J2ME itself speaks volumes about the magnitude of the task at hand. We will not be covering any details here. For the curious here is the link 'http://www.jcp.org/en/jsr/tech?listBy=1&listByType=platform'.

The rest of the chapter focuses on the details. We begin with an introduction to J2ME. This is followed by sample exercises in application development.

Readers are encouraged to read the Java 2 Platform Micro Edition (J2ME) Technology for Creating Mobile Devices white paper @ http://es.sun.com/wireless/descarga/kvmwp.pdf

15.3 JAVA 2 MICRO EDITION (J2ME) TECHNOLOGY

The coffee cup for the small devices is christened Java 2 Micro Edition (J2ME). J2ME was conceived from the need to define a computing platform that could accommodate consumer electronics and embedded devices. As we saw in the chapter on 'Introduction to Client Programming' the handheld gadgets comprise a whole gamut of devices that come in varied configurations in terms of resources and capabilities. The low-end PDAs may offer only offline data storage with a serial cable to sync with the PC while the high-end communicators would be microcomputers. Mobile phones are likely to have low bandwidth intermittent connectivity while the set top boxes would have uninterrupted connectivity. It was not practical to attempt to define a single J2ME platform for all of these. The biggest challenge for J2ME was to specify a platform that could support a consistent set of services across a broad spectrum of devices with a large multitude of capabilities. To be able to support the large brood of devices, a modular structure was essential. The designers of J2ME came up with a concept of configurations and profiles towards achieving this goal.

In practice the primary differentiators are, computing power, power supply and I/O capabilities. Moore's law however makes this differentiation quite fuzzy, because technology enables additional capability to be placed in smaller devices.

A Configuration defines the lowest common denominator or the minimum capabilities that will be available across a given range of devices. It is a complete Java runtime environment, consisting of:

- A JVM (Java virtual machine).
- A set of Core Java runtime classes.
- A set of supported API (Application Programming Interface)

Configuration specify classes and methods that are inherited from Java 2 Standard Edition (J2SE) classes. That means, J2ME is a subset of J2SE. However they are generally not complete subsets of J2SE. Configurations also include additional classes to adapt to device capabilities and constraints.

In short, a configuration can be defined as a specification that identifies the system-level facilities available; for example, the characteristics and features of the virtual machine present and the minimum Java libraries that are supported. Members of a given genera are considered to have similar capabilities in memory and processing power and can be expected to provide a certain level of system support. Although a configuration does provide a complete Java environment, the set of core classes is normally quite small and must be enhanced with additional classes supplied by J2ME profiles or by configuration implementer. In particular, configurations do not define any user interface classes.

To avoid fragmentation and a deluge of incompatible platforms, J2ME defines only two configurations. They represent the two distinct categories of devices.

The first category is devices that have superior UI facilities higher computing power and are constantly connected. These implement the Connected Device Configuration (CDC) E.g. set-top boxes, Internet TVs, Internet-enabled screen phones, high-end communicators, and car entertainment/navigation systems.

The second being personal, mobile information devices that are capable of intermittent communications. These implement the Connected, Limited Device Configuration (CLDC) e.g., mobile phones, two-way pagers, personal digital assistants (PDAs), and organizers.

While the configuration concept is helpful, it's still broad and incomplete. A profile takes the configuration a step further by defining the libraries used to create applications. The profile specifies the application-level interface for a particular class of devices representing a vertical market. Profiles are built on top of and utilize the J2ME configurations. Event handling, I/O functions, User interface APIs (for specific device categories,

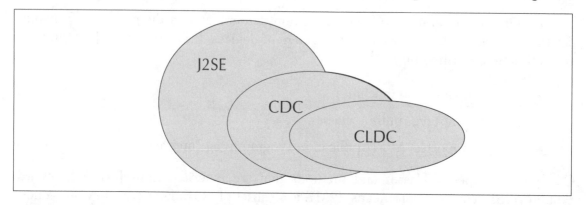

Figure 15.4 How the three fit-in together

like mobile phones, and PDAs) and application lifecycle management lie in the purview of the profile. Profiles are also a means to guarantee interoperability—between all devices of the same category or vertical market.

Applications are built on top of the profile; they can use only the class libraries provided by these two lower-level specifications. Profiles can be built on top of one another. A J2ME platform however, can contain only one configuration.

Together the configurations and profiles enable creation of 'customized flavors' of Java for different device categories. Java has already been accepted as the de-facto standard for devices. A large number of manufacturers are adopting J2ME technology, and as the range of devices using J2ME increases, newer profiles will need to be implemented. By providing a common minimum platform J2ME configurations and profiles enable creation of these custom editions in a structured way.

15.3.1 CDC

CDC is a configuration for high end devices with larger memory, in the range of 2 MB or more (at least 512K for the runtime environment, plus another 256Kfor applications), providing connectivity through some kind of network and some UI support. The CDC is a superset of the Connected Limited Device Configuration and is quite close to a conventional Java 2 Standard Edition (J2SE) runtime environment.

The CDC specification defines

- A full-featured JVM, called the CVM.
- A subset of the J2SE 1.3 classes.
- APIs introduced in the CLDC—i.e., the Generic Connection Framework.

Eric Giguere, an undisputed guru on the subject has a lot of useful articles on his site, http://www.ericgiguere.com.

The CDC supports the following standard Java packages:

File I/O	java.io
Core java system classes	java.lang
Networking support	java.net
Security framework	java.security
Text, dates, numbers, internationalization	java.text
Utility classes	java.util

The CDC defines three profiles. They are the Foundation Profile which is essentially the worker that provides support to other profiles, the Personal Profile which is personal java retrofitted to J2ME and the Remote Method Invocation (RMI) profile for RMI support. We will now briefly look at these.

Foundation Profile

The Foundation profile is targeted at devices supporting a strong network, but do not provide any UI (User Interface). Essentially it does the "useful" work. It is also used as a base by other profiles, which then build on its functionality by adding UI and other components. Typically Foundation Profile has the following requirements:

- Memory minimum 1024k ROM and 512k RAM for the profile and configuration.

- Stable Network Connectivity.

Personal Basis Profile

The J2ME Personal profile is a reincarnation of the Personal Java Application Environment suitably modified to fit into the J2ME environment. Built on top of the Foundation Profile, it is backward compatible with Personal Java 1.1 and 1.2. It caters to devices that enjoy reliable and constant Internet connectivity and rich GUI. These devices are usually characterized by

- A Minimum of 2.5 M of ROM and 1 M RAM.

- Robust internet connectivity.

- Rich (GUI), with browser like internet support.

J2ME RMI Profile

The J2ME RMI (Remote Method Invocation) profile, as the name suggests provides support for RMI across applications. It is built on top of the Foundation profile and uses TCP/IP as the underlying connection protocol. It is interoperable with the Java 2 Standard Edition (J2SE) RMI API 1.2 and higher. Details of the RMI Profile Specifications can be found at JSR-66 home.

15.3.2 CLDC

As noted earlier the Connected Limited Device Configuration (CLDC) is meant for low-end, intermittently connected, battery-operated devices. A CLDC device is characterized by the following capabilities:

- A minimum of 128 to 512 KB for the platform.

- A 16-bit or 32-bit low end processor

- A low bandwidth network with Intermittent connectivity (mostly wireless GSM/GPRS/CDMA)

CLDC runs on KVM, which is a highly optimized JVM for resource constrained devices. It includes just the basic classes from the java.lang, java.io and java.util packages, with a few additional classes from the new javax.microedition.io package. In particular the CLDC specification does not support the following Java language features.

- floating point calculations

- object finalization

- custom class loader

- error classes

For a detailed index of available classes please refer to the CLDC specifications which can be freely downloaded from the sun's site.

A notable difference between the CLDC and J2SE is in the process of class verification. Class files are required to be verified by a class verifier, through a process called preverification. This is done before the application can be loaded to the device. The J2ME VM is similar to the JavaCard VM in this respect. This allows the VM to be leaner. At runtime, the VM uses information inserted into the class files by the preverifier to perform the final verification steps.

Though CLDC is a subset of the CDC, yet the two are quite different. The two configurations are independent of each other, and cannot be used together to define a platform. Figure 15.4 shows the relationship between the two configurations and the parent J2SE platform.

Sun has released two profiles that sit atop the CLDC. They are the Mobile Information Device Profile (MIDP) and the Personal Digital Assistant Profile.

MIDP

J2ME Mobile Information Device Profile is by far the most popular and widely supported one. It was also the first profile to be released. Currently it is in the 2.0 Version. The MIDP specs were written by an expert group, the Mobile Information Device Profile Expert Group, an international forum represented by several leading companies in the mobile device industry.

It provides classes for writing downloadable applications and services that are of interest to the consumer for example games, commerce applications, personalization services etc.

The MIDP profile requires the devices to have the following capabilities:

- A minimum of 512K for the platform.
- Intermittent connectivity to some type of wireless network.
- Limited User interfaces.
- Some kind of input mechanism.

For further details please refer to the specification document available at JSR 37 / JSR 118 home.

The MIDP specifies the following APIs:

Application control	javax.microedition.midlet
User interface	javax.microedition.lcdui
Persistent storage	javax.microedition.rms
Networking and local IO	javax.microedition.io, java.io
System classes and interfaces	java.lang
Utility objects	java.util

PDAP

J2ME Personal Digital Assistant (PDA) Profile is a recently released profile catering specifically to the PDA market. It lies on top of the CLDC specification. It provides user interface and data storage APIs for devices with the following resource constraints:

- Minimum 1000K for the platform.

- Battery powered low power devices.

- Good UI capability with a resolution of 128x128 pixels and dept of 1 pixel, a touch screen, and or character input mechanism in the form of a T keypad or a full "qwerty" keyboard.

- An Intermittent and low bandwidth two way connectivity.

For further details please refer to the specification document available at JSR 000075 home.

15.4 PROGRAMMING FOR CLDC

Covering the entire gamut of profiles mentioned above is beyond the scope of this book. We will take one profile from the CLDC group the MIDP. MIDP has been chosen as it is the most widely used as also has created a good amount of noise with the release of its version 2.0. Here unless specifically referred to as introduced by MIDP 2.0 they are applicable to both the versions. The reader however is encouraged to study the vendor specific features too as they allow the developer to use the support of the underlying native environment.

A MIDP application is called a MIDlet. It is a take-off on the traditional Applet, as the two are similar in some ways. We now look at the MIDlet Model.

15.4.1 The MIDlet Model

A typical MIDP application or MIDlet sits atop the MIDP which in turn requires the services of the CLDC and the VM below it. Finally it is the device hardware that executes instructions on behalf of the software layers above. Figure 15.5 shows a top down view of a MIDP application.

Like an applet a MIDlet needs an execution environment. The browser's equivalent in the MIDP world is called Application Management Software or AMS. It is a device resident software (normally provided by the device vendor) and all MIDlets run within the context of an AMS. This is also required because the handset needs to respond

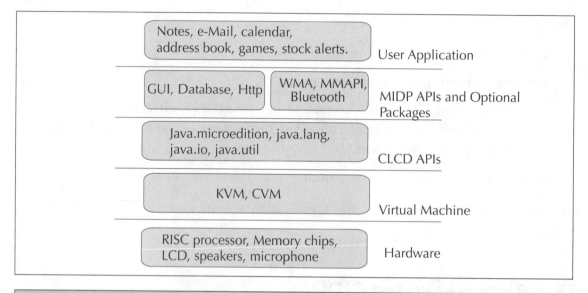

Figure 15.5 A top down view of a typical MIDP application

to events outside the MIDlet scope; for example, a call might need to be answered while reading an e-mail. All MIDlets are registered with the AMS during installation. A set of MIDlets can be grouped together into a MIDletsuite. Other than managing the MIDlet, the AMS is also responsible for Application Provisioning and Application removal.

15.4.2 Provisioning

Provisioning is the process of application discovery, download and installation. PDAs allowed this by downloading the applications on to a pc and then using a serial cable to transfer it to the device. The solution defeats the very purpose of mobility.

Provisioning includes

- Search: which can be performed by the user manually entering the URL where the application is hosted or using a device resident browser step 1 in Figure 15.6 above.

- Next step 2 is to retrieve the descriptor file, which includes the application details, checks for version, and compatibility issues.

- In step 3 it proceeds to download the appropriate jar file. (All middle suites are packaged into a jar file).

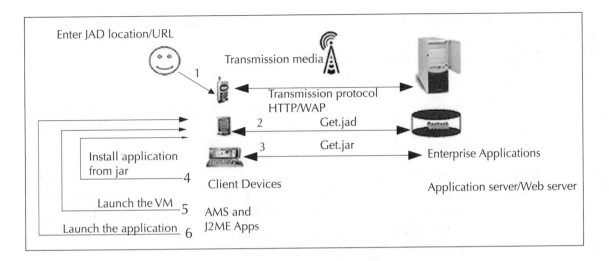

Figure 15.6 Provisioning a MIDP application (OTA)

- Once the download is completed the AMS is called to install (step 4).

- Finally the VM is launched (step 5) and the

- Applications are launched (step 6).

The process includes appropriate user interactions, including payment authorizations. Figure 15.6 above depicts the entire process.

MIDP 2.0 has introduced a more interesting and unique capability in the form of a push registry. (MIDP 1.0 specifications allowed only pull applications, ie., the applications had to initiate the transaction.) These features are discussed further a little later.

15.4.3 The MIDlet Life-Cycle

As shown in Figure 15.7, a MIDlet has three states. A MIDlet class extends javax.microedition.midlet.MIDlet which defines the corresponding life-cycle notification methods. These Life Cycle Methods allows the AMS to notify and request MIDlet state changes.

- Applications are launched either by a user selection or in response to an external event from the push registry. On being activated by the AMS, the MIDlet is constructed but is still inactive. This is the reason why resource allocation is not advisable in the constructor. Now the MIDlet is in a *paused* state.

- Once constructed, the AMS initializes and activates the MIDlet by invoking its startApp() method. The MIDlet's now changes to active state. If the initialization fails a javax.microedition.midlet.MIDletStateChangeException is thrown and the state is changed to the *destroyed.*

- A transition from *active* state to the paused state is initiated by the AMS by calling the MIDlet's pauseApp() method or by the MIDlet itself through the MIDlet context. In this state a MIDlet should release all its resources.

- A MIDlet can be *destroyed* from either *active* or *paused* state, If destruction is initiated by the AMS, the MIDlet's destroyApp() method is invoked with a boolean parameter true/false for optional or forced destruction. The Optional destruction can result in a MIDletStateChangeException. The possible state changes and the transitions are shown in Figure 15.7

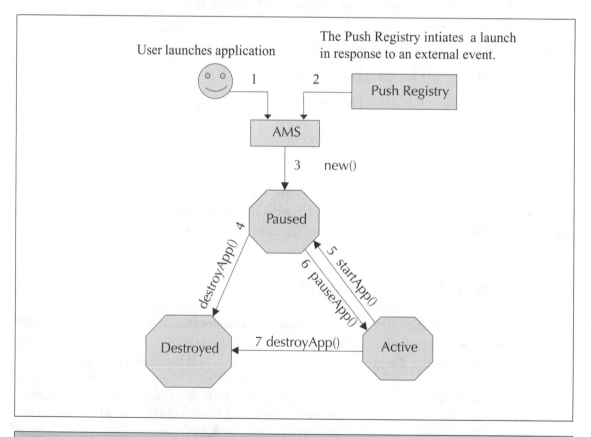

Figure 15.7 MIDlet lifecycle

Briefly then the three states of a MIDlet are

- *paused:* The MIDlet is constructed but inactive, transition occurs by a call to pauseApp()

- *active:* The MIDlet is active and can process requests, transition occurs by a call to startApp()

- *destroyed:* The MIDlet has been destroyed and is ready for garbage collection, transition occurs by a call to destroyApp().

Apart from the life cycle methods javax.microedition.midlet.MIDlet also defines some MIDlet Context methods. These are as follows.

- getAppProperty() which retrieves properties from the *application descriptor.* Properties are name-value pairs in the JAD file. The contents of a sample JAD and Manifest files are shown below.

- resumeRequest() is a request to the AMS to reactivate the MIDlet. However it the prerogative of the AMS to decides if and when to reactivate the MIDlet. Reactivation make a call to the MIDlet's startApp () method.

- notifyPaused() is an alert to the the AMS that the MIDlet is transitioning to a paused state;

- notifyDestroyed() informs the AMS that the MIDlet will now destroy itself.

Sample JAD file contents

MIDlet-1: GUI, GUI.png, FirstMIDlet

MIDlet-2: SimpleTextBox, , SimpleTextBoxMIDlet

MIDlet-3: TextBoxWithCommandListener,
TextBoxWithCommandListenerMIDlet

MIDlet-4: CompleteTextBox, CompleteTextBoxMIDlet

MIDlet-5: SimpleList, SimpleListMIDlet

MIDlet-6: CompleteList, CompleteListMIDlet

MIDlet-Jar-Size: 5698

MIDlet-Jar-URL: GUI.jar

MIDlet-Name: GUI

MIDlet-Vendor: Sun Microsystems

MIDlet-Version: 1.0

MicroEdition-Configuration: CLDC-1.0

MicroEdition-Profile: MIDP-2.0

JAD file is a text file that lists important information about a set of MIDlets packaged together into a single JAR file (*a MIDlet suite*)

Sample MF file contents

MIDlet-1: GUI, GUI.png, FirstMIDlet

MIDlet-2: SimpleTextBox, SimpleTextBoxMIDlet

MIDlet-3: TextBoxWithCommandListener,

TextBoxWithCommandListenerMIDlet

MIDlet-4: CompleteTextBox, CompleteTextBoxMIDlet

MIDlet-5: SimpleList, SimpleListMIDlet

MIDlet-6: CompleteList, CompleteListMIDlet

MIDlet-Name: GUI

MIDlet-Vendor: Sun Microsystems

MIDlet-Version: 1.0

MicroEdition-Configuration: CLDC-1.0

MicroEdition-Profile: MIDP-2.0

The *manifest* is the standard JAR manifest packaged with the MIDlet suite.

15.4.4 First MIDLet

The code for our FirstMIDlet would look as below:

```java
import javax.microedition.MIDlet.*;
import javax.microedition.lcdui.*;
/**
 * @author  ryavagal
 * @version
 */
public class FirstMIDlet extends MIDlet
{
    private Display display;

    public void startApp() throws MIDletStateChangeException
{
            if( display == null ){
               init(); //  one-time initialization
          }
}
public void pauseApp() {
     }

public   void   destroyApp(boolean   unconditional)   throws
MIDletStateChangeException   {
        exit();
    }

    private void init(){
        display = Display.getDisplay( this );
         //  initialization stuff goes here

    }
```

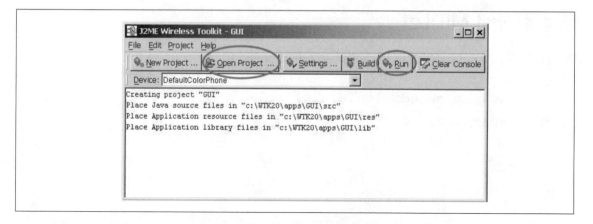

Figure 15.8 Running samples from WTK

```
public void exit(){
        // It is a good practice to release all
resources and cleanup
        notifyDestroyed();
    }
}
```

But before you can see this in action you will need to setup your development environment. The easiest way to set-up the development environment is to visit http://java.sun.com/j2me/index.jsp detailed download and installation instructions are given. Integration with the IDE of our choice will involve further investigation on our part. For our purpose we will use the WTK2.0 running on a windows platform. (Details are excluded here as it will only be duplication of content from the Sun site.)

Once we have successfully installed the WTK. We can start it from the start menu. This will look as shown in Figure 15.8 Click on "Open Project" to view any of the sample programs. To run the applications click "Run".

15.4.5 Creating a new application

To create our own application we need to do the following

1. Click on "New Project".

2. Enter Appropriate Names for the Application and the Application class and click Create Project. The resulting window is shown in Figure 15.9.

3. By default WTK will create the folder Structure under the Apps folder of the WTK root (Figure 15.10).

4. Use any editor, (Even Notepad will do) to enter the program above and save it under the source folder of the application we created in 2 above.

5. Go back to the WTK and click Build. If everything is OK we will get a Build complete message (Figure 15.11).

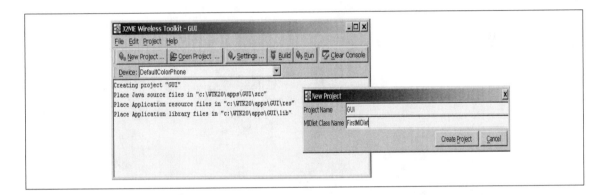

Figure 15.9 Creating a new application

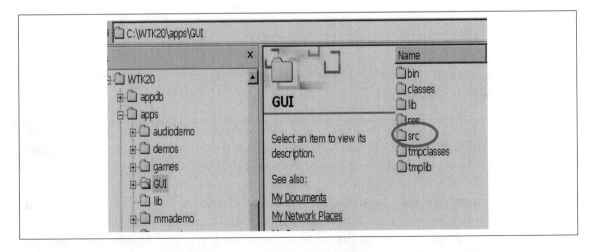

Figure 15.10 Application Project directory structure

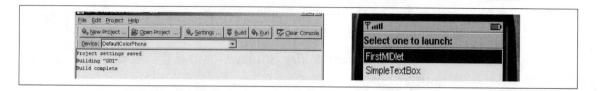

Figure 15.11 Building the First MIDlet

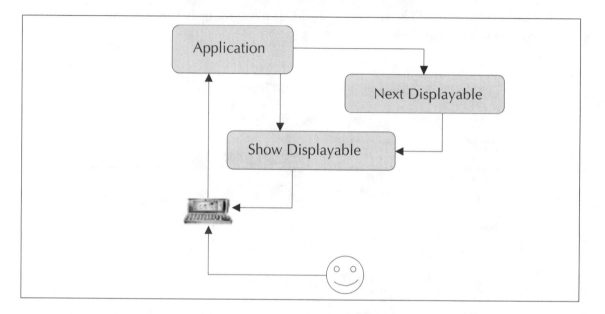

Figure 15.12 Application control flow

6. Clicking on run will launch the emulator which lists the Midlet currently registered with the WTK AMS. For steps 4, 5 and 6 refer to Figure 15.11.

But selecting the FirstMidlet and clicking Launch does nothing! That is because we have not done anything yet. If we take a closer look at the code above there is variable called Display. What is it? And what does it do? And the ;lcdui' package?

We guessed right. We will be using it to display elements on the screen. GUI in MIDP has two core concepts the *Display* and *Displayable*. In short the MIDP's display is represented by the *Display* class. All displayable elements are called *Displayable*. To show an element we use the setCurrent method of the *Display* class. The lifecycle of GUI interactions is shown in Figure 15.12.

15.4.6 MIDlet event handling

User interactions generate events. These could be:

- Screen inputs.
- Item state change
- Handset data update.

MIDP event handling mechanism is based on a listener model. It provides interfaces for each of the events mentioned above. These interfaces implement callback methods, which in turn invoke application-defined methods. These methods perform the desired functions in response to events. The three interfaces provided are: ItemStateListener, CommandListener, and RecordListener. Let's look at each in some detail.

Command Listener

The CommandListener as the name implies is responsible for notifying the MIDlet of any commands or events generated by the user. Objects extending it, implement the commandAction method. This method takes two parameters a Command object and a displayable (Command c, Displayable d). This method implements the functionality that needs to be executed in response to the command event on the associated Displayable.

- Displayable.setCommandListener(CommandListener L)sets the listener L to a Displayable.

Item State Listener

An ItemStateListener informs the MIDlet of changes in the state of an interactive item. It calls the itemStateChanged(Item I) method in response to an internal state change.

- Form.setItemStateListener(ItemStateListener L)sets the item state listener L for the given displayable.

Record Listener

RecordListener is related to database events which are discussed in the later section on RMS.

15.5 GUI IN MIDP

As shown in Figure 15.13 the *"Displayable"* has two main subclasses Screen and Canvas. The screen is a super class for a set of predefined UI elements. The predefined UIs are called High Level UI elements and the canvas elements are called Low Level Items. It is always preferable to use these as they involve less coding and are more portable. The Canvas allows the developer to have low level control on the screen. Games normally use the Canvas. The Figure 15.13 below shows the subclasses of the *"Displayable"*.

15.5.1 High Level UI

We will begin with the High Level UIs and then proceed to the Low Level UIs Let us begin by adding some of the displayables to our applications. Copy the FirstMidlet into a new program called SimpleTextBoxMIDlet.

To the init method, add the following code:

```
Display display = Display.getDisplay(this);
TextBox text = getTextBox();
display.setCurrent(text);
```

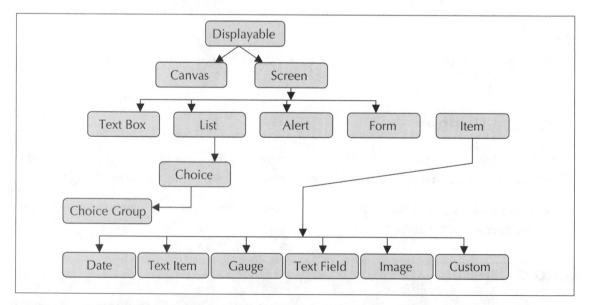

Figure 15.13 UI elements in MIDP

Now add this new Method

```
public static TextBox getTextBox()

{

    TextBox textBox = new TextBox("Hello Midlet", "", 50, 0);
    return textBox;

}
```

Save the file Build and Run. Wait a min why can't we see it in the list? We need to add the MIDlet to out suite first. Click on settings. We see the DialogBox showing the currently available MIDlets. Refer to Figure 15.14.

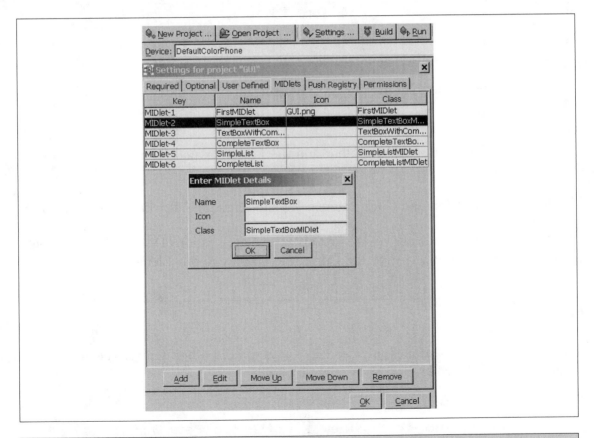

Figure 15.14 Adding a new MIDlet to the suite

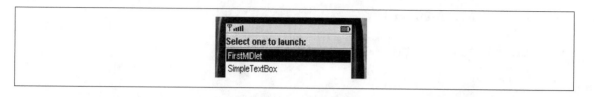

Figure 15.15

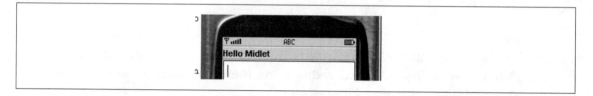

Figure 15.16

- Click on Add in this dialog to add a new MIDlet to our suite.
- Enter The Name to be Displayed and associate it with the Midlet class.
- Click OK.

Now say Run from the WTK toolbar and the newly created SimpleTextBox should appear (Figure 15.15).

Select SimpleTextBox and Click Launch. We should see the TextBox (Figure 15.16).

Lets the revisit the code above. We first obtain the applications Display, Creat the element we want to use and then set the Current to Display to our Element.

Forms

A form is similar to its html counterpart. It is used to hold other items and elements. In its simplest form the code for a form might look like

```
display = Display.getDisplay( this );
Form myform = new Form("Hello Form");
display.setCurrent(myform);
```

Figure 15.17 Form

If we put it in the init method of the MIDlet above, build and run we should get the window or form shown in Figure 15.17.

Items

To do some useful work we have put some items into the form we created above. MIDP provides a set of predefined items. Table 15.1 shows a listing of all Items available in MIDP.

Table 15.1 Subclasses of Item

Item	Description
ChoiceGroup	Allows user to select one or more elements from a group.
DateField	Counterpart of Java date field. Used for date and time values.
Guage	A bar graph representation used for integer values.
ImageItem	To display an Image.
StringItem	Equivalent of html's label widget and used for non-interactive text.
TextField	For text input.
CustomeItem	A user defined Item (MIDP2.0).

ChoiceGroups

The ChoiceGroup Item allows users to select one or more elements from a group. These groups are similar to the "radio button", "check boxes" and "drop down" elements in the html parlance. AchoiceGroup item contains a simple String and an optional image per

Table 15.2 Choice Type Constants

Constant	Value
EXCLUSIVE	Allows only one element to be selected at a time.
MULTIPLE	Allows the selection of multiple elements.
POPUP	Is a dropdown like construct that allows a single option selection

member in the group. ChoiceGroups are of three different types. ChoiceGroup and List have a lot of similarity in the options that they support. The three types are listed in Table 15.2.

The ChoiceGroup constructor takes a label and a type value. Optionally images and hover text can also be added. Members can be added after creation using the append() method, it takes a String for the label and an Image or null.

```
ChoiceGroup grp = new ChoiceGroup("Select One",Choice.POPUP);
        grp.append("Implicit", null);
        grp.append("Explicit", null);
        grp.append("Multiple", null);
        grp.setLayout(Item.BUTTON);
```

grp.setLayout (Item.BUTTON) allows us to specify the layout, in this case in a button format. To add this to the form use myform.append(grp);

The resulting output would be as shown in Figure 15.18. The other options of exclusive and multiple are common with another component, the List and will be discussed in the section that pertains to LIST.

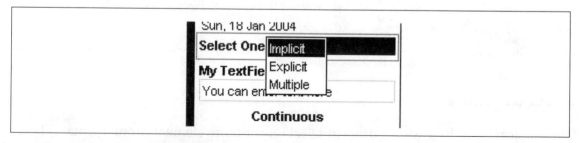

Figure 15.18 ChoiceGroup of type POPUP.

DateField

The DateField is an interactive item to enter/retrieve date and time information. DateField extends the Item class so can be placed on Form objects. As usual we will create the date component and append it to the form. Display option for a DateField is specified using input mode constants in the constructor. Possible DateField mode constants are listed in Table 15.3.

Table 15.3 DateField Constants

Constant	Value
DATE	For date only.
DATE TIME	For both date and time information.
TIME	For time only.

So here goes:

```
DateField dtf = new DateField("Today",DateField.DATE);
        Date date = new Date();
        dtf.setDate(date);
        dtf.setLayout(Item.HYPERLINK);
```

And say myform.append(dtf); we should get the output as shown in Figure 15.19 and Figure 15.20.

TimeZone and locale issues complicate date and time formatting. MIDP does not include the java DateFormat, only a subset of the Calendar, Date and TimeZone classes. The two-argument constructor above uses the default time zone of the device.

To explicitly specify a time zone MIDP provides an constructor with the following format.

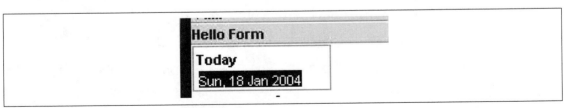

Figure 15.19 Date object

Figure 15.20 Date selection dialog

public DateField(String label, int mode, java.util.TimeZone zone); where, *zone* specifies the TimeZone. What we saw above was an un-initialized date:

To initialize the field to a particular date or time, we will have to use setDate and pass in a java.util.Date object initialized to the correct value:

```
Calendar cal = Calendar.getInstance();
    cal.set( Calendar.MONTH, Calendar.FEBRUARY );
    cal.set( Calendar.DAY_OF_MONTH, 06 );
    c.set( Calendar.YEAR, 1977 );
    cal.set( Calendar.HOUR_OF_DAY, 04 );
    cal.set( Calendar.MINUTE, 00 );
    cal.set( Calendar.SECOND, 00 );
    cal.set( Calendar.MILLISECOND, 0 );
    Date currenttime = cal.getTime();
    DateField df = new DateField( null, DateField.DATE_TIME
    );
    df.setTime(currenttime);
```

A Date object represents the number of milliseconds since midnight, 1 January 1970.

Note: it is always wiser to use Calendar class for date manipulation and not the raw milliseconds value stored in a Date object.

To obtain or update date/time, we use getDate method:

```
DateField df = ....;
Date editedDate = df.getDate();
```

Gauge

The Gauge item can be thought of as a progress bar. The constructor takes a label, a Boolean flag, where true indicating interactive and false indicating non-interactive, an upper limit and an initial or starting value. An interactive Guage as the name implies, allows the user to change the value using some device-dependent input method.

The following code snippet shows the construction of different types of Gauge. The current value of the Gauge can be set using the method setValue() and read using the method getValue(). Analogous setMaxValue() and getMaxValue() methods let we access the maximum value of the Gauge.

We will take a look at the timers in a later section the code is otherwise self-explanatory.

```
Gauge gug = new Gauge("Continuous Running Guage",false,
Gauge.INDEFINITE,Gauge.CONTINUOUS_RUNNING);
    gug.setLayout(Item.LAYOUT_CENTER);

    Gauge gug1 = new Gauge("I do nothing. Use a timer
to upgrade me",false,10,0);
    gug1.setLayout(Item.LAYOUT_EXPAND);

    //Task and Timer for a progress bar.
    MyTask task = new MyTask();
    timer.schedule(task,500,1000);

    Gauge gug2 = new Gauge("Interactive Guage Click to
inc or dec me",true,10,0);
    gug2.setLayout(Item.LAYOUT_EXPAND);
```

The output of the code is shown in Figures 15.21 and 15.22.

StringItem

`StringItems` are similar to labels in functionality. They have a name or label and the display string. These are extremely important for I18N or internationalization. The following code snippet shows the creation of a simple text label.

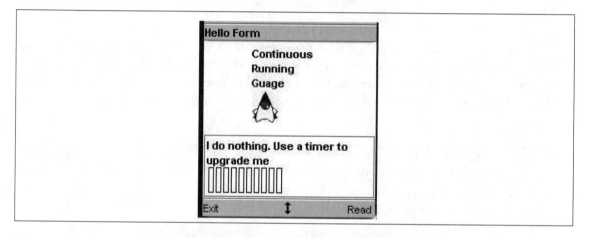

Figure 15.21 Guage Non interactive

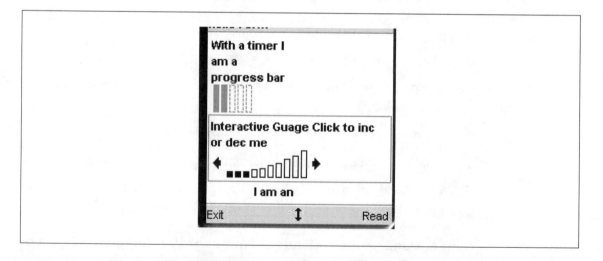

Figure 15.22 Interactive Guage

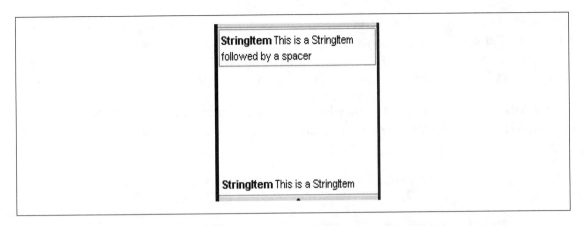

Figure 15.23 String Item

```
StringItem stringitm = new StringItem("StringItem","This is
a StringItem followed by a spacer");
StringItem stringitm1 = new StringItem("StringItem","This
is a StringItem after a spacer");
```

The label of the `StringItem` can be accessed using the `setLabel()` and `getLabel()` methods inherited from `Item`. To access the text, we can use the methods `setText()` and `getText()`.

```
myform.append(stringitm);
myform.append(stringitm1);
```

The resulting UI looks like Figure 15.23. Now where did that big space come from? This is a spacer Item introduced in MIDP2.0. It is a non-interactive item used to correctly position the elements on the screen. Normally we define the size of the empty space as `minWidth` and `minHeight`, both integers. Its usage is very simple.

```
Spacer spacer = new Spacer(100,100);
myform.append(spacer);
```

ImageItem

The Image Item is also a non-interactive Item, its constructor takes an Image object, a layout parameter, and an alternative text string to be displayed if the image cannot be displayed. The image source is the location of the image file. The layout parameter can take one of the following   constant values LAYOUT_CENTER,   LAYOUT_DEFAULT, LAYOUT_LEFT, LAYOUT_NEWLINE_AFTER, LAYOUT_NEWLINE_BEFORE, LAYOUT_RIGHT. The values are self explanatory.

The following code snippet shows how a center aligned ImageItem is added to the sample MIDlet: The Duke comes with the WTK, but we could use any image of our choice

```
Image img = Image.createImage("/Duke.png");
     ImageItem imgItem =
     new ImageItem("I am an ImageItem", img,
          Item.LAYOUT_CENTER, null,
          Item.BUTTON);
     imgItem.setDefaultCommand( new Command("Image Command",
Command.ITEM, 1));
          imgItem.addCommand(CMD_IMG);
          imgItem.setItemCommandListener(this);
```

The output is as shown in Figure 15.24 and Figure 15.25.

We have set the command listner. So what happens when we click the corresponding Image Command shown in Figure 15.25.

We should see the following lines printed on our console.

I am here now Image Command

I am out

We will see more about the command listener soon.

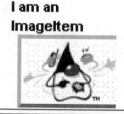

Figure 15.24 Image Item

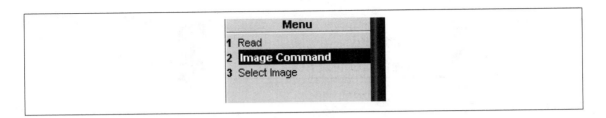

Figure 15.25 Image Command

Table 15.4 TextField Constraint Constant Values

Constant	Value
ANY	Any text can be entered
EMAILADDR	A valid e-mail id of type userl@domainname.com.
NUMERIC	Numeric values.
PASSWORD	A masked text is allowed. It displays all "*".
PHONENUMBER	A valid phone number
URL	A valid http URL

TextFields

The TextField class handles all text input. The constructor takes four parameters a label or the name of the Item, initial text (this can be an empty string if we don't want to display anything), the input length, and constants that indicate the type of input allowed.

```
TextField txtfld = new TextField("My TextField","You can enter
text here",50,TextField.ANY);
```

Valid constraint values are listed in Table 15.4.

Refer to the java docs for more options.

A textfield is added to a form using the append function. Example: myform.append(txtfld); Sample output is shown in Figure 15.26.

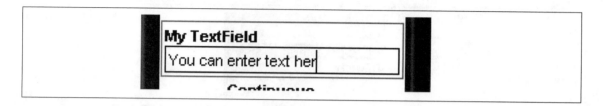

Figure 15.26 Text Field

To read the values from an updated textfield we first need to attach a command listener to it and then check in our command action class. A sample code block might look like

```
if(itm instanceof TextField)
{

        TextField txt = (TextField)itm;
        buf.append(delim);
        buf.append(txt.getString());
}
```

The output of a read command will be the newly entered string displayed on the console.

Custom Element

The new user interface API in MIDP 2.0 includes a class specifically intended for building custom UI elements. CustomItem extends the Item class, offers a simple template for developing custom items. The bonus is, its extensibility that allows for building on top of the existing Item base class features. We will now look at how to use this exciting new item.

Our custom Item is the Directory Item which most of are familiar with. In the most simplified form this will display a + symbol. When selected, it will expand to display a sub-level item. The process involves two steps. First we need to extend the item class and create our new Item. Then we will use this item in a form.

Figure 15.27 side shows the output of our custom item which we have titled the DirListing.

The code for the DirListing begins by extending the CustomItem class. A new feature introduced in MIDP 2.0 is the `ItemCommandListener` class. This allows a

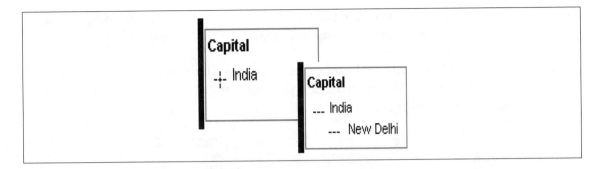

Figure 15.27 Custom item (Directory item)

command listener to be specified for an item rather than for the whole form. We will use it here to implement the expand function.

```
public class DirListing

extends CustomItem

implements ItemCommandListener {
```

Custome Item is an abstract class hence we have to implement all the methods therein. These are given below. The values returned are optimized based on a trial- error method for my display settings and may not hold good for ours.

```
protected int getMinContentHeight() {     }
protected int getMinContentWidth() {    }
protected int getPrefContentHeight(int width) {      }
protected int getPrefContentWidth(int height) {      }
protected void sizeChanged(int w, int h){       }
```

In a real world application these functions use the values based on the screen co-ordinates. We will see the details of that in the section on Low level APIs.

The sizeChanged() method take height and width parameters that are used by the application to redraw an item. Size changes occur due user actions that change the screen size such as minimize or reduce etc.

paint() as the name implies does the actual job of painting the item.

Curser movements cause a call to the traverse() method. It will return true if the traversal is internal this will then activate the Item, or false when another item in the form is selected. Other information provided by traverse()the direction of traversal, available area etc. We will now see the implementation details of the CheckItem

The DirListing item has two basic functions: it responds to the movement of the cursor and expands the selected value (clicking on the country gives the capital). The listing below gives us the constructor of the item.

```
private    final    static    Command    CMD_EXPAND    =    new
Command("Expand", Command.ITEM, 1);

public DirListing(String root, String Country, String
Capital) {
     super(root);
     this.Country = Country;
     this.Capital= Capital;
     setDefaultCommand(CMD_EXPAND);
     setItemCommandListener(this);
}
```

The constructor takes the country and capital and sets the Expand command. The next piece is the paint() method which is responsible for drawing the item on the screen.

```
protected void paint(Graphics g, int w, int h) {
     g.setStrokeStyle(Graphics.DOTTED);
     if (UnExpanded==1)    // The item is unexpanded so
draw the + and the associated text is the country.
        {
            g.drawLine(10,10,10,20);
            g.drawLine(5,15,15,15);
        g.drawString(Country, 22, 4, Graphics.TOP|Graphics.LEFT);
        }
     else {       // The state is expanded so the symbol
is - and both the country and capital need to be shown
            g.drawLine(5,15,15,15);
```

```
g.drawString(caption1, 22, 4, Graphics.TOP|Graphics.
LEFT);
g.drawLine(20,30,30,30);
g.drawString(Capital, 40, 20, Graphics.TOP|Graphics.
LEFT);
        }
    }
```

Note for the sake of simplicity we have used absolute values. But in a live application it is advisable to use relative co-ordinates. Finally the Command handler and the traverse() method.

protected boolean traverse(int dir, int viewportWidth, int viewportHeight,   int[] visRect_inout)

```
    {
        switch (dir) {
            case Canvas.DOWN:
                repaint();
                break;
            case Canvas.UP:
                repaint();
                break;
        }
        return false;
    }
```

The traverse is quite simple we just repaint the item. In the command we first toggle the state between UnExpanded=1 and 0 and call repaint().

```
        public void commandAction(Command c, Item i) {
            if (c == CMD_EDIT) {
                if(UnExpanded ==1 ) UnExpanded =0;//location;
                else UnExpanded =1;
                repaint();
```

```
                notifyStateChanged();
        }
    }
```

Now our item is ready to be used. We have a new class, the DirListingDemo. The following code gives a sample form that includes the DirListing item.

```
Form mainForm = new Form("DirListingDemo");
mainForm.append(new DirListing("Capital", "Pakistan",
"Islamabad"));
mainForm.append(new DirListing("Capital", "India", "New
Delhi"));
```

List

We will continue with our discussion of the GUI components. We take up a list. A list, as the name suggests, is a group of options that the user can select. Lists are basically of three types.

- IMPLICIT: This type is associated with the select command. When selected the event can be trapped in the command listener.

- EXCLUSIVE: This allows the user to select any one option. Similar to the choice group item.

- MULTIPLE: As the name suggests it allows the user to select multiple values.

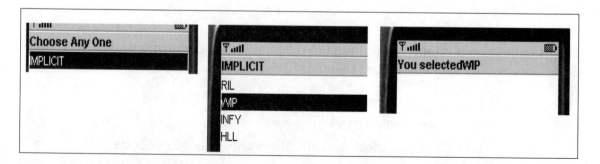

Figure 15.28 Choice group (Implicit)

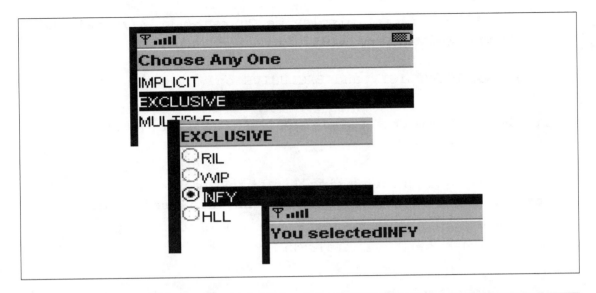

Figure 15.29 Choice group (Exclusive)

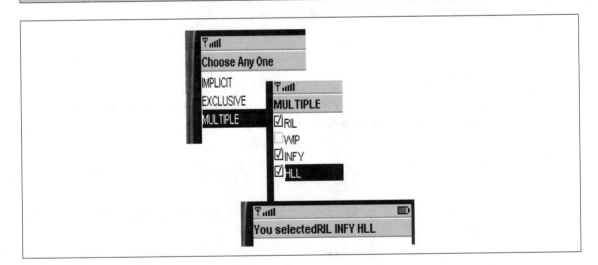

Figure 15.30 Choice group (Multiple)

The constructor would be given by List list = new

List("IMPLICIT",List.IMPLICIT); where the arguments to the constructor are the label and the type. To read the selected value from a list we can use the following piece of code.

```
List list = ((List)display.getCurrent()); //Obtain the
currently displayed list.

// For both IMPLICIT and EXCLUSIVE only one option can be
selected.
    if(list.getTitle().equals("IMPLICIT") ||list.getTitle().
    equals("EXCLUSIVE"))
    {
    String msg = list.getString(list.getSelectedIndex()); //
    Get the label from the index of the value selected.
        Alert alrt = new Alert("You selected" + msg);
        display.setCurrent(alrt);
    }

//For MULTIPLE more than one options can be selected the
use age
//therefore is slightly different.
else
{
        boolean[] sel = new boolean[list.size()];
        list.getSelectedFlags(sel); //Read all the selected
        values into a Boolean array
        StringBuffer buf = new StringBuffer();
        for(int i=0;i<list.size();i++)
        {
            if(sel[i]) //if the option has been selected
            store it some where
            {
              buf.append(list.getString(i));
              if(i!= list.size())
              buf.append(" ");
            }
        }
        String msg = buf.toString();
        Alert alrt = new Alert("You selected" + msg);
```

```
//Display the values selected.
    display.setCurrent(alrt);
  }
```

The output of the sample code above is shown in Figures 15.28, 15.29 and 15.30. What we saw so far were high-level APIs. What that means is, the users only had to instantiate and use these objects. The actual implementation and on screen rendering was the responsibility of the OEM or the MIDP implementer. Though easy to use and highly portable they are very restrictive for applications like games that need more control on the display. For these we have a set of low-level APIs derived from the canvas class.

15.5.2 Low-level GUI Components

Low-level APIs, as the name suggests, allow the application to control displayable on the screen and allow the program to directly draw on the screen using the screen co-ordinates. This is specifically needed for applications like games, business tools that need to show graphs, bars, pie charts etc. The package to be used is the javax.microedition.lcdui. To implement such control we need another class, mentioned above the Canvas class. The canvas is a blank screen. We can draw lines, text and shapes on this screen. To show the low-level capability of J2ME we will take two examples here.

Canvas

The first is our good old custom Item DirListing. Yes we know we discussed it in the section on High Level UI APIs. The noticeable difference between the CustomeItem and other items in that section is the fact the custom item is defined by the item developer and the item code is responsible for handling its display response to cursor movements and other user interactions; while in the case of other items, all the responsibility of display and user interaction was the responsibility of the system. In essence, a CustomItem uses some of the low-level APIs but this implementation is abstracted from the user of the item. The low level implementation of DirListing will use Canvas and not the Item class.

The second is a simple application that will move the Duke across the screen. Obviously, after implementing these examples we cannot stake claim to fame as game developers. We, however, will have made a beginning. The WTK comes with some excellent examples for game implementations. We urge the readers to work through them.

The first sample is very simple. The DirListing class subclasses the Canvas class, which is an abstract class that extends Displayable, and overrides the paint( ) method. The traverse methods are used to detect cursor movements. The paint( ) method uses the drawing methods of the javax.microedition.lcdui.Graphics class.

For the purpose of this example we will enhance the paint() method to show Unexpanded values in red color and the Expanded versions in blue. The updated code is listed below.

```
public void paint(Graphics g) {

    g.setColor(255, 255, 255);

    g.fillRect(0, 0, getWidth(), getHeight(  )); // Gives the
    coordinates of the screen and paints it white.

    g.setColor(255, 0, 255);     if (checked==1)

    {
        g.drawLine(10,10,10,20);

        g.drawLine(5,15,15,15);

        g.drawString(Country, 22, 4, Graphics.TOP|Graphics.LEFT);

    }
    else {  // The state is expanded so the symbol is - and both
    the country and capital need to be  shown
        g.setColor(0, 0, 255);

        g.drawLine(5,15,15,15);

        g.drawString(Country, 22, 4, Graphics.TOP|Graphics.LEFT);

        g.drawLine(20,30,30,30);

        g.drawString(Capital, 40, 20, Graphics.TOP|Graphics.LEFT);

    }
```

As seen above, setColor takes parameters (red,green,blue) in that order. An important difference here is the event handler. The canvas class provides for different events like pointer movements, keyPressed events, notify events etc. Here we show a simple implementation of the keyPressed event.

```
protected void keyPressed(int keyCode)
{
    if (this.getKeyName(keyCode).equals("DOWN")) // Each key is
    associated with a unique id and a corresponding name.
        {
            checked = 0;
            System.out.println(" down");
            repaint();
        }
    if (this.getKeyName(keyCode).equals("UP"))
        {
            checked = 1;
            System.out.println("up");
            repaint();
        }
}
```

The code to invoke the canvas implementation is as follows.

```
public void startApp(   )
{
    Canvas canvas = new MyCanvas(   );
    Display display = Display.getDisplay(this);
    display.setCurrent(canvas);
}
```

An interesting technique that is often used in gaming context is Double Buffering. It is usually used where smooth animation effects are desirable. The idea is to create an off-screen buffer maintained in memory. This buffer is used in lieu of the screen. Once the buffer is ready, it is copied to the display. As is evident this process is faster and hence provides smoother effects.

To implement double buffering, first create a Image buffer equal to the size of the screen:

```
int width = getWidth(   );
int height = getHeight(   );
```

```
Image buffer = Image.createImage(width, height);
Next, attach a graphics context to the buffer:
Graphics gc = buffer.getGraphics(   );
Now, we can draw to the buffer:
// normally an animation
// ..
```

Finally copy it to the screen by overriding the paint( ) method

```
   public void paint(Graphics g)
   {
      g.drawImage(buffer, 0, 0, 0);
   }
```

Note some MIDP implementations implicitly support double-buffering. To check we can use isDoubleBuffered( ) method.

Game API

An interesting enhancement to the low-level API in the MIDP 2.0 is the Game APIs. We will use these for our second example while briefly describing some of these. The intention is to introduce the process of game development. All game APIs are in the package javax.microedition.lcdui.game

The first class in this package is the GameCanvas a subclass of javax.microedition. lcdui. Canvas. As implied it inherits from canvas and basically provides the screen for a game. The GameCanvas provides an additional facility to query the current state of the game keys and synchronous refresh. There is a provision for a unique buffer. (The concept is analogous to an implicit double buffering.) A game class using the GameAPI would typically be defined as

```
   class GameTest extends GameCanvas{ ….. }
```

A typical game logic works in a loop till EOG. Within the loop the application checks for user response, update the game appropriately and display the new game configuration. In adition to the methods inherited from the cavas, GameCanvas provides methods to obtain the buffer and flush the buffer. It also provides for gamekeys up down right and left. The implementation however is device/OEM dependent. The GameAPI provides constructs that help in drawing and rendering images. The First of such a construct is a layer.

Layer

A Layer can be considered as an application element's visual part, a player or a field for example. It provides basic attributes like location, size, and visibility. As the above statement implies a game application can use several layers. A LayerManager is a construct that can automate the rendering process for multiple layers. The layer manager uses the concept of a user view which represents the current user window.

```
LayerManager layer;

Create a new layer:  Layer = new LayerManager();

Insert a new layer obj into the layer: layers.insert(layer);

Remove an existing layer from the layer.layers.remove(layer,
index);

Transfer the layer image to the game graphics: layers.paint
(g, 0, 0);

To render the layer on to the screen: flushGraphics();
```

Sprite

Another important construct is the Sprite. This is mainly used for animation purpose. The Sprite provides various operations including flip, rotation, collision detection and more. In short it simplifies the implementation of a game's logic.

We will now see a simple example of an implementation of a Sprite to move and rotate the duke. The output is shown in Figure 15.31. More complex examples are available from Sun within the samples with WTK.

```
Image img = Image.createImage("/Duke.png");
    sprite = new Sprite(img);
    g = getGraphics();
    sprite.paint(g);
flushGraphics();
```

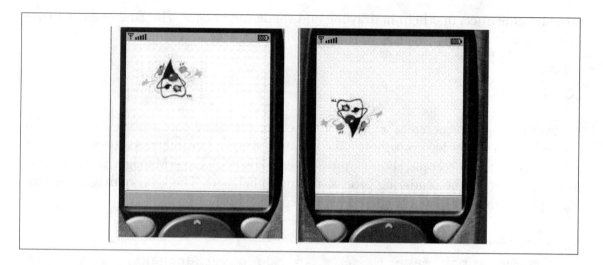

Figure 15.31 Low level UI

```
sprite.move(Xcord, Ycord);
    sprite.setTransform(Sprite.TRANS_ROT180);
    sprite.paint(g);
flushGraphics();
```

Tiled Layer

The last of the constructs is a Tiled layer. A tiled layer can be pictured as a brick wall in which each brick represents a tile. Multiple tiles together make up a single image object. Tiles can also be filled with animated objects whose corresponding pixel data can be changed very rapidly. Imagine a puppet dancing, where multiple tiles make up the puppet, we can change only the corresponding tile to say shake its hand. This allows the application to render large images without actually using the resources required for one.

```
TiledLayer tiles = new TiledLayer(xcolumns, yrows, Image,
tileW, tileH);
```

The call above creates a tiledlayer that is a (xcolumns * tileW) by (yrows * tileH) grid. This grid consists of cells or tiles that are tileW by tileH each. Note that the image dimensions must be within the tile dimensions or its multiples theirin. Initially, the layer, is empty. To modify it we use

```
setCell(int, int, int) or fillCells(int, int, int, int, int).
```

We have used tiles.fillCells(0, 0, Image.getWidtht, Image.getHeight, 1); To provide a background filler. The background tileset can now be initialized using updateTile(x, y); Wher x and y are th cell co-ordinates. The final task is to load this on to the layer we created earlier.

```
layers.append(tiles);
```

Both sprite and Tiledlayer extend layer and hence can be added to and removed from a layer. With this we come to the end of our discussion on GUI.

At this point it is appropriate to include a discussion on certain important considerations in mind while designing application UI:

15.6 UI Design issues

- Entering text through the T keypad is not very attractive nor is filling out long forms, UI should be small simple and easy to use.

- High-level API should be used where ever possible as these are portable across different handsets.

- While using low-level API, it is advisable to restrict to elements defined in the Canvas class.

- Device capabilities like screen width, height, resolution vary from device to device and hence applications should adapt to the LOW level objects accordingly.

15.7 Multimedia

Another new entry in MIDP 2.0 is the Media API, a direct subset of the Mobile Media API (JSR-135) specification. This library provides support for audio capability. Note: There is no support for video or graphics formats. As the Figure 15.32a alongside shows, a j2ME audio application consists of three main parts.

A Manager that is responsible for creating and controlling the audio resources.

A player, the workhorse that does the actual job of playing music.

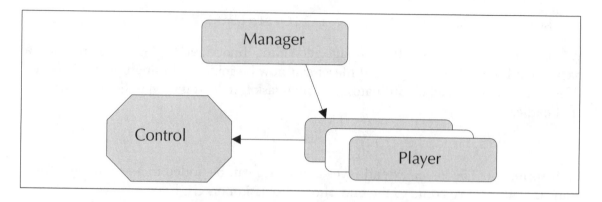

A control to regulate the features of the player. The manager also provides support information on property, content and protocol support. To see how the manager can be used to play a simple tone, just include the following code in any MIDlet.

```
Manager.playTone(ToneControl.C4, 100, 100)
```

ToneControl is an interface that applications implement to play a user defined tone sequence, ranging from C1 to G9. We recommend that the user refer to the javadocs for more information. The ToneControl interface defines two constants Middle C (C4) and SILENCE. The second parameter is the duration for which the tone is to be played and the last is a volume control.

Of course the major purpose of a manager is to create a player. This can be done through a call to createPlayer. But before we can make this player do something we need to identify what we want it to play. Sound inputs can be in three forms.

- Make a request to a .wav file over the network,
- play a file stored in the .jar
- create a tune or a sequence of tunes.

A tune is essentially a byte array that contains information about the tune. A simple tune can be created using the following

```
Private byte[] MyTune =
{
ToneControl.VERSION, 1,
```

```
    C4,D4,E4,G4
};
```

```
    Where
    byte C4 = ToneControl.C4;;
    byte D4 = (byte)(C4 + 2); // a whole step
    byte E4 = (byte)(C4 + 4); // a major third
    byte G4 = (byte)(C4 + 7); // a fifth
```

Creating a tune sequence is left as an exercise for our audience. To play this tune we first create a player

```
Player player = Manager.createPlayer(Manager.TONE_DEVICE_
LOCATOR);
```

Alternate calls can be of the form

```
Player player = Manager.createPlayer("http://webserver/file.
wav");
```

A `Player` plays the tune/tune sequence. *A Player's lifecycle has five states.* As shown in Figure 15.32b, a player begins its life in the *UNREALIZED* state. Here it is still unformed and requires additional information to acquire resources. At this point a call to realize() transitions it to the next state which is *REALIZED*. This method essentially allows the player to collect information by which it can obtain resources. For example which media device audio or video or both would be required to play the content

In the *REALIZED* state the player is ready to acquire the required resources. From here a call to *prefetch()* trasitions to the *PREFETCHED* State.

In the *PREFETCHED* state all initializations, for example, acquire audio device, happen.

A call to the *start()* method will now start the player. The *start()* method also causes a *STARTED* event to be fired. It is now in the *STARTED* state.

From here once a player reaches the end of the file it will automatically stop. Alternatively if the *stop()* method is called it will stop and transition to the *STOPPED* state. A corresponding *STOPPED* or *END_OF_MEDIA* event will be triggered.

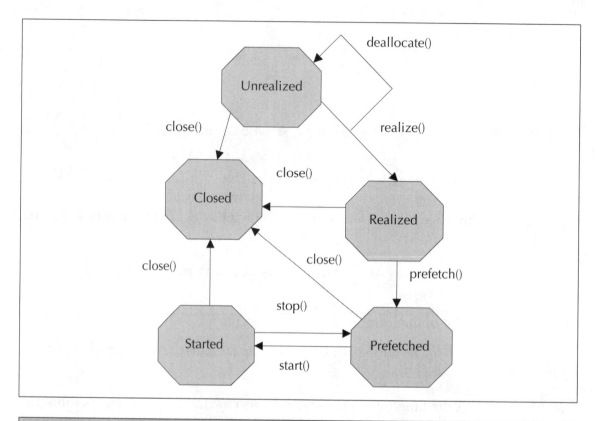

Figure 15.32b States in a player's life cycle

When a Player stops, it returns to the *PREFETCHED* state. The following code snippet shows the transition through all five states.

```
player.realize();
ToneControl  toneControl =
(ToneControl)tonePlayer.getControl("javax.microedition.media.
control.ToneControl");
toneControl.setSequence(MyTune);
player.start();
```

The last state in the lifecycle is the CLOSED state. Here it releases all the resources and must not be used again.

With this we conclude our discussion of media API. Our next discussion concerns the need for persistent storage.

15.8 RECORD MANAGEMENT SYSTEM

Most application data needs to be stored for future use. For example a PIM application needs to store the appointments information on the device. All devices provide some kind of persistent storage. However heavy duty APIs to manage persistent data provided in Java 2 Standard Edition (J2SE), such as JDBC, are not suitable for handheld devices. MIDP provides a light weight persistent storage in the form of RMS (Record Management System). Record stores are essentially flat files in the binary format. Similar to the Palm Os RMS is also a record store consisting of a collection of records. In the WTK you will find these under the folder appdb. Record stores are platform-dependent and its definition lies in the domain of the OEM. Record stores are identified by names that are

- Unique to the MIDlet Suite(no two record stores in the same application can have the same name)
- All characters
- Case sensitive
- Cannot be more than 32 characters long
- Stored as *"DBNAME.db"*.

Note: records associated with the application are deleted when the application is uninstalled or deleted from the device.

Each record store has a version number as well as a date/time stamp. Both values are updated whenever a operation is performed on the Store.

Till MIDP1.0 there was no mechanism to share the record stores across MIDlet Suites. MIDP2.0 provides new API to share record stores between MIDlet Suite. This accomplished by

"RecordStores are uniquely named using the unique name of the MIDlet suite plus the name of the Record Store. MIDlet suites are identified by the MIDlet-Vendor and MIDlet-Name attributes from the application descriptor. Sharing is accomplished through the ability to name a RecordStore created by another MIDlet suite. Access controls are defined when Record Stores to be shared are created. Access controls are enforced when RecordStores are opened. The access modes allow private use or shareable with any other MIDlet suite"

Java Docs API documentation WTK2.0

- Explicit permission granted by MIDlet that owns (creates) the store.

Within the Record Store, individual records can be considered to be organized as rows, with each row consisting of two columns: the first, a unique integer row identifier and the second, a series of bytes representing the data in the record. The row identifier is referred to as the *Record ID* and is the primary key for the record. RecordIDs are sequential values i.e., the first entry will have the ID of 1, the next of 2 and so on.

The RMS APIs provide the following functionality:

- Create and delete record stores
- Mechanism for sharing data across MIDlet suites.
- Allow application to add update and delete records within a record store.

The RMS are Defined in the 'javax.microedition.rms' package. This package contains:

Interfaces:

- *RecordComparator* : To create a comparator object used for searching and sorting.
- *RecordEnumeration* : An enumeration Object to traverse through the records.
- *RecordFilter*: To create a Filter object to retrieve data based on filter criteria.
- *RecordListener*: To trap modify operations on the record store.

Classes:

RecordStore: Functionality is to

- Create /delete a record store.
- Add/update/delete records in the store.

Exceptions:

- *InvalidRecordIDException*: Thrown when the application tries to access a recordID that does not exist.
- *RecordStoreException* : A general exception.

- *RecordStoreFullException* : Thrown when there is no more storage space for RMS.

- *RecordStoreNotFoundException*: Thrown when the record store with the specified name does not exist.

- *RecordStoreNotOpenException*: Thrown when a modify operation is tried on a store that is closed.

We will see these in detail now. Any class wishing to implement DBfunctions has to implement the RecordListner interface. As a good coding practice this class should be separate from the main MIDlet class. In our case the RMS class implements the DB functions needed by the TestStore class.

```
public class RMS implements RecordListener {
  public RMS() throws RecordStoreException
  {
    MakeNewRecordStore(DefaultStoreName);
  }
public RMS(String RecStoreName)  throws RecordStoreException
  {
    MakeNewRecordStore(RecStoreName);
  }
```

15.8.1 Create a new Record Store

Before any operation on a record store is possible, we need to create and open it. This is the functionality implemented in the MakeNewRecordStore. The sample code is

```
recordStore = RecordStore.openRecordStore(fileName, true);
    recordStore.addRecordListener(this);
```

MIDP defines a dual purpose API

```
'recordStore = RecordStore.openRecordStore(fileName, create
IfNecessary);'
```

The first argument to this call is the name of the store to be created or opened. The second is a *boolean* parameter which, if set to true, will cause the record store to be created if it does not exist. The call returns a reference to the RecordStore. If an application tries to open a store that is already open then a reference to the store is returned.

MIDP2.0 has introduced two new API for sharing record stores across the MIDlet Suites.

15.8.2 To create a sharable Store

'openRecordStore(fileName, createIfNecessary, authmode, writable)'

The first two are the same as above while *authmode* defines the *authorization mode or permissions* on the store. Currently two modes are defined

- AUTHMODE_PRIVATE: Only the creator can access it. Behaves the same as API above
- AUTHMODE_ANY: Any MIDlet can access the RecordStore. This should be used carefully as it could have privacy and security implications.

The last is a Boolean parameter writable it defines the access permissions on the store as read-only if false else read/write.

Note: Both the argument above are ignored if the RecordStore exists.

The second API is

```
'openRecordStore(fileName, vendorName, suiteName)'
```

This call is used to open a shared store belonging to another MIDlet/MIDlet suite, where

```
vendorName is the vendor of the owning MIDlet suite and
suiteName, the name of the MIDlet suite that created and owns
the store.
```

In response to this call the following may happen.

- Access is granted if the authorization mode of the RecordStore allows access i.e., authorization mode set to AUTHMODE_ANY.
- If the record store is already open, a reference to the same RecordStore is returned.

- If the caller is also the owner of the record store, the call is same as openRecordStore(recordStoreName, false).

Before the application quits, it should close the RecordStore to avoid data corruption. This is accomplished by a call to

```
recordStore.closeRecordStore();
```

15.8.3 To delete a RecordStore use

We use recordStore.deleteRecordStore(*recordStoreName*);

The code snippet implementing the method above is given below. Note the function throws RecordStoreNotOpenException exception.

```
public void deleteRecordStore()
  {
    try
    {
      if (recordStore.getNumRecords() == 0)
      {
        String fileName = recordStore.getName();
        recordStore.closeRecordStore();
        recordStore.deleteRecordStore(fileName);
      }
      else
      {
        recordStore.closeRecordStore();
      }
    }
  catch(RecordStoreNotOpenException rsNopEx){ rsNopEx.printStack
  Trace();}
  catch(RecordStoreException rsEx){ rsEx.printStackTrace();}
```

Now we will see how add update and delete records to the RecordStore.

15.8.4 To add records

```
recordStore.addRecord(b, 0, b.length); // where b is the record
data in a byte array.
```

```java
public synchronized void addNewRecord(String record)
    {
        byte[] b = record.getBytes();
        try
        {
            recordStore.addRecord(b, 0, b.length);
        }
        catch (RecordStoreException rse)
        {
            System.out.println(rse);
            rse.printStackTrace();
        }
    }
```

The corresponding call to this function would look like

```java
System.out.println("Adding records to the store");
testRMS.addNewRecord("Pakistan";);
testRMS.addNewRecord("India";);
testRMS.addNewRecord("China";);
```

Since our RMS implementation has registered a record listener that prints the event triggered to the console we should see the following

MSG From RecordListener: Record added Store Name " Countries" ID 1 Value Pakistan

MSG From RecordListener: Record added Store Name " Countries" ID 2 Value India

MSG From RecordListener: Record added Store Name " Countries" ID 3 Value China

15.8.5 To update record

```
"recordStore.setRecord(id,byteInp,0,byteInp.length);"
```

Where

- id is the RecordID of the record to be updated.
- byteInp is the byte array for the input
- *startat* is the offset in the byte array from where the data starts
- *byteInp.length* is the length of the input.

```
public synchronized void UpdateRecord(int id, String record)
    {
        byte[] byteInp = record.getBytes();
        try
        {
            recordStore.setRecord(id,byteInp,0,byteInp.length);
        } catch (RecordStoreException e)
        {
            e.printStackTrace();
        }
    }
```

Note: that there is no way to modify part of the data in a record. It can only be replaced with a new byte array.

A call to the update method above is given below.

```
System.out.println("Now let us update a record");
testRMS.UpdateRecord(1,"Bangaladesh");
```

Our RecordListner now tells us
MSG From RecordListener: Record changed Store Name " Countries" ID 1 Value Bangaladesh

To read records in the RecordStore

```
resp = recordStore.getRecord(Id)
```

where *Id* is the unique RecordID of the record and response is the data in the record returned as a byte array. We have written a small utility called dumpRecordStore that reads all the records in the store and prints the contents to the console. The method is listed below.

```
public void dumpRecordStore(RMS rms )
    {
    System.out.println("******** This is a record store Dump
    *******");
    System.out.println( rms.getRecord(1));
    System.out.println( rms.getRecord(2));
    System.out.println( rms.getRecord(3));
    }
```

The output to this would be

******** This is a record store Dump *******

Bangaladesh

India

China

As is very apparent, this is bad programming. A read utility should be able to traverse the entire length of the store and print all the contents. This is where enumeration comes into picture.

15.8.6 To delete a record

```
"recordStore.deleteRecord(id);"
```

where *Id* is the unique RecordID of the record and response is the data in the record returned as a byte array. It is interesting to note that there is no method to obtain the RecordID of a record. Developers will have to device their own work around.

```
public boolean deleteRecord(int id)
    {
    try
```

```
    {
      recordStore.deleteRecord(id);
    }
  catch (RecordStoreException e)
    {
      e.printStackTrace();
      return false;
    }
      return true;
  }
```

To call this method we use the following listing

```
  System.out.println("Trying to delete a record");
  if(testRMS.deleteRecord(1))
    {
      System.out.println("Successfully deleted record 1");
      dumpRecordStore(testRMS);
    }
      else
        System.out.println("Couldnot delete record 1");
```

Output is

Trying to delete a record

MSG From RecordListener: Record deleted Store Name " Countries" ID 1 Value □□□□
□□□□□□□□□□□

Successfully deleted record 1

Note: some points about the delete options that need to be remembered are:

- RecordID of the deleted record is not reused (So the last RecordID of the store does not indicate the number of records in the store).

- Even if the record is deleted, the space is not reused. i.e., the space is not freed-up for reuse by another record. This is possible only by deleting RecordStore. So if our application is deleting very frequently we might want to perform a clean-up during init/destroy by writing a new and current Store. Of course our RecordID will also be updated.

To work around the missing getRecordID we need to use the other constructs RecordEnumerator, RecordComparator and RecordFilter.

We can test this by calling a delete record for an ID that has already been deleted

```
System.out.println('Now lets try to delete the same record
again');

    if(testRMS.deleteRecord(1))
    {
        System.out.println("Successfully deleted record 1");
        dumpRecordStore(testRMS);
    }
    else
        System.out.println("Couldnot delete record 1");
```

The output this time will show.

******** This is a record store Dump *******

Now let's try to delete the same record again

javax.microedition.rms.InvalidRecordIDException

Could not delete record 1

To see if the ID of the deleted record is reused, we can add a new record and loop through to see the ID associated with it. To get a list of all the records in a record store, we use an enumeration.

15.8.7 The Enumerator

The enumerateRecords method returns a RecordEnumeration. This allows a bidirectional movement through the set of records.

```
"recordStore.enumerateRecords(null, null, false)"
```

The first two are used to filter and sort the records in an enumeration (we will discuss filtering and sorting shortly). The last argument if set to true, causes the enumeration to update itself as the record store is modified. It is, however, important to remember that it's an expensive operation and should be used judiciously.

To traverse the resulting enumerator we can use the following:

```
while( enum.hasNextElement())
  {
     int id = enum.nextRecordId();
     // Do something here whit the ID
     System.out.println(testRMS.getRecord(id));
  }
```

The output of this loop is

Now we will loop through the Store and print them out

China

India

Another method defined by `RecordEnumeration` is `getNextRecord()` to obtain the record in bytearray format. To traverse backwards we use `getPrevious RecordId()` and `getPreviousRecord()`. An application can use hasNextElement() or hasPreviousElement() to break the loop. An enumeration can be reset to the beginning by calling `reset`.

Note: It is a good idea to `destroy` the enumeration after use. This frees up any resources used by the enumeration.

Passing null as first two arguments causes all the records in the record store to be returned in an undefined order. But it is not prudent to extract all the records for every operation. RMS provides two additional methods to get a subset of the records or to return the records in a specific order. They are filter and comparator. Developers can use one or both depending on their requirement.

15.8.8 Filter

We will first see a filter. For example to create a filter for string that have length > 0 and begin with the alphabet 'I', the filter defined will resemble

```
public class MyFilter implements RecordFilter
  {
    public boolean matches( byte[] recordData )
    {
    return( recordData.length > 0 && recordData[0] == 'I' );
    }
  }
```

Code listing below shows the use of this filter

```
System.out.println('Let's try a filter Filtering for records
begining with I');
  try
    {
        enum = testRMS.enumerate(new MyFilter());
        while( enum.hasNextElement())
        {
          int id = enum.nextRecordId();
          System.out.println(testRMS.getRecord(id));
        }
    }
  catch( RecordStoreException e )
    {
      e.printStackTrace();
    }
  finally
    {
      enum.destroy();
    }
```

The corresponding output is

Let's try a filter Filtering for records begining with 'I'

India

What happens when we have multiple entries that match the filter? Let's try and see. So here's the repeat code.

```
System.out.println("Will it work for multiple record begin-
ing with 'I'");
System.out.println("Add IceLand");
testRMS.addNewRecord("IceLand");
System.out.println("Filtering again for records begining
with 'I'");
try
   {
     enum = testRMS.enumerate(new MyFilter());
     while(enum.hasNextElement())
      {
       int id = enum.nextRecordId();
       System.out.println(testRMS.getRecord(id));
      }
   }
catch( RecordStoreException e )
   {
     e.printStackTrace();
   }
   finally
   {
     enum.destroy();
   }
```

The output now is as below.

Will it work for multiple record begining with 'I'

Add IceLand

MSG From RecordListener: Record added Store Name 'Countries' ID 4 Value IceLand.

Filtering again for records begining with 'I'

India

IceLand

As we can see the filter class has to implement the RecordFilter interface. This interface has one method *matches* that performs the matching operation on the input bytearray. The filter's matches method is called for each record in the record store. If the matches method returns true, the record is included in the enumeration. Use the filter like this:

recordStore.enumerateRecords(fltr, null, false);
Where fltr is an instance of Myfilter defined above.
Note: that the filter only gets the contents of a record.
To compare my Record with other records in the RecordStore or to search for a particular record the application must implement the Comparator.

15.8.9 Comparator

The Comparator interface is used to compare two records. The return value indicates the ordering of the two records. RecordComparator defines the following constants

- FOLLOWS: Value = 1 parameter on LHS follows the right parameter in terms of search or sort order.

- PRECEDES: Value = -1 parameter on LHS precedes the right parameter in terms on search or sort order.

- EQUIVALENT: Value = 0 Both parameters are the same.

The RecordComparator interface again defines a single method: compare We will now define a comparator that will, when used with an enumerator, return a list of strings in lexical sorted order.

```
public class MyComparator implements RecordComparator
    {
    public int compare( byte[] rec1, byte[] rec2 )
      {
        String s1 = new String(rec1);
        String s2 = new String(rec2);
        int cmp = s1.compareTo(s2);
        System.out.println( s1 + " compares to " + s2 +"
        gives " + cmp );
        if( cmp != 0 ) return ( cmp < 0 ? PRECEDES : FOLLOWS );
```

```
      cmp = s2.compareTo(s2);
      if( cmp != 0 ) return ( cmp < 0 ? PRECEDES : FOLLOWS );
      return EQUIVALENT;
   }
```

The call is made as below:

    "recordStore.enumerateRecords(null, cmptr, false);"

where cmptr is an instance of MyComparator. Again, only the contents of the records to compare are passed to the comparator.

The following snippet shows the use of the comparator above.

```
System.out.println('How does Does a comparator work?');
   testRMS.addNewRecord("Bangaladesh");
   try
   {
      enum = testRMS.enumerate(new MyComparator());
      System.out.println('Now what does the resulting enumara-
      tor look like?')
      while( enum.hasNextElement())
      {
         //int id = enum.nextRecordId();
         System.out.println( new String(enum.nextRecord()));
      }
   }
   catch( RecordStoreException e )
      {
         e.printStackTrace();
      }
      finally
      {
         enum.destroy();
      }
```

The output is

How does Does a comparator work?

Bangaladesh compares to China gives -1

China compares to China gives 0

IceLand compares to China gives 6

India compares to China gives 6

China compares to China gives 0

India compares to India gives 0

IceLand compares to India gives -11

Now what does the resulting enumarator look like?:

Bangaladesh
China
IceLand
India

Note: an enumeration can take both a filter and a comparator. The filter is called first to determine which records are in the enumeration. The comparator is then called to sort those records. If tracking is set to true, filtering and sorting occurs whenever the record store is modified. To refresh data selectively we have a `keepUpdated` method. This is used to enable or disable an enumeration's tracking of the underlying record store. We can keep tracking disabled but 'refresh' the enumeration from the record store, by calling the Enumeration `rebuild` method.

15.8.10 The RecordListener

The RecordListener interface defines event handlers for record change events from a recordstore associated with the application. An application overrides the following methods to trap and use these events.

- `recordAdded()`: When a new record is added to the record store.
- `recordDeleted()`: When a record in the record store is deleted.
- `recordChanged()`: When a record in the store has been updated.

Any application that wants to listen to the RecordListener has to implement the RecordListener interface and override the three functions above. The next step is to associate listener with the object. The sample implementation just shows the methods and a simple print statement for the function activated. We have already seen the output from these methods above.

```
public void recordAdded(RecordStore recordStore,int recordId)
  {
    try
    {
      System.out.println("MSG    From    RecordListener:    Record
      added " + "Store Name \" " + recordStore.getName() + "\"
      ID " + recordId + " Value " + getRecord(recordId) );
    } catch (RecordStoreNotOpenException e) {
      e.printStackTrace();
    }
  }
public void recordChanged(RecordStore recordStore, int
recordId)
  {
    try
    {
      System.out.println("MSG From RecordListener: Record
      changed  " + "Store Name \" "+ recordStore.getName() +
      "\" ID " +recordId + " Value " + getRecord
      (recordId));
    } catch (RecordStoreNotOpenException e)
    {
      e.printStackTrace();
    }
  }
public void recordDeleted(RecordStore recordStore,int
recordId)
  {
    try
```

```
    {
        System.out.println("MSG  From  RecordListener:  Record
        deleted  " + "Store Name \" " + recordStore.getName()
        + "\" ID  " + recordId + " Value " + getRecord
        (recordId));
    } catch (RecordStoreNotOpenException e) {
        e.printStackTrace();
    }
}
```

With this we will conclude our discussions of RMS. Before we continue however we will take a Quick peek at good programming practices.

A Warning about using RMS with threads: *No locking operations are provided in this API. Record store implementations ensure that all individual record store operations are atomic, synchronous, and serialized, so no corruption will occur with multiple accesses. However, if a MIDlet uses multiple threads to access a record store, it is the MIDlet's responsibility to coordinate this access or unintended consequences may result. Similarly, if a platform performs transparent synchronization of a record store, it is the platform's responsibility to enforce exclusive access to the record store between the MIDlet and synchronization engine.*

JavaDocs API MIDP2.0

15.8.11 Good programming practices

Our mantra is

- Don't create Object unless necessary
- REUSE as far as possible.
- Every byte saved is a byte gained.

Specific to RMS

- Read and write operations should be optimized such that they are minimal.

- Writes are expensive and hence should be kept to a minimal

- Know our data well and use the storage optimally

- Keep functionality in mind, flush and recreate the RecordStore periodically.

- Handle exceptions properly so that users do not end up with frustrating blank or frozen screens.

An excellent article titled 'Record Management System Basics' by Eric Giguere discusses optimization techniques for RMS in greater details. At the time of this writing the article can be found at http://developers.sun.com/techtopics/mobility/midp/ttips/rmsefficient/

WTK20 provides a monitoring facility. Among other options is the possibility of observing the runtime usage of memory. Also important is the VM emulation feature which allows the developer to limit the resources available to the application and simulate a device like environment. This is very imp to ensure proper performance for applications.

15.9 COMMUNICATION IN MIDP

One of the unique selling points of j2me was its claim of supporting multiple modes of connections while at the same time achieving a small footprint. Since its inception J2Me has mandated that http must be supported on all j2ME handsets. There are other protocols too that are supported and some new introductions in the MIDP 2.0. All connection APIs come under the broad scope of a Generic Connection Framework provided by CLDC. The GCF provides a single consistent set of abstractions for the developer. The Figure 15.33 below provides a look at the structure of the GFC. At the top we have the Connection the base interface as represented by the connector class. A static method open defines creation of all types of connections. If a call to open is successful it returns the object that implements the desired protocol. {This is in essence a factory class, however we will not get into discussions on design here.}

In the Figure 15.33 all blocks numbered 1 represent CLDC 1.0 interfaces.It's important to remember that the CLDC only defines the classes and interfaces that make up the GCF. It is the profiles built on top of the CLDC which define the specific protocols an implementation must support. The HttpConnection interface (labeled 2) was added by MIDP 1.0, while SeverSocketConnection UDPDatagramConnection and Socket Connection (numbered 3) were added by MIDP2.0.

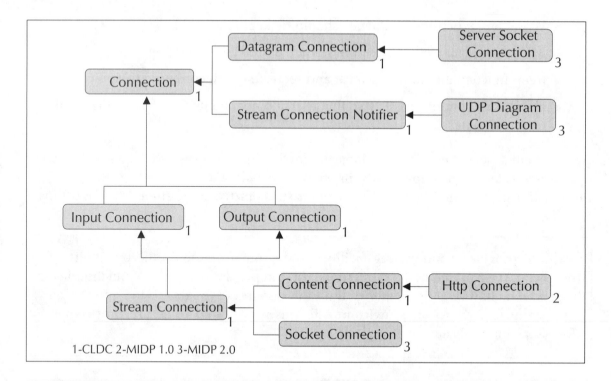

Figure 15.33 Generic Connection Framework

A call to the connection interface would look like

```
ConnectionType conn = (ConnectionType) Connector.open( URI );
```

where the URI is typically of the form : <protocol> :< address> :< parameters>

- `protocol` specifies the protocol to be used it is simply string-like while 'http', 'mailto', 'ftp' etc. are the names used to identify specific protocols. As mentioned above, MIDP mandates the implementation of some protocols, for example, the HTTP protocol (MIDP 1.0). While OEM implementations can add support for other protocols these should fit into the framework.

- `address` specifies the destination of the connection.

- `parameters are` connection parameters as required by the specified protocol.

The Connector class uses the protocol parameter to

- Find and load the appropriate connection class

- Makes the connection
- Returns the newly created Connection object.

This design ensures that the application code need not be changed even if the underlying protocol implementation changes. We will look at some details below.

Sample URI and the resulting Connection calls are given below.

CLDC interfaces.

- urI = "file:e:/myj2me/readme.txt";

InputConnection conn = (InputConnection) Connector.open(uri, Connector.READ);

- urI = "file:e:/myj2me/readme.txt append=true";

OutputConnection conn = (OutputConnection) Connector.open(uri, Connector. WRITE);

- urI = "socket://www.mysite.com:80";

StreamConnection conn = (StreamConnection) Connector.open(uri);

- urI = "http://www.mysite.com/index.html";

ContentConnection conn = (ContentConnection) Connector.open(uri);

Further examples will however concentrate only on MIDP implementations namely HTTP, Socket, ServerSocket and Datagram protocols. A discussion about the various protocols though helpful is beyond the scope of this text. Links for articles dealing with details are provided at the end of this chapter.

MIDP 2.0 supports four protocols: HTTP, Socket, ServerSocket and Datagram. We will

1. HTTP Connection
   urI = http://www.yahoo.com ;
   HttpConnection conn = (HttpConnection) Connector.open(urI);

2. urI = 'datagram://mysite.com:79' ;
   DatagramConnectio conn = Connector.open(urI)

3. urI = "socket://mysite.com:port";
   SocketConnection client = (SocketConnection) Connector.open('socket://' + hostname + : + port);

4. urI= socket://:2500;
   ServerSocketConnection server = (ServerSocketConnection) Connector.open-(urI);

Note though we say MIDP supports HTTP protocol, its implementation is not restricted to IP protocols. Non-IP protocols such as WAP and i-Mode are also supported. The non tcp/ip mechanisms usually involve an edge gateway that does the translation between the client and the HTTP server. The control flow is as depicted in Figure 15.34. In such cases security lies in domain of the network provider and hence needs to be accounted for.

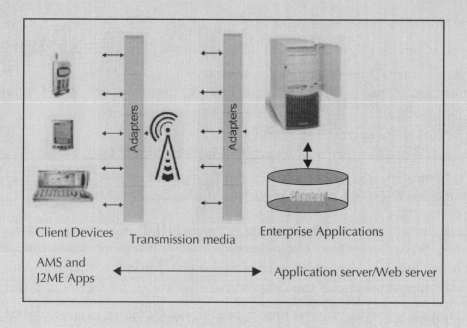

The open method throws a ConnectionNotFoundException if the requested protocol handler is unavailable. It is interesting to note that while MIDP 2.0 *mandates* support for HTTP 1.1 as well as HTTPs connections, support for low-level IP networking is optionall which means that the decision is finally in the hands of the OEMs. However for the applications to be portable it is advisable to stick to http(s) and do the heavyweight stuff on the server end. A chat or multi user game using a server on one of the handsets is fun, but GSM's data bandwidth makes the response time so poor that the users will soon lose interest. The decision regarding the constructs to be used, vary depending on the application's requirements and should be made judiciously.

From an application perspective, a communication mechanism conceptually has three parts to it: Open a channel, Read/Write to the channel and close the channel. We have already seen the first. We will now see the last two. As shown above http inherits from both the stream and content connection while socket connections are inherited from stream connection.

15.9.1 HTTP Connections

http connections leverage on the protocol implementation to provide information regarding the headers. This allows us to set the user-agent (to identify the device), the user language the version of MIDP?CLDC supported as also application specified variables. This feature is extremely useful when applications use server processing capabilities of a server that supports various formats including web wap J2ME etc. For this purpose MIPD provides setRequestPropertyand getRequestProperty *methods.* An example might look like:

User-Agent: Profile/MIDP-2.0 Configuration/CLDC-1.0 Accept-Language: en-US

All Connections are associated with an underlying InputStream OutputStream. To read, we open an input stream using the connection we obtained earlier and then continue to read and store the contents into a buffer till we receive a -1 that indicates a EOS. The loop would look something like below.

```
InputStream is = conn.openInputStream();
// Get the length and process the data
//Read the data.
while ((ch = s.read()) != -1)
  {
     //Store into a buffer that can be used later.
  }
//A better way of doing it would be read chunks of data
instead of just one character something like this.

int len = (int)conn.getLength();
byte[] data = new byte[len];
//Read the entire content of the buffer.
dis.readFully(data);

//Sometimes depending on their implementation the web-
servers do not return the complete buffer in one go but
```

return multiple chunks of fixed lengths or less as may be available. In such cases we need to read chunks of data. The logic then looks something like

```
int datalen = (int)conn.getLength();
int read = 0;
int totalread = 0;
byte[] data = new byte[datalen];
while ((totalread != len) && (read != -1))
{
    read = is.read(data, totalread, datalen - totalread);
    totalread += read;
}
```

To write to the channel we use the output stream. The sample implementation might look as below.

```
//Open an output connection
OutputStream os = conn.openOutputStream();
String content = "My first connection";
//Write to the connection
os.write(content.getBytes());
//flush it so that it is sent out.
os.flush();
```

Finally we use conn.close() to close the channel.

```
HTTP supports some constants as defined in the protocol. These include
```

- HTTP_OK to indicate a 200 or correct response,
- HTTP_NOT_FOUND indicates a 404 condition of server not found etc.

The java docs accompanying the WTK give an extensive explanation and details of the API available.

The only difference between HTTP and HTTPs from the application perspective is the implementation class and the urI which now become

url = https://mysite.com
HttpsConnection conn = (HttpsConnection)Connector.open(url);

The implementation details are abstracted from us.

15.9.2 The SocketConnection Interface

The SocketConnection interface as shown above returns a socket connection and is normally used to access TCP/IP servers. The difference between a server socket and an inbound client is the host name. If a host name is specified, we get a client; else we get a server. Also if the port number is omitted then by default an available port number is assigned to the socket. This can then be retrieved by a call to getLocalPort() and getLocalAddress(). The following snippet shows a simple example.

```
SocketConnection client = (SocketConnection) Connector.open
("socket://" + myhost + ":" + 7001);

    InputStream is = client.openInputStream();
    OutputStream os = client.openOutputStream();

    // send a request to server
    os.write(content.getBytes());

    // read server response
    int c = 0;
    while((c = is.read()) != -1)
    {
        // store for later use.
    }
    // close all
    is.close();
    os.close();
    client.close();
```

We can set user defined options on the socket using Call setSocketOption. SocketConnection provide constants like DELAY, KEEPALIVE etc[]

15.9.3 The `ServerSocketConnection` Interface

The `ServerSocketConnection` interface defines the server socket stream connection. It is used when we need to establish a server on the device. Multiuser game applications are ideal candidates here.

```
// create a server to listen on port 7001
ServerSocketConnection server = (ServerSocketConnection)
Connector.open ("socket://:7001");

// wait for a connection
SocketConnection client = (SocketConnection) server.
acceptAndOpen();

// open streams

DataInputStream dis = client.openDataInputStream();

DataOutputStream dos = client.openDataOutputStream();

int datalen =0;

// read client request refer to the WTK javadocs for other
related APIs

while(datalen)

{

datalen= dis.available()

c = dis.read();

if( c == -1) break;

}

// process request and send response

os.write(...);


// close streams and connections

is.close();

os.close();

client.close();

server.close();
```

Similar to https, we also have secure socket connection which is the same as the socket connection except in the class to be used and the URI.

```
urI = ssl://host.com:79;
SecureConnection sc = (SecureConnection)Connector.open(urI)
```

Both secure protocols provide information regarding the security implementation in terms of the protocol, version, certificate information and the cipher suite used.

With this we conclude our discussion of the communication facilities provided by MIDP. The one topic remaining is security concerns. These are enumerated in the section on security.

Till MIDP 2.0 came along, applications were all pull based i.e., the application had to be running and connected to receive any input from the external world. No messages or network activity could be pushed to the application, essentially forcing applications to function only in a synchronous or polling mode. This was a major restriction on network aware applications. MIDP 2.0 rectified this by introducing the push registry. Our next topic deals with the details of this.

15.9.4 Push Registry

The push registry is actually an addition to the AMS. As we have seen above the AMS is responsible for managing the application lifecycle. Given the sandbox model followed in MIPD, it would not have been possible to implement it else-where. This module allows applications to register for incoming connection or timer-bound requests. The registry itself is a list of all timer events and inbound connections. To utilize this facility the applications need to register to the AMS. It can be done in two ways: one, statically by defining the connections they need to access in the app descriptor. While installing the AMS, check if it supports all the mentioned events; if not, the user is informed and the application will not install. Two dynamically using registerConnection() API of the Class javax.microedition.io. PushRegistry, Once registered, it is the responsibility of the AMS to launch the application in response to the event that it has registered for. This is made available through Serial Port Communications, a CommConnection newly available in the MIDP 2.0.

Static registrations are defined by listing one or more MIDlet-Push attributes in the JAD file or JAR manifest. When the MIDlet suite is uninstalled, the AMS automatically unregisters all its associated push registrations. This entry is the

```
MIDlet-MyPushReg-1: socket://:7001, PushMIDlet, * in the
sample jad file above.

// MIDlet-Permissions: javax.microedition.io.PushRegistry,

// javax.microedition.io.Connector.serversocket
```

The format of the MIDlet-Push attribute is...

MIDlet-Push-<n>: <ConnectionURL>, <MIDletClassName>, <AllowedSender>

- **MIDlet-Push-<n>** a unique name number combination to identify the registration e.g., MIDlet-MyPushReg-1

- *<ConnectionURL>* string that identifies the urI. It is the same URI that we have defined in the section on connections above.

- *<MIDletClassName>* fully qualified class name of the MIDlet that is requesting the registration.

- *<Allowed-Sender>* is a filter that defines the origin server of this registration. This is infact a security mechanism to regulate external access to the application.

Note also the MIDlet-Permissions property, which is used to request permissions for the MIDlet suite. This example requests permission to use the push registry and socket APIs. The 'Security Considerations' section will cover permissions in more detail.

In the dynamic mode the application makes a call to registerConnection(). A sample code snippet would look like:

```
// Open the connection.
   ServerSocketConnection conn  = (ServerSocketConnection)
   Connector.open(urI);

// Register the connection to the AMS
   PushRegistry.registerConnection(url, MIDletClassName,
   filter); //filter is a string variable that can include
   wild characters.

// The rest of the connection code is as before
// wait for a connection
```

```
SocketConnection client = (SocketConnection) server.
acceptAndOpen();
```

Push implementation is the combined responsibility of the MIDlet and the AMS. Post registration we can have two scenarios.

1. When the MIDlet is *not* active, the AMS monitors registered push events on behalf of the MIDlet. When a push event occurs, the AMS activates the appropriate MIDlet to handle it. Figure 15.35 illustrates this sequence for a network activation:

2. If the MIDlet *is* active (running), the MIDlet itself is responsible for all push events. It must set up and monitor inbound connections, and schedule any cyclic tasks it needs–basically standard networking and timer processing.

The push registry can also be used for timer based alerts or events. With this we conclude our discussion of Push. The Push facility gives rise to many exciting applications. For the interested is an excellent discussion on using the Push registry for various inbound and outbound connection including socket, server socket, datagram and http connections while Extending this idea further working with the (WMA) to incorporate the SMS feature of GSM networks.is available at the sun's site under the section technical articles.

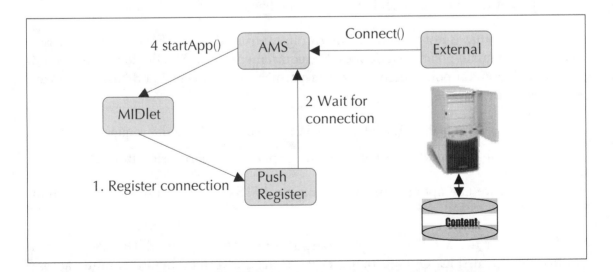

Figure 15.35 Application activation sequence through Push Registry

15.10 Security considerations in MIDP

Security considerations are of different types namely application security, data security and communication security. MIDP 1.0 implemented a strict sandbox for all applications. Thus APIs prevented access to sensitive device capability assuring application security. Only MIDlets within the same MIDlet suite could access record stores belonging to the suite, i.e. created by the MIDlets in the suite. These restrictions clubbed with the absence of any push facility ensured a very safe environment for MIDlets. MIDP 2.0 expanded the scope of MIDlets by having possibility of inter suite access to record stores and push facility. Therein lay the need for security. Following is a discussion on the security features built into MIDP 2.0 to protect the users from rogue applications while at the same time, allow access to sensitive data.

15.10.1 Application security

The MIDP 2.0 security model introduces the concept of security domain. A security domain defines the access permissions for an application.

Applications belong to either

- the trusted domain i.e. permitted to use APIs that are considered sensitive.
- the un-trusted domain i.e. having restricted access.

A MIDlet suite that is not trusted will be run in un-trusted mode. Each domain is associated with a domain policy that defines the requirements for a MIDlet to be considered as trusted. A domain policy defines a set of permissions "allowed" and "user" that can be granted to the MIDlet.

- "allowed" grants the MIDlet permission to access and use the requested API.
- "user" means the MIDlet has to obtain explicit user permission through UI.

For example if we attempt an http connection the AMS throws up a screen as shown in Figure 15.36.

Another way for the MIDlet to request permissions is through its MIDlet-Permissions and MIDlet-Permissions-Opt attributes. Permissions are verified before granting access to the restricted APIs. For maximum security it is advisable to use signed MIDlet suites. Essentially this process uses the X.509 PKI security from a trusted source. The

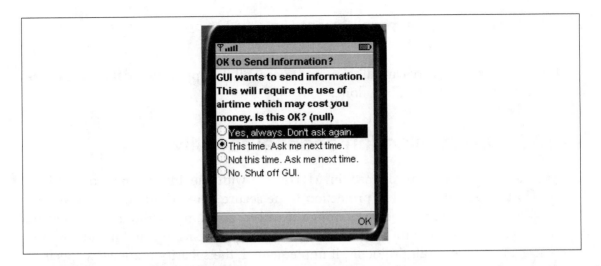

Figure 15.37 Security domain permissions in MIDP 2.0

provisioning systems play a very important role in this. This ensures that only authenticated MIDlets are registered as members of the trusted domains. For further details please refer to the documentation accompanying the WTK.

15.10.2 Recordstore security

MIDP 2.0 allows for shared record stores. Sharing is defined by the MIDlet suite at the time of creation. Access controls are enforced when RecordStores are opened. We have already seen the access modes in the section on RMS. Briefly then

```
openRecordStore(recordStoreName,

      createIfNecessary,

      AUTHMODE_ANY,

      writable) throws Exception ........
```

Creates a RecordStore that allows other MIDlets access to this RecordStore.

```
Now openRecordStore(recordStoreName,

      vendorName,
```

```
suiteName) throws Exception, will open the store cre-
ated above.
```

But these must be used very carefully as this exposes all the application data to the caller and no further verification will be done.

15.10.3 Communication channel security

Secured networking was introduced in MIDP 2.0 with the introduction of HTTPS (secure HTTP) . Another level of protection is the secure domains discussed above. All protocols have to initiate a request through a call to javax.microedition.io.Connector .open(...). The permissions are granted individually to protocols. It is not mandatory for all devices to implement all protocols, if implemented, the security framework specifies the naming of permissions based on the package and class name.

15.10.4 Security of PushRegistry

The PushRegistry is also uses the security framework and permissions. Only MIDlet suites having the javax.microedition.io.PushRegistry permission can register for a Push based launch. A detailed discussion on the protection domains and related development is beyond the scope of this book.

15.11 OPTIONAL PACKAGES

To enhance J2ME capability we have optional packages. An optional package is a set of APIs to support additional features that don't really belong in one specific configuration or profile. Bluetooth support, for example, is defined as an optional package.

Optional packages prescribe their own minimum requirements and are dependent on a particular configuration and one or more profiles There are many optional packages in development, including

- JDBC Optional Package a subset of JDBC (database access APIs)

- Wireless Messaging API (WMA), for sending and receiving Short Message Service (SMS) messages.

- The Mobile Media API (MMAPI) for the capture and playback of multimedia content.

- RMI Optional Package remote method invocation.

- APIs for Bluetooth to integrate into a Bluetooth environment

- Mobile Game API   game development.

- Location API build location-aware applications

- Mobile 3D Graphics API for interactive 3D manipulations.

- Event Tracking API tracking of application events and the submission of these event records to an event-tracking server via a standard protocol.

- Advanced Graphics and User Interface to migrate the core Swing, Java 2D Graphics and Imaging, Image I/O, and Input Method Framework for advanced graphics and user interface facilities

- Content Handler API to handle multi-media and web content.

- This is by no means either a complete or exhaustive list. For more details the readers are referred to "http://www.jcp.org/en/jsr/tech"

15.11 CONCLUSION

J2ME capabilities are quite restrictive from a java developer's perspective, but to someone who has worked with Sim Tool Kit (STK) or JavaCard technologies it is a luxury. The configurations have been formed keeping in mind the limitations of the devices. Before we move over to other technologies let us take a quick look at the list of 'Will and Wonts' for J2ME.

Will

- Basic Java: Object, Class, Runnable, String, System,Thread,Throwable java.lang.

- Data Types: Int, char, String, Stringbuffer

- Utility classes Stack, Vector, Hashtable, Enumeration, Date,Calendar, random numbers java.util

- I/O Stream classes (java.io)

- Operations: integer math,abs, min, max java.lang.Math.

- Exceptions limited list of standard exception classes.

Won't

- Floating point: Most small CPUs don't have on-chip floating point and MIDP doesn't include it.

- Class loader: There is no option for the developer to choose the class loader. It is the responsibility of the default loader.

- Finalization doesn't exist. The developer is responsible for cleaning up before the object deletion takes place.

- JNI: There is no standardized calling convention or hardware profile and hence doesn't make much sense to native support. This is also a security threat.

- Reflection is not included in J2ME.

- Error handling: a specific set of exceptions is generated. Other errors are left to the device manufacturer to handle as they deem fit.

- High-level APIs: Heavy weight GUI APIs, Swing and AWT, are replaced by more appropriate APIs.

- File handling: Due to memory constraints, there is no interface for reading or writing persistent files. We have only a mechanism called Record Management System (RMS) which provides a means for creating persistent data records.

With this we conclude the chapter on J2ME. Next on our agenda is WINce the Microsoft offering for the handheld devices.

REFERENCES/FURTHER READINGS

1. A historical perspective of JAVA is here
   http://www.wired.com/wired/archive/3.12/java.saga.html?topic=&topic_set.

2. Official account of the first five years of java is here http://java.sun.com/features/ 2000/06/time-line.html.

3. JCP overview http://www.jcp.org/en/introduction/overview.

4. JSR for CLDS J2ME is here http://www.jcp.org/en/jsr/detail?id=30.

5. Introduction to various configurations and profiles http://developers.sun.com/ techtopics/mobility/getstart/articles/intro.

6. J2me home http://java.sun.com/j2me/index.jsp.

7. A great article on provisioning
   http://developers.sun.com/techtopics/mobility/midp/ttips/provisioning/index.html.

8. MIDP Game API article
   http://www.microjava.com/articles/techtalk/game_api?content_id=4271.

9. Network programming with MIDP 2.0
   http://developers.sun.com/techtopics/mobility/midp/articles/midp2network/.

10. RMS usage http://www-106.ibm.com/developerworks/java/library/wi-rms/.

11. A good source for all mobile technologies http://www.devx.com/DevX/Door/16148.

12. A detailed article for using Communication APIs in MIDP
    http://developers.sun.com/techtopics/mobility/configurations/ttips/cldcconnect.

13. Push registry in MIDP http://developers.sun.com/techtopics/mobility/midp/articles/pushreg/.

14. Security in J2ME http://www-106.ibm.com/developerworks/library/j-midpds.html.

15. An article on MIDP push using SMS is here http://weblogs.java.net/blog/billday/archive/2004/02/midp_push_using.html.

16. Links to lots of free online articles and resources
    http://www.j2meolympus.com/freebooks/freej2mebooks.jsp.

17. Sun has some excellent articles at http://www-106.ibm.com/developerworks/wireless/library/wi-midlet2.

18. Vartan Piroumian, Wireless J2ME Platform Programming, Prentice Hall, 2002.

19. Yu Feng, Jun Zhu, Wireless Java Programming with J2ME, SAMS, 2001.

REVIEW QUESTIONS

Q1: What are the differences between J2ME and other flavors of Java for example J2SE or J2EE?

Q2: What is J2ME MIDP?

Q3: What is CLDC? How do you program for CLDC?

Q4: Explain MIDlet lifecycle?

Q5: How do you program for multimedia in J2ME?

Q6: What is Record management in J2ME? How do you handle records in J2ME?

Q7: What are the different security considerations in J2ME?

CHAPTER 16

Wireless Devices with Windows CE

16.1 INTRODUCTION

The current trend in electronics is to get more for less. Digital world is becoming more and more miniature. Starting from PDA to cellular phones, all are having more power in lesser space and obviously at lesser cost. In Chapter 13, we have studied PalmOS in PDAs. We have studied Symbian OS for cellular phones in Chapter 14. We have also studied Java (J2ME) in micro devices starting from cellular phones to PDAs in Chapter 15. One question comes in mind, what about Microsoft? Are they not present in this space? Well, Microsoft is not lagging behing. Microsoft has an OS solution for all these small devices. This OS is Windows CE, first released in early 1997. Windows CE is the Windows operating environment for the small devices. The word CE does not have any full form; some people claim that CE stands for Compact Edition though. Some people also claim that Windows CE is a scaled down version of Windows 95. Does not matter what the history is, Microsoft claim that Windows CE has been developed from scratch. Some of the major differences of Windows CE with respect to desktop Windows NT are listed in Table 16.1.

Windows CE is available from various OEM (Original Equipment Manufacturer). An OEM is a company, which manufactures the device that will use the Windows CE operating system. This includes both PDA and cellular phone manufacturers. Following is a list of some of the Windows CE Palm Size devices:

- o Casio E-15, E-100, E-105
- o HP Jornada 420 and 428
- o Compaq Aero A2xxx
- o Philips Nino 520
- o Everex FreeStyle 540

Table 16.1

Feature	Windows CE	Windows NT
Minimum Memory Usage	512K for OS, 4K for RAM	6MB for OS 24MB for RAM
Available memory	~1MB-64MB	~64MB-1GB
Instant ON (no boot)	Yes	No
Networking support	Limited	Full
COM support	Limited	Full
Transaction Server Support	No	Yes
Message Queuing Support	Client	Client+server
Support for SQL DB	Limited	Full
Maximum number of Process	32	Unlimited
Server Services Support	HTTP	HTTP, FTP, File server, Print server, SQL server
Hard Disk	No	Full
Processor	Various	Intel i386
Multimedia support	Limited	Full
Touch screen support	Built in	Third party
Unicode supported	Yes	Yes
Development platform	Limited Win32 APIs	Full Win32 API
Power	Optimized for low power	No power optimization

16.2 DIFFERENT FLAVORS OF WINDOWS CE

Windows CE operating system is available for the entire range of handheld devices. This includes the smallest device like a phone to slightly larger device like PDA or Pocket PC. These devices can be grouped into two major categories. One is a handheld device and the other one is an embedded device. In the case of handheld devices, the user has high interaction with the OS, whereas in the case of the embedded, the interaction is less. An embedded version can run without a keyboard and display. Different flavors of Windows CE are available today. These are:

- **Windows Mobile for Pocket PC:** A Windows Mobile-based Pocket PC is a miniature of a PC, which can fit the size of a pocket. Windows Mobile software for Pocket PC enables mobile computing devices to run applications suitable for small devices by optimizing the user interface, applications size and corresponding feature sets around mobile personal information management and connectivity scenarios. By standardizing core resource requirements and providing a consistent set of programming APIs, Windows Mobile environment provides a consistent application development environment across devices. Windows mobile for pocket PC enables the user to store and retrieve e-mail, contacts, appointments, play multimedia files, games, exchange text messaging, browse the Web, etc. The user can also exchange or synchronize information with a desktop computer.

- **Windows CE .NET:** Windows CE .NET and Windows XP Embedded belong to the Microsoft family of embedded operating systems. Windows CE .NET combines an advanced, real-time operating system with tools for rapid development for the next generation of smart, connected and small footprint devices. Windows CE .NET provides a componentized, customizable, embedded OS. It offers rich configuration and application options for a broad range of embedded devices. Such devices range from enterprise tools such as industrial controllers, communications hubs and Windows-based thin clients to consumer products such as digital cameras, voice-over Internet protocol devices and IP-based set-top boxes. Platform Builder is the integrated development environment for building, debugging and deploying a customized embedded OS based on Windows CE .NET. Windows CE 3.0 serves as the underlying OS for Windows for Pocket PC 2002, while Windows CE .NET 4.2 serves as the underlying OS for Windows Mobile 2003 for Pocket PC.

- **Windows Mobile for Phones:** This version of Windows CE comes in two flavors. These are Pocket PC Phone and Smartphone. The basic difference between the two is that Pocket PC phone is basically a PDA combined with phone features. These are also known as communicators. Whereas, a Smartphone is a phone combined with PDA features. Windows Mobile software for Pocket PC Phone offers functions like dial from contacts database, send SMS messages, identify incoming callers or take call notes. Using a Windows Mobile-based Pocket PC Phone and wireless service through GPRS, the user can access the Internet, send and receive e-mail, and access corporate intranet applications while on the move. A Windows Mobile-based Smartphone integrates PDA-type functionality into a voice-centric handset comparable in size to today's mobile phones. A Windows Mobile-based Smartphone is designed for one-handed operation with keypad access for both voice and data features. It is optimized for voice and text communication, wireless access to Outlook information, and wireless access to corporate and Internet information and services.

16.3 WINDOWS CE ARCHITECTURE

Figure 16.1 depicts the architecture for the Windows CE. It is a layered architecture. At the bottommost layer we have the hardware layer with all the hardware. Next layer is the OEM layer followed by the Operating system layer. At the top of this stack is the Application layer. Application layer interfaces with the user, whereas the hardware layer interfaces with the hardware resource. All other layers are the facilitators for the user to access the resource.

16.3.1 OEM Layer

The OEM layer is responsible for getting a Windows CE–based OS to run on a new hardware platform. OEM layer is a layer of functions which is developed and maintained by an OEM. For example, HP is an OEM who offers Windows CE with the HP Jornada. Within the OEM layer there is an OEM adaptation layer (OAL). This layer

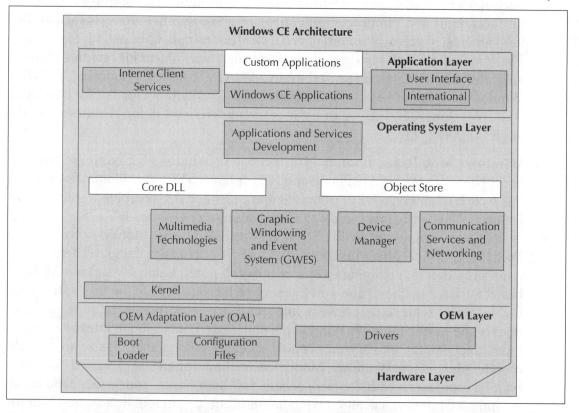

Figure 16.1 Windows CE Architecture

resides between the Windows CE kernel and the hardware of the device. It facilitates communication between the operating system (OS) and the target device and includes code to handle interrupts, timers, generic IOCTLs (I/O control codes), etc. Physically, the OAL is linked with the kernel libraries to create the kernel executable file.

In any device we have a layer of software called device drivers. This is no different in the case of Windows CE. In Windows CE, different hardwares interact with the kernel through respective device drivers.

The boot loader is a piece of software that is required to boot the device. The boot loader generally resides in nonvolatile storage on the device. It is executed at system power-on/reset (POR). To get the boot loader on the target device for the first time, some special program is used. However, updates of boot loaders are handled by boot loader itself flash the new OS images. The platform initialization code for the device is shared between the boot loader and the OAL. The boot loader provides a menu that allows the user to set different configuration options, such as DHCP or static IP information. These configuration parameters are stored in a configuration file.

16.3.2 Operating System Layer

The operating system layer contains all the software supplied by Microsoft as a part of the operating system. The main component in this layer is the kernel. Along with the kernel, this layer includes functions like Applications and Services Development, Core DLL, Object Store, Multimedia Technologies, Graphics Windowing and Event System (GWES), Device Manager, Communication Service and Networking. Several of these modules are divided into components. Components helped Windows CE to become very compact (less than 200K of ROM), using only the minimum ROM, and RAM. These are explained in the following sections.

Kernel

The kernel is the core of the OS, and is represented by the **Coredll.dll** module. It provides the base operating system functionality that is common to all devices. Like any other operating system, Windows CE kernel is responsible for memory management, process management, and certain required file management functions. It manages virtual memory, scheduling, multitasking, multithreading and exception handling. There are some optional kernel components that are needed to include features like telephony, multimedia and graphics device interface (GDI).

Windows CE maps the bottom section of memory into 33, 32Mb slices called "slots". The lowest slot is used for the currently running process (the process at slot 0), and other low slots are used for system processes as:

Slot 0: current running process

Slot 1: kernel (NK.EXE)

Slot 2: File system – object store, registry, CeDB etc. (Filesys.exe)

Slot 3: Device manager (Device.exe)

Slot 4: Windows CE shell (Shell32.exe).

5 slots are used, leaving 28 remaining slots for user processes.

Being a 32-bit machine, the Windows CE address space is 4GB. The top 2Gb address space is used by the operating system, which includes hardware, object store and ROM. The bottom 2Gb address space is used for processes and application shared space.

FFFF FFFF: Top of memory

BF00 0000: Beginning of ROM area

AB00 0000: I/O Registers on a MIPS R4xxx

AA00 0000: Beginning of screen memory

A000 0000: Beginning of DRAM area

8000 0000: Beginning of kernel memory space

[…]

4200 0000: begin shared app memory (memory mapped files)

4000 0000: slot 32 space

[...]

0800 0000: slot4 – shell32.exe

0600 0000: slot3 – device.exe

0400 0000: slot 2 - filesys.exe

0200 0000: slot 1 - nk.exe

0000 0000: slot 0 - current process space

Graphic Windowing and Event System (GWES)

Graphic Windowing and Event System, commonly known as GWES is the graphical user interface between a user and the OS. GWES handles the user input/output. It provides controls, menus, dialog boxes and resources for devices that require a graphical display. It manages user input by translating keystrokes, stylus movements and control selections into messages that convey information to applications and the OS. GWES handles output to the user by creating and managing the windows, graphics and text that are displayed on display devices and printers. All applications need windows in order to receive messages from the OS. This is true even for those devices that lack graphical displays.

Device Manager

In desktop operating systems we load all the device drivers during the startup. However, in Windows CE we do not do so, we use the Device Manager to load the drivers as and when necessary. The Device Manager is launched from the kernel and runs continuously. Device Manager provides the following functions:

- Loads drivers by reading and updating registry keys
- Unloads drivers when a device no longer needs them
- Manages device interfaces and interface notifications
- Manages resources relevant to device drivers, such as I/O space and interrupt requests.

The Device Manager searches the **HKEY_LOCAL_MACHINE\Drivers*RootKey*** registry key to determine the key to begin the driver loading process. Device Manager tracks loaded drivers and their interfaces. It can notify the user when device interfaces appear and disappear. Additionally, it sends power notification callbacks to device drivers and provides power management services.

Windows CE Storage

Windows CE offers different types of storage. These are registry, file system, object store, and databases. Object store is a new concept and available only in Windows CE. Most of these storages in Windows CE are RAM-based. As they are RAM-based, these storages are kept alive through the internal battery of the device. In the event of a cold reset, all data in these storage are lost. It is therefore advised to synchronize these data with a

PC. The Windows CE file system can also use Flash-RAM cards to store data. Data on Falsh-RAM are saved even during cold-reset.

The registry: This storage is similar to the registry on Windows desktop computers. The system DLL managing registry is **coredll.dll**. The respective include files and library files are **coredll.h** and **coredll.lib**. Windows CE supports only three root keys. These are:

- HKEY_LOCAL_MACHINE
- HKEY_CURRENT_USER
- HKEY_CLASSES_ROOT

The file system: Like any other computer system, Windows CE has a file system for persistent storage. However, this persistent storage is not on a hard disk, but on a RAM. The file system RAM is protected by battery. As the file system is not on a disk drive, the file system on Windows CE has no concept of the drive letter like "c:\". All drives on Windows CE are mounted under the root directory. There are different APIs to handle the Windows CE file system.

The object store: This is a storage that is unique to Windows CE. This is similar to a database but for some special types of information. This database stores all PIM (Personal Information Manager) information. The system DLL handling PIM is **coredll.dll**. The respective include files and library files are **coredll.h** and **coredll.lib**. Accessing the object store involves 4 steps. These are:

- Mounting of the database volume
- Create a new database or open an existing database–functions to create a database is CeCreateDatabase(), whereas to open an existing database, the function is CeOpenDatabase().
- Read or write a record–for reading the database, we may need two functions. These are CeSeekDatabase() and CeReadRecordPropsEx(). For writing into a database we use CeWriteDatabaseProps(). These functions look a little strange as these are not just read or write; they are read or write properties. This is because we store properties and information together in an object store. Let us explain it further. In a standard file, we define the property during the creation of the file. However, in case of object store or a PIM database, it is not so. In case of object store, every record contains the property of the record embedded as a part of the record. The number of properties for different record can be different. Even a

property in one records can be a different type in the next record, though it can use the same property ID. If we need to use the CeSeekDatabase() call, all records in the database must share at least one common property. For convenience this could be the sort order on some primary information field.

- Close the database–for this we use the CloseHandle()

Object storage can be accessed through VB, C/C++ or MFC. There are three MFC classes to do these functions. These are:

- CCeDBDatabase class contains all necessary methods to access the database in the object store

- CCeDBRecord class relates to all record–related methods like adding, deleting, reading of records.

- CCeDBProp class offers all the necessary methods to manipulate all fields and properties of a record. Properties supported by Windows CE are Short, Unsigned Short, Long, Unsigned Long, File-Time, Unicode String, Binary Blob, Boolean, and Double.

These are databases as available in desktop computers. Examples are

Database: Pocket-Access or SQL for Windows CE. These are databases that are based on ADO (Active Data Object) technology. The DLL required for this subsystem are adoce.dll, adocedb.dll, adoce30.dll, adocedb30.dll, adoceoledb30.dll, adosync.dll.

ADO is part of Microsoft's universal data access (UDA) strategy. ADO is a database middleware and offers a complete datastore abstraction. Using ADO, an application can be made completely datastore independent. For Windows CE, the ADO version is known as ADOCE. Through ADOCE, an application can access different types of datastore like:

- SQL datastore: like any formal databases, which can be accessed through structured query language. Examples could be SQL server, Oracle etc

- Non-SQL datastore: These could be datastores like email, directory, text files, streams, excel, documents etc.

- Mainframe and legacy data: These datastores would be required by a client application on a Windows CE device to access a database on the remote mainframe.

16.3.3 Communication Services and Networking

The communications component in Windows CE provides support for various types of media access. This varies from device to device. However, the most important ones are communication through the serial port and infrared. Also there is support for Internet and remote access. Windows CE is capable of connecting to the enterprise local area network through both Ethernet and WiFi wireless LAN. In addition to built-in communication hardware, such as a serial cable or infrared (IR) transceiver, PCMCIA support permits a wide variety of aftermarket communications devices to be added to the basic package. Finally, Windows CE supports voice, SMS (Short Message Service) and other telephony services. Following is the list of communications hardware and data protocols supported by Windows CE:

- o Serial I/O support

- o RAS

- o TCP/IP

- o LAN

- o Wireless Services for Windows CE

- o Telephony API (TAPI).

Windows CE Communications Architecture

The underlying assumption for Windows CE device is that it is a mobile device. Therefore, essentially a Windows CE device needs to communicate to the external world. Windows CE supports two basic types of communications. These are serial communication and communication over a network. Most devices feature built-in communications hardware, such as a serial port or an infrared (IR) transceiver. The NDIS implementation on Windows CE supports Ethernet (802.3), Token Ring (802.5), IrDA, and Internet through WAN (Wide Area Network).

Windows CE communications model is designed to function on a variety of devices. Applications and devices in the handheld and mobile space differ in their communications needs. Therefore, Windows CE supports a diverse variety of communication options and associated APIs. It provides an OEM with a diverse set of options to choose from. For application developers, Windows CE supports most of the common types of communication. They are accessible via familiar Win32-based APIs, allowing developers to readily implement communications capability in their applications. In many cases, existing code from other flavors of Windows CE can be used with little or no modification.

Serial I/O

Serial I/O is the most fundamental feature of the Windows CE communications model and is available virtually for all devices. The serial communication would be accessed via a cable or through the IR transceiver. A cable connection is handled with the standard API for serial and file system functions. They can be used to open, close, and manipulate COM ports and read from and write to them. The IR transceiver is also assigned a COM port. Therefore, direct serial I/O is available on IrDA port using the usual serial communications functions.

Networking and Communication Support

Networking support includes primarily the Socket programming interface. This includes different APIs and application interfaces to user programs along with WinSock for normal sockets and IrSock for infrared sockets.

- WinSock and IrSock: WinSock is the socket implementation for the standard TCP/IP. IrSock is an extension of WinSock implementation over the IrDA interface. The Windows CE TCP/IP stack is designed quite efficiently so that it can be configured to effectively support WiFi wireless networking. Windows CE also supports Secure Sockets Layer 2.0, 3.0 and PCT1.0 security protocols. IrSock enables socket-based communication via an infrared transceiver. It is designed to support the industry-standard IrDA protocols. Applications implement IrSock in much the same way as conventional WinSock, although some of the functions are used somewhat differently.

- Browser support (WinINET API): Windows CE supports subsets of the WinINET and Wnet APIs, and an SMB (Service Message Block) redirector. The WinINET API provides support for Internet browsing protocols, including FTP and HTTP 1.0. Only one proxy is supported, and there is no caching. It also provides access to two internet security protocols, Secure Sockets Layer (SSL), and Private Communication Technology (PCT).

- Remote file access (Wnet API): The Wnet API provides access to an SMB redirector for remote file access. Currently only Microsoft Windows 95 and Windows NT operating system connections are supported.

Remote Access and Networking

This includes remote access where the Windows CE device is a client.

- Windows CE supports a remote access services (RAS) client. RAS is multi-protocol router used to connect remote devices. The Windows CE RAS client supports one point-to-point connection at a time.

- NDIS 4.0 for local area networking: For local area networks (LANs), Windows CE includes an implementation of NDIS 4.0. At present, only Ethernet miniport drivers are supported. Wide area networks (WANs) are not supported.

- Windows CE-based devices will connect to their network via a serial communications link, such as a modem. To support this type of networking, Windows CE implements the widely used Serial Line Interface (SLIP), and point to point (PPP) protocols. Authentication is provided via password authentication protocol (PAP) and challenge authentication protocol (CHAP).

Telephony API (TAPI)

For smart phones we need various telephony supports on the device. These telephony supports are generally required for voice communication in GSM or CDMA cellular networks. These telephony interfaces are abstracted and function as an independent isolated interface within the same device. However, there will be instances when the telephony interface needs to integrate with the applications. For example, a calendar application may like to send a reminder to a few people through SMS. Also, there could be need to initiate a GPRS data call to connect to the intranet application. The connection will be established by the telephony interface within the device. These are done through the TAPI (Telephony API). TAPI is a collection of utilities that allows applications to take advantage of a wide variety of telephone and communications services. Windows CE includes TAPI service for AT command-based modems (Unimodem). AT stands for attention and a command level interface; through this interface, we can tell the telephony to do certain tasks for us. In the SMS chapter we have seen some of the AT examples. TAPI can be used with either attached or PCMCIA modems. The Windows CE TAPI supports outgoing calls with outbound dialing and address translation services. TAPI does not support inbound calls yet.

Multimedia support module

Windows CE supports audio and multimedia technologies. Within Windows CE this is achieved through the high-performance DirectX. DirectX provides low-level access to audio and video hardware in a device-independent manner. DirectX delivers a consistent set of capabilities across a variety of hardware configurations. This is achieved through the use of a hardware abstraction layer (HAL) and a hardware emulation layer (HEL).

Windows CE also supports Windows Media technologies. These are designed to provide audio and video playback support for a wide variety of streaming and non-streaming media formats. The Windows Media Player control allows a developer to add playback support to web pages or other applications. The following multimedia technologies are supported in Windows CE:

- o Microsoft Direct 3D
- o DVD-Video API
- o Windows Media audio and video codes
- o A new unified audio model that uses waveform audio drivers for waveform audio, the audio mixer, and DirectSound
- o Microsoft DirectDraw
- o Microsoft DirectSound
- o Microsoft DirectShow
- o Windows Media technology
- o Windows Media Player control.

All the above multimedia components other than the Direct3D are supported on all Windows CE devices. Direct3D feature requires floating point support on the device.

COM Support Module

The Component Object Model or COM is available across all Windows operating systems. COM is used by applications for object-oriented inter-application and intra-application communicates. The desktop version of COM offers different methods for communication. However, Windows CE offers following subset:

- • ActiveX control: Reusable code and objects, specially for user interface and other invocations can be implemented quite efficiently through ActiveX controls.

- In-process activation: In-process COM servers are practically ActiveX controls without a user interface.

- EXE to EXE communication: Using EXE to EXE communication, applications can communicate over the process boundaries. Two applications can have peer-to-peer communicate with each other through this technology without using shared memory, sockets or temporary file.

- MTS (Microsoft Transaction Server)–client only: MTS is used for application offereing transaction processing paradigm. However, a Windows CE device can be a client to a MTS-hosted application running on a desktop.

- DCOM (Distributed COM): DCOM is the implementation of COM across machine boundaries.

- Message Queue–client only: Message queues are very useful for asynchronous communication between peers. A Windows CE device can act as a message queue client.

Windows CE Shell Module and User interface

We have discussed that in Windows CE, GWES is the interface between the user, the applications and the operating system. Windows CE shell module and UI combine the Microsoft Win32 API (application programming interface), UI (user interface) and GDI (graphics device interface) libraries into the GWES module (Gwes.exe). GWES supports all the windows, controls and resources that make up the Windows CE user interface. GWES also includes support for user input and output, through support for keyboard input, fonts, text drawing, line and shape drawing, palettes and printing.

GWES supports various types of resources. Resources are objects that are used within an application but are defined outside an application. GWES includes support for the following resources:

- o Keyboard accelerators
- o Menus
- o Dialogs boxes and message boxes
- o Bitmaps
- o Carets
- o Cursors

o Icons

o Images

o Strings

o Timers.

GWES supports various types of controls. A control is a child window that an application uses in conjunction with another window to perform input/output tasks. Common controls are a set of windows that are supported by the common control library. Common control is a DLL that is included with Windows CE. GWES includes support for the following common controls:

o Command band

o Command bar

o Header control

o Image list

o List view

o Month calendar control

o Pocket PC-style ToolTips

o Progress bar

o Property sheet

o Rebar

o Status bar

o Tab control

o Toolbar

o ToolTip

o Trackbar

o Tree view

o Up-down control

o Date and time picker.

GWES supports various types of window controls. A window control is a predefined child window that allows a user to make selections, carry out commands, and perform input/output tasks. GWES includes support for the following window controls:

o Check boxes

o Push buttons

o Radio buttons

o Group boxes

o Combo boxes

o Edit controls

o List boxes

o Scroll bars

o Static controls.

GWES also supports the CAPEDIT Control and SBEDIT Control for use in edit controls. Following are some additional GWES features that developers can add to display-based platforms.

o Accessibility: Options that allow persons with disabilities to use the device more easily.

o Fonts: Provides 60 different fonts for displaying and printing text.

o Mouse/Pen: Allows users to provide input through a pointer device like mouse or pen.

o Stylus: Allows users to provide input through stylus and touch screen.

o Multiple Screens: Enables a device to connect to multiple screens.

o Printing: Supports the ability to print.

o Input Panel: Allows users to provide input through a input panel displayed on a touch screen.

Developers of applications for Windows CE can customize the UI by creating a skin. Platform developers can change the behavior of menus. In Windows CE menus by default contain only one level; which means that menu items do not open submenus.

Platform developers can choose the Overlapping Menus feature to provide support for cascading, overlapping menus to enable submenus.

The shell architecture in Windows CE allows developers to implement a wide variety of shells. All the source code for the presentation and user interface aspects of the standard shell is available to developers. This allows fully customized shells built for individual platforms to be completely integrated into the OS. Developers can add the following shell features to display-based platforms.

  o Standard Shell: Provides a shell that is similar to the shell on the Windows-based desktop platforms. The source code for this shell is available for customization.

  o Command Processor: This is used for a command-line-driven shell interface for console input and output and a limited number of commands.

  o Windows Thin Client Shell: Provides a Windows Thin Client user interface

  o API compatibility support: These extensions to the standard shell are achieved through AYGShell API extensions. AYGShell support allows most Pocket PC based applications to run on a Windows CE-based device after being recompiled for an OEM's Windows CE-based platform.

Embedded Windows CE platforms do not require full GWES features because such platforms do not require an interactive display or an input device like a keyboard or a mouse. Therefore, for embedded devices we can use the Minimal GWES Configuration Features. These GWES support are the glue between the embedded platforms and display-based platforms that Windows CE provides.

Following are the Minimal Configuration features in Windows CE.

  o Minimal GWES Configuration: Provides basic windowing and message queue support.

  o Minimal Input Configuration: Provides support for keyboard input.

  o Minimal GDI Configuration: Provides GDI support, including TrueType fonts, text drawing, and palette support.

  o Minimal Notifications Configuration: Provides support for notifications.

  o Minimal Window Manager Configuration: Provides support for window managemen.

16.3.4 Application Layer

The last and final layer is the custom application layer. All the Windows CE user applications form this layer. Subsystems for user interface and internationalization are part of this layer. Windows CE applications available from Microsoft, applications developed by OEM or application developed by third parties are part of this layer.

16.4 WINDOWS CE DEVELOPMENT ENVIRONMENT

The application for Windows CE is developed on a desktop environment. It is tested through a simulator on the desktop environment. Once the application is found working on the simulator it is loaded on the physical Windows CE device for testing. Therefore, Windows CE development environment comprises different systems and subsystems in the desktop. Also, it requires some components on the Windows CE device. Following are the components:

Windows 2000 operating system on the workstation: Windows CE development environment can run on desktop OS like Windows 95, Windows 98 or other flavors of Windows. The facility available in Windows 95 and 98 are quite restricted. Therefore, it is recommended that Windows 2000 is used as the host OS on the workstation for the Windows CE development.

- **Visual Studio:** The Windows CE development environment is an add-on on the desktop development environment. Visual Studio for Windows CE extends the Visual Studio for desktop.

- **eMbedded Visual Tools:** This set of software comprises of different tools for the Windows CE development and debugging. The set includes:

   o Remote file viewer

   o Remote Heap Walker

   o Remote Process Viewer

   o Remote Registry Editor

   o Remote Spy++

   o Remote Zooming

   o Control Manager.

- **SDK (Software Development Kits):** for Palm-size PC and Handheld/PC SDK can also be used. This can give a better control over the device.

- **ActiveSync:** Like we archive our files in the desktop, we need to archive our files and data in the Windows CE device. Also, as the data is stored in RAM, it is advised to save the data from time to time. ActiveSync is used to do all these functions. All Windows CE devices come with a cradle that attaches to a desktop or a laptop PC via ActiveSync. It is a system that manages the connection between a desktop computer and the Windows CE device. ActiveSync can be configured to synchronize e-mail, calendar appointments, contacts and many more applications.

Once the above software are installed, we need to connect the workstation with the Windows CE device. This is required for multiple functions starting from installing of the Windows CE software to the debugging of the Windows CE system. This will be used as the PC-Link.

16.4.1 Windows C++ Development

One of the most common platforms for developing applications for Windows CE is Visual C++. All application developed using Visual C++ will work on all Windows CE platforms. For C++ development we need Visual C++ Toolkit for Windows CE. We can develop application for Windows CE using the MFC classes. To develop applications using Visual C++ we will need the following:

o Visual C++ Professional or Enterprise edition (latest release)

o HPC or PSPC SDK, downloadable for free from the Microsoft web site.

o The Visual C++ Toolkit for Windows CE (latest release).

16.4.2 VB Development

Visual Basic is another platform for developing Windows CE applications. To develop a Windows CE application using VB we need the following:

o The Visual Basic Toolkit for Windows CE (latest release)

o Visual Basic, Professional or Enterprise edition (latest release)

o The HPC SDK, HPC Pro SDK or the PSPC SDK v. 1.2.

16.4.3 Windows CE Programming

As discussed in previous sections, for Windows CE programming all the popular programming languages from Microsoft are available. These are VB (Visual Basic) and Visual C/C++ with MFC. Programming of Windows CE is no way different from standard Windows on any desktop PC. Though the underlying hardware and operating system are different in Windows CE, the programming interfaces through VB and C/C++/MFC are the same as in desktops.

REFERENCES/FURTHER READING

1. Chris Muench: The Windows CE Technology Tutorial, Addison Wesley, ISBN 0-201-61642-4.

2. David J. Kruglinski, George Shepherd, Scot Wingo: Programming Microsoft Visual C++, Microsoft Press.

3. http://www.microsoft.com.

4. http://www.msdn.com.

5. http://msdn.microsoft.com/library/default.asp?url=/library/en-us/wceintro5/html/wce50conIntroducingWindowsCE.asp.

REVIEW QUESTIONS

Q1: Explain different layers in the WindowsCE architecture?

Q2: Explain how communications and networking are handled in WindowsCE. How does one program these interfaces?

Q3: How does one program the telephony interfaces in WindowsCE?

Q4: How are multimedia handled in WindowsCE?

Q5: Explain the WindowsCE development environment?

CHAPTER 17

Voice over Internet Protocol and Convergence

17.1 VOICE OVER IP

Traditionally, for decades circuit-switched technologies were in use for voice communications. In a circuit-switched technology, a channel (timeslot in Time Division Multiplexing, a frequency in Frequency Division Multiplexing, or space in Space Division Multiplexing etc), is reserved to establish an end-to-end circuit. The channel is reserved for the connection, and users pay for the entire length of the circuit (in space and time) irrespective of whether users are talking or thinking. The circuit could carry voice traffic, which could be either a digitized voice or analogue voice. While circuit switching provides good voice quality, it may not be efficient in channel utilization. In contrast, packet-switched networks carry data in packets from multiple sources and destinations over one channel. Such networks are better in channel utilization but suffer from delays and jitters. For realtime traffic like voice, delays and jitters are not nice qualities to have. IP (Internet Protocol) is one such packet-switched network which is efficient for data communication but not suited for realtime voice.

In 1995 some hobbyists in Israel made an attempt to send voice over IP network between two PCs. Later in the same year, Vocaltec, Inc. released Internet Phone Software. By 1998 few companies started setting up gateways to allow PC-to-Phone and later Phone-to-Phone (over private corporate IP networks) connections. Technology to enable such voice communication over the IP network became known as Voice over Internet Protocol or VoIP in short. By 2000, VoIP traffic exceeded 3% of voice traffic. Most of these VoIP technologies were proprietary and did not interoperate. To ensure interoperability between protocols and equipments from different vendors, standards started emerging. These standards were from two major camps viz., the telecommunication camp and the data camp. Today there are two sets of standards for VoIP switching, media, and gateways. These are H.323 from ITU (International Telecommunications

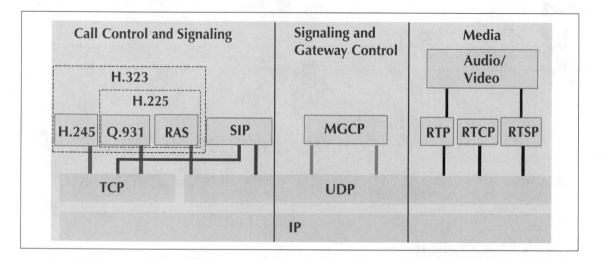

Figure 17.1 H.323, SIP and MGCP

Union) and SIP (Session Initiation Protocol) from IETF (Internet Engineering task Force). Figure 17.1 depicts the H.323, SIP and MGCP (Media Gateway Control Protocol) and connection among them.

17.2 H.323 FRAMEWORK FOR VOICE OVER IP

The H.323 is a set of protocol standards that provide a foundation for multipoint conferencing of audio, video, and data communications over IP networks standardized by the ITU. It is used for peer-to-peer, two-way delivery of realtime data. The scope of H.323 (Figure 17.2) includes parts of H.225.0–RAS, Q.931, H.245 RTP/RTCP and audio/video codecs, such as the audio codecs G.711, G.723.1, G.728, etc. and video codecs like H.261, H.263 that compress and decompress media streams. It includes codecs for data conferencing through T.120 and fax through T.38. H.235 Specifies security and encryption for H.323 and H.245 based terminals. H.450.N recommendation specifies supplementary services such as call transfer, call diversion, call hold, call park, call waiting, message waiting indication, name identification, call completion, call offer, and call intrusion. H.246 specifies internetworking of H Series terminals with circuit switched terminals.

In a H.323 implementation, along with the end-user devices three logical entities are required as depicted in Figure 17.3. These are Gateways, Gatekeepers and Multipoint Control Units (MCUs). Terminals, Gateways, and MCUs are collectively known as endpoints. It is possible to establish an H.323-enabled network with just terminals, which are H.323 clients. Yet for more than two endpoints, a MCU is required.

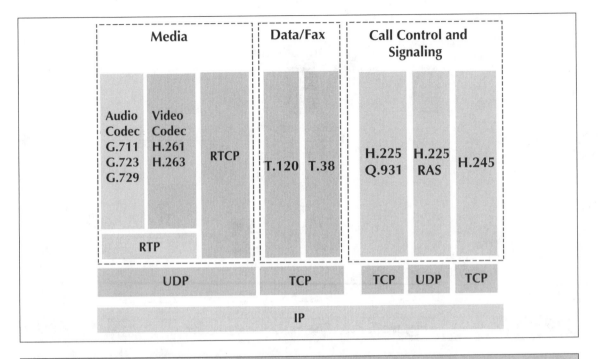

Figure 17.2 H.323 Umbrella specification

17.2.1 Gateway

The purpose of the gateway is to do the signal and media translation from IP to circuit-switch network and vice versa. This includes translation between transmission formats, translation between audio and video codecs, call setup and call clearing on both the IP side and the circuit-switched network side. The primary applications of Gateways are:

- Establishing links with analog PSTN terminals.

- Establishing links with remote H.320-compliant terminals over ISDN-based switched-circuit networks.

- Establishing links with remote H.324-compliant terminals over PSTN networks

17.2.2 Gatekeeper

A gatekeeper acts as the central point of control for all calls within its zone for all registered endpoints. A gatekeeper is not mandatory in an H.323 system. However,

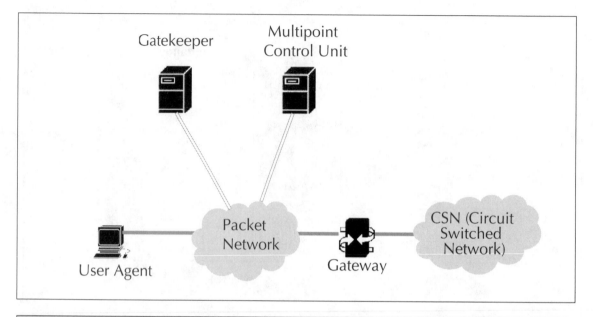

Figure 17.3 The H.323 architecture

if a gatekeeper is present, terminals must use the services offered by gatekeepers. Gatekeepers perform functions like address translation and bandwidth management. For example, if a network has a threshold for the number of simultaneous conferences on the LAN, the gatekeeper can refuse to make any more connections once the threshold is reached. An optional feature of a gatekeeper is its ability to route H.323 calls. By routing a call through a gatekeeper, service providers can meter a call with an intention of charging. Routing of call through gateway can also be used for call forwarding to another endpoint. The gatekeeper plays a major role in multipoint connections by redirecting the H.245 Control Channel to a multipoint controller. A gateway could use a gatekeeper to translate incoming E.164 addresses into IP addresses.

17.2.3 Multipoint Control Unit

The Multipoint Control Unit (MCU) supports conferences between three or more endpoints. An MCU consists of a Multipoint Controller (MC) and a Multipoint Processor (MP). The MC handles H.245 negotiations between all terminals to determine common capabilities for audio and video processing. An MCU optionally may have one or more MPs to deal with the media streams. MP mixes, switches, and processes audio, video, and/or data bits.

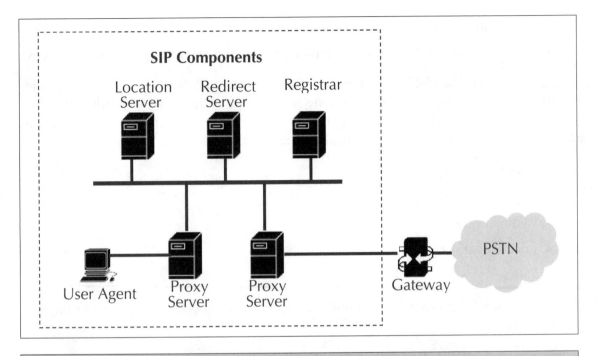

Figure 17.4 The SIP architecture

17.3 SESSION INITIATION PROTOCOL (SIP)

Session Initiation Protocol (SIP) is a signaling protocol for telephone calls over IP. SIP is defined by the IETF and is gaining popularity. Unlike the H.323, SIP is designed specifically for the Internet. SIP defines interfaces for establishing, modifying and terminating sessions with one or more participants in the VoIP environment. It facilitates development of telephony application. These facilities also enable personal mobility of users. SIP supports five facets of establishing and terminating multimedia communications:

- User location: Determine the location and end systems to be used for communication

- User capabilities: determination of the media and media parameters to be used for the communication

- User availability: determining the called parties' willingness to engage in communication

- Call setup: "ringing," establishing call parameters at both calling and called party

- Call handling: manage the transfer of data (voice)

- Call teardown: at the end of the call, terminate the call and release all resources

Figure 17.4 depicts the VoIP architecture with respect to SIP. In such a VoIP setup, the end-user device can be either an IP phone or a computer in an IP network. The conversation can be IP-to-IP, PSTN-to-IP, IP-to-PSTN. In a SIP environment along with the endpoint devices, five entities are required. These are:

- Proxy server

- Registrar server

- Redirect server

- Location server

- Gateways

We describe the functions of these entities and the mode of communications in the following sections.

17.3.1 Proxy Server

According to the SIP standard (RFC3261), a proxy server is defined as: "SIP proxies are elements that route SIP requests to user agent servers and SIP responses to user agent clients. A request may traverse several proxies on its way to a UAS. Each will make routing decisions, modifying the request before forwarding it to the next element. Responses will route through the same set of proxies traversed by the request in the reverse order." In the SIP context user agent client (UAC) is the endpoint initiating a call and user agent server (UAS) is the endpoint receiving the call.

SIP proxies function similar to routers and make routing decisions, modifying the request before forwarding it to the next element. SIP standard make provision for proxies to perform actions such as validate requests, authenticate users, resolve addresses, fork requests, cancel pending calls. The versatility of SIP proxies allows the operator to use proxies for different purposes and in different locations in the network. Proxies could be deployed as edge proxy, core proxy or even enterprise proxy. This versatility also allows for the creation of a variety of proxy policies and services, such as routing calls on various intelligent rules. The 3GPP IMS architecture (section 17.9 below) for example, uses proxies known as Call State Control Functions of different kinds for various purposes.

17.3.2 Registrar Server

The Registrar server in a VoIP network can be defined as the server maintaining the whereabouts of a domain. It accepts REGISTER requests from nodes in the VoIP network. It places the information it receives in as a part of those requests into the location service for the domain it handles. REGISTER requests are generated by clients in order to create or remove a mapping between their externally known SIP address and the IP address they wish to be contacted at. It uses the location service in order to store and retrieve location information. The location service may run on a remote machine and may be contacted using any appropriate protocol (such as LDAP).

17.3.3 Redirect Server

Redirect server does similar functions as done in case of call forwarding in a PSTN or cellular network. A redirect server receives SIP requests and responds with redirection responses (3xx). This enables the proxy to contact an alternate set of SIP addresses. The alternate addresses are returned as contact headers in the response SIP message.

17.3.4 Presence Server

Presence is a service that allows the calling party to know the ability and willingness of the other party to participate in a call. A user interested in receiving presence information for another user (Presentity) can subscribe to his/her presence status and receive Presence status notifications from the Presence system. This is achieved through an Event Server. An Events Server is a general implementation of specific event notification, as described in RFC3265. RFC3265 provides a framework that allows an entity to subscribe for notifications on the state change of other entities. The IETF SIPMPLE Working Group is developing a set of specifications for the implementation of a Presence system using SIP. They are working within a general IETF requirements framework for Presence and Instant Messaging, which is called Common Presence and Instant Messaging (CPIM).

17.3.5 SAP/SDP

Session Announcement Protocol (SAP) is an announcement protocol that is used by session directory clients. A SAP announcer periodically multicasts an announcement packet to a known multicast address and port. The scope of multicast announcement is same as the session it is announcing. This ensures that the recipients of the announcement can also be potential recipients of the session the announcement describes.

The Session Description Protocol (SDP) describes multimedia sessions for the purpose of session announcement, session invitation and other types of multimedia session initiation. SDP communicates the existence of a session and conveys sufficient information to enable participation in the session. Many of the SDP messages are sent using SAP. Messages can also be sent using email or the WWW (World Wide Web).

17.3.6 Quality of Service and Security

In any network, quality of service and security are very important. In Internet protocols, RSVP (Resource ReSerVation Protocol) protocol is designed for quality integrated services. RSVP is used by a host to request specific quality of service (QoS) from the network. This could be a SIP service or any other service. RSVP requests the quality results in resources being reserved in each node along the data path.

COPS (Common Open Policy Service) protocol is a simple query and response protocol that can be used to exchange policy information between a policy server (Policy Decision Point–PDP) and its clients (Policy Enforcement Points–PEPs). The model does not make any assumptions about the methods of the policy server, but is based on the server returning decisions to policy requests. The policy could be related to security, authentication, or even QoS. One example of a policy client is an RSVP router that must exercise policy-based admission control over RSVP usage.

17.4 COMPARISON BETWEEN H.323 AND SIP

Functionally, SIP and H.323 are similar. Both H.323 and SIP support call control, call setup and call teardown. H.323 and SIP support basic call features such as call waiting, call hold, call transfer, call forwarding, call return, call identification, or call park. H.323 defines sophisticated multimedia conferencing like whiteboarding, data collaboration, or video conferencing. SIP supports flexible and intuitive service creation using SIP-CGI and CPL. Both H.323 and SIP support capabilities exchange. Third party call control is currently only available in SIP. Though SIP's deployment started later, it seems to gain momentum. SIP is adopted by 3GPP. The primary factors that encourage SIP's adoption are, SIP is simple, it is scales, and it is flexible. Table 17.1 summarizes some of these features of SIP and H.323.

Table 17.1 Comparison between SIP and H323

	SIP	H.323
Standard Body	IETF	ITU
Relationship	Peer-to-peer	Peer-to-peer
Client	Intelligent User Agent	Intelligent H.323 terminal
Core Servers	SIP Proxy server, Redirect server, Location server, and Registration servers	H.323 Gatekeeper, Gateway, Multipoint Control Unit
Current Deployment	SIP is new with less installations but gaining interest	Widespread
Capabilities Exchange	SIP uses SDP protocol for capabilities exchange; but not as extensive capabilities exchange as H.323.	H.245 provides structure for detailed and precise information on terminal capabilities.
Control Channel Encoding Type	Text based UTF-8 encoding.	Binary ASN.1 PER encoding.
Server Processing	Stateless or stateful.	Version 1 or 2–Stateful. Version 3 or 4–Stateless or stateful.
Quality of Service	SIP relies on other protocols such as RSVP, COPS to implement or enforce quality of service.	Bandwidth management/control and admission control is managed by the H.323 gatekeeper. The H323 specification recommends using RSVP for resource reservation.
Security	Registration–User agent registers with a proxy server. Authentication–User agent authentication uses HTTP digest or basic authentication. Encryption–The SIP RFC defines three methods of encryption for data privacy.	Registration–If a gatekeeper is present, endpoints register and request admission with the gatekeeper. Authentication and Encryption–H.235 provides recommendations for authentication and encryption in H.323 systems.

Table 17.1 (*Continued*)

Endpoint Location and Call Routing	Uses SIP URL for addressing. Redirect or location servers provide routing information.	Uses E.164 or H323ID alias and an address mapping mechanism if gatekeepers are present in the H.323 system. Gatekeeper provides routing information.
Conferencing	Basic conferencing without conference or floor control.	Comprehensive audiovisual conferencing support. Data conferencing or collaboration defined by T.120 specification.
Service or Feature Creation	Supports flexible and intuitive feature creation with SIP using SIP-CGI and CPL. Some example features include presence, unified messaging, or find me/follow me.	H.450.1 defines a framework for supplementary service creation.
Instant Messaging	Supported	Not Supported

17.5 Real Time Protocols

We have discussed that for good quality voice we need realtime support. To allow realtime data transmission over TCP/IP, various protocols have been developed. These include protocols for realtime data, audio, video, movie, and streaming data in a unicast or multicast situation. Examples of such protocols are Realtime Transport Protocol (RTP – RFC1889), Real Time Control Protocol (RTCP – RFC3605), and Real Time Streaming Protocol (RTSP – RFC 2326). RTP is a transport protocol for the delivery of real-time data, including streaming multimedia, audio and video. RTCP helps with lip synchronization and QoS management for RTP. RTSP is a control protocol for managing delivery of streaming multimedia from media servers. RTSP can be considered as the "Internet VCR remote control protocol".

17.5.1 Real Time Transport Protocol

The Realtime Transport Protocol (RTP) is both an IETF and ITU standard (H.225.0). It defines the packet format for multimedia data. RTP is used by many standard protocols, such as RTSP for streaming applications, H.323 and SIP for IP telephony applications, and by SAP/SDP for pure multicast applications. It provides the data delivery format for all of these protocols.

17.5.2 Real Time Control Protocol

The Real Time Control Protocol (RTCP) is based on the periodic transmission of control packets to all participants in the session. RTCP uses the same distribution mechanism as the RTP data packets. RTCP can deliver information such as the number of packets transmitted and received, the round-trip delay, jitter delay, etc. that can be used to measure Quality-of-Service in the IP network. This facility allows monitoring of the data delivery in a manner scalable to large multicast networks, to provide minimal control and identification functionality. For RTCP to work effectively, the underlying protocols must provide multiplexing of the data and control packets.

17.5.3 Real Time Streaming Protocol

Real Time Streaming Protocol (RTSP) is a client-server protocol, designed to address the needs for efficient delivery of streamed multimedia over IP networks. Interoperability on streaming media systems involves many components. These are players in the client device, servers that stores the content, encoders that transform or compress the data, tools that create the content. All these must share common mechanisms for interoperability. Encoders and tools must store data types in files in formats that will be understood by players. Encoders and content-creation tools must be able to store content in files that servers can read. Servers must be able to stream content using protocols that players in the client device can understand.

17.6 CONVERGENCE TECHNOLOGIES

To make convergence and interworking between PSTN and IP networks possible, three functional gateway elements are defined. Two of these are interface elements: the Media Gateway and the Signaling Gateway. The third element is the Media Gateway Controller. Signaling gateway (SG) is responsible for interfacing to the SS#7 network and forwarding signaling message to the IP network. The Media Gateway (MG) is responsible for packetization of voice and other realtime traffic (media). The Media Gateway Controller (MGC) plays the role of the mediator to enable and control access

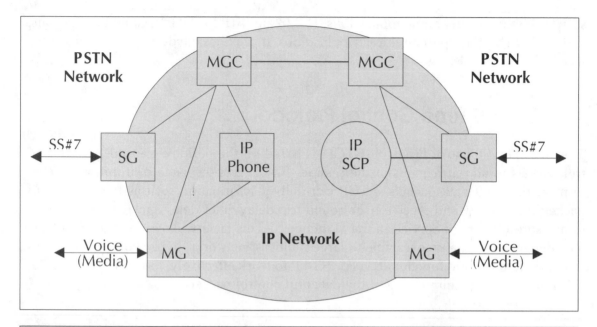

Figure 17.5 Interfaces between IP and PSTN networks

and resource usage between the IP and PSTN network. Together, these elements form the building blocks for a distributed architecture approach to providing voice, fax and a set of digital data services over IP networks. Figure 17.5 depicts the convergence architecture. In this architecture we can see an IP SCP (Service Control Point). The functionality of the SCP is similar to those we have described in Intelligent Networks (IN) in chapter 11. However, an IP SCP is addressable from the SS#7 network.

17.6.1 Media Gateway

The primary responsibilities of the Media Gateway (MG) are to allow media of various types e.g., voice, fax, video, and modem data to be transported from one type of network to another network. These media must be transportable, both as packets in the IP network and as digital or analog streams in the circuit switched network. They must also be able to move without loss of integrity or degradation of quality. These criteria are met through the use of various coding, compression, echo cancellation, and decoding schemes. The Media Gateway function provides a bi-directional interface between a circuit switched network and media-related elements in an IP network. Typically, Media Gateways will interact either with IP Telephony end-user applications residing in computers attached to the IP network, or with other Media Gateways. The technology necessary to implement Media Gateways is evolving at a very rapid pace. Media Gateways can implement a variety of physical interfaces to the PSTN. For example, highly scalable Media Gateway systems can implement high speed Time Domain Multiplexing (TDM) trunk interfaces, which are commonly used between switching elements in the circuit switched network.

17.6.2 Media Gateway Controller

The key responsibilities of the Media Gateway Controller (MGC) are to make decisions based on flow-related information, and to provide associated instructions on the interconnecting of two or more IP elements so that they can exchange information. Media Gateway Controllers maintain current status information of all media flows, and they generate the administrative records necessary charging and billing. Typically, Media Gateway Controllers instruct Media Gateways on how to set up, handle and terminate individual media flows. A media gateway controller exchanges ISUP (ISDN User Part) messages with central office switches via a signaling gateway. They also provide the parameters associated with bandwidth allocation and, potentially, quality of service characteristics. Media Gateway Controllers can be used by sophisticated end-user interface applications. In H.323, significant Media Gateway Controller functions are performed in network elements called Gatekeepers. Because media gateway controllers is built primary through software using off-the-shelf computer platforms, a media gateway controller is sometimes called a **softswitch**.

17.6.3 Signaling Gateway

The Signaling Gateway (SG) function implements a bi-directional interface between an SS7 network and various call control-related elements in an IP network. The key responsibilities of the Signaling Gateway are to repackage SS#7 information into formats understood by elements in each network, and to present an accurate view of the elements in the IP network to the SS#7 network. Typically, the associated IP network elements will implement Media Controller functions, database storage, or query functions. The SS#7 network has stringent reliability constraints on all devices directly attached to it. By definition, Signaling Gateways needs to implement reliable SS#7 messaging that obeys all the rules of the SS#7 network, while also accommodating a variety of behaviors in the IP network. Many of these behaviors are entirely appropriate in an IP world, but not acceptable by PSTN standards. To actually enable IP elements, like Media Controllers, to perform their designated administrative functions, the Signaling Gateway repackages the information contained in various high level SS#7 message protocols such as ISUP and TCAP into formats that can be understood by IP elements. It is necessary for Signaling Gateways to understand all of SS#7 protocols and messaging standards. Finally, since an IP network is a shared medium lacking physical security, Signaling Gateways must filter out the inappropriate traffic that shows up at the Signaling Gateway. It is essential that the Signaling Gateway function protect the SS#7 network from malicious intrusion or accidentally induced undesirable traffic.

17.6.4 Megaco/H.248: Media Gateway Control Protocol

Megaco or Media Gateway Control Protocol is defined in RFC 3015. It is adapted by ITU as H.248 recommendation. Megaco defines the protocol for control of different elements in a physically decomposed multimedia gateway. There are two basic components in Megaco. These are Terminations and Contexts. Terminations represent streams entering or leaving the MG. Example could be analog telephone lines, RTP streams, ATM stream, or MPEG (Moving Picture Experts Group) stream. Terminations may be placed into contexts (Fig 17.6), which are defined as two or more termination streams are mixed and connected together. Contexts are created and released by the MG under command of the MGC. A context is created by adding the first termination, and it is released by removing (subtracting) the last termination. There is a special Context called the null Context. It contains Terminations that are not associated to any other Termination.

Figure 17.6 depicts an example of one way call waiting scenario in a decomposed access gateway. In Figure 17.6 (a), Terminations T1 and T2 in Context C1 are engaged in a two-way audio call. While T1 and T2 are in middle of the call, a second call arrives for T1 from Termination T3. T1 has call waiting facility; therefore, T3 stands alone (null Context) in Context C2 at waiting state. In Figure 17.6 (b), T1 accepts the call from T3. This places T2 on hold (parked call). This action results in T1 moving into Context C2.

17.6.5 Sigtran and SCTP

The Signaling Transport (SIGTRAN) group of the IETF defines Sigtran Protocol Architecture through RFC2719 and SCTP (Stream Control Transmission Protocol) standards through RFC2960. SCTP is an end-to-end, connection-oriented protocol that transports data in independent sequenced streams. SCTP was designed to provide a general-purpose transport protocol for message-oriented applications, as is needed for the

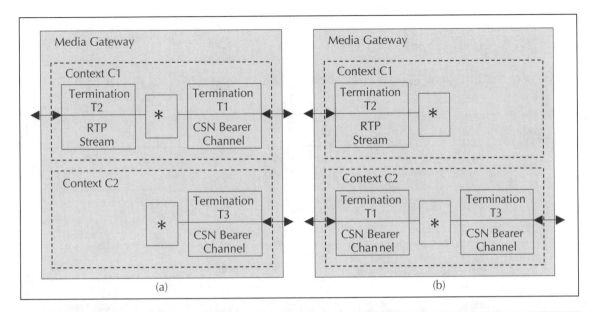

Figure 17.6 Example Call waiting scenario in Megaco

transportation of signalling data. In the TCP/IP network model, SCTP resides in the transport layer, alongside TCP and UDP.

There are two main differences between SCTP and TCP. These are Multihoming and Multistreaming. In chapter 11 we have seen that in the signalling network all nodes have sufficient redundancy. For signaling data transfer over IP, similar facility needs to be provided by the Sigtran. Through *Multihoming* SCTP supports multi-homed nodes, i.e. nodes which can be reached under several IP addresses. If we allow SCTP nodes to support more than one IP address, during network failure data can be rerouted to alternative destination IP addresses. This makes the nodes fault tolerant. In TCP if a packet is lost, the connection effectively stops while it waits for the retransmission to happen. This phenomenon where packets are blocked by a packet in front which has been lost is known as Head-of-Line Blocking. *Multistreaming* is an effective way to limit Head-of-Line Blocking. The benefit in having multiple independent data streams is, if a packet is lost in one stream, while that stream is on wait for the retransmission, the remaining unaffected streams can continue to send data.

One of the main roles of Sigtran is to tunnel SS#7 signaling traffic. Therefore, to have a proper convergence, all the functions of SS#7 stack as discussed in section 11.4.1 need to be emulated in the IP network. Figure 17.7 is the protocol stack of

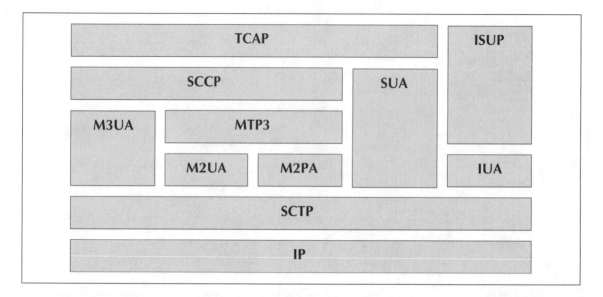

Figure 17.7 Sigtran Protocol Stack

Sigtran. This stack is the SS#7 replica in the IP domain. In Sigtran, User Adaptation layers (UA) are defined to identify SS#7 services. They are named according to the services they replace. For example M2UA (MTP2 User Adaptation layer) in Sigtran will provide the same services, look and feel to its users as MTP2 (Message Transfer Part 2) does in the SS#7 network. M2PA (MTP2 Peer-to-peer Adaptation layer) on the contrary, will provide MTP2 services in a peer-to-peer manner, such as transparent SG to SG connection over the IP network. The role of M3UA is to provide MTP3 user adaptation APIs. IUS offers ISUP User Adaptation APIs. SUA (SCCP User Adaptation layer) provides the means by which an application part (such as TCAP) on an IP SCP may be accessed by a SCP in the PSTN side via an IP SG. To access a SG, we need a pointcode attached to the SG. However, to access an SUA on the IP SCP, we do not need a pointcode for the IP SCP. Functions of TCAP, SCCP, ISUP, MTP are already defined in chapter 11.

17.7 CALL ROUTING

In VoIP, call routing can be divided into four groups. These will be IP to IP, IP to PSTN, PSTN to IP and PSTN to PSTN via IP. Hardware elements, nodes, and protocols used by these groups are not same. For example over the LAN, VoIP (IP to IP) may not need a signaling gateway at all. In following section we discuss the call flow for some of these protocols.

17.7.1 SIP to SIP Call flow

Figure 17.8 depicts a typical example of a SIP message exchange taken from RFC3161. This relates to a communication between two users "A" (Alice in RFC3161) and "B". (Bob in RFC3161). In this example each message is labeled with the letter "F" and a number for reference by the text. In this example, "A" uses a softphone to call "B" over the IP network. Also there are two SIP proxy servers in the system that act on behalf of "A" and "B" to facilitate the session establishment. "A" calls "B" using B's SIP URI (Uniform Resource Identifier). It has a similar form as an email address, typically containing a username and a host name. In this case, it is sip:bob@biloxi.com, where biloxi.com is the domain of B's SIP service provider. "A" has a SIP URI of sip:alice@atlanta.com. A's URI could be sips:bob@biloxi.com to signify a secured URI.

In this example, the transaction begins with A's softphone sending an INVITE request addressed to B's SIP URI. INVITE is a SIP method indicating a connection request. The INVITE request contains a number of header fields. The INVITE (message F1 in Figure 17.8) might look like this:

```
INVITE sip:bob@biloxi.com SIP/2.0
Via: SIP/2.0/UDP pc33.atlanta.com;branch=z9hG4bK776asdhds
Max-Forwards: 70
To: Bob <sip:bob@biloxi.com>
From: Alice <sip:alice@atlanta.com>;tag=1928301774
Call-ID: a84b4c76e66710@pc33.atlanta.com
CSeq: 314159 INVITE
Contact: <sip:alice@pc33.atlanta.com>
Content-Type: application/sdp
Content-Length: 142
```

The first line of the text-encoded message contains the method name INVITE. The lines that follow are a list of header fields. To contains a display name (Bob) and a SIP or SIPS URI (sip:bob@biloxi.com) towards which the request was originally directed. SIPS is used for secured transfer like HTTPS. The from field contains a display name (Alice) and a SIP URI (sip:alice@atlanta.com) that indicate the originator of the request.

Since the softphone does not know the location of "B" or the SIP server in the biloxi.com domain, the softphone sends the INVITE to the SIP proxy server that serves A's domain, atlanta.com. The proxy server receives the INVITE request and sends a 100 (Trying) response back to A's softphone. The 100 (Trying) response indicates that the

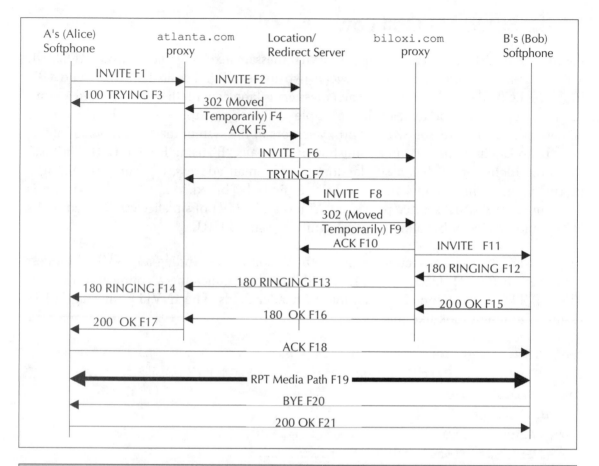

Figure 17.8 SIP session setup example with SIP trapezoid

INVITE has been received and that the proxy A (atlanta.com) is working on her behalf to route the INVITE to the destination. The proxy A sends the INVITE to the location server to determine whether the user B is available or not. If B has moved to a new place that information will be communicated to proxy A. Proxy A then sends the INVITE to the proxy B (biloxi.xom) at the other end. The biloxi.com proxy server receives the INVITE and responds with a 100 (Trying) response back to the atlanta.com. The proxy server consults the location service, that contains the current IP address of B. B's SIP phone receives the INVITE and as a result B's phone rings. B's SIP phone indicates this in a 180 (Ringing) response, which is routed back to A through both the proxies in the reverse direction. Each proxy uses the Via header field to determine where to send the response and removes its own address from the top. B decides to answer the call. When he picks up the handset, his SIP phone sends a 200 (OK) response to indicate that the call has been answered. A and B enters into conversation. When they are done, they hang-up, resources are released.

17.7.2 SIP to PSTN Call flow

In case of SIP to PSTN call let us take the case of party A (Alice) calling party B (Bob) from an IP phone with address 192.168.3.123 to a PSTN phone with phone number 011-31313131 using a user agent from her computer. This is depicted in Figure 17.9. She dials in 011-31313131. This number gets converted to enum e.164 format i.e +13131313110. Last three digits of this are used as the domain name for the SIP server to be searched to route the message to. In this case, the domain name turns out to be 1.1.0, which is the SIP proxy server. Before starting to route this message, the local SIP proxy queries local database. The database can be an equivalent of HSS (Home Subscriber Subsystem) or HLR. The database will have routing information along with personalization and provisioning information of the user. The proxy finds out if the user has the facility of calling the person or not.

Foreign SIP server on receiving this request, queries its database and finds out that the number belongs to the PSTN. Then SIP server, triggers a call agent/ *media gateway controller.* The type of trigger/event will depend upon the type of message the SIP server has received. Depending on the type of trigger received by call agent, it will either contact

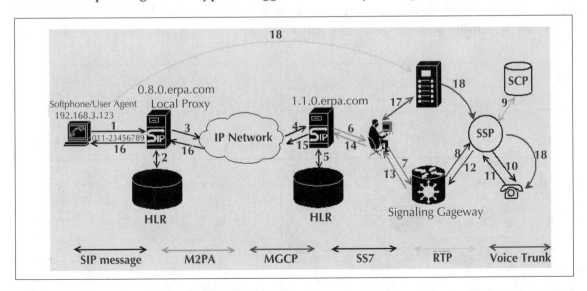

Figure 17.9 SIP to PSTN call flow

signaling gateway or media gateway. If the trigger is corresponding to some call establishment/maintenance/tearing then signaling gateway will be queried. This Signaling gateway is responsible for converting message in to PSTN understandable format. On reaching the destination SSP (Service Switching Point), this request will get replied and travel back to the originating MGC. This MGC's Signaling Gateway again converts this PSTN signal to SIP signal in the opposite direction.

The Call Agent/Media Gateway control also contacts Media Gateway using MGCP for establishing an RTP path and for Codec Conversion. So a RTP path is established between caller (Party A) in IP and callee (Party B) in PSTN (through media gateway).

17.7.3 PSTN to IP Call flow

In this section we look at the PSTN to IP call flow (Figure 17.10). When Party B (PSTN) tries to call party A (IP), assume that party A has a number which can be dialled from a normal DTMF telephone. Now as the idea is to have maximum traffic transferred on IP network, the SSP to which the calling party is linked or associated should transfer SS#7 messages onto the IP network via the signaling gateway associated with it. But for this every SSP should be having a Media/Signaling gateway linked to itself. However, this doesn't seem practical. To make the design a little easier and practical, in case the SSP doesn't have a media/signaling gateway associated with it, it handles those messages to some SSP which has a media/signaling gateway.

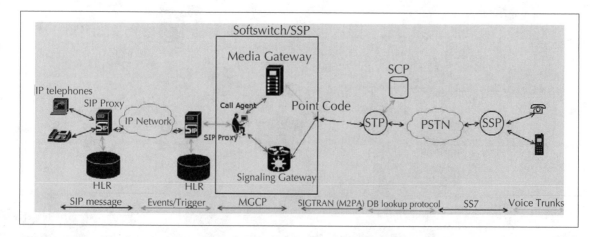

Figure 17.10 PSTN to SIP call flow

Let us suppose that the number dialed by the person is 011-23567890, and then at the first signaling gateway, the SIP proxy would know that the destination SIP proxy is in Delhi (following the same convention as the earlier case). So as the normal SIP message routes through IP network, this message also routes till it reaches the SIP proxy at Delhi.

After the SIP invitation has been received by the destination, acknowledgement follows the same way back. Now an RTP path is established between the source and destination and voice packets flow by. Here again protocol to control Media Gateway is MGCP and the protocol to carry SIP messages is M2PA.

17.8 VOICE OVER IP APPLICATIONS

In addition to voice communication, PSTN offer many other services like Caller Identification, 800 number translation, 900 premium services, and various Intelligent Networks services. These types of services are not offered by Internet telephony. If these Intelligent Network (IN) services are integrated within the Internet Telephony then a whole new range of services can be offered. Therefore, to create new applications Internet telephony needs to be programmed.

The key to programming Internet telephony services with SIP is adding logic that controls behavior at each of the system elements. In a SIP proxy server, this logic determines how the packet should be formatted, where the requests will be proxied to, how

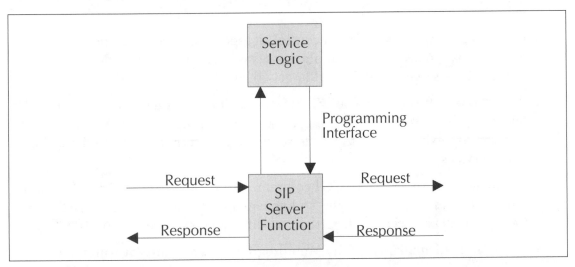

Figure 17.11 Model for programming SIP services

the responses should be processed, so on and so forth. A simple service such as call forwarding based on time of day will require logic in the SIP server to obtain the time when a call setup request arrives. Based on the time, the proxy will forward request to a particular destination. The logic can direct the server's actions based on many parameters defined by the subscriber. These could be time of day, caller, call subject, session type, call urgency, location of the subscriber, media composition, data obtained from directories, data obtained from Web pages, etc. Based on some conditions, the logic may also instruct the server to generate new requests or responses. The basic model for providing logic for SIP services is shown in Figure 17.11.

17.8.1 SIP CGI

In the Web, CGI (Computer Gateway Interface) is the most flexible mechanism for creating dynamic content. As SIP's functionality is similar to that of HTTP, some researchers applied CGI interfaces to Internet telephony. This is because CGI possesses the following characteristics:

- Language Independence: CGI works with Perl, C, VisualBasic, Tcl, and many other languages.

- Exposure of All Headers: CGI exposes the application to all header content in an HTTP request through environment variables. This approach can be directly applied to SIP because its methods of encoding messages are similar to those in HTTP.

- Creation of Responses: CGI can control all aspects of a response, including headers, response codes, and reason phrases, as well as content. This flexibility helps in SIP where services are defined largely through response headers.

- Access to Any Resources: The CGI script is an ideal starting point for creating IP telephony services because it is a general-purpose program whose flexible interface can use existing APIs to let the service logic access an unlimited set of network services.

- Component Reuse: Much of CGI components provides easy reading of environment variables and easy parsing and generation of header fields. As SIP reuses the basic syntax of HTTP, these tools are immediately available to SIP CGI.

- Environment Familiarity: Many Web programmers are familiar with CGI.

- Easy Extensibility: Because CGI is an interface rather than a language, it is easy to extend and reapply to other protocols, such as SIP.

17.8.2 Call Processing Language

While SIP CGI is an ideal service creation tool for trusted users, it is too flexible for service creation by untrusted users. Therefore a new scripting language has been developed. This is called the Call Processing Language (CPL), which allows untrusted users to define services. Users can upload CPL scripts to network servers. The logic can be read in and verified and the service instantiated instantly.

17.9 IP Multimedia Subsystem (IMS)

IP Multimedia Subsystem (IMS) is an emerging international standard, which looks at total convergence of voice and multimedia. Some literatures even refer IMS as "All IP network". IMS was specified by the Third Generation Partnership Project (3GPP/3GPP2) and now being embraced by other standards bodies including ETSI. It specifies interoperability and roaming. It provides bearer control, charging and security. It is well integrated with existing voice and data networks. This makes IMS a key enabler for fixed-mobile-multimedia convergence with value-based charging. For a normal user, IMS-based services enable person-to-person and person-to-content communications in a variety of modes that include traditional telephony services and non-telephony services such as instant messaging, unified messaging, push-to-talk, video streaming, multimedia messaging, text, fax, pictures and video, or any combination of these in a personalized and controlled way. For a network operator, IMS takes the concept of layered architecture one step further by defining a horizontal architecture, where service enablers and common functions can be reused for multiple applications. IMS will meet the following requirements:

- Separation of the access and transport layer from the services layer
- Consistent mechanisms for authenticating and billing end users
- Consistent mechanisms for sharing user profile information across services
- Session management across multiple real time communication services
- Compatibility with existing Advanced Intelligent Networks (AIN) services (Toll free 800, Premium Service 900, Calling name, Local Number Portability, Customized Applications for Mobile Networks Enhanced Logic (CAMEL), American National Standards Institute-41, etc.)
- Coarse-grained bearer QoS control
- Transparent interworking with legacy TDM networks, which will support numbering plans, progress tones etc.

- Convergence of wireline and wireless services

- Authentication, user management and charging based on existing 2G functions

- Blending of voice and real time communications services including Instant Messaging

- Consistent and blended graphical user interface

- Open standard interfaces and APIs for new services by service providers and 3rd parties

IMS will offer following converged services:

- Advanced service components like Presence, Instant Messaging, Push-to-talk over Cellular (PoC), etc

- Multimedia call like VoIP, PoC, etc)

- Multimedia messaging (MMS, etc)

- Group services (Collaboration, Buddies list, PoC, etc)

- Infotainment (Audio Visual distribution, Interactive Gaming, etc)

The IMS services architecture is a unified architecture that supports a wide range of services enabled by the flexibility of SIP. As shown in Figure 17.12, the IMS architecture is a collection of logical horizontal functions, which can be divided into three major layers:

- Communication Layer

- Session Control Layer

- Applications or Service Layer

In the IMS architecture, Call Session Control Function (CSCF) is used in the session control layer. The session control layer contains the Call Session Control Function (CSCF), which provides the registration of the endpoints and routing of the SIP signaling messages to the appropriate application server. The session control layer includes the Home Subscriber Server (HSS) database that maintains the unique service profile for each end user. The end user's service profile stores all of the user service information and preferences in a central location. This includes an end user's current registration information (i.e., IP address), roaming information, telephony services (i.e., call forwarding information), instant messaging service information (i.e., buddies list), voice mail box options (i.e., greetings), etc. Media resource function (MRF) includes functions related to

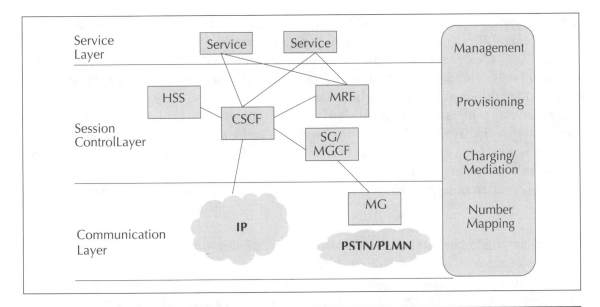

Figure 17.12 Simplified architecture of IMS

conference booking and floor management. Conference booking provide booking information like start time, duration, list of participants etc. Through floor control, end users (participants, or chairman of the conference) can influence floor and provide information to the MRF Controller on how incoming media streams should be mixed and distributed.

17.10 MOBILE VoIP

Mobility and wireless issues have not been considered till date in detail within the scope of IP telephony. In particular, while mobility has been considered to some extent within SIP, it has not been addressed comprehensively within H.323 or Megaco either. In a VoIP application, mobility may include terminal mobility, user mobility, and service mobility. Terminal mobility refers to the ability for a terminal to change physical location while the ongoing voice connection is maintained. User mobility is defined as the ability for communications of the mobile user irrespective of the terminal type in use. Service mobility is the ability for a user to access a particular service independent of user and terminal mobility.

In the context of VoIP, roaming refers to the ability that connectivity between endpoints are assured even while one or both endpoints are moving. Such reachability can either be discrete or continuous. Discrete reachability is service portability, implying no on-line reachability and communications taking place while moving. Continuous reachability is the service mobility allowing seamless communication continuity while roaming. Obviously, mobility encompasses portability, and requires the on-going connection to be handed off when a mobile terminal is on the move. Upon crossing a region boundary, a handoff must be initiated; otherwise, the connection is broken and the ongoing conversation is interrupted. Mobility management is the key to enabling mobile Internet telephony service over connectionless IP networks. The core operations include registration, call establishment, roaming, and handoff. In H.323 or SIP there is no provision for support for roaming or handoff handling, and callee location tracking and location update.

When an H.323 terminal moves across different subnets during a call, it causes the IP address to change. This results in on-going connection to be broken. In the intrazone handoff shown in Fig. 17.13 (a), both subnets are under the management of the same GK (Gatekeeper), whereas in the interzone roaming shown in Fig. 17.13 (b), they are under the management of different GKs within the same ITG (Internet Telephony Gateway).

The existing activities in the international standards bodies toward VoIP mobility include efforts made by ETSI TiPHON (Telecommunications and Internet Protocol Harmonization over Networks) Working Group 7, and ITU-T Study Group 16 H.323 Mobile Annex. TiPHON's mandate is to develop specifications (protocol profiles and test suites) to enable end-to-end telephony and multimedia communications services

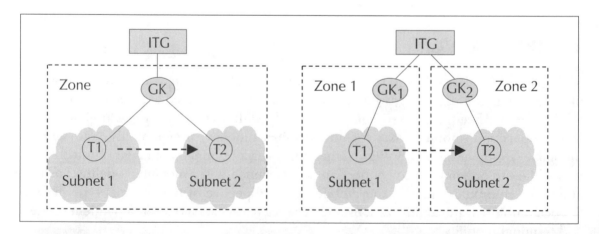

Figure 17.13 Handoff/Roaming scenario (a) Intra-zone handoff, (b) Inter-zone roaming

over Next Generation Networks. The standards will cover VoIP mobility including roaming and handoffs.

References/Further Reading

1) Source for Open Source Communication: www.Vovida.org.

2) Performance Technologies Learning Center, http://www.pt.com/learning.html.

3) Voice over IP, www.protocol.com.

4) SIP Server Technical Overview; SIP Server version 2.0, April, 2004, http://www.radvision.com.

5) RTP: A Transport Protocol for Real-Time Applications, IETF RFC1889, January 1996, www.ietf.org.

6) Real Time Control Protocol (RTCP) attribute in Session Description Protocol (SDP), RFC3605, October 2003, www.ietf.org.

7) Real Time Streaming Protocol (RTSP), RFC 2326, April 1998, www.ietf.org.

8) SIP: Session Initiation Protocol, RFC3261, June 2002, www.ietf.org.

9) Framework Architecture for Signaling Transport, IETF RFC2719, October 1999, www.ietf.org.

10) Stream Control Transmission Protocol, IETF RFC 2960, October 2000, www.ietf.org.

11) Megaco IP Phone Media Gateway Application Profile, RFC3054, January 2001, www.ietf.org.

12) Instant Messaging / Presence Protocol Requirements, RFC2779, February 2000, www.ietf.org.

13) Session Initiation Protocol (SIP)-Specific Event Notification, RFC3265, June 2002, www.ietf.org.

14) Resource ReSerVation Protocol (RSVP), RFC2205, September 1997, www.ietf.org.

15) The COPS (Common Open Policy Service) Protocol, RFC2748, January 2000, www.ietf.org.

16) All-IP Core Network Multimedia Domain, IP Multimedia Subsystem–Stage 2, 3GPP2 X.S0013-002-0 1, Version 1.0 2, December 2003.

17) Vijay K. Gurbani and Xian-He Sun, Terminating Telephony Services on the Internet, IEEE/ACM Transactions On Networking, Vol. 12, No. 4, p571, August 2004.

18) Jonathan Rosenberg, Jonathan Lennox and Henning Schulzrinne, Programming Internet Telephony Services, IEEE Network, May/June 1999, p42.

19) Wanjiun Liao, Mobile Internet Telephony: Mobility Extension to H.323 IEEE Transaction on Vehicular Technology, Vol. 50, No. 6, November 2001, p1403.

20) Roch H. Glitho, Ferhat Khendek, Alessandro De Marco, Creating Value Added Services in Internet Telephony: An Overview and a Case Study on a High-Level Service Creation Environment, IEEE Transaction On Systems, Man, And Cybernatics–Part C: Applications And Reviewes, Vol. 33, No. 4, November 2003.

REVIEW QUESTIONS

Q1: What is SIP? How does SIP handle call setup and teardown of calls?

Q2: What are the basic differences in functionality between H.323 and SIP ?

Q3: What are the different realtime protocols available for realtime data transmission over IP ?

Q4: What are the different elements in a VoIP architecture where a call is originated in IP through SIP and terminating into an IP phone in an IP network?

Q5: What is Sigtran ? What is it used for?

Q6: Describe Voice over IP applications. What is SIP CGI?

CHAPTER 18

Security Issues in Mobile Computing

18.1 INTRODUCTION

Mobile computing is pervading our society and our lifestyles with a high momentum. Mobile computing with networked information systems help increase productivity and operational efficiency. This however, comes at a price. Mobile computing with networked information systems increase the risks for sensitive information supporting critical functions in the organization which are open to attacks.

The fundamental premise of mobile computing is that the information will be accessed from outside of the organization. As long as the information is within the four walls, the environment will be better known. It may be easier to control this environment and make it secure. When the information or computing environment is outside the controlled environment we do not have much control either from its users or usage patterns. Today, all the computers of the world are interconnected through extranet. Moreover, in a majority of cases, mobile computing uses wireless networks. Wireless media works on the principle of broadcast; information is radiated to everyone within the radio wave range thus increasing the security threats. Unlike a physical attack, cyber attacks can be replicated quite easily. Therefore, unless special care is taken, all systems are open to attack. This chapter discusses different techniques to secure information over mobile computing environment.

18.2 INFORMATION SECURITY

In any defense system, we need to know our enemy. We also need to determine possible areas—weak points, vulnerabilities—where the enemy may attack. We need to build a defense system around these vulnerabilities. To build an information security system, we need to answer the following questions:

- Who is the enemy?

- What are the vulnerabilities? What are the weak links in the system?

- What could be the possible exploitation of these vulnerabilities by the resulting attacks?

- What needs special protection?

- To protect our assets from attack, we need to build a security system. How much does the security system cost in terms of money, resource and time?

- When the security system is deployed, to what extent will it affect the openness and add to inconvenience?

- Is prevention better that cure? If prevention is expensive or impractical, what is the strategy to recover from the loss following an attack?

There is no absolute security. What may appear to be absolute security in one context may not be absolute security in another context. Therefore, while building a security system, we need to arrive at a proper balance amongst the answers emerging from the above questions. In a mobile environment, the user roams through different networks with heterogeneous security infrastructure. In such an environment where device mobility and network mobility is a necessity, offering homogenous service over heterogeneous devices and networks is the key. In such an environment weak security link from a wireless network could become a point of vulnerability for the entire system. Therefore, in a mobile computing environment, it is necessary to have a robust security and trust infrastructure.

18.2.1 Attacks

A security system is a system to defend our assets from attacks. In the physical world, these attacks are carried out at the weak points in the defense system. Likewise in the electronic world, attacks are carried out at the point of vulnerability. When the vulnerability is exploited for some interest or selfish motive, it is an attack on the system. Of course there could be occasions where the vulnerability is exposed by accident as well. Where the vulnerability is exploited, there is a loss. This loss can be either of static information asset (static asset) or an information asset in transit (dynamic asset). If we look at an information system, static assets cover a large portion of the asset base. All the databases, files, documents, etc. in the computers fall in this category. Examples of attacks on static asset are virus deleting files in a computer or jamming a network. An example of

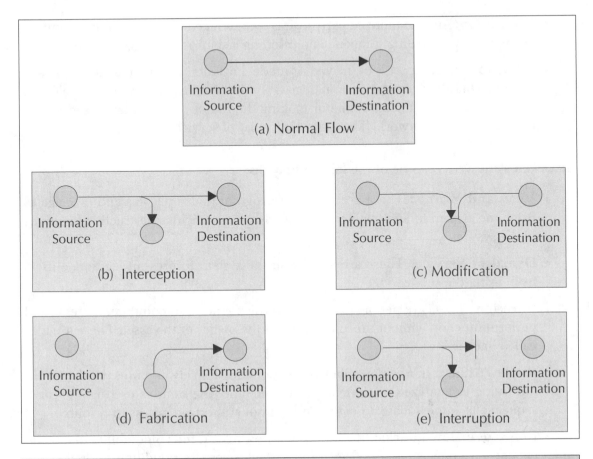

Figure 18.1 Types of attacks

an attack on a dynamic asset is the theft of a credit card number while a user is doing a credit card transaction on the web.

Attack on dynamic assets can be of the following types (Figure 18.1):

- **Interception:** An unauthorized party gaining access to an asset will be part of this attack. This is an attack on **confidentiality** like unauthorized copying of files or tapping a conversation between parties. Some of the sniffing attacks fall in this category.

- **Modification:** An unauthorized party gaining control of an asset and tampering with it is part of this attack. This is an attack on **integrity** like changing the content of a message being transmitted through the network. Different types of man-in-the-middle attacks are part of this type of attack.

- **Fabrication:** An unauthorized party inserts counterfeited objects into the system; for example, impersonating someone and inserting a spurious message in a network.

- **Interruption:** An asset is destroyed or made unusable. This is an attack on **availability**. This attack can be on a static asset or a dynamic asset. An example could be cutting a communication line or making the router so busy that a user cannot use a server in a network. These are all Denial of service attack.

Attacks on static assets can be of the following types:

- **Virus and Worms:** These are a type of program that replicates and propagates from one system to another. Most of the virus do malicious destructive functions in the system.

- **Denial of Service:** These are attacks on the system to prevent legitimate users from using the service.

- **Intrusion:** These are people or software, which enter into computer systems and perform function without the knowledge of the owner of the asset. These are also called hackers.

- **Replay Attack:** In a replay attack the opponent passively captures the data without trying to analyze the content. At a later time, the same is used in the same sequence to impersonate an event and gain unauthorized access to resource.

- **Buffer overflow attacks:** In a buffer overflow attack, the vulnerability of an executable program is exploited to force a stack overflow condition, inducing the program counter of the process to change. The program counter is then manipulated to do the work for the attacker.

- **Trapdoor attacks:** These are exploitations of some undocumented features of a system. Undocumented functionality are designed to debug, service, support or take control of the system.

A security system needs to be so designed that the system is able to counter and recover from attacks.

18.2.2 Components of Information Security

For centuries, information security was synonymous with secrecy. The art of keeping a message secret was to encrypt the message and thus hide it from others getting to know of it. However, in today's netcentric electronic world, the taxonomy of information security is much beyond encryption. Information security needs to cater to all the possible attacks related to confidentiality, integrity, availability, non-repudiation, authorization, trust and accounting (CIANATA). **Confidentiality** is the property where the information is kept secret so that unauthorized persons cannot get at the information. **Integrity** is the property of keeping the information intact. **Availability** is the property of a system by which the system will be available to its legitimate users. **Non-repudiation** is the property by which the identity of both sender and receiver of the message can be identified and verified. **Authorization** is the property by which the user's properties can be associated to the information access. **Trust** is the property of expectation, confidence, and belief over time. **Accounting** is the property of calculating the fee for a service rendered.

Confidentiality

Confidentiality is ensured through encryption of the data. To a person a comprehensible message is written in a particular language. The language can be English, Hindi, French or any other language. These messages are called plaintext or cleartext messages. Through encryption (or encipher) we disguise this message in such a fashion that it is no longer understandable by either a person or a machine. An encrypted message is called ciphertext. The process of converting a ciphertext back into plaintext is called decryption (or deciphering). Plaintext need not be a written text. It can even be an audio or video message as well. When leaders of two countries talk, the message is encrypted so that a man eaves dropping cannot make any sense of the conversation. The plaintext message can also be a data file in the computer disk. Figure 18.2 depicts the process of encryption and decryption.

In cryptography there are two components, viz., **algorithms** and **protocols**. A cryptographic algorithm is a mathematical function used for encryption and decryption, and

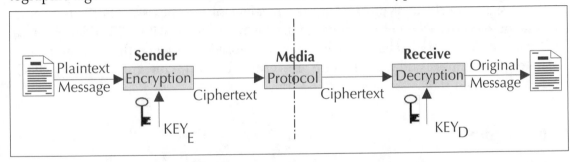

Figure 18.2 Encryption and decryption with a key

protocol relates to the process and procedure of using algorithms. A protocol is the way algorithms are used to ensure that the security is ensured and the system is less prone to attack. In a security system the plaintext message is encrypted by using a key KEY_E. The encrypted message is then sent from the sender to the received through a media (wired, wireless, or even postal) using some protocol. The encrypted message is then decrypted using a key KEY_D to extract the original message. A cryptographic key is generally a large number. The range of possible values of a key is called **keyspace**. The larger the keyspace is, the more difficult it is for an attacker to guess the key and restore the original message. Therefore a larger keyspace makes a ciphertext more secure. This is similar to a lock. A conventional lock of 11 levers is more secure compared to a 7-lever lock.

The art of keeping message secure using the science of encryption and decryption is called cryptography. People who practise cryptography are called **cryptographers**. There are people who try to break the secrecy of encryptions. These are for many purposes; some are for research purposes to measure the strength of the security and some, for stealing the information. Some are hackers who try to break the security for fun or for a price. These people who try to break the secrecy of the cryptography are called **cryptanalysts**. The practice of cryptanalyst is called **cryptanalysis**. There is another science in security engineering. This is called **steganography**. Steganography is the science of hiding secret message in other messages so that the existence of the secret message is concealed; for example, sending some secret message by changing some bits in a large picture message. By looking at the picture, others will not be able to guess that in reality the picture is carrying a secret message.

Integrity

Integrity is to ensure the integrity of the message. Integrity is achieved by adding additional information in to the message. This is done through checksums, message digests or digital signature. In a crypto system, the receiver of the message checks this extra information to verify whether the message has been tampered with. This is similar to a bank cheque. A cheque issued to a customer is honored only when the customer signs it. The cheque number and the signature are verified to ensure integrity. Integrity check is advised for both static asset and asset on transit.

Authorization

Authorization deals with privileges. In any transaction, there is a subject (a person) and an object (data items or file). The subject wants some function to be performed on the object. The privilege to an object is defined through ACL or Access Control List. ACL is used while allowing access to the object. The privilege on an object can be read, write, or execute. Besides objects there need to be privilege-based type of subjects. This is done through authorization.

Authorization is implemented through policy-based resource accessibility. In an organization (or society) where there is a hierarchy, there will be certain functions allowed to certain levels in the hierarchy. A clerk in a corporation may have authorization to approve an expense claim less than a specified threshold, supervisors might have a higher limit, and vice-presidents might have a still higher limit. Similarly, role-based security will be used when an application requires multiple layers of authorization and approvals to complete an action. Privilege management infrastructure together with the role-based authorization allows the administration and enforcement of user privileges and transaction entitlements. In the authorization process, users are checked to see if they have the required rights to access the resource. If they have been granted the required rights, they can access the resource, otherwise they are denied access.

Non-repudiation

Authentication and **Non-repudiation** have some overlapping properties. Authentication is a process by which we validate the identity of the parties involved in a transaction. In non-repudiation we identify the identity of these parties beyond any point of doubt. Non-repudiation can be considered as authentication with formal record. These records will have legal bindings. Like a signature in a cheque, using digital signature we achieve non-repudiation.

Availability

Media management is not within the scope of security protocols and algorithms. However, media management is part of the larger security framework. Media management is needed to ensure **availability** of service. For a message a confidentiality may be maintained; also, the integrity is intact but an attacker can manipulate the media to make sure that the message does not reach the destination. This is like there is no theft of power, power quality is good, but someone blows the transmission line of the power grid.

Attack on availability happens for industrial espionage or from political motivation. During a festive season, one company may target to block the e-commerce site of a competition. In a social framework, someone may try to stop people's voice by using threats or other means of intimidation to compel the author to remove the web page. If these methods prove unsuccessful, various denials of service attacks can be launched against the site to make it impossible to access. In less high-profile cases, people often enjoy far less support for exposing corruption or criticizing employers and particularly litigious organizations. Also, there need to be some way where terrorist organizations or dictators cannot block the mass opinion. This field of research area is called **Censorship-resistant Publishing**. Censorship-resistant publishing is achieved through document entanglement.

Trust

Computers rely on user authentication and access control to provide security. Within a network, it may be safe to assume that the keyholder is authentic, and the keyholder is using the key assigned to him or her. However, these strategies are inadequate for mobile computing environments with high level of flexibility. Mobile computing lacks centralized control and its users are not all predetermined. Mobile users expect to access resources and services anywhere and anytime. This leads to serious security risks and access control problems. To handle such dynamic everchanging context, **trust**-based security management is necessary. Trust involves developing a security policy, assigning credentials to entities, verifying that the credentials fulfill the policy. Also, we need delegation of trust to third parties, and reasoning about users' access rights.

Accounting

For any service, the service provider needs to be paid. The service can be either a content service or a network service. Accounting and billing is a very critical aspect in mobile computing environment. **Accounting** is the process by which the usage of the service is metered. Based upon the usage, the service provider collects the fee either directly from the customer or through the home network. This will be true even if the user is roaming in a foreign network, and using the services in the foreign network.

RADIUS (Remote Authentication Dial In User Service) protocol (RFC 2865) has been in use for a long time for the AAA (Authentication, Authorization, and Accounting) functions in Internet. With the demanding service requirement of mobile computing, it is now apparent that RADIUS is incapable of supporting all these complexities. A new protocol called Diameter (RFC 3588) has been released to address the AAA needs for data roaming and mobile computing. Diameter can work in both local and roaming AAA situations.

18.3 SECURITY TECHNIQUES AND ALGORITHMS

Generally the encryption algorithms are divided into two main groups. These are symmetric key encryption and public key encryption. In a symmetric key encryption, the key used for decryption is the same as the key for encryption. In some cases of symmetric encryption, even the algorithm used for encryption and decryption is the same. In the case of public key algorithms, the key used for decryption is different from the key used for encryption.

18.3.1 Stream Ciphering and Block Ciphering

In stream cipher, a bit or a byte is taken at a time and encrypted. The algorithm looks at the input plaintext as a stream of bits and encrypts them one bit (or byte) at a time as the stream progresses. In this technique, the length of the plaintext and the key size will be same. Wireless LAN (WiFi) uses stream cipher. In this methodology, the key has to be unique for every encryption. If the same key is used for multiple packets, and these packets can be captured, there is vulnerability. The other technique is block cipher. In a block cipher, one block of plaintext is taken as a whole and used to produce a ciphertext block of equal length. Typically a block of 64 bits (8 octets) or 128 bits (16 octets) is used for block cipher. Majority of cryptosystems use block cipher.

18.3.2 Symmetric Key Cryptography

In a symmetric key cryptography, the same key is used for both encryption and decryption. This is like a lock where the same key is used to lock and unlock. In cryptography, symmetric key algorithms are in use for centuries; that is why symmetric key algorithms are called conventional or classical algorithms as well. In this type of encryption, the key is secret and known only to the encrypting (sender) and decrypting (receiver) parties. Therefore, it is also known as a secret key algorithm. Some authors refer to symmetric key cryptography as shared key cryptography as well. This is because the same key is shared between the sender and the receiver of the message. The unique key chosen for use in a particular transaction makes the results of encryption unique. Selection of a different key causes the cipher that is produced for any given set of inputs to be different. The cryptographic security of the data depends on the security of the algorithm used and the key used to encipher the data. The strength of the security depends on the size of the key. Unauthorized recipients of the cipher who know the algorithm but do not have the correct key cannot derive the original data algorithmically. However, anyone who does have the key and the algorithm can easily decipher the cipher and obtain the original data. Symmetric key algorithms are much faster compared to their asymmetric (public key) counterparts.

In a symmetric key cryptography, there are four components. These are **plaintext**, **encryption/decryption algorithm**, **secret key** (key for encryption and decryption), and the **ciphertext**. In Figure 18.2, if we make $Key_E = Key_D$, this becomes a symmetric key algorithm. There are many symmetric key algorithms. The most popular symmetric key algorithms are:

DES: Data Encryption Standard, this algorithm is the most widely used, researched and has had the longest life so far.

3DES: This is a modification of DES. In this algorithm, DES is used 3 times in succession.

AES: Advances Encryption Standards, this is the current accepted standard for encryption by FIPS (Federal Information Processing Standards) of USA.

Skipjack/FORTEZZA: This is a token-based symmetric algorithm used by defense personnel in the US.

DES (Data Encryption Standard)

In the late 1960s, IBM set up a research project in computer cryptography led by Horst Feistel. In 1971, the project concluded with an outcome of an algorithm named Lucifer. The original algorithm used 64-bits block and 128-bits key. IBM reduced the length of the key to fit the algorithm into a single chip. This algorithm was adopted in 1977 by NIST (National Institute of Standards and Technology) as the data encryption standard (DES). A DES key consists of 64 bits of which 56 bits are randomly generated and used directly by the algorithm. The other 8 bits are used for error detection and not for encryption.

DES employs the principle of scrambling and substitution. These processes are repeated a number of times with keys to ensure that the plaintext is completely transformed into a thoroughly scrambled bit stream. The DES can be divided into the following major functions. These are:

- Permutations of bits in a block. This is the first and last step in DES. In this step the 64-bit plaintext block is rearranged through Initial Permutation **IP**. This is done through a 64-bit register where the bits of the input block are scrambled in a particular fashion. As the last step, the reverse permutation is done through $\mathbf{IP}^{-1}$.

- A key dependent computation. This includes multiple rounds (iteration) of transformation through combination of permutation and substitution. This is in the core of the encryption function.

- Swapping of half blocks of data in each round.

- Key schedule; this breaks the 56-bit key into two 28-bit subkeys and use them to compute the bits in data blocks. In each iteration, the bits within the subkey are shifted to generate a new subkey.

- The key-dependent computation is run through 16 rounds. Each round uses the data from the previous round as input.

The beauty of DES algorithm is that the same algorithm is used for both encryption and decryption. DES demonstrates a very high avalanche effect. In an avalanche effect one bit of change in either the input data or the key changes many bits in the output. For example, in DES one bit of change in the input data changes 34 bits, whereas one bit change in the key affects 35 bits.

3DES (Triple DES): With the increase of processing power available in PC, 56 bits of key became vulnerable for attack. Therefore, to protect the investment and increase security 3DES (commonly known as Triple DES) was proposed. 3DES uses the same DES algorithm three times in succession with different keys. This increases the keysize resulting in higher security. Also, as the fundamental algorithm in 3DES is practically the DES, it is easily adaptable without additional investment. There are two different flavors of 3DES. One uses two 56-bit key and the other uses three 56-bit key. By using three 56-bit key, the effective security can be increased to the key size, to 168 bits. Till today 3DES is the most widely used algorithm for symmetric cryptography.

AES (Advanced Encryption Standard)

We have discussed that the strength of security of a cryptographic algorithms depends on the size of the key. The larger the size of the key, the longer it takes to decipher the encrypted data through brute force. With GHz of computing power easily available, 56-bit key size is found to be unsafe today. To overcome these challenges, 3DES became popular. However, 3DES was quite slow. Also, scientists found that the 64-bit block which both DES and 3DES use, may not be the best. A higher block size is desirable from efficiency and security point of view.

To overcome these drawbacks, in 1997 NIST (National Institute of Standards and Technology) in US issued a call for algorithms for advanced encryption standard or AES. According to the call for proposal, the AES standard was to have equal or better security compared to 3DES and more efficient than the 3DES. NIST also specified that AES had to be a symmetric cipher with block size of 128 bits. Also, it has to support keys of size 128-bits, 192-bits, and 256-bits. Many algorithms competed for the AES standard. Following a rigorous evaluation process in November 2001, NIST selected the Rijndael as the AES algorithm. Rijndael is named after two researchers from Belgium who developed the algorithm. They were Joan Daemen and Vincent Rijmen. Rijndael was designed to have the following characteristics:

- Resistance against all known attacks
- Design simplicity
- Speed and code compactness on a wide range of platforms.

Like DES, AES also uses permutation and substitution. However, AES does not use Feistel structure. In a Feistel structure, one half of the data block is used to modify the other half of the data and then swapped.

18.3.3 Public Key Cryptography

In symmetric key encryption we use the same key for both encryption and decryption. In public key cryptography we use two different keys, one key for encryption and a different key for decryption. As there are two different keys used, this is also called asymmetric key cryptography. The development of public key cryptography can be considered as the greatest advance in the history of cryptography. Public key cryptosystem is based on mathematical functions rather than permutation and substitution. However, it is not true that the public key cryptosystem is more secure for general purpose. There is nothing in principle, which makes one algorithm superior to another from the point of view of resisting cryptanalysis. It is computationally infeasible to derive the decryption key given only the encryption key and knowledge of the cryptographic algorithm. The encryption key and the decryption key together form a key pair. One of these keys from the key pair is made public and the other one kept private or secret. That is why this algorithm is called public key cryptosystem.

Whitfield Diffie and Martin Hellman in 1976 came up with the principle of asymmetric key or public key cryptography. Public key cryptography proposed by Diffie and Hellman solved two difficult problems of **Key distribution** and **digital signature** in

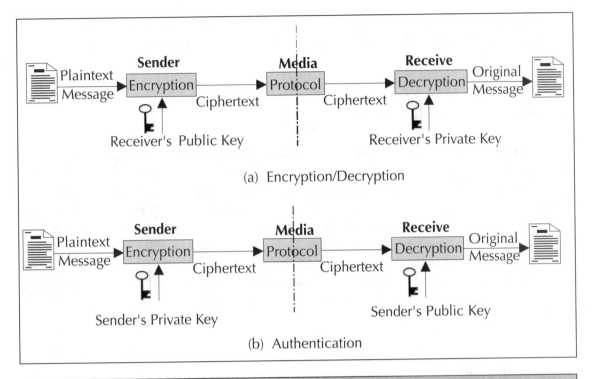

Figure 18.3 Public key cryptography

cryptography. In public key cryptography, there are six components (Figure 18.3). These are:

- **Plaintext:** This is the human readable message or data given to the public key algorithm as input for encryption.

- **Ciphertext:** This is the scrambled data produced as output of the encryption algorithm. This is a unique data and depends only on the unique key used for encryption.

- **Encryption algorithm:** This is the algorithm that does computation and various transformations on the input plaintext. The output of the transformation is too garbled to be decipherable for an intruder.

- **Decryption algorithm:** This algorithm does the reverse function of the encryption algorithm. This function accepts the ciphertext as input and does some transformation on the data so that the original data is recovered.

- **Public key:** This is one of the keys from the key pair. This key is made public for anybody to access. This key can be used either for encryption or decryption.

- **Private key:** This is the other key from the key pair. This key is called the private key, because this is kept secret. This can be used either for encryption or decryption.

There are three public key cryptosystems most widely used today. These are **Diffie Hellman, RSA**, and **Elliptic carve**.

The methodology used for encryption of data and the digital signature is different. During the encryption, the sender uses the public key of the receiver. This is because only the receiver should be able to decrypt the message using his or her own secret private key. If there is a surrogate who is able to intercept the encrypted message, he will not be able to decrypt the message, as the key required to do so is the private key. Receiver's private key is kept secret with the receiver. The methodology used for authentication or digital signature is just reverse. In case of signing the transaction, the private key of the sender is used by the sender. The receiver uses the public key of the sender to read the signature. This authenticates that the transaction was indeed done by the sender.

Diffie Hellman

Whitfield Diffie and Martin Hellman first introduced the notion of public key cryptography in 1976. In Diffie Hellman technique, secret keys are never exchanged. However, the technique allows two parties to arrive at a secret key through the usage of public keys. Communicating parties select a pair of private and public keys. Public keys are exchanged. The shared secret key is generated from the private key and the public key of the other party.

Let us assume that there are two parties A and B. A and B choose some prime number p and another number g less than p. These numbers are selected and made available to both A and B in advance. The steps followed in Diffie Hellman algorithms for key generation are as follows:

1. Let these p and g be: $p = 13$ and $g = 3$;

2. A chooses a random number S_A. This number is kept secret as a private key with A. Let this number be 5.

3. B chooses a random number S_B. This number is kept secret as a private key with B. Let this number be 7.

4. A takes g and raises it with his secret key S_A modulo p. This will be $T_A = (g \char`\^ S_A)$ mod $p \Rightarrow (3 \char`\^ 5)$ mod $13 = (243)$ mod $13 = 9$. This number 9 is A's public key. A already has chosen 5 as his private key.

5. B takes g and raises it with his secret key S_B modulo p. This will be $T_B = (g \char`\^ S_B)$ mod $p \Rightarrow (3 \char`\^ 7)$ mod $13 = (2187)$ mod $13 = 3$. This number 3 is B's public key. B has already chosen 7 as his private key.

6. Public keys of A and B are exchanged. This means A send the public key 9 to B and B send his public key 3 to A over a public channel like Internet.

7. A takes B's public key and raises it with his own private key mod p. Therefore, we now have $K_A = (T_B \char`\^ S_A)$ mod $p \Rightarrow (3 \char`\^ 5)$ mod $13 = (243)$ mod $13 = 9$.

8. B now takes A's public key and raises it with his own private key mod p in a similar fashion as A. The result will be $K_B = (T_A \char`\^ S_B)$ mod $p \Rightarrow (9 \char`\^ 7)$ mod $13 = (4782969)$ mod $13 = 9$.

9. The value of $(T_A \char`\^ S_B)$ mod $p = (T_B \char`\^ S_A)$ mod $p = 9$. Though K_A and K_B have been calculated by A and B independently; it will always be equal. Therefore, these keys K_A and K_B can now be used by A and B as the shared key for payload encryption.

Neither A nor B shared their secret key for use in symmetric encryption, but arrived at that using some properties of modulo arithmetic with prime numbers. The example above may look trivial. However, when these numbers are large, nobody can calculate the key just by knowing p, g and S_x in a reasonable period of time. An eavesdropper could not compute discrete logarithm, i.e., figure out K_A based on seeing S_B.

RSA

RSA is named after its inventors R.L. Rivest, A. Shamir and L. Adleman. It is a public key algorithm that does encryption/decryption, authentication, and digital signature. The key length is variable and the most commonly used key size is 512 bits. The key length used In India by CCA (Controller of Certifying Authorities) is 2048 bits. Key length can be large for higher security; the key length can be smaller for better efficiency. The plaintext data block is always smaller than the key length. However, the ciphertext block is the same as the key length. RSA is much slower than symmetric key encryption. That is why RSA is generally not used for payload encryption. RSA is used primarily for encrypting a secret key for key exchange.

The RSA algorithm works as follows:

1. Choose two prime numbers p and q.

2. Multiply p and q to generate n. n will be used as the modulus.

3. Calculate $\Phi(n) = (p - 1) * (q - 1)$. $\Phi(n)$ is the Euler's totient function. $\Phi(p)$ is the number of positive integers less than p and relatively prime to p.

4. Choose a number e such that it is relatively prime to $\Phi(n)$.

5. Find d such that it is multiplicative inverse of e; $d = e^{-1} \bmod \Phi(n)$.

6. (e, n) is the public key and (d, n) is the private key

7. To encrypt we use the formula (Ciphertext block) = (Plaintext block)^e mod n

8. To decrypt we use the formula (Plaintext block) = (Ciphertext block)^d mod n

Let us take an example where we choose two prime numbers $p = 7$ and $q = 17$.

Calculate $n = p * q = 7 * 17 = 119$

Find the value of $\Phi(n)$ using the formula $\Phi(n) = (p - 1) * (q - 1) = (7 - 1) * (17 - 1) = 6 * 16 = 96$.

Now we need to select an e. e will be relatively prime to $\Phi(n)$ and less than $\Phi(n)$. We can see that 2, 3, 4 have factors with 96, therefore, are not relatively prime. Whereas, 5 is relatively prime to 96. Therefore, we can choose e to be 5.

We know that $d * e = 1 \bmod \Phi(n)$, which in other words $d * e = ((Y * \Phi(n) + 1) \bmod \Phi(n)$. To find the value of d, we use the formula $((Y * \Phi(n)) + 1)/e$. Replace Y with 1 then 2 then 3 and so on until we get an Integer. When we set $Y = 4$, the equation evaluates:

$$d = (4 * 96 + 1)/5 = (384 + 1) / 5 = 385/5 = 77$$

Therefore, we get d = 77. We have just generated our key pair. The public key is (5, 119) and private key is (77, 119). We can now use this to encrypt and decrypt values.

To encrypt we use the formula

(Ciphertext block) = (Plaintext block)^e mod n. Assuming that the plaintext block is 8 bits long and the value is 65. Therefore, the ciphertext will be $(65 \text{ ^ } 5) \bmod 119 =>$ (1160290625) mod 119 = 46. To decrypt, we use the formula (Plaintext block) = (Ciphertext block)^d mod $n => (46 \text{ ^ } 77) \bmod 119 = (1.07734063167916956809383545838 5e+128)$ mod 119 = 65

The example above may look trivial and someone may think that by knowing (5, 119) one can easily find out d. This is almost impossible if the numbers are large, for example 128 bits long. Also, to know the private key, the eavesdropper needs to evaluate *p* and *q* from *n*. The eavesdropper has to factorize the number *n* to get the two large prime numbers, which is extremely hard even in a huge timeframe. RSA uses the complexity in prime factorization.

Elliptic Carve

A majority of the products and standards that use the public key cryptography use RSA for encryption, authentication, and digital signature. Due to extensive research in cryptanalysis in RSA and increase in availability of computing power, some vulnerabilities of RSA have been discovered. There are subexponential algorithms available today for breaking RSA and Diffie-Hellman algorithms. To overcome these threats, the size of the RSA key has been increasing over time. This puts a tremendous demand on computing power. Elliptic Carve Cryptography (ECC) has shown a lot of promise for higher security with lesser resource. Elliptic curve cryptography was proposed by Victor Miller and Neal Koblitz in the mid 1980s. Till date there is no subexponential algorithms available to break ECC. An elliptic curve is the set of solutions (x,y) to an equation of the form $y^2 = x^3 + ax + b$, together with an extra point O which is called the point at infinity.

ECC is believed to offer a similar level of security with a much smaller size of key. For example, it is claimed that the level of security that 1024 bits of RSA provide can be achieved by 160 bits of ECC. A 210-bit key of ECC is equivalent to 2048 bits of RSA. This makes ECC very attractive for small footprint devices like cell phones or PDAs.

18.3.4 Hashing Algorithms

Hashing functions are one-way functions used for message digests. Hash function takes an input data of any size and produces an output stream of some fixed size. The outputs are collision free. This means that two different inputs will not produce the same output. It is also not possible to derive the input from a known output. This means that if we have a message digest, it is impossible to derive the original message. The most commonly used hash functions are MD5 and SHA-1.

MD5

MD5 (Message Digest version 5) hashing algorithm is described in RFC 1321. The MD5 algorithm is an extension of the MD4 message-digest algorithm and is slightly slower than MD4. The MD5 algorithm takes a message of arbitrary length as input and produces a 128-bit 'message digest' as output. The algorithm processes 512 bits of the input message in blocks. The digest produced by the algorithm can also be considered as a 'fingerprint' of the message. It is conjectured that it is computationally infeasible to produce two messages having the same message digest. It is also conjectured that it is computationally infeasible to produce any message having a given message digest. The MD5 algorithm is intended for digital signature applications in a public key cryptosystem.

SHA

The Secure Hash Algorithm (SHA) was developed by the NIST (National Institute of Standards and Technology). SHA was first published in 1993. Later in 1995, a revised version of the algorithm was published as SHA-1. SHA processes input in 512 bits block and produces 160 bits of output. Like MD5, SHA-1 is also based on MD4 algorithm. As both MD5 and SHA-1 are based on MD4, they are quite similar in nature. However, as SHA-1 generates a longer digest of 160 bits compared to 128 bits by MD5, it is considered to be more secure.

MAC

MAC stands for Message Authentication Code. MAC is used to do the integrity check on the message. A secret key is used to generate a small fixed size data block from the message. This is similar to a checksum of the message. Both the sender and the receiver share the same secret key for MAC. When the sender has a message to be sent to the receiver, the message is sent along with the MAC. The receiver receives the message; and calculates the MAC from the message and the shared key. The receiver checks the MAC received from the sender. If they are the same, the message is considered to be in perfect state. HMAC is another mechanism for message authentication using cryptographic hash functions like MD5, or SHA-1, in combination with a secret shared key. HMAC has been defined in RFC 2104. The cryptographic strength of HMAC depends on the properties of the underlying hash function.

18.4 SECURITY PROTOCOLS

To provide confidentiality, integrity etc. we need to use different algorithms. However, we need to device protocols that will use these algorithms in such a fashion that vulner-

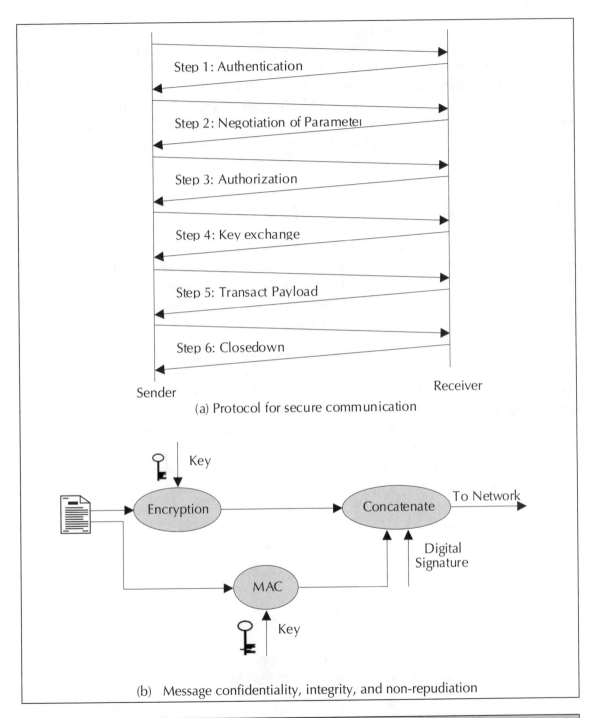

(a) Protocol for secure communication

(b) Message confidentiality, integrity, and non-repudiation

Figure 18.4 Security protocols

abilities are eliminated and security is ensured. The protocol needs to be so robust that a masquerader is unable to get the message being sent. The protocols need to ensure that if the masquerader is able to modify the message, we can detect it. There are many protocols for secured communication. One such protocol is depicted in Figure 18.4. However, the most popular protocol is SSL (Secured Socket layer–section 18.4.1). SSL was originally developed by Netscape. The Internet standards for TLS (Transport Layer-Security–section 18.4.2) and WTLS (Wireless Transport Layer Security–section 18.4.3) have been derived from the SSL protocol.

18.4.1 Secured Socket Layer (SSL)

The Secured Socket Layer or SSL protocol is used to provide security of data over public networks like Internet. It runs above the TCP/IP protocol layer and below higher level protocols such as HTTP or IMAP (Figure 18.5). SSL allows both machines (server and the client) to establish a secured encrypted channel so that all the data transacted between them are confidential and tamper-resistant.

Public-key encryption provides better authentication techniques. On the other hand, symmetric key encryption is much faster than the public key encryption. The SSL protocol uses a combination of both public key and symmetric key encryption. An SSL session begins with **SSL handshake.** SSL handshake allows the server to authenticate itself to the client using public-key techniques. Optionally, the handshake also allows the client to authenticate itself to the server. It then allows the client and the server to cooperate in the creation of symmetric key. It then uses this shared key for payload encryption, decryption, and tamper detection during the session that follows.

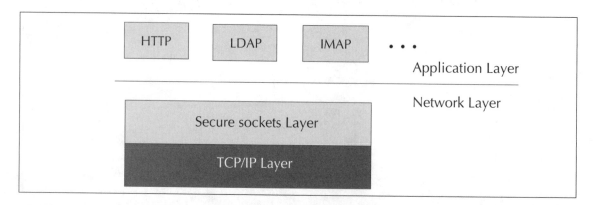

Figure 18.5 SSL layer

18.4.2 TLS

Transport Layer Security or TLS in short is a security protocol to offer secured communication at the transport layer. TLS protocol is the Internet standard and based on the SSL 3.0 protocol specification. According to RFC 2246 (TLS Protocol Version 1.0), the primary goal of the TLS protocol is to provide privacy and data integrity between two communicating applications. At the lower levels, TLS uses TCP transport protocol. The TLS protocol is composed of two layers: the TLS Handshake Protocol and the TLS Record Protocol.

The TLS Handshake Protocol provides connection security that has three basic properties:

1. Peer's identity can be authenticated using asymmetric or public key cryptography (e.g., Diffie-Hellman, RSA, etc.).

2. The negotiation is reliable: no attacker can modify the negotiation communication without being detected by the parties in the communication.

3. The negotiation of a shared secret is secured: the negotiated secret is unavailable to anybody eavesdropping in the middle of the connection.

TLS Record Protocol provides connection security that has two basic properties:

1. Privacy: The confidentiality of the data is maintained through encryption. Symmetric cryptography is used for data encryption (e.g., AES, DES, RC4, etc.). Keys for symmetric encryption are generated uniquely for each connection. These encryption algorithms are negotiated by the TLS Handshake Protocol.

2. Integrity: The connection is reliable. Message transport includes a message integrity check using a keyed MAC. Secure hash functions (e.g., SHA, MD5, etc.) are used for MAC computations.

18.4.3 WTLS

The transport layer security protocol in the WAP architecture is called the Wireless Transport Layer Security or WTLS in short. WTLS provides functionality similar to TLS 1.0 (see Chapter 8) and incorporates new features such as datagram support, optimized handshake and dynamic key refreshing. The WTLS layer operates above the transport protocol layer similar to TLS. WTLS provides the upper-level layer of WAP with a secure transport service interface that preserves the transport service interface below it. In addition, WTLS provides an interface for creating and terminating secure connections. The primary goal of the WTLS layer is to provide privacy, data integrity and authentication between two communicating applications. The WTLS protocol is optimized for low-bandwidth bearer networks with relatively long latency.

18.4.4 Multifactor Security

In a security system larger key implies higher security. This is simply because, larger key means larger lock. However, it may not be always possible to keep on increasing the size of the key. Therefore, we keep on looking for alternate methods of increasing security. One such method is splitting the key and distributing it. For example, in a bank, a locker cannot be opened with one key. It requires multiple keys. One key belongs to the customer; the other key is with the bank employee. Both the keys need to be used to open the locker. Take the example of ATM, where an ATM card and the PIN are required to withdraw cash. This technique is called multifactor security. These factors are generally a combination of 'what you have', 'what you know', and 'what you are'. Multifactor security can be a combination of any of the following factors.

What You Have

- Magnetic stripe card
- Private key protected by password
- Smart card
- Hardware token
- RF badge
- Physical key.

What You Know

- Password
- Pass Phrase
- PIN (Personal Identification Number)
- Answer to some personal questions
- Sequence of numbers
- Predetermined events.

Who You Are

- Fingerprint
- Voice Recognition
- Retinal Scan
- Hand Geometry
- Visual Recognition
- Face (picture in passport)
- Other biometric identities.

Most of the multifactor security systems in use today are two-factor ones. However, for defense systems and high security establishment three-factor securities are used. In a two-factor security any two of the above factors are used. In a three-factor security, one each from the above factors are used.

18.4.5 Digital Watermark

Watermarks are being used for a long time as a security measure. If we take a 100-rupee currency note of Reserve Bank of India and hold it in front of light, we can see Gandhi's face on the white circle. This is called the watermark in the currency note. If we photocopy a currency note using a color photocopier, we will not be able to copy the watermark. The term 'digital watermark' refers to a pattern of information inserted into a file. The file can be a digital audio file, digital video file, or a data file that identifies the file's copyright information (author, rights, etc.). The purpose of digital watermark is to provide copyright protection for intellectual property that is in digital format. Unlike printed watermark which are intended to be visible, digital watermarks are designed to be completely invisible. In the case of audio clips, the watermarks are inaudible. The information representing the watermark is scattered throughout the file in such a way that it cannot be identified, manipulated or reproduced.

18.4.6 Key Recovery

Encryption is an important tool for protecting the confidentiality of data. This data can be either data on transit over a network or a static data in a file. When suitably strong encryption algorithms are employed and implemented with appropriate assurance, encryption can prevent the disclosure of data to unauthorized parties. However, the unavailability, loss or corruption of the keys may prevent legitimate parties from accessing the data. For law enforcement agencies it may be sometime necessary to decrypt encrypted data. To facilitate authorized access to encrypted data in the face of such situations, there are needs for key recovery procedures and standards. This type of systems is called Key Recovery System (KRS). KRS will enable authorized persons to recover plaintext from encrypted data when the decryption key is not otherwise available. Key recovery is achieved through different key recovery techniques and key recovery information (KRI). Key recovery information refers to the aggregate of information needed by a key recovery technique to recover a target key. In third party systems like a Certification Authority (CA), the KRI is securely stored with the CA.

18.5 PUBLIC KEY INFRASTRUCTURE

Public Key Infrastructure or PKI in short consists of mechanism to securely distribute public keys. PKI is an infrastructure consisting of certificates, a method of revoking certificates, and a method of evaluating a chain of certificates from a trusted root public key. The framework for PKI is defined in the ITU-T X.509 Recommendation. PKI is also defined through RFC3280. In RFC3280 the goal of PKI is defined as 'to meet the needs of deterministic, automated identification, authentication, access control, and authorization functions. Support for these services determines the attributes contained in the certificate as well as the ancillary control information in the certificate such as policy data and certification path constraints.'

PKIX is the Internet adaptation for PKI and X.509 recommendation suitable for deploying a certificate-based architecture on the Internet. PKIX also specifies which X.509 options should be supported. RFC2510, RFC2527 and RFC3280 define the PKIX specifications.

18.5.1 Public Key Cryptography Standards

Public-key Cryptography Standards or PKCS in short comprises of standards proposed and maintained by RSA lab. These standards are accepted as de-facto standards for public key cryptography helping interoperability between applications using cryptography for security. Most of the crypto libraries available today support PKCS standards. PKCS standards consist of a number of components, which are defined through PKCS #1, #3, #5, #6, #7, #8, #9 #10, #11, #12, #13 and #15.

- **PKCS #1, RSA Encryption Standard:** PKCS #1 describes a method for encrypting data using the RSA public-key cryptosystem. Its intended use is in the construction of digital signatures and digital envelopes as described in PKCS #7. Digital enveloping is a process in which someone 'seals' a plaintext message in such a way that no one other than the intended recipient can open the sealed message. PKCS #1 also describes syntax for RSA public keys and private keys.

- **PKCS #2:** Incorporated as part of PKCS #1.

- **PKCS #3, Diffie-Hellman Key Agreement Standard:** PKCS #3 describes a method for implementing Diffie-Hellman key agreement whereby two parties, without any prior arrangements, can agree upon a secret key that is known only to them.

- **PKCS #4:** Incorporated as part of PKCS #1.

- **PKCS #5, Password-Based Encryption Standard:** PKCS #5 describes a method for encrypting an octet string with a secret key derived from a password. PKCS #5 is generally used for encrypting private keys when transferring them from one computer system to another, as described in PKCS #8.

- **PKCS #6, Extended-Certificate Syntax Standard:** PKCS #6 describes syntax for extended certificates. An extended certificate consists of an X.509 public-key certificate and a set of attributes, collectively signed by the issuer of the X.509 public-key certificate.

- **PKCS #7, Cryptographic Message Syntax Standard:** PKCS #7 describes a general syntax for data that may have cryptography applied to it, such as digital signatures and digital envelopes.

- **PKCS #8, Private-Key Information Syntax Standard:** PKCS #8 describes a syntax for private-key information. PKCS #8 also describes syntax for encrypted private keys.

- **PKCS #9, Selected Attribute Types:** PKCS #9 defines selected attribute types for use in PKCS #6 extended certificates, PKCS #7 digitally signed messages and PKCS #8 private-key information.

- **PKCS #10, Certification Request Syntax Standard:** PKCS #10 describes a syntax for certification requests. A certification request consists of a distinguished name, a public key and optionally a set of attributes, collectively signed by the entity requesting certification. Certification authorities may also require non-electronic forms of request and may return non-electronic replies.

- **PKCS #11, Cryptographic Token Interface Standard:** This standard specifies an API, called Cryptoki to devices which hold cryptographic information and perform cryptographic functions.

- **PKCS #12, Personal Information Exchange Syntax Standard:** This standard specifies a portable format for storing or transporting a user's private keys, certificates, miscellaneous secrets, etc.

- **PKCS #13, Elliptic Curve Cryptography Standard:** It will address many aspects of elliptic curve cryptography, including parameter and key generation and validation, digital signatures, public-key encryption, and key agreement.

- **PKCS #15, Cryptographic Token Information Format Standard:** PKCS #15 is intended at establishing a standard which ensures that users, in fact, will be able to use cryptographic tokens to identify themselves to multiple, standards-aware applications, regardless of the application's cryptoki provider.

18.5.2 Storing Private Keys

For optimum security, some security information needs to be stored in tamper-resistant storage. This will help in protecting some of the sensitive data like private keys. Also, to provide application level security, part of the security functionality needs to be performed through this tamper-resistant device. The WAP Identity Module (WIM) is designed to address all these needs. WIM will be used to perform WTLS and application level security functions. In a GSM, GPRS or 3G phones, it can be the SIM (Subscriber Identity Module) or USIM (Universal SIM) card containing additional functionality of the WIM or an external physically separate smart card. Use of generic cryptographic features with standard interfaces like PKCS#15 makes it possible to use the WIM for non-WAP applications like SSL, TLS, S/MIME etc.

18.6 TRUST

A portable computer never connected to a network, a standalone computer, never exposed to any unknown environment, can be assumed to be safe and secure. What happens to the security if we connect the same computer to a small private network? What happens if we connect the same computer to the Internet? What happens if we take this computer out in a football stadium and connect to the Internet over WiFi? The question is, can we trust these environments?

In early days, business was always face-to-face. In those days business used to be carried out amongst people who knew each other and in close physical proximity. In those days, one handshake literally closed the deal. The problem posed by mobile computing today is very much like that faced by business in the second half of the nineteenth century. During that time, the growth of transportation and communication networks in the form of railroads and telegraphs formed national markets and people were forced to do business with people whom they had never met. Let us take some examples. When a person searches the web for some authentic information on earthquake, what are the options? The obvious answer is to use an Internet search engine like Google. There are shops, forums, music groups with name earthquake. How do we know out of a few million hits, which are authentic information on earthquake? It may be relatively easy for a human being to determine whether or not to trust a particular web page. But is it that easy for software agents in our computers? Like in a database, can we form a SQL like query to extract an authentic technical research paper on earthquake from the Internet? In another example, let us assume for the moment that you are 55 years old and having a chest pain with sweating and vomiting; will you go to Google and give a keyword 'chest pain doctor' to look for medical help? The question, therefore, is 'Which information sources should my software agent believe?' This is equally important like the question "Which agent software should I believe and allow to access my information source?' If we look into these questions carefully we will find that first question is about trust and the second question is about security. In mobile computing, we need to address both.

We said the question, 'Which agent software should I believe and allow to access my information source?' relates to security. However, there is a catch. Suppose a person by name Anita tries to access my information source. My agent denies access to her. She then produces a certificate that she is a student in my mobile computing course, what action is expected from my agent? Of course, the agent should allow her to access my information source. This is an example of trust. The person who was not trustworthy becomes trustworthy when she produces a certificate. It is interesting to note that this certificate is not the conventional certificate as issued by a CA. Trust is explained in terms of a relationship between a trustor and a trustee. Trustor is a person who trusts a certain entity, whereas, trustee is the trusted entity. Based on the trust in the trustee, a trustor can decide whether the trustee should be allowed to access her resources and what rights should be granted. Therefore, trust plays an important role in deciding both the access rights as well as provenance of information. Trust management involves using the trust information, including recommendations from other trustees. There are different models of trust. These are direct trust, hierarchical trust and web of trust.

Direct Trust: In a direct trust model, parties know each other. This is like early days where everyone personally knew others in the line of business. A user trusts that a key or certificate is valid because he or she knows where it came from. Every organization today uses this form of trust in some way. Many companies today do business through Internet. However, before they start doing business over Internet, a due diligence and audit is done. Following this they do business over Internet with proper trust using trusted certificates and known key source.

Hierarchical Trust: In a hierarchical system, there are a number of 'root' certificates from which trust extends. This is like the holding company establishing a trust and then member companies use this trust and key (certificate). These root certificates may certify certificates themselves or they may certify certificates that certify still other certificates down the chain. This model of trust is used by conventional CA.

Web of Trust: A web of trust encompasses both of the above models. A certificate might be trusted directly or trusted in some chain going back to a directly trusted root certificate or by some group of introducers. The web of trust uses digital signatures as its form of introduction. When any user signs another's key, he or she becomes an introducer of that key. As this process goes on, it establishes a web of trust. PGP (Pretty Good Privacy) uses this model of trust. PGP does not use the CA in its conventional sense. Any PGP user can validate another PGP user's public key certificate. However, such a certificate is only valid to another user if the relying party recognizes the validator as a trusted introducer.

18.6.1 Certificate

Digital certificate plays a significant role in establishing trust. Through a digital certificate, we can associate a name with a public key. Certificate is a signed instrument vouching that a particular name is associated with a particular public key. It is a mapping between a domain name (like mybank.co.in for example) and a public key. The structure of certificates is hierarchical originating from a trusted root certificate. For example, the root certification authority in India is called Controller of Certification Authority (CCA–http://cca.gov.in). CCA is responsible for generating the key pair using SHA-1 and 2048 bit RSA algorithm. CCA issues these certificates to users through different RAs (registration authority). An RA is an organization to which a CA delegates administrative functions of creation, distribution, and bookkeeping of the public-private key pair.

Here are the data and signature sections of a certificate in human-readable format taken from an example sited in the Netscape site:

```
Certificate:
        Data:
                Version: v3 (0x2)
                Serial Number: 3 (0x3)
                Signature Algorithm: PKCS #1 MD5 With RSA Encryption
                Issuer: OU=Ace Certificate Authority, O=Ace Industry, C=US
                Validity:
                        Not Before: Fri Oct 17 18:36:25 1997
                        Not  After: Sun Oct 17 18:36:25 1999
                Subject: CN=Jane Doe, OU=Finance, O=Ace Industry, C=US
                Subject Public Key Info:
                        Algorithm: PKCS #1 RSA Encryption
                    Public Key:
                        Modulus:
                            00:ca:fa:79:98:8f:19:f8:d7:de:e4:49:80:48:e6:2a:2a:86:
                            ed:27:40:4d:86:b3:05:c0:01:bb:50:15:c9:de:dc:85:19:22:
                            43:7d:45:6d:71:4e:17:3d:f0:36:4b:5b:7f:a8:51:a3:a1:00:
                            98:ce:7f:47:50:2c:93:36:7c:01:6e:cb:89:06:41:72:b5:e9:
                            73:49:38:76:ef:b6:8f:ac:49:bb:63:0f:9b:ff:16:2a:e3:0e:
                            9d:3b:af:ce:9a:3e:48:65:de:96:61:d5:0a:11:2a:a2:80:b0:
                            7d:d8:99:cb:0c:99:34:c9:ab:25:06:a8:31:ad:8c:4b:aa:54:
                            91:f4:15
                        Public Exponent: 65537 (0x10001)
                    Extensions:
                        Identifier: Certificate Type
                            Critical: no
                            Certified Usage:
                                SSL Client
                        Identifier: Authority Key Identifier
                            Critical: no
                            Key Identifier:
                                f2:f2:06:59:90:18:47:51:f5:89:33:5a:31:7a:e6:5c:fb:36:
                                26:c9
        Signature:
```

```
Algorithm: PKCS #1 MD5 With RSA Encryption
Signature:
        6d:23:af:f3:d3:b6:7a:df:90:df:cd:7e:18:6c:01:69:8e:54:
        65:fc:06:
        30:43:34:d1:63:1f:06:7d:c3:40:a8:2a:82:c1:a4:83:2a:fb:
        2e:8f:fb:
        f0:6d:ff:75:a3:78:f7:52:47:46:62:97:1d:d9:c6:11:0a:02:
        a2:e0:cc:
        2a:75:6c:8b:b6:9b:87:00:7d:7c:84:76:79:ba:f8:b4:d2:62:
        58:c3:c5:
        b6:c1:43:ac:63:44:42:fd:af:c8:0f:2f:38:85:6d:d6:59:e8:
        41:42:a5:
        4a:e5:26:38:ff:32:78:a1:38:f1:ed:dc:0d:31:d1:b0:6d:67:
        e9:46:a8:
        dd:c4
```

Here is the same certificate displayed in the 64-byte-encoded form interpreted by software:

```
-----BEGIN CERTIFICATE-----

MIICKzCCAZSgAwIBAgIBAzANBgkqhkiG9w0BAQQFADA3MQswCQYDVQQGEwJVUzERMA8G
A1UEChMITmV0c2NhcGUxFTATBgNVBAsTDFN1cHJpeWEncyBDQTAeFw05NzEwMTgwMTM2
MjVaFw05OTEwMTgwMTM2MjVaMEgxCzAJBgNVBAYTAlVTMREwDwYDVQQKEwhOZXRzY2Fw
ZTENMAsGA1UECxMEUHViczEXMBUGA1UEAxMOU3Vwcml5YSBTaGV0dHkwgZ8wDQYJKoZI
hvcNAQEFBQADgY0AMIGJAoGBAMr6eZiPGfjX3uRJgEjmKiqG7SdATYazBcABu1AVyd7c
hRkiQ31FbXFOGD3wNktbf6hRo6EAmM5/R1AskzZ8AW7LiQZBcrXpc0k4du+2Q6xJu2MP
m/8WKuMOnTuvzpo+SGXe1mHVChEqooCwfdiZywyZNMmrJgaoMa2MS6pUkfQVAgMBAAGj
NjA0MBEGCWCGSAGG+EIBAQQEAwIAgDAfBgNVHSMEGDAWgBTy8gZZkBhHUfWJM1oxeuZc
+zYmyTANBgkqhkiG9w0BAQQFAAOBgQBtI6/z07Z635DfzX4XbAFpj1R1/AYwQzTSYx8G
fcNAqCqCwaSDKvsuj/vwbf91o3j3UkdGYpcd2cYRCgKi4MwqdWyLtpuHAH18hHZ5uvi0
0mJYw8W2wUOsYORC/a/IDy84hW3WWehBUqVK5SY4/zJ4oTjx7dwNMdGwbWfpRqjd1A==

-----END CERTIFICATE-----
```

18.6.2 Simple PKI

It was thought that digital certificate would address issues related to trust. However, finally, certificates emerged as an instrument for authentication. PKCS also made some attempt to address the need of trust through PKCS#6 and PKCS#9. IETF developed yet another standard called Simple PKI or SPKI (RFC 2692, RFC 2693) in short. SPKI defined a different form of digital certificates whose main purpose is authorization in addition to authentication. Purpose of SPKI is to define a certificate structure and operating procedure for trust management in the Internet.

18.7 SECURITY MODELS

We have discussed different types of security algorithms. We have also discussed security protocols. These algorithms and protocols are used to protect our assets. These assets can be either static assets in the form of priceless data in a database, files, or documents within a computer or assets in transit. The security and trust model we choose should be able to secure our assets and protect our interests. To protect ourselves from different threats, we need to look at security and trust at system level and application level.

18.7.1 Infrastructure Level Security

Infrastructure level security offers security at the perimeter of the system. This will primarily include networks and the infrastructure. We can call this **Network security** as well. Infrastructure level security will include protecting the infrastructure or the network so that attacks from worms, viruses, Trojan horses can be prevented. Prevention from other forms of attacks like intrusion etc. different firewalls in the network are all part of the infrastructure-level security. Virtual private network (VPN) is part of infrastructure security as well.

Infrastructure level security for mobile-computing environment need to handle some additional threats compared to a wired network. For mobile-computing network, the last mile access network will be wireless in most of the situations. Therefore, at the access level, additional infrastructure security is necessary. An example is encryption in GSM using A5 algorithm. WiFi/wireless LAN networks use WEP. Some vulnerabilities have been identified in infrastructure security for WiFi; therefore, new security protocols like 802.1x and 802.11i have been proposed to take care of the over the air interface in the access network.

18.7.2 System Level Security

In system level security we secure our systems to protect our assets. In the security framework provided by the operating system, shells will be part of system level security. In authentication challenges during Unix login, or login into a mainframe computer through username, password is the system level security. Access control list (ACL), file system security, memory security etc will also be part of the system level security. It protects the system from worms, viruses and Trojan horses. Prevention from other forms of attacks like buffer overflow attacks, intrusion etc can also be part of system level security as also security protocols like SSL and TLS. There is a concept of capability-based system, where security is policy-driven and managed through capability. Even if a virus enters into such a system, or an intrusion happens, it will not be able to damage any asset in the system. One of such operating system is EROS.

Database security is part of system level security. In database security, data in the database is secured by the database software. This can be encrypting a column in a row or some special check based on ACL and capability. Most of the database software today offer security at this layer. This will be over and above the security offered by the operating system.

18.7.3 Policy Based Security

Security systems implemented for wired networks in organization are primarily policy based. Effective security policies make frequent references to standards and guidelines that exist within an organization. Policy is a written down rules about what is allowed or what is not allowed in the network. Policies are usually area-specific, covering a single area. According to the RFC 2196, security policy is defined as "A security policy is a formal statement of the rules by which people who are given access to an organization's technology and information assets must abide." For example, there could be a rule in the organization that nobody will be allowed to have a global IP address. To stop spam and mail bound viruses, there may be another rule that prevents access to external email systems (like hotmail) from the corporate network. To stop the possibility of espionage, there could be a rule that FTP from the Internet is not allowed to any machine within the intranet. A standard is typically collection of system-specific requirements that must be met by everyone. Standards are necessary when we need interoperability. A guide line is typically collections of system specific procedural specific "suggestions" for best practice. Guidelines are not requirements to be met, but are strongly recommended.

In a wired network where systems are stationary, and the network structure is static, it is possible to define security policies. In such networks, it is possible to enforce these policies or rules. However, things are different when we move to a mobile computing environment. In a mobile computing environment, user will move from one device to another device, from one network to another network. These devices or networks may be of similar type or different types. For example user moves from a WiFi network to a CDMA2000 network or from a PalmOS to a WindowsCE. In a network with static nodes, it is possible to define a security policy. It may be possible to enforce such policies as well. However, in case of mobile computing where nodes are roaming from one network to another, it may not be practical to define a security policy and implement it. Therefore, over and above policy based security, for mobile nodes, we need object security. In object security, objects will carry their security signatures and capabilities. This is achieved through the concept of principal. Therefore, when a device moves from network to network, the device carries the security requirement and security signature with it. Principal based security system is in the process of maturing. OMA DRM (discussed in section 8.4.1) is an example of security principal.

18.7.4 Application Level Security

Infrastructure and system level security take care of security at the infrastructure and system level respectively. The parameters for these securities are not very flexible; most of the time vendors define them. In a mobile-computing environment we need security at the application level. Application security looks at the security at the content level. This can also be termed as **Peer to peer security**. The application at the client device will talk to the application at the server and handle security requirements end-to-end as the content may demand.

In a mobile computing environment, we cannot make any assumption related to the client context or the network context. Therefore, the security needs to be addressed at the content level, using the context awareness, J2ME, .NET, WIM or MExE (Mobile Execution Environment) environment. Using cryptographic libraries, we can build security at the application level. This security will be custom-built and can use standard algorithms or new algorithms as agreed by the peer nodes. Of course the system/infrastructure level security, if any, will be over and above the application level security.

18.7.5 Java Security

Security model provided by Java covers both system level security and application level security. Java system level security is provided through the 'sandbox' model. Sandbox provides a restricted environment for code execution through Java virtual machine. In the sandbox model, local code is trusted to have full access to system resources like file system, memory etc. However, downloaded code from a remote site as an applet is not trusted. Therefore, applet can access only the limited resources provided inside the sandbox. Java supports digitally signed trusted applet. A digitally signed applet is treated like local code, with full access to resources. Digitally signed applets use public key infrastructure. Prior to transmission, the applet server signs an applet JAR file using its digital certificate. Upon receipt, the client side Java security manager verifies the signature and decides whether the origin and integrity of the application is trusted. Once the authentication is successful, the application code is delivered to the client for execution.

Java offers tools to facilitate various security-related operations. These are:

- **Keytool:** This is a command line tool. Keytool is used to manage keystore, which includes the following functions.

    o Create public/private key pairs

    o Issue certificate requests (which will be sent to the appropriate Certification Authority)

    o Import certificate replies (obtained from the Certification Authority)

    o Designate public keys belonging to other parties as trusted keys and certificates are used to digitally sign applications and applets. A **keystore** is a protected database that holds keys and certificates for an enterprise. Access to a keystore is guarded by a password. In addition, each private key in a keystore can be guarded by its own password.

- **Jar:** This is a command line tool to create JAR (Java Archive) files. The JAR file format enables users to bundle multiple files into a single archive file. Typically a JAR file will contain the class files and auxiliary resources associated with applets and applications. After importing appropriate keys into the keystore, **jarsigner** tool is used to digitally sign the JAR file.

- **Jarsigner:** This is a command line tool to sign JAR files. This is also used to verify signatures on signed JAR files. The jarsigner tool accesses a keystore that is created and managed by **keytool**, when it needs to find the private key and its associated certificate chain. The jarsigner tool prompts for needed passwords.

- **Policytool:** Unlike the other tools, this tool has a graphical user interface. Policytool is used to create and modify the external policy configuration files that define installation's security policy.

As a part of application level security, Java framework supports cryptographic library through Java cryptography architecture (JCA). JCA refers to a framework for accessing and developing cryptographic functionality for the Java platform. These cryptographic services are:

- Symmetric key encryption algorithms
- Public key encryption algorithms
- Digital signature algorithms
- Message digest algorithms
- Message authentication code generation
- Key generation algorithms
- Key exchange algorithms
- Keystore creation and management
- Algorithm parameter management
- Algorithm parameter generation
- Key factory support to convert between different key representations
- Certificate factory support to generate certificates and certificate revocation lists (CRLs) from their encodings
- Random-number generation (RNG) algorithm
- Support of SSL and TLS through http support.

Cryptographic library is available for the entire Java framework. This includes J2EE (Java 2 Enterprise Edition), J2SE (Java 2 Standard Edition), J2ME (Java 2 Micro Edition), and JC (Java Card). However, due to security reasons and resource constraints J2ME and JC functionalities are restrictive. Some of the APIs, which are available in J2EE and J2SE are not supported in Java card.

18.8 SECURITY FRAMEWORKS FOR MOBILE ENVIRONMENT

Mobile applications usually span over several networks. One of these networks will be a wireless radio network. Others will be wired networks. At the boundary of any of these networks, there is a need for protocol conversion gateways. These gateways run either at the transport layer or at the application layers. Moreover, while the user is roaming in foreign networks, there will be multiple wired networks (PLMNs) managed and controlled by different network operators. Multiple gateways and multiple networks make security challenges in mobile environments complex.

In a security system, authentication, and non-repudiation are meaningful only when these are implemented end-to-end between parties that need to authenticate each other. Authorization is a direct function of authentication; therefore, it is also an end-to-end function. Authentication, authorization, and non-repudiation must therefore be implemented at the application layer. Confidentiality and integrity on the other hand can be implemented at any layer or through multiple layers. Confidentiality can be realized by encrypting isolated legs between gateways or even end-to-end. When confidentiality is realized in isolated legs, the gateways or nodes between the legs need to be secured and trusted.

Therefore, to offer secured environment in a mobile environment, security procedures will be a combination of many procedures and functions. Following sections cover some of the vulnerabilities and techniques to offer security in Mobile environment.

18.8.1 3GPP Security

In a mobile computing environment inside a campus the access network is likely to be WiFi. However, outside of the campus, access network will be one of the cellular wide area wireless networks like GPRS, CDMA, or GSM. It could also be WiMax. We have discussed WiFi security in section 10.8. We have discussed GSM security in section 5.9. We have discussed the GPRS security in section 7.3.4. We have also discussed the security issues of CDMA networks in section 9.3.5.

WiFi security is an extension of the LAN security and primarily designed for data and applications. However, security procedures for wireless wide area networks GSM, GPRS, CDMA, are designed primarily keeping the operator in mind. All these security principles mainly try to protect an operator from fraud and network misuse. None of these procedures address the security concerns of user information or the application. Current security procedures in all these wide area wireless networks failed to provide trusted environment where mobile users felt confident enough to place sensitive information over these networks.

In order to perform an attack in a wireless wide area network, the adversary has to possess one or more of the following capabilities:

- **Eavesdropping.** This is the capability through which the adversary eavesdrops signalling and data traffic associated with a user. Equipment required for such attack is a modified mobile station or a radio receiver.

- **Impersonation of a user.** This is the capability whereby the adversary sends signaling and user data to the network, in an attempt to make the network believe that they originate from a genuine (target of the impersonation) user. Equipment required for such attack is a modified mobile station or a radio transmitter/receiver.

- **Compromising authentication vectors in the network.** The adversary possesses a compromised authentication vector, which may include challenge/response pairs, cipher keys and integrity keys. This data may have been obtained by compromising network nodes or by intercepting signalling messages on network links and then through brute force attack.

- **Impersonation of the network.** This is the capability whereby the adversary sends signalling and user data to the target user, in an attempt to make the target user believe that the data originate from a genuine network. Equipment required for such attack is a modified base station or a radio transmitter/receiver.

- **Man-in-the-middle.** This is the capability whereby the adversary puts itself in between the target user and a genuine network and has the ability to eavesdrop, modify, delete, re-order, replay, and spoof signalling and user data messages exchanged between the sender and the receiver. The required equipments in such attack are modified base station in conjunction with a modified mobile station.

3GPP looked into these concerns and come up with changes in the security architecture of the current wireless wide area networks. 3GPP proposed a new architecture (Figure 18.6) through following important changes:

- Changes were made to defeat the false base station attack. The extended security mechanism is now capable of identifying the network.

- Key lengths are increased to allow stronger algorithms for encryption and integrity.

- Mechanisms are included to support security within and between networks.

- Security is based within the switch rather than the base station to ensure that links are protected between the base station and switch.

- The authentication algorithm has not been defined, but guidance on choice will be given.

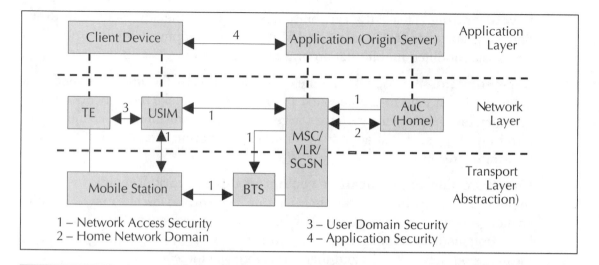

Figure 18.6 3GPP Security Architecture

- Integrity mechanisms for the terminal identity (IMEI) have been included.

18.8.2 MOBILE VIRTUAL PRIVATE NETWORK

Virtual Private Network (VPN) provides an end-to-end security infrastructure. This generally deals with authentication, non-repudiation, integrity, and confidentiality at a layer between the transport layer and the application layer. In section 10.8.8 we have discussed Wireless VPN with respect to WiFi. Like wireless VPN, mobile VPN is a private network over a public network (usually the Internet) to connect two endpoints. Instead of using a dedicated physical connection such as leased line, a VPN uses "virtual" connections routed through the Internet from the enterprise's private network to the remote mobile device. VPN implements this through an encrypted private connection between nodes. It generally uses IPSec and other PKI frameworks to offer confidentiality, authentication, non-repudiation (through digital signature), and integrity. With mobile VPN, mobile workers have the freedom to safely use wireless applications on their PDAs, smart phones and other handheld devices in the field as if they are in a private network.

18.8.3 Multifactor Security

In section 18.4.4, we have discussed multifactor security where factors could be "what you have", "what you know", "what you are". In mobile network multifactor security can be extended to multiple networks. For example the security key (session key) in a GPRS network can be split into multiple parts and sent through Internet (TCP/IP) and AMS network. By now we know that SMS uses SS#7 network for its traffic. SS#7 network is closed and protected. SS#7 is physically more secured that the IP network.

18.8.4 Smartcard Security

Smart cards offer data encryption and the ability to store secret information for the purpose of authenticating the cardholder. There are various types of smartcards used in different application scenarios. One of such examples is the SIM card on a mobile phone. ETSI standard 03.48 specifies procedures for SIM card to be used as a security engine. SIM cards used in mobile phones are processor cards. Processor cards are smart cards with an inbuilt processor and memory. This local processor protects the content in the SIM and makes it tamper resistant. The 03.48 standard specifies the interoperability standards for cryptographic functions. Also, many of the SIM cards have RSA, DES, 3DES, AES algorithms implemented within the card. There are different file systems within a SIM card. These files have very stringent security controls. Files can be protected through passwords known to the user or operator. Private key and many other secret keys can be stored in these files. As these files are protected through password, even if the card is lost, these information are protected. To counter brute force attack, a smartcard processor does not allow more than 10 attempts to read a file data with wrong password. Therefore, a user can make use of the card as a security factory. Using this security factory is quite easy through Java card interfaces. In Java card technology, a Java interface is provided on the SIM card. Java cryptographic architecture (JCA) and Java programs running on the smartcard can do all these things quite easy.

18.8.5 Mutual and Spatial Authentication

In section 18.8.4 we have discussed how a SIM card can be used as a security factory. The SIM card can store secured information like private keys, wireless identity module, and many other private secured information. It also has various algorithms implemented. This can facilitate mutual authentication. In SSL or TLS over Internet generally client authentication is not done. However, using SIM card, we can do client authentication over wireless wide area networks. This is called mutual authentication; because, using GSM 03.08 procedures, a client can authenticate the server, also the server can authenticate the client.

SIM cards in a GSM/GPRS network store location information. This includes country, network, and base station information. This information can be obtained and sent from the mobile phone using GSM 03.48 standards specification. Location information then can be used to implement spatial authentication. For example, if the user is in a neighborhood which is insecure, access to some critical applications can be prevented.

18.8.6 RFID Security

In section 4.3 we have discussed about RFID. Application areas for RFID is increasing. However, it has certain vulnerabilities. For example using the RFDump tool (http://www.rf-dump.org/), an adversary can detect an RFID-Tag and extract its meta information like Tag ID, Tag Type, manufacturer etc. It can even be used to rewrite the data stored in some RFID tags using either a Hex or an ASCII editor. All these vulnerabilities pose a serious threat toward RFID based systems starting from merchandise in a store to the e-passports. The US governments sometime ago decided to issue passports with RFIDs. It is nicknamed as "e-passports". However, the concerns over RFID security delayed this plan. According to the specification of e-passport, there will be 64-bit RFID tags attached in the passport that will contain name, date of birth, place of birth, a digital photograph and a digital face recognition template of the passport holder. This RFID is supposed to work only in a very close proximity. A RFID reader placed beyond the distance of more than 10 centimeters should not be able to read the content of the e-passport. However, in reality it was found that the radio tags' readable distance is as large as 30 feet. This makes the security information in the e-passport available over radio for an adversary to grab.

18.8.7 Mobile Agent Security

Mobile agents are processes that can autonomously migrate from one networked computer to another. Mobile agents can be useful for many applications, especially these in Internet. For example I give my weekly shopping list to my mobile agent, which visits the web sites of all the stores in 3 kilometer radius and tells me which shop is having which fish at what price, where tomato is cheapest, where can I get my favorite pickle etc. The mobile visits all these store's web site does a shopping plan for me.

Despite its many practical benefits, mobile agent technology results in significant security threats from both malicious agents and malicious hosts. For examples, as the mobile agent traverses multiple hosts that are trusted to different degrees, its state may be changed in a way that an adversely can impact the decision making process of the agent.

18.8.8 Mobile Virus

Viruses are common in the PC and desktop environment. However, they were not common in mobile environment. However, things are changing; as the mobile device become more intelligent with more flexibility and higher capabilities, viruses are surfacing. Already some of them are in the wild. In June 2004, as a proof concept a virus called Cabir was developed to exploit Bluetooth vulnerability. In early November 2004 a mobile virus called "Skull.A" was reported for some models of Nokia phones. A new version named as "Skull.B" emerged in the wild, which combines Skull.A and Cabir. One more virus identified as Commwarrior.A surfaced in March 2005. This virus uses a combination of Bluetooth and MMS (Multimedia Messaging Service) to propagate. The principles these viruses use are similar in concept as the desktop viruses do.

18.8.9 Mobile Worm

A worm needs to propagate, execute, and reproduce in an automated fashion. To reproduce and then propagate, the worm needs to execute a piece of code (designed by the worm writer) on the target system. Therefore, it is necessary to have an execution environment available to the worm code on the target mobile device. On a mobile equipment today we have various execution environments like:

1. WAP/WML Script (MExE Classmark I)
2. JavaPhone/Personal Java (MExE Classmark II)
3. J2ME (MExE Classmark III)
4. Symbion
5. WindowsCE
6. PalmOS
7. Linux

These can access both the TCP/IP and SMS interfaces. Therefore, worms can replicate and propagate through both TCP/IP and SMS interfaces of JavaPhone, PersonalJava or J2ME framework.

Along with JavaPhone, technology on the mobile equipment, JavaCard facility is also available on the SIM cards. Using all these technologies, it will be possible to develop viruses, worms, and Trojan horses for mobile phones. These viruses and worms will be able to replicate, access the address book, use the network facility and propagate.

References/Further Reading

1. 3G TR 33.900 V1.2.0 (2000-01) 3rd Generation Partnership Project; Technical Specification Group SA WG3; A Guide to 3rd Generation Security (3G TR 33.900 version 1.2.0), January 2000.

2. S Gindraux, From 2G to 3G: A Guide to Mobile Security, Proceedings of Third International Conference on 3G Mobile Communications Technologies, 2002.

3. Yang Kun, Guo Xin, Liu Dayou, Security in Mobile Agent System: Problems and Approaches, ACM SIGOPS Operating Systems Review, 2000.

4. Katherine M. Shelfer and J. Drew Procaccino, Smart Card Evolution, Communications Of The ACM July 2002, Vol. 45, No. 7, p83.

5. Thomas S. Messerges, Ezzat A. Dabbish, Digital Rights Management in a 3G Mobile Phone and Beyond, DRM'03, October 27, 2003.

6. Asoke K Talukder, Debabrata Das, Artificial Hygiene: Non-Proliferation of Virus in Cellular Network, Journal of Systems and Information Technology, Volume 8, 1st December 2004, pp 10–22.

7. 3GPP TS 03.48, Digital cellular telecommunications system (Phase 2+); Security mechanisms for SIM application toolkit; Stage 2, 1999.

8. Zhiqun Chen, Java Card Technology for Smart Cards, Addison Wesley, 2000.

9. Wireless Transport Layer Security, Version 06-Apr-2001: WAP-261-WTLS-20010406-a; http://www.wapforum.com.

10. Transport Layer Security Protocol: RFC 2246.

11. Internet X.509 Public Key Infrastructure Certificate and Certificate Revocation List (CRL) Profile: RFC3280.

12. Controller for Certification Authority; Government of India: Digital Certificate, http://cca.gov.in.

13. Shashi Kiran, Patricia Lareau, Steve Lloyd: PKI Basics-A Technical Perspective, November 2002, http://www.pkiforum.org.

14. Data Encryption Standard (DES); Federal Information Processing Standard Publication, 1999 October 25, U.S. Department Of Commerce/National Institute of Standards and Technology.

15. Elliptic Curve Cryptosystem: http://www.certicom.com.

16. Charlie Kaufman, Radia Perlman, Mike Speciner: Network Security Private Communication in a Public World 2nd Edition; Prentice Hall of India, 2002.

17. William Stallings, Network Security Essentials: Applications and Standards; Pearson Education. 2000.

18. William Stallings: Cryptography and Network Security Principles and Practices; Pearson Education, Third edition, 2003.

19. R. Rivest, The MD5 Message-Digest Algorithm, RFC 1321, April 1992.

20. H. Krawczyk, M. Bellare and R. Canetti: HMAC: Keyed-Hashing for Message Authentication, February 1997, RFC 2104.

21. Secured Socket Layer (SSL): http://developer.netscape.com/docs/manuals/security/pkin/contents.html.

22. Report of the Technical Advisory Committee to Develop a Federal Information Processing Standard for the Federal Key Management Infrastructure, http://csrc.nist.gov/keyrecovery/.

23. Marc Waldman and David Mazieres Tangler: A Censorship-Resistant Publishing System Based On Document Entanglements, December 8, 2001.

24. WPKI, WAP-217-WPKI, Wireless Application Protocol Public Key Infrastructure Definition, Apr-2001.

25. Wireless Application Protocol Identity Module Specification, Version 05-Nov-1999, WAP Forum.

26. Burton S. Kaliski Jr.; An Overview of the PKCS Standards, published by RSA Lab http://www.rsasecurity.com/rsalabs/pkcs/.

27. C. Rigney, S. Willens, A. Rubens and W. Simpson, RADIUS (Remote Authentication Dial In User Service), RFC 2865, June 2000.

28. P. Calhoun, J. Loughney, E. Guttman, G. Zorn and J. Arkko: Diameter Base Protocol, RFC 3588, September 2003.

29. Whitfield Diffie and Martin Hellman: New Directions in Cryptography, 1976.

30. Francis Fukuyama, The Virtual Handshake: E-Commerce and the Challenge of Trust, http://www.ml.com/woml/forum/ecommerce1.htm.

31. Tim Finin and Anupam Joshi, Agents, Trust and Information Access on the Semantic Web.

32. Henry M Levy, Capability-Based Computer Systems, Digital Press, 1984.

33. Jonathan S. Shapiro and Norm Hardy: EROS: A Principle-Driven Operating System from the Ground Up, IEEE Software Magazine, January 2002.

34. Java Security: http://java.sun.com/products/jdk/1.2/docs/guide/security/CryptoSpec.html.

35. http://www.sans.org/resources/policies/.

36. http://www.tbs-sct.gc.ca/pubs_pol/gospubs/TBM_12A/gsp-psg1_e.asp#poli.

REVIEW QUESTIONS

Q1: What are the different components of information security?

Q2: What do you understand by security algorithms and security protocols? What are the differences between them? How are they related?

Q3: Describe symmetric key and Public key encryption. If you are required to design a security system when will you use which algorithms?

Q4: Explain the security framework for mobile computing. How do we ensure security in a mobile environment through Mobile VPN?

Q5: Give examples of RFID security vulnerability?

List of Abbreviations

1G	First Generation
2G	Second Generation
2.5G	2.5 Generation
3G	Third Generation
3GPP	Third Generation partnership Project
3GPP2	Third Generation partnership Project 2

A

AABS	Automatic Alternative Billing Service
AAS	Adaptive Antenna System
AC	Admission Control
AC	Authentication Centre
ACL	Asynchronous Connectionless Link
ACL	Access Control List
ACK	Acknowledgement
ACM	Address Complete Message
AES	Advanced Encryption Standard
AI	Application Interface (prefix to interface class method)
AMPS	Advanced Mobile Phone System
AMR	Adaptive Multi Rate
ANM	Answer Message
ANSI	American National Standard Institute
AP	Access point
API	Application Programming Interface
APN	Access Point Name
APN-NI	APN Network Identifier
APPUI	Application User Interface

ARFCN	Absolute Radio Frequency Channel Numbers
ARIB	Association of Radio Industries and Business
ARP	Address Resolution Protocol
ARPA	Advance Research Project Agency
ARQ	Automatic Repeat Request
ASCII	American Standard Code for Information Interchange
ASP	Application Service Provider
ASP	Active Server Page
AT	Attention
ATM	Asynchronous Transfer Mode
ATD	Absolute Time Difference
AUC	Authentication Center

B

B2B	Business to Business
BASIC	Beginners All purpose Symbolic Instructional Code
BCCH	Broadcast Control Channel
BER	Bit Error Rate
BG	Border Gateway
BIB	Backward indicator bit
BMP	Bit Map
BPSK	Binary Phase Shift Keying
BS	Base Station
BSA	Basic Station Area
BSC	Base Station Controller
BSN	Backward Sequence Number
BSS	Base Station Subsystem
BSS	Basic service set
BSSAP	BSS Application Part
BT	Busy Tone
BTS	Base Transceiver Station

C

CA	Content Aggregator
CA	Certification Authority
CAMEL	Customised Application for Mobile Network Enhanced Logic
CAP	CAMEL Application Part
CAS	Call Associated Signalling
CC	Country Code
CC/PP	Composite Capabilities/Preference Profiles
CCK	Complementary Code Keying
CCSSO	Common Channel Signaling Switching Office
CDC	Connected Device Configuration
CDMA	Code Division Multiple Access
CDPD	Cellular Digital Packet Data
CDR	Call Detail Record
CEK	Content Encryption Key
CEPT	Conference of European Posts and Telegraphs
CF	Contention free
CFU	Call Forwarding Unconditional
CFB	Call Forwarding Busy
CFNA	Call Forwarding Not Answered
CFNR	Call Forwarding Not Reachable
CFP	Contention-Free Period
CGI	Computer Gateway Interface
cHTML	compact Hyper Text Markup Language
CI	Call Identifier
CICS	Customer Information Control System
CID	Cell ID
CIMD	Computer Interface to Message Delivery
CLDC	Connected Limited Device Configuration
CLI	Caller Line Identification
CM	Connection Management

CN	Core Network
CORBA	Common Object Request Broker Architecture
CP	Content Provider
CP	Contention Period
CPI	Capability and Preference Information
CPU	Central Processing Unit
CRC	Cyclic Redundancy Code
CRP	Customer Routing Point
CS	Carrier Sense
CS	Capability Set
CSD	Circuit Switched Data
CSE	CAMEL Service Environment
CSMA/CA	Carrier Sense Multiple Access with Collision Avoidance
CSMA/CD	Carrier Sense Multiple Access with Collision Detection
CTI	Computer Telephony Interface/ Computer Telephony Integration
CTS	Clear To Send
CUG	Closed User Group
CWTS	China Wireless Telecommunication Standard group
Cyborg	Cyber Organism

D

DCE	Data Circuit terminating Equipment
DCF	DRM Content Format
DCF	Distributed Coordination Function
DFP	Distributed Functional Plane
DFRD	Device Family Reference Designs
DHCP	Dynamic Host Configuration Protocol
DIFS	Distributed Inter Frame Space
DL	Downlink
DLL	Data Link Layer
DLL	Dynamic Link Library

DMA	Direct Memory Access
DMH	Data Message Handler
DNS	Domain Name Server
DoCoMo	DO (Everywhere) + COMO (Communication)
DoD	Department of Defense
DPC	Destination Point Code
DRM	Digital Rights Management
DRNC	Drift RNC
DS	Direct Sequence
DS	Distribution System
DSSS	Direct Sequence Spread Spectrum
DSL	Digital Subscriber Link
DSP	Digital Signal Processing
DT	Dial Tone
DTE	Data Terminal Equipment
DTMF	Dual Tone Multi Frequency
DUP	Data User Part

E

E-OTD	Enhanced Observed Time Difference
EAP	Extensible Authentication Protocol
EDGE	Enhanced Data rate for GSM Evolution
EGPRS	Enhanced GPRS
EIA	Electronic Industries Alliance
EIFS	Extended Inter Frame Space
EIR	Equipment Identity Register
EMS	Extended Message Service
ENIAC	Electronic Numerical Integrator and Computer
ENUM	Electronic Numbering
ERP	Enterprise Resource Planning
ESME	External Short Message Entity

ESN	Electronic Serial Number
ESS	Electronic Switching System
ESS	Extended Service Set
ETSI	European Telecommunication Standards Institute
EU	End User

F

FDD	Frequency Division Duplex
FDMA	Frequency Division Multiple Access
FE	Functional Entity
FEA	Functional Entity Action
FEC	Forward Error Correction
FER	Frame Error Rate
FH	Frequency Hopping
FHSS	Frequency Hopping Spread Spectrum
FIB	Forward Indicator Bit
FISU	Fill-In Signal Units
FNC	Federal Networking Council
FSN	Forward Sequence Number
FTP	File Transfer Protocol
FWA	Fixed Wireless Access

G

GERAN	GSM EDGE Radio Access Network
GFP	Global Functional Plane
GGSN	Gateway GPRS Support Node
GIF	Graphics Interchange Format
GIWU	Gateway Inter Working Unit
GMLC	Gateway MLC
GMSC	Gateway MSC

GPRS	General Packet Radio Service
GPS	Global Positioning System
GSM	Global System for Mobile communications
GT	Global Title
GTT	Global Title Translation
GUI	Graphical User Interface

H

HCI	Human Computer Interface
HDLC	High level Data Link Control
HDML	Handheld Device Markup Language
HDTP	Handheld Device Transport Protocol
HE	Home Environment
HLR	Home Location Register
HPLMN	Home Public Land Mobile Network
HTML	Hyper Text Markup Language
HTTP	Hyper Text Transfer Protocol

I

I/O	Input/Output
IAM	Initial Address Message
IAPP	Inter-Access Point Protocol
IBSS	Independent Basic Service Set
IC	Integrated Circuit
ICAP	Internet Content Adaptation Protocol
ICCC	International Computer Communication Conference
ICL	International Computers Limited
ICMP	Internet Control Message Protocol
ICT	Information and Communication Technology
IDE	Interactive Development Environment

IED	Information Element Data
IEDL	Information Element Data Length
IEEE	Institute of Electrical and Electronics Engineers
IEI	Information Element Identifier
IETF	Internet Engineering Task Force
IMAP	Internet Message Access Protocol
IMEI	International Mobile Equipment Identity
IMSI	International Mobile Subscriber Identity
IMT	International Mobile Telecommunications
IN	Intelligent Networks
INAP	Intelligent Network Application Part
INCM	IN Conceptual Model
IP	Internet Protocol
IPCP	Internet Protocol Control Protocol
IPDL	Idle Period Downlink
IPNG	Next Generation Internet Protocol
IR	Infra Red
IrDA	Infrared Data Association
IrMC	Infrared Mobile Communication
ISDN	Integrated Services Digital Network
ISM	Industrial, Scientific, and Medical
ISP	Internet Service provider
ISO	International Organization for Standardization
ISUP	ISDN User Part
ISV	Independent Software Vendor
ITTP	Intelligent Terminal Transfer Protocol
ITU	International Telecommunication Union
ITU-T	International Telecommunication Union–Telecommunication Standardization
IVR	Interactive Voice Response
IWF	Inter Working Function
IWMSC	Inter Working MSC

J

J2EE	Java 2 Enterprise Edition
J2ME	Java 2 Micro Edition
J2SE	Java 2 Standard Edition
JDBC	Java Data Base Connector
JFIF	JPEG File Interchange Format
JPEG	Joint Photographic Experts Group
JSP	Java Server Pages

L

L2CAP	Logical Link Control and Adaptation Protocol
LA	Location Area
LA	Location Application
LAF	Location Application Function
LAI	Location Area Identifier
LAN	Local Area Network
LAP	LAN Access Point
LAP	Link Access Procedure
LAPD	Link Access Procedure - D
LBS	Location Based Services
LCAF	Location Client Authorization Function
LCCF	Location Client Control Function
LCCTF	Location Client Co-ordinate Transformation Function
LCD	Liquid Crystal Diode
LCF	Location Client Function
LCP	Link Control Protocol
LCS	LoCation Services
LCZTF	Location Client Zone Transformation Function
LDR	Location Deferred Request
LIR	Location Immediate Request

LLC	Logical Link Control
LMP	Link Manager Protocol
LMSI	Local Mobile Subscriber Identity
LMU	Location Measurement Unit
LNP	Local Number Portability
LPC	Linear Prediction Coding
LSAF	Location Subscriber Authorization Function
LSBcF	Location System Broadcast Function
LSBF	Location System Billing Function
LSCF	Location System Control Function
LSOF	Location System Operation Function
LSPF	Location Subscriber Privacy Function
LSSU	Link Status Signal Units

M

M2M	Machine to Machine
MAC	Media Access Control
MAN	Metropolitan Area Network
MAP	Mobile Application Part
MCC	Mobile Country Code
ME	Mobile Equipment
MExE	Mobile Execution Environment
MGCP	Media Gateway Control Protocol
MIB	Management information base
MIDP	Mobile Information Device Profile
MIN	Mobile ID Number
MIS	Management Information Systems
MLME	MAC sublayer Management Entity
MLC	Mobile Location Center
MM	Mobility Management
MMI	Man Machine Interface
MMS	Multimedia Message Service

MMSC	MMS Controller
MMSE	MMS Environment
MMU	Memory Management Unit
MNC	Mobile Network Code
MO	Mobile Originated
MO-LR	Mobile Originated Location Request
MOM	Message Oriented Middleware
MPDU	MAC Protocol Data Unit
MPEG	Moving Picture Experts Group
MS	Mobile Station
MSC	Mobile Switching Centre
MSDU	MAC Service Data Unit
MSIN	Mobile Subscriber Identification Number
MSISDN	Mobile Station ISDN
MSRN	Mobile Station Roaming Number
MSU	Message Signal Units
MT	Mobile Terminated
MT-LR	Mobile Terminated Location Request
MTP	Message Transfer Part
MVC	Model-View-Controller
MVNO	Mobile Virtual Network Operator

N

NA-ESRD	North American Emergency Service Routing Digits
NA-ESRK	North American Emergency Service Routing Key
NAV	Network allocation vector
NDC	National Destination Code
NI-LR	Network Induced Location Request
NIC	Network Interface Card
NID	Network Identification
NO	Network Operator
NSF	National Science Foundation

NSS	Network and Switching Subsystem
NTT	Nippon Telegraph and Telephone Corporation

O

OBEX	Object Exchange Protocol
OCC	Occasionally Connected Computing
OCR	Optimal Call Routing
ODBC	Open Data Base Connectivity
OEM	Original Equipment Manufacturer
OFDM	Orthogonal Frequency Multiplexing
OFDMA	Orthogonal Frequency Multiple Access
OMA	Open Mobile Alliance
OMAP	Operations, Maintenance and Administration Part
OMC	Operation and Maintenance Center
OOPS	Object Oriented Programming
OPC	Originating Point Code
OPL	Organiser Programming Language
OS	Operating System
OSA	Open Service Architecture
OSA	Open Service Access
OSS	Operation and Support Subsystem
OTA	Over-The-Air
OTDOA	Observed Time Difference Of Arrival

P

P3P	Platform for Privacy Preference Project
PAN	Personal Area Network
PBX	Private Branch Exchange
PC	Point Coordinator
PC	Power Control

PCF	Point Coordination Function
PCF	Power Calculation Function
PCH	Paging Channels
PCI	Peripheral Component Interface
PCM	Pulse Coded Modulation
PCMCI	Personal Computer Memory Card International Association
PCN	Personal Communication Networks
PCS	Personal Communications System
PDA	Personal Digital Assistant
PDC	Personal Digital Cellular
PDN	Packet Data Network
PDP	Packet Data Protocol
PDTCH	Packet Data Traffic Channel
PDU	Protocol Data Unit
PE	Physical Entity
PEAP	Protected EAP
PHP	Hypertext Preprocessor
PHS	Personal Handyphone System
PHY	Physical (layer)
PICS	Platform for Internet Content Selection
PIFS	Point (coordination function) Inter Frame Space
PIM	Personal Information management
PKI	Public Key Infrastructure
PLCP	Physical Layer Convergence Procedure
PLL	Physical Link Layer
PLMN	Public Land Mobile Network
PLW	PSDU Length Word
PMD	Physical Medium Dependent
PN	Pseudo random Noise
POI	Point Of Initiation
POI	Privacy Override Indicator
POP	Post Office Protocol

POR	Point Of Return
POS	Point Of Sale
POTS	Plain Old Telephone Service
PP	Physical Plane
PPDU	PLCP Protocol Data Unit
PPG	Push Proxy Gateway
PPP	Point-to-Point Protocol
PRCF	Positioning Radio Co-ordination Function
PRNG	Pseudo-Random Number Generator
PRRM	Positioning Radio Resource Management
PS	Power Save (mode)
PSDU	PHY Sublayer Service Data Unit
PSE	Personal Service Environment
PSF	PLCP Signaling Field
PSMF	Positioning Signal Measurement Function
PSPDN	Public Switched Packet Data Networks
PSTN	Public Switched Telephone Network
PTM	Point-To-Multipoint
PTP	Point-To-Point

Q

QPSK	Quadrature Phase Shift Keyed
QoS	Quality of Service

R

RA	Routing Area
RACH	Random Access Channel
RADIUS	Remote Authentication Dial In User Service
RAM	Random Access Memory
RAN	Radio Access Network

RANAP	RAN Application Part
RAND	Random Number
RDF	Resource Description Framework
REL	Release
RF	Radio Frequency
RFC	Request For Comments
RFCOMM	Radio Frequency Communication
RFID	Radio Frequency Identifiers
RFL	Radio Frequency Layer
RISC	Reduced Instruction Set Computer
RLC	Radio Link Control
RLC	Release Complete
RLP	Radio Link Protocol
RIS	Radio Interface Synchronization
RNC	Radio Network Controller
ROM	Read Only Memory
RPC	Remote Procedure Call
RPE-LPC	Regular Pulse Excited - Linear Predictive Coder
RRM	Radio Resource Management
RSA	Rivest, Shamir, Adelman
RSACi	Recreational Software Advisory Council - Internet
RT	Ring Tone
RTD	Real Time Difference
RTS	Request To Send
RTT	Radio Transmission Technology

S

SAP	Service Access Point
SAT	SIM Application Toolkit
SBS	Switched Beam System
SC	Service Centre
SCCP	Signalling Connection Control Part

SCF	Service Capability Feature
SCO	Synchronous Connection-Oriented link
SCP	Service Control Point
SCS	Service Capability Servers
SCUA	Service Control User Agent
SDK	Software Development Kit
SDP	Service Discovery Protocol
SF	Service Feature
SGSN	Serving GPRS Support Node
SI	Service Interface (prefix to interface class method)
SIB	Service Independent Building Block
SIBO	Single Board Organizer
SID	System Identification
SIF	Service Information Field
SIFS	Short Inter Frame Space
SIM	Subscriber Identity Module
SIO	Service Indicator Octet
SIP	Session Initiation Protocol
SIS	Symbian OS Installation
SIR	Signal Interference Ratio
SLPP	Subscriber LCS Privacy Profile
SME	Short Message Entity
SME	Station Management Entity
SMIL	Synchronization Multimedia Integration Language
SMLC	Serving Mobile Location Center
SMMO	Short Message Mobile Originated point-to-point
SMPP	Short Message Peer-to-Peer
SMMT	Short Message Mobile Terminated point-to-point
SMS	Short Message Service
SMS	Service Management System in SMS/800
SMSC	SMS Centre
SMTP	Simple Mail Transfer Protocol

SN	Subscriber Number
SN	Service Node
SNA	System Network Architecture
SNDCF	Sub Network Dependent Convergence Function
SNDCP	Sub Network Dependent Convergence Protocol
SNR	Signal to Noise Ration
SOAP	Simple Object Access Protocol
SoD	Session oriented Dialogue
SP	Service Point
SP	Service Plane
SPC	Signaling Point Code
SQL	Structured Query Language
SRC	Short Retry Count
SRES	Signature Response
SRNC	Serving RNC
SS	Station Service
SS	Signalling System
SS7	Signaling System No 7
SSID	Service Set Identifier
SSL	Secured Socket Layer
SSP	Service Switching Point
STA	Station
STP	Signaling Transfer Point
SWAP	Shared Wireless Access Protocol

T

TA	Timing Advance
TACS	Total Access Communication System
TCAP	Transaction Capabilities Application Part
TCP	Transmission Control Protocol
TCP/IP	Transmission Control Protocol/Internet Protocol

TCS	Telephony Control Specification
TDD	Time Division Duplex
TDMA	Time Division Multiple Access
TIA	Telecommunication Industries Association
TINA	Telecommunications Information Network Architecture Consortium
TLS	Transport Layer Security
TKIP	Temporal Key Integrity Protocol
TMSI	Temporary Mobile Subscriber Identity
TN3270	Telnet protocol for IBM 3270
TN5250	Telnet protocol for IBM 5250
TOA	Time Of Arrival
TP	Transaction Processing
TPMS	Transaction Processing Management System
TTA	Telecommunications Technology Association-Korea
TTC	Telecommunication Technology Committee-Japan
TTML	Tagged Text Mark-up Language
TTS	Text To Speech
TUP	Telephone User Part

U

UAProf	User Agent profile
UDH	User Data Header
UDHI	User Data Header Indicator
UDP	User Datagram Protocol
UDT	Unit Data message
UE	User Equipment
UL	Uplink
UMTS	Universal Mobile Telecommunication System
URI	Universal Resource Identifier
URL	Universal Resource Locator
USIM	Universal Subscriber Identity Module

USSD	Unstructured Supplementary Service Data
UTRAN	Universal Terrestrial Radio Access Network

V

VAS	Value Added Service
VASP	Value Added Service Provider
VDU	Visual Display Unit
VHE	Virtual Home Environment
VLSI	Very Large Scale Integration
VLR	Visitor Location Register
VME	Virtual Machine Environment
VPN	Virtual Private Network
VPS	Voice Processing System
VRU	Voice Response Unit
VSAT	Very Small Aperture Terminal
VT3K	Visual Terminal for HP 3000

W

W3C	WWW Consortium
WAE	Wireless Application Environment
WAP	Wireless Application Protocol
WBMP	Wireless BMP
W-CDMA	Wideband CDMA
WCDMA	Wideband CDMA
WDP	Wireless Datagram Protocol
WiFi	Wireless Fidelity
WiLL	Wireless in Local Loop
WLAN	Wireless LAN
WLL	Wireless Local Loop
WM	Wireless Medium
WML	Wireless Markup Language

WSP	Wireless Session Protocol
WTA	Wireless Telephony Applications
WTAI	Wireless Telephony Application Interface
WTLS	Wireless Transport Layer Security
WTP	Wireless Transaction Protocol
WWAN	Wireless Wide Area Network
WWW	World Wide Web

X

XML	eXtensible Markup Language

Index